FUNDAMENTALS
OF SOLID-STATE
ELECTRONICS

FUNDAMENTALS OF SOLID-STATE ELECTRONICS

ROBERT D. PASCOE
Jefferson County Technical Institute

JOHN WILEY & SONS, INC.
New York · London · Sydney · Toronto

Library of Congress Cataloging in Publication Data:

Pascoe, Robert D   1943-
 Fundamentals of solid-state electronics.

 Includes index.
 1. Semiconductors. 2. Electronic circuits.
I. Title.

TK7871.85.P36   621.3815′3   75-25818
ISBN 0-471-66905-9

Printed in the United States of America

10 9 8 7 6 5 4

To Professors J. Hattman and R. W. Schramm,
two excellent teachers

PREFACE

This book gives the student a foundation in solid-state devices and circuitry. It is written on the engineering technician level, and it assumes that the student has a background in ac and dc theory.

Unique features of the book are:

1. Chapter 20 ties together the preceding material by demonstrating simple systems that incorporate material discussed in the first nineteen chapters. The problems in Chapter 20 ask the student to design the component values used in the systems discussed.

2. The amount of material is appropriate for two courses in solid-state theory; possibly three courses can be covered. The purpose of including too much material is to make the text flexible so that it will "fit in" with many existing courses.

3. Almost all of the circuitry is pretested, and the illustrations can be incorporated into laboratory experiments. Suggestions for laboratory use of this material are in the *Instructor's Manual*.

4. Problems are solved frequently to demonstrate the practical relationship between equations and circuit configurations.

5. Objectives are given at the beginning of each chapter, and a short paragraph is included that explains why the student should study this material.

6. Chapter 11 emphasizes chips that are available with applications instead of discussing, in detail, manufacturing techniques.

7. Appendixes are included to explain commonly used abbreviations, specifications for devices, and derivations.

The material, as arranged, can be covered in two quarter or semester courses in electronics. Course One: Chapters 1 to 10. Course Two: Chapters 11 to 20.

If the material is carefully selected, the book can be altered in presentation without losing its continuity. Chapter 12, "Power Supply Regulation," was purposely organized so that the diode used as a rectifier was discussed immediately before regulating circuitry. I found this approach very effective because the student does not have to "brush up" on rectifying circuitry before he is confronted with regulators. However, sections 1 to 17 of Chapter 12 may follow Chapter 2, "The Junction Diode and Applications," if the more traditional approach is used. Also, Chapter 13, "Thyristors," may follow Chapter 17, "Timers." This alteration will not break the sequence of chapters devoted to the amplifier. I have discovered that a "break" in the amplifier chapters (Chapter 13 discussed before Chapter 14) works very well in keeping students' interest.

With the organization described above, the two courses can be outlined as follows. Course One: Chapters 1, 2, 12, 3, 4, 5, 6, 7, 8, and 9—Course Two—Chapters 10, 11, 14, 15, 16, 17, 13, 18, 19, and 20.

If time is limited, Chapter 7, "Quiescent Point Stabilization," and Chapter 8, "Circuit Models," may be presented briefly. Chapter 16, "Digital Electronics," and the examples in Chapter 20, "Electronic Systems," may be eliminated without losing continuity.

For a single course in electronics, assuming that the student has a background in BJTs and FETs, the course may begin with Chapter 8, "Circuit Models," and proceed to Chapter 20, "Electronic Systems." This course consists of Chapters 8, 9, 15, 12 (sections 18 to 23) and Chapters 11, 16, 17, 13, 18, 19, and 20. The remaining material can be used as reference material.

Chapter 1 reviews semiconductor theory. It gives a general review of the theory by considering the makings and workings of solid-state devices. Chapter 2 discusses the junction diode and its applications. Many applications, except power supplies, are dealt with in this chapter. The power supply is discussed in Chapter 12. Chapter 3 presents the transistor and circuit applications, and discusses definitions of circuit configurations and terms. Chapter 4 considers the common-emitter amplifier, input resistance, gain calculations, and supply requirements.

Chapter 5 covers common-collector and base amplifiers with circuit applications. Chapter 6 reviews the various biasing networks and lists advantages and disadvantages of these methods. The selection of biasing resistors also is explored. Chapter 7 discusses quiescent point stabilization and defines stability factors. Chapter 8 presents circuit models for devices and defines (with examples) the parameters commonly found in publications. Chapter 9 deals with feedback principles, including both current and voltage feedback (with applications).

Chapter 10 considers the junction and metal oxide semiconductor field effect transistors. Integrated circuits are dealt with in Chapter 11, which emphasizes chips that are on the market with applications rather than manufacturing techniques. Chapter 12 covers power supplies, rectifying circuitry, regulation, and filtering. Chapter 13 presents thyristors. The theory of the various devices is discussed with dc and ac applications. Chapter 14 discussess multistage and power amplifiers with common circuit configurations.

Chapter 15 presents operational amplifiers. Circuit theory and applications are discussed and demonstrated. Chapter 16 gives a cursory presentation of digital electronics. Chapter 17 discusses applications of PUT, UJT, bipolar, and IC timers commonly found in circuitry. Chapter 18 presents the photoelectronic devices found in circuitry. Chapter 19 considers LC, RC, and crystal oscillators. Chapter 20 discusses electronic systems. These systems incorporate the devices and circuitry given in the first 19 chapters of the text. Chapter 20 ties together the material presented in the rest of the text to give the student an overview of the entire subject.

Robert D. Pascoe

ACKNOWLEDGMENTS

I am grateful to the following companies for their permission to reproduce data sheets.

Burroughs Corp.
P. O. Box 1226
Plainfield, N.J. 07061

Datel Systems
1020 Turnpike St.
Canton, Me. 02021

Hamlin Inc.
Lake & Grove Streets
Lake Mills, Wi. 53551

Howard W. Sams & Co., Inc.
4300 West 62nd Street
Indianapolis, In. 46206

Intersil Corp.
10900 North Tantau Ave.
Cupertino, Ca. 95014

ISE Corp.
1472 West 178th Street
Gardenia, Ca. 90248

Motorola Corp.
5005E McDowell Rd.
Phoenix, Az. 85008

RCA Corp.
Box 3200
Somerville, N.J. 08876

Signetics Corp.
2460 Lemoine Ave.
Fort Lee, N.J. 07024

Texas Instruments Corp.
P. O. Box 5012
Dallas, Tx. 75222

R. D. P.

CONTENTS

1. SEMICONDUCTOR THEORY

2. THE JUNCTION DIODE AND APPLICATIONS

3. THE TRANSISTOR AND CIRCUIT CONFIGURATIONS

4. COMMON-EMITTER AMPLIFIER

5. COMMON-COLLECTOR AND COMMON-BASE AMPLIFIERS

6. dc BIASING TECHNIQUES

7. QUIESCENT POINT STABILIZATION

8. CIRCUIT MODELS

9. FEEDBACK PRINCIPLES

10. THE FIELD EFFECT TRANSISTOR

11. INTEGRATED CIRCUIT APPLICATIONS

12. POWER SUPPLY REGULATION

13. THYRISTORS

14. MULTISTAGE AND POWER AMPLIFIERS

15. OPERATIONAL AMPLIFIERS

16. DIGITAL ELECTRONICS

17. TIMERS

18. PHOTOELECTRONIC DEVICES

19. OSCILLATORS

20. ELECTRONIC SYSTEMS

APPENDICES

INDEX

FUNDAMENTALS
OF SOLID-STATE
ELECTRONICS

OBJECTIVES FOR CHAPTER 1, SEMICONDUCTOR THEORY

At the completion of this chapter you can demonstrate your understanding of the material by answering these questions:

- What is the structure of matter?
- What are insulators, semiconductors, and conductors?
- What are energy levels?
- What is meant by the terms *intrinsic semiconductors* and *extrinsic semiconductors*?
- How and why does temperature affect silicon and germanium?
- What are solid-state devices?

1 SEMICONDUCTOR THEORY

Semiconductor material is used in the construction of all solid-state devices (discrete and integrated). You should study semiconductor material in order to comprehend the workings of, the effects of temperature on, and the dissipation ratings of such devices as diodes, bipolar and unipolar junction transistors, thyristors, and integrated circuits.

The by-products of the advancements in semiconductor technology are such devices as bipolar and unipolar transistors, metal oxide semiconductor field effect transistors, thyristors, and integrated circuits. Some applications of solid-state devices are hand-held calculators, electronic digital watches, mini-computers, and microprocessors on a chip. This chapter discusses semiconductor material that is used in all solid-state devices.

THE STRUCTURE OF MATTER

In 1913, Bohr introduced a model for the atom using the quantum theories of Planck and Einstein. Bohr's model for the atom was very successful in explaining the absorption and emission spectrum for hydrogen. This model is called the planetary model for the atom because of its resemblance to the planets revolving around the sun.

Bohr pictured the atom as having a nucleus in the center with electrons revolving around the nucleus (see Figure 1.1). The levels at which the electrons travel are assigned specific energies. Bohr labeled these *energy levels* or *shells L, M, N,* and *O.* The maximum number of electrons that can exist in the *L* shell is 2; in the *M* shell, 8; in the *N* shell, 18; and in the *O* shell, 32.

This model for the atom shows the shells or energy levels at which electrons can revolve. The closer the electrons are to the nucleus, the greater the influence

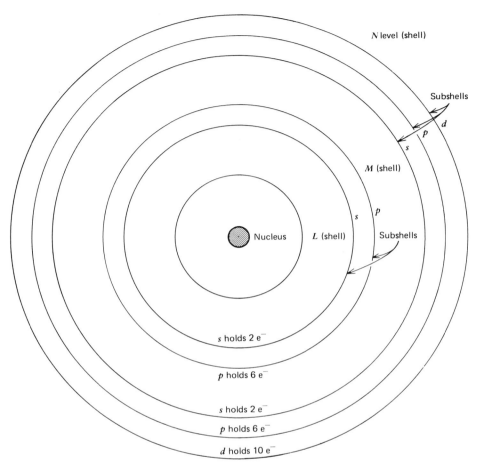

Figure 1.1
A model of the atom.

the nucleus has on the electrons. The outermost shell containing electrons is called the *valence shell*; the electrons in the outermost shell are called the valence electrons. The number of valence electrons determines the chemical and electrical properties of the element. As an example, if the outermost shell is full, the element is called *inert* or an *insulator*. If the outermost shell contains only a few electrons, the element is called a *conductor*. The amount of energy needed to extract an electron from the valence shell determines whether the element is a conductor, an insulator, or a semiconductor.

Bohr's model has been modified to explain certain experimental data. One modification of the model was the addition of subshells because of Pauli's Exclusion Principle, which states that no two electrons in an atom can have exactly the same quantum state (energy). The subshells, labeled *s*, *p*, and *d*, are located within each of the shells. The *L* shell has a maximum of 2 electrons and no subshells. The *M* shell has a maximum of 8 electrons, with a maximum of

2 electrons in the *s* subshell and a maximum of 6 electrons in the *p* subshell, and so on.

INSULATORS, SEMICONDUCTORS, AND CONDUCTORS

If it takes a great deal of energy, expressed in electron volts (1 eV = 1.6 × 10⁻¹⁹ joules), to extract electrons from the valence shell for conduction, the element is called an *insulator*. If it takes very little energy to extract electrons from the valence shell, the element is called a conductor.

Figure 1.2*a* illustrates an energy level diagram for an insulator. Note that several electron volts of energy separate the conduction and valence bands. The valence electrons all have slightly different amounts of energy according to Pauli's Exclusion Principle. Thus the electrons in the valence shell are represented by a band of energy called the *valence band*. If an electron is extracted from the valence band, it is "free" and goes into the conduction band. The magnitude of energy necessary to extract an electron from the valence band determines its energy in the conduction band. The energy level between the valence and conduction band is called the *forbidden energy gap* because no electrons can exist in this region. For an insulator, the energy gap is considered greater than a few electron volts.

Figure 1.2*b* illustrates an energy level diagram for a conductor. Note that the conduction and valence bands overlap. This overlapping implies that it takes very little or no energy for electrons to jump from the valence band to the conduction band. An example of a good conductor is copper. Copper has 1 valence electron that is readily given up for conduction.

Figure 1.2*c* illustrates an energy level diagram for a *semiconductor*. The energy level of the forbidden gap is approximately 1 eV. Some examples of semiconductors are silicon (Si), which has an energy gap of 1.1 eV at room temperature; germanium (Ge), which has an energy gap of 0.72 eV at room temperature; and gallium arsenide (GaAs), which has an energy gap of 1.3 eV at room temperature.

Figure 1.2
Energy level diagrams: (a) insulator, (b) conductor, and (c) semiconductor.

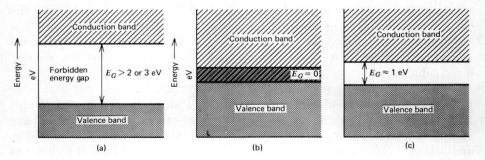

INTRINSIC PROPERTIES OF SILICON AND GERMANIUM

Silicon and germanium are the most commonly used semiconductors in the manufacturing of solid-state devices. *Intrinsic* silicon (pure silicon) is a crystalline structure having 4 valence electrons (see Figure 1.3). The crystalline structure of silicon has the atoms stationary with the valence electrons being shared by adjacent atoms. The valence shell for silicon is stable with 8 electrons in it. This sharing of the valence electrons is called *covalent bonding*.

A block of silicon has an overall charge of zero. However, at room temperature a specific number of electron-hole pairs are formed. If a potential difference is

Figure 1.3
Intrinsic semiconductor material: (a) silicon material with covalent bonding and (b) current carriers through silicon —holes = electrons.

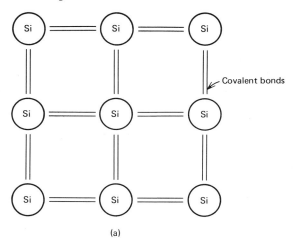

(a)

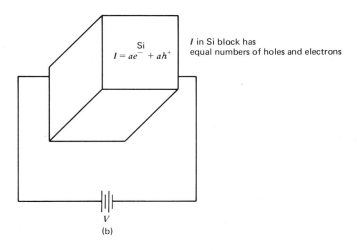

(b)

placed across the face of the silicon block, a current flows (see Figure 1.3b). The current flow through the block of silicon is made up of equal numbers of electrons and holes.

A hole is the absence of an electron in the covalent bonding. The electrons in the block of silicon shown in Figure 1.3b are flowing from left to right, while the holes are flowing from right to left. As an electron leaves a covalent bond, it leaves behind a hole; therefore, the current carriers of holes and electrons move in opposite directions. Thus as the electrons drift toward the positive side of the potential difference, the holes move toward the negative side of the potential difference. It is important to realize that the number of holes and electrons in the intrinsic semiconductor is equal.

EXTRINSIC SEMICONDUCTORS

If to the pure silicon block a selected impurity is added, the current flow through the block of semiconductor material can be made up of majority current carriers of either electrons or holes. If impurities called dopant atoms (a *donor*) with a valence number of 5 is added or mixed with the pure silicon semiconductor, the majority current carriers through the material are electrons. With the majority current carrier being electrons, the material is called N-type semiconductor material (see Figure 1.4a).

Normally, the energy gap for intrinsic silicon is 1.1 eV. But for extrinsic silicon doped with atoms having a valence of 5, the extra electron is held loosely to the parent nucleus with only 0.05 eV of energy required to move the electron into the conduction band (see Figure 1.4b). The dopant atom has 5 electrons in its valence shell. The covalent bonding is very satisfied with sharing only 4 electrons; thus the fifth electron is held to the parent nucleus with very little energy. The electron is considered "free" for conduction.

It is important to realize that the N-type semiconductor material has an overall charge of zero. Even though the donor atom may lose its electron and become an ion (positive), the block of N-type material will have an overall charge of zero. If a potential difference is applied across the face of the N-type material, the current flow through the block is made up of a majority of electrons; thus the term N-type semiconcudtor material (see Figure 1.4c). The majority current carriers for N-type material are electrons. The ratio of dopant atoms to silicon (or germanium) atoms is small. As an example, if one dopant atom is added for every million silicon atoms, the number of current carriers at room temperature changes from 10^{13} to 10^{17}.

If the silicon semiconductor material is doped with atoms that have a valence number of 3, the semiconductor material is called P-type, semiconductor material. The dopant atoms that have a valence number of 3 are called *acceptor* atoms (see Figure 1.5a). Because the acceptor atoms have 3 valence eletcrons, they leave a hole in the covalent bonding. The dopant atom is called an acceptor atom because the hole in the covalent bond readily accepts other electrons.

Figure 1.5b illustrates the energy level diagram for P-type material. The energy necessary for the hole to move into the valence band is only 0.05 eV. Because the hole can move in the valence band, this movement is considered

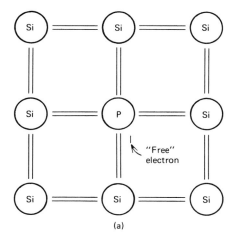

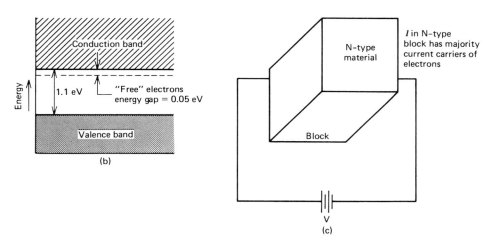

Figure 1.4
Extrinsic N-type semiconductor material: (a) donor atom among silicon atoms, (b) energy level diagram of N-type material, and (c) majority current carriers in N-type material—electrons.

current flow. If a potential difference is placed across the face of a block of P-type semiconductor material, the majority current carriers through the material are made up of holes (see Figure 1.5c). Note that when the acceptor atom gains an electron, it becomes a negative ion, but that the block of P-type material has an overall charge of zero.

TEMPERATURE EFFECTS ON SILICON AND GERMANIUM

It was stated earlier that the energy gap for silicon is 1.1 eV at room temperature and that for germanium the energy gap is 0.72 eV. Because silicon has a wider

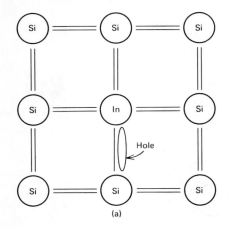

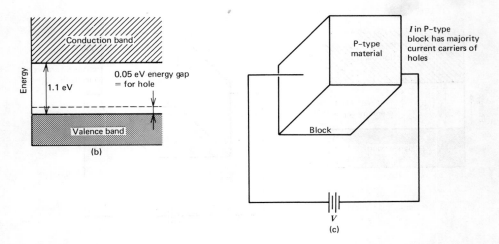

Figure 1.5
*Extrinsic P-type semiconductor material: (a) acceptor atom among silicon atoms, (b) energy
level diagrams for P-type material, and (c) majority current carriers for P-type material—
holes.*

energy gap, it is less affected by temperature than germanium. Thus, at a given
temperature, silicon has fewer current carriers available than does germanium.
The resistance of a cubic centimeter of silicon is much greater than the resistance
of a cubic centimeter of germanium at the same temperature. Many solid-state
devices use silicon at the semiconductor material, with donor or acceptor atoms
added because of the lower number of current carriers available with temperature.

Both silicon and germanium have negative temperature coefficients with
respect to their resistances. As the temperature increases, covalent bonds are
broken that free holes and electrons for conduction. Thus the resistance decreases as the temperature increases.

SOLID-STATE DEVICES

Semiconductor material is used in the manufacturing of solid-state devices. As examples, there are PNP and NPN transistors, PNPN switches, P-channel and N-channel field effect transistor, unijunction transistors (UJTs), programmable unijunction transistors (PUTs), light-emitting diodes (LEDs), silicon controlled rectifiers (SCRs), triacs, and light-sensitive solid-state devices. The remaining chapters of this text are devoted to the discussion of the solid-state devices and circuits already mentioned and many more.

REVIEW QUESTIONS AND PROBLEMS

1-1 In terms of valence electrons, what elements are (a) good conductors, (b) good insulators, and (c) semiconductors?

1-2 What is meant by the term *covalent bond*?

1-3 How are the electrical properties of an element changed when electrons move from the valence band into the conduction band?

1-4 How are the electrical properties of a semiconductor changed when a hole moves from the conduction band into the valence band?

1-5 Define (a) valence band and (b) conduction band.

1-6 What significance does Pauli's exclusion principle have in defining the energy levels for the conduction and valence bands?

1-7 Discuss the meaning of the terms *intrinsic* and *extrinsic*.

1-8 Explain why an N-type semiconductor material has an overall charge of zero.

1-9 What are the majority current carriers in (a) N-type material, (b) P-type material, and (c) intrinsic semiconductor material?

1-10 What is meant by the term dopant atom?

1-11 What effect does a dopant atom have on the energy required for conduction in silicon material?

1-12 What is meant by the term *acceptor atom*?

1-13 What is meant by the term *donor atom*?

1-14 Using the periodic chart of the elements, list some possible donor and acceptor atoms.

1-15 Explain why semiconductor materials have a negative temperature coefficient.

OBJECTIVES FOR CHAPTER 2, THE JUNCTION DIODE AND APPLICATIONS

At the completion of this chapter you can demonstrate your understanding of the material by answering these questions:

- How is a diode constructed?
- What is the circuit model for the diode?
- What are the forward and reverse characteristics of the diode?
- What differences are there between a real and an ideal diode?
- What is a zener diode and what are some of its characteristics?
- What is a voltage-variable capacitance diode?
- What are some of the applications of the diode?

2 THE JUNCTION DIODE AND APPLICATIONS

The diode has many interesting and useful circuit applications. In order to comprehend the applications, you must know the theory and workings of the many types of diodes that are available. The diode should be studied because of its use in such circuits as power supply, electronic tuning, voltage regulation, and voltage protection.

The workings of the PN junction or diode, some of the various types of diodes commonly found in circuitry, and applications are discussed in this chapter. Conventional diode applications in rectifying circuitry are presented in Chapter 12.

THE DIODE

A solid-state diode is constructed of P- and N-type semiconductor material. The device has one junction and is considered nonlinear and unidirectional (i.e., passes current in only one direction). The schematic symbol and characteristic curve for the diode are shown in Figure 2.1.

With a positive potential at the anode (P-type material) with respect to the cathode (N-type material), the diode is *forward-biased*. The positive potential at the anode repels the majority current carriers of holes, causing the holes to move toward the junction. The negative potential at the cathode repels the majority current carriers of electrons, causing the electrons to move toward the junction. With the holes and electrons meeting at the junction, annihilation occurs, causing energy to be released at the junction. When the diode is forward-biased, its internal resistance is low, allowing current flow through the device.

With a negative potential at the anode with respect to the cathode, the diode is *reverse-biased*. The negative potential at the anode attracts holes away from

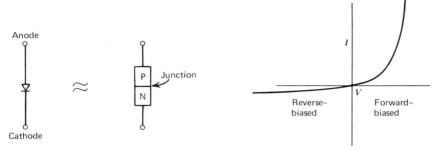

Figure 2.1
Schematic symbol for the diode with its characteristic curve.

the junction; the positive potential at the cathode attracts electrons away from the junction. In the reverse-biased mode, the junction area has very few holes and electrons in it. The area around the junction is called the *depletion area* when the diode is reverse-biased. The small magnitude of current in the reverse-biased mode of the diode is called *leakage current*. This small magnitude of leakage current is caused predominantly by the minority current carriers in the P- and N-type material.

In Figure 2.1, the diode has very low internal resistance (ideally 0Ω) when forward-biased and very high resistance (ideally infinite ohms) when reverse-biased.

With the diode in the forward-biased mode, an approximate circuit model can be shown. A circuit model is a means of explaining the characteristics of the device. Figure 2.2 shows the circuit model for a forward-biased diode. The circuit model consists of an ideal diode D, a resistance R_f, and a potential V.

The ideal diode has 0Ω of resistance when forward-biased and functions as a switch in the circuit model. The resistance R_f is the ohmic value of resistance representing the internal resistance of the diode when it is forward-biased. The value of R_f is found from the $v\text{-}vs\text{-}i$ curve for the diode. Notice that the resistance is considered a constant in the circuit model. The voltage V_{FB} represents the

Figure 2.2
Characteristic curve and equivalent circuit of a forward-biased diode.

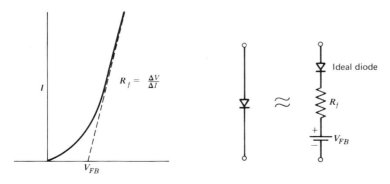

necessary potential to forward bias the ideal diode; V_{FB} is approximately 0.3 V for germanium and approximately 0.6 V for silicon. The power dissipated when the diode is forward-biased is given by

$$P = I_D{}^2 R_f + I_D V_{FB} \tag{1}$$

where I is the current flow through the forward-biased diode.

The voltage V_{FB} represents the barrier potential of the diode; its magnitude depends on whether the semiconductor material is silicon or germanium. When the diode is first fabricated, electrons from the N-type material are attracted by the holes in the P-type material. These electrons and holes "migrate" to the junction. At first, one would think that this migration of current carriers would continue indefinitely. However, migration does not continue indefinitely. As was pointed out in Chapter 1, the doping elements are fixed or immobile in the crystalline structure of the semiconductor material. When migration of the holes and electrons first starts, ions are formed near the junction because the doping atoms lose electrons in the N-type material and holes in the P-type material.

The loss of an electron for the N-type dopant causes the atom to become a positive ion fixed or immobile in the crystalline structure of the semiconductor material. The loss of a hole for the P-type dopant causes the atom to become a negative ion fixed or immobile in the crystalline structure of the semiconductor material. Because the migration occurs near the junction of the P- and N-type material, ions are "uncovered" near the junction. The positive ions are in the N-type material and the negative ions are in the P-type material. The potential difference generated by the uncovering of ions is called the *barrier potential* of the diode and is shown in Figure 2.3. This barrier potential is the V_{FB} labeled in the circuit model for the diode. This barrier potential has to be overcome by the biasing voltage across the diode before the diode can become forward-biased. Once the diode is forward-biased, it can conduct current.

Figure 2.4 shows a circuit model for the diode when it is reverse-biased. This circuit model has only an ideal diode and a resistance R_r. The resistance value of R_r is obtained from the curves for the diode and is a very high value.

Figure 2.5 shows how the reverse- and forward-biased circuit models just discussed are connected to show the workings of a real diode. The need for ideal

Figure 2.3
Barrier potential for a PN junction.

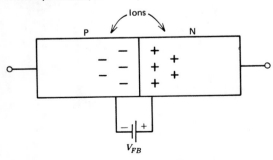

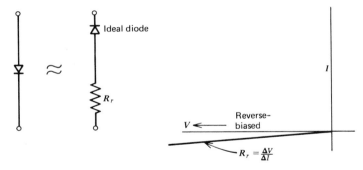

Figure 2.4
Characteristic curve and equivalent circuit of a reverse-biased diode.

diodes is made clear in Figure 2.5. The ideal diodes isolate each circuit element of the model until the correct polarity of potential is across the real diode. A positive potential at the anode of the real diode will cause the forward-biased ideal diode to be "on" (zero ohm) and the reverse-biased ideal diode to be "off" (infinite ohms). If a negative potential is at the anode of the real diode, the forward-biased ideal diode is "off" and the reverse-biased ideal diode is "on."

Power diodes have a low value of R_f, and signal diodes have a relatively high value of R_f compared to power diodes. With power diodes, the forward-biased conducting current is in the high milliampere or ampere range, depending on the power rating of the device. With signal diodes, the forward-biased conducting current is less than 50 mA.

EXAMPLE

Calculate the forward-biased resistance R_f for a small signal diode 1N914 and for a power diode 1N4007 having the current-voltage measurements shown in Table 2.1.

Figure 2.5
Circuit model for a real diode.

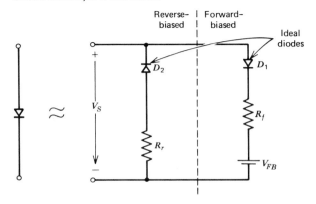

Table 2.1

1N914		1N4007	
I_D	V_{FB}	I_D	V_{FB}
5.0 mA	0.663 V	30.0 mA	0.715 V
14.4 mA	0.714 V	58.3 mA	0.747 V

The forward-biased resistance R_f is found by calculating the slope of the v-i curve, or

$$R_f = \frac{V_{FB1} - V_{FB2}}{I_{D1} - I_{D2}} = \frac{\Delta V_{FB}}{\Delta I_D}$$

Thus, for the 1N914,

$$\Delta I_D = 14.4 \text{ mA} - 5.0 \text{ mA} = 9.4 \text{ mA}$$

$$\Delta V_{FB} = 0.714 \text{ V} - 0.663 \text{ V} = 0.052 \text{ V}$$

and

$$R_f = \frac{\Delta V_{FB}}{\Delta I_D} = \frac{0.052 \text{ V}}{9.4 \text{ mA}} = 5.5 \ \Omega$$

and for the 1N4007,

$$\Delta I_D = 58.3 \text{ mA} - 30.0 \text{ mA} = 28.3 \text{ mA}$$

$$\Delta V_{FB} = 0.747 \text{ V} - 0.715 \text{ V} = 0.032 \text{ V}$$

$$R_f = \frac{\Delta V_{FB}}{\Delta I_D} = \frac{0.032 \text{ V}}{28.3 \text{ mA}} = 1.1 \ \Omega$$

TEMPERATURE EFFECTS ON THE DIODE

Figure 2.6 shows the effects of temperature on the forward-biased diode. Note that the barrier potential *changes* with temperature. This voltage change decreases with an increase in temperature. The effect is approximately given by

$$\Delta V_{FB} \approx -2.5 \ \frac{\text{mV}}{{}^\circ\text{C}} \tag{2}$$

Equation 2 states that the barrier potential (V_{FB} in the circuit model) decreases 2.5 mV for every 1.0 °C increase in temperature.

In the reverse-biased mode, the diode has a small leakage current, which is predominantly caused by the minority current carriers in the P- and N-type semiconductor material (see Chapter 1). The reverse-biased voltage across the junction of the diode permits these minority current carriers (electrons in the P-type material and holes in the N-type material) to cross the junction. The leakage current is a function of temperature. For germanium semiconductor

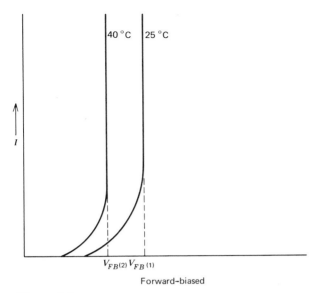

40 °C   25 °C

I

$V_{FB(2)}$ $V_{FB(1)}$

Forward-biased

Figure 2.6
Effects of temperature on the barrier potential of a diode.

material, the leakage current approximately doubles for every 10 °C increase in temperature. For silicon semiconductor material, the leakage current approximately doubles for every 6 °C increase in temperature. At room temperature (25 °C), the leakage current for germanium is of the order of microampere (10^{-6} A) and for silicon is of the order of nanoampere (10^{-9} A).

PEAK INVERSE VOLTAGE

The peak inverse voltage (PIV) is the maximum reverse-biased voltage that can be maintained across the PN device. If the PIV is exceeded, the diode will break down. Once the diode is broken down in the reverse-biased region, a great deal of power is dissipated in the junction of the diode. Therefore, conventional diodes are not made to operate in the reverse-biased voltage region. Special diodes that are made to work in this region are called zener diodes.

ZENER DIODES

Figure 2.7 shows the schematic symbol and characteristic curve for a zener diode. As can be determined from the figure, a zener diode can be used as a voltage source if it is operated in the zener region. If the current through the zener diode is increased or decreased, the zener voltage V_Z is changed very little.

The magnitude of V_Z is controlled by the concentration of dopant atoms in the N- and P-type material by the manufacturer. The zener diode breaks down as a result of the strong electric field across the junction of the diode caused by

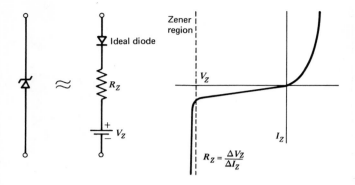

Figure 2.7
Circuit model and characteristic curve for a zener diode.

the reverse-biased voltage. This electric field accelerates the minority current carriers in the N- and P-type material, gaining kinetic energy. The minority carriers collide with the dopant atoms and give up some of their kinetic energy in the collision to other dopant atoms. The excited dopant atoms free additional current carriers that are accelerated in the electric field and collide with other dopant atoms. The effect avalanches to such a degree that many current carriers cross the junction, causing the diode to break down. As can be reasoned, the number of donor and acceptor atoms per unit area of the semiconductor material determines the potential at which this zener effect takes place. Generally, the greater the concentration of donor and acceptor atoms, the lower the voltage at which this zener breakdown occurs.

As a point of interest, most zener diodes should be called avalanche diodes because of the mechanism by which the diode breaks down. If the energy of the current carriers of the dopant atoms is great enough to cause additional current carriers from other dopant atoms, the breakdown is called the avalanche effect. The potential generally required to initiate the avalanche effect is equal to or greater than 6 V. The breakdown of diodes below 6 V is the zener effect. The zener effect is caused by the collision of the minority carriers with the dopant atoms, but the energy of the released current carriers of the dopant atoms is not great enough to cause other dopant atoms to release additional current carriers.

Figure 2.8 shows the circuit model of a zener diode. In the model, V_Z is the zener voltage, and R_Z is the resistance of the device when operated in the zener region; the diode shown is an ideal diode. The ideal diode does not conduct until the potential across the diode is equal to or greater than V_Z. The power dissipated in the diode is equal to

$$P_Z = I_Z^2 R_Z + I_Z V_Z \qquad (3)$$

where I_Z is the zener current, V_Z the zener voltage, and R_Z the zener resistance. The value of R_Z ranges from less than 1 Ω to 20 Ω, depending on the power rating of the zener diode.

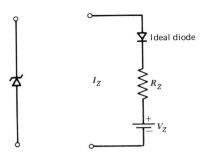

Figure 2.8
Circuit model for a zener diode.

Figure 2.9 shows a diagram of a zener diode being used as a voltage regulator. Ideally, with changes in the load current I_L, the voltage across the load V_Z should not change. The function of R_S in the circuit is to insure that the diode is always operated in its zener region.

In Figure 2.9, the current I_S is equal to the sum of the load current I_L and the zener current I_Z, or

$$I_S = I_L + I_Z \tag{4}$$

If R_S is chosen so that it is at least 10 times the value of R_Z, I_S is then considered constant. If the load current increases, the zener current decreases; if the load current decreases, the zener current increases. Thus V_S with R_S functions as a current pump, pumping a constant current into the zener and load resistance. The voltage regulation of the zener can now be seen. If I_Z is maintained between some maximum and minimum value in order to keep the voltage V_Z in the zener region, the voltage across the load resistance, with varying current changes, will remain constant.

The value of R_S is the difference between the source voltage V_S and the zener voltage V_Z divided by the sum of the currents $I_{Z(min)}$ and $I_{L(max)}$. As a rule of thumb, the minimum zener current is 10 percent of the maximum load current. The above relationship is given by

$$R_S = \frac{V_S - V_Z}{I_{Z(min)} + I_{Z(max)}} \tag{5}$$

where

$$I_{Z(min)} = 0.1 \times I_{Z(max)}$$

If the load is removed from across the zener diode and if R_S is $\gg R_Z$, the maximum zener current is given by

$$I_{Z(max)} \approx \frac{(V_S - V_Z)}{R_S} \tag{6}$$

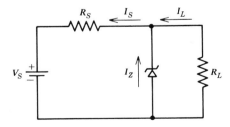

Figure 2.9
A zener diode voltage regulator.

The maximum zener current for a specific power rating of the zener diode is given by the manufacturer and should not be less than the value given in Equation 6.

EXAMPLE

A 9.1-V zener diode is to be used with an 18-V source and a load current that varies from 0 to 200 mA. Calculate the necessary value of R_S, the maximum value of R_Z, and the power rating of the zener diode needed.

Using Equation 5, we have

$$I_{Z(min)} = 0.1 \times I_{Z(max)} = 0.1 \ (200 \ \text{mA})$$

$$= 20 \ \text{mA}$$

$$R_S = \frac{V_S - V_Z}{I_{Z(min)} + I_{Z(max)}} = \frac{(18 - 9.1) \ V}{20 \ \text{mA} + 200 \ \text{mA}}$$

$$= 40.45 \ \Omega \qquad (\text{use } 39 \ \Omega)$$

Now R_Z should be much less than R_S or

$$R_Z \leq \frac{R_S}{10} = \frac{39}{10}$$

$$\leq 3.9 \ \Omega \tag{A}$$

The maximum zener current, when the load resistor is removed, is found with Equation 6, or

$$I_{Z(max)} = \frac{V_S - V_Z}{R_S} = \frac{(18 - 9.1)}{39 \ \Omega} \ V$$

$$= 228 \ \text{mA}$$

Using Equation 3, we have the maximum power dissipated by the zener diode, or

$$P_Z = I_{Z(max)}^2 \ R_Z + I_{Z(max)} \ V_Z$$

$$= (0.228)^2 \ R_Z + (0.289) \ 9.1$$

$$= 0.052 \ R_Z + 2.08 \ W \tag{B}$$

The result of Equation B indicates that a zener diode of greater than 2 W should be used. Because zener diodes are commonly manufactured in ½-W, 1-W, 5-W, 10-W, and 50-W power dissipations, a look through a parts catalog in the 5-W section of zener diodes shows a 1N5346B, 9.1 V zener having an R_Z of 2 Ω. The 2 Ω of R_Z satisfies our requirement for R_Z in Equation A. The power dissipated by the 1N5346B is then

$$P_Z = 0.052 \times 2 \ \Omega + 2.08 \ W$$

$$= 2.184 \ W$$

VARIABLE-VOLTAGE CAPACITOR

As was stated early in the chapter, a diode in the reverse-biased mode has a depletion area within the junction region. The depletion area has very few current carriers and increases in area as the reverse voltage increases. This effect can be used as a capacitor because the depletion area acts as an insulator, and the ions on each side of the depletion area act as conductors. Diodes that are manufactured to operate in this manner are called variable-voltage capacitors. Their capacitance varies as the voltage across the diode is increased and decreased. Some of the trade names for these variable-voltage capacitors are Varicap, Tuning Capacitor, and `Capacitance Diode.

Figure 2.10 shows the characteristic curve for a variable-voltage capacitor. Note that as the reverse-bias voltage increases, the capacitance decreases. The range over which the capacitance is varied is small (in pF).

The major use for the variable-voltage capacitor is in tuning tank circuits electrically. Such an arrangement is shown in Figure 2.11. The variable-voltage capacitor is connected across the tank circuit via the 0.01 μF capacitor. The series capacitance of the variable-voltage capacitance and the 0.01 μF is approximately the capacitance of the diode. The RFC (radio frequency choke) and resistor R isolate the diode from the low resistance of the variable-voltage sourve V_{RB}. The value of R is generally in megohms (MΩ). This high resistance has a very small voltage drop because the diode, when it is reverse-biased, has a very small leakage current (in nanoamperes). As V_{RB} is varied, the capacitance of the diode is changed, causing the resonant frequency of the tank circuit to change. The circuit can be modified to make a simple but effective frequency modulator (see Chapter 20).

Figure 2.10
Characteristic curve for a variable-voltage capacitor (VVC) diode.

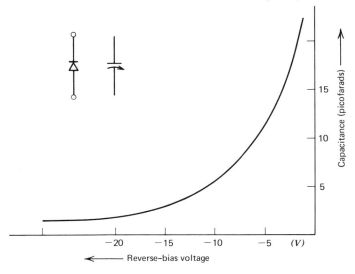

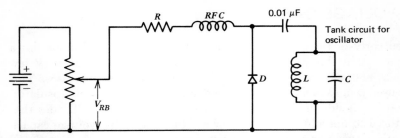

Figure 2.11
Simple application of a VVC.

LIGHT-EMITTING DIODE

When a diode is forward-biased, holes and electrons meet at the junction of the diode and annihilation occurs. When an electron and hole meet, energy is given off in the junction of the diode. Specially constructed diodes allow this energy in the form of photons to be emitted from the junction area of the diode. These specially constructed diodes are called light-emitting diodes (LED). The LED is supplied a fixed amount of current to emit light in the visible spectrum. Some of the applications for the LED are pilot lamps, display devices, and light sources for photooptical coupling devices. For a detailed discussion and applications of the LED, see Chapter 18.

APPLICATIONS

Diode Bias Clamping. Figure 2.12a shows a circuit configuration that biases an ac signal with a resistor, a capacitor, and a diode. Figure 2.12b shows the ideal output waveform across the diode. From the figure it can be seen that when v_s is positive, diode D is forward-biased, causing the capacitor C to charge up to the peak voltage value of v_s. When v_s goes negative, its value is in series with the voltage across the capacitor C. Thus the ac signal v_s is biased and has a dc value of V_s, as shown in Figure 2.12b.

For the circuit to function as stated, the time constant of the resistor and capacitor (RC) has to be at least 10 times the period of the ac voltage v_s. The long time constant of R and C, compared to the period of v_s, allows the dc voltage that the capacitor C charges up to (V_s) to remain constant.

EXAMPLE

In Figure 2.12, the frequency of v_s is 1 kHz. What should the value of R and C be? The period of the source v_s is calculated by

$$T = \frac{1}{f} = \frac{1}{10^3 \text{ Hz}}$$

$$= 1 \text{ msec}$$

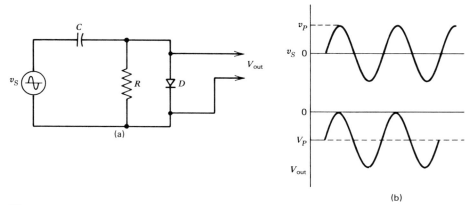

Figure 2.12
Diode bias clamping: (a) circuit configuration and (b) waveforms.

Because RC is 10 times the period T, we have

$$RC = 10T = 10 \times 1 \text{ msec}$$

$$= 10 \text{ msec}$$

If $R = 10 \text{ k}\Omega$, then C is equal to

$$C = \frac{10 \text{ msec}}{R} = \frac{10^{-2} \text{ sec}}{10^4 \ \Omega}$$

$$= 1.0 \ \mu\text{F}$$

The amplitude of v_s can be varied and then V_p follows v_s. If the diode were to be turned around in Figure 2.12a, the biased ac signal would be on the positive axis, as shown in Figure 2.12b. The diode used in Figure 2.12 was assumed to be ideal. With a real diode in the circuit, the peak output voltage across the diode would be slightly positive. This type of biasing method can be used in nonlinear amplifiers.

cemf PROTECTION

Figure 2.13 shows a circuit arrangement for a diode protecting a transistor (or other devices) against the effect of the cemf voltage of a relay. When the relay is pulled in (Q is turned "on"), current flows through the relay and the diode D is reverse-biased. When the relay is turned "off" (Q is turned "off"), the collapsing magnetic field of the relay coil induces a voltage across the relay of opposite polarity. This voltage is called the cemf voltage. Because the cemf is opposite in polarity, the diode D will become forward-biased. The energy of the collapsing field is then dissipated in the dc resistance of the relay coil.

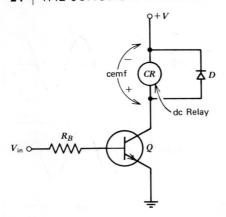

Figure 2.13
*Diode used to suppress the cemf
generated by a dc relay.*

CLIPPING CIRCUITS

Figure 2.14 shows a diode connected to clip the negative voltage spike of a differentiating circuit. The voltage across the resistor R in the figure without the diode D would be both a positive and negative spike. With the diode D connected as shown in the figure, the diode is reverse-biased for positive voltage spikes and forward-biased for negative voltage spikes. Thus the output waveform consists of only positive voltage spikes. If the diode were to be reversed, only negative voltage spikes would appear across R.

Figure 2.15 shows a circuit configuration used to clip the peaks of an ac sinewave voltage. If V_{in} is positive, diode D_2 will be forward-biased if the amplitude of V_{in} is greater than V_1. If V_{in} positive is not greater than V_2 and V_{in} negative is not greater than V_1, the output voltage will be a sinewave, unclipped. The circuit shown in the figure could be used to clip noise spikes that might appear on an audio signal. For this application, V_1 and V_2 would be adjusted to a low "normal amplitude" audio signal so that it would not be clipped. Because the noise spikes are greater than the normal amplitude signal, they would be clipped.

Figure 2.16 shows two zener diodes connected back-to-back in series. This combination would serve the same function as the circuit shown in Figure 2.15. A zener diode, when forward-biased, behaves as a conventional diode forward-

Figure 2.14
Differentiating circuit: (a) diode used to clip negative spikes and (b) waveforms.

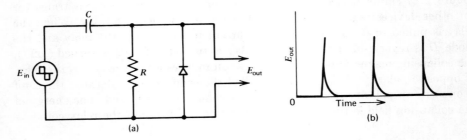

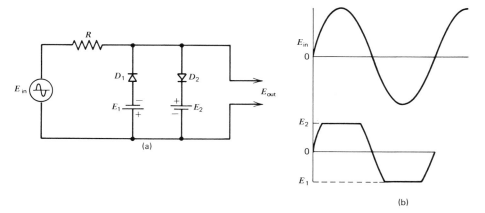

Figure 2.15
Negative and positive diode clipper: (a) circuit configuration and (b) waveforms.

biased. If V_{in} is positive and greater than V_{Z1}, D_{Z1} will conduct in the zener region and D_{Z2} will be forward-biased. If V_{in} is negative and greater than V_{Z2}, D_{Z2} will conduct in the zener region and D_{Z1} will be forward-biased.

SWITCHING

Figure 2.17 shows a circuit arrangement using diodes to control relays. If the switch S is in position A, diode D_1 will be forward-biased, causing relay CR_1 to be pulled in. If the switch S is in position B, diode D_2 will be forward-biased, causing relay CR_2 to be pulled in. As can be seen from the figure, the position of the switch determines which relay is picked up.

Figure 2.16
A clipper using zener diodes: (a) circuit and (b) waveforms.

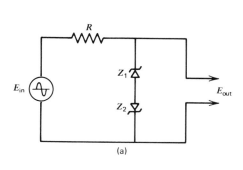

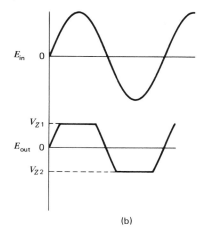

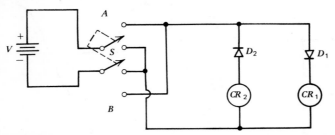

Figure 2.17
Polarity sensitive relays.

CORRECT POLARITY CIRCUIT

Portable solid-state equipment requires some external power source such as a battery. A sample circuit arrangement of diodes can be used to make sure that the portable equipment receives the correct polarity from the battery. Figure 2.18 shows such an arrangement. With the circuit shown, no matter which way the battery is connected, the equipment will always have the correct polarity. If point A of the battery is positive and B is negative, diodes D_1 and D_3 will conduct. If point A of the battery is negative and B positive, diodes D_2 and D_4 will conduct.

TEMPERATURE DETECTION

Figure 2.19 shows several diodes connected in series. The battery V forward-biases the diodes, and the resistor R limits the conducting current of the diodes. The output voltage V_0 can be used to monitor temperature changes because the diode voltage will be affected by temperature. Each diode has a voltage change of approximately -2.5 mV/°C. The four diodes connected in series will have a voltage change of -10 mV/°C. Assume that the current through the diodes will cause 2.4 V for V_0 at 25 °C. If the temperature of the environment that the

Figure 2.18
Circuit for a simple correct polarity device.

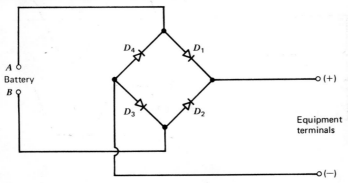

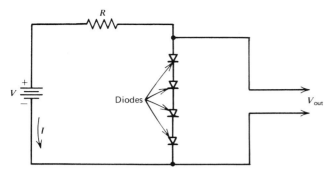

Figure 2.19
Diodes used as a temperature probe.

diodes are in changes to 35 °C, the output voltage V_0 would be 2.3 V. Thus the voltage V_0 would be a linear function of temperature.

COUNT-RATE METER

Figure 2.20 shows a diagram for a simple count-rate meter. The square-wave source V_{in} has a frequency F_{req} and a peak voltage amplitude V_p. When V_{in} is at its positive value, diode D_1 conducts, causing capacitor C_1 to charge to V_p with the polarity shown in the figure. When V_{in} goes to zero volts, diode D_1 becomes reverse-biased and diode D_2 becomes forward-biased. With D_2 forward-biased, capacitor C_1 is placed in parallel with capacitor C_2. The charge of C_1 is then dumped into capacitor C_2 if C_2 is much greater than C_1. When V_{in} goes positive again, D_2 becomes reverse-biased and D_1 conducts, charging C_1 again. In this state, the charge of C_2 is dumped into the meter movement M. The average current indicated by M is given by

$$I_{ave} = V_p C_1 F_{req} \tag{7}$$

(See Appendix 3 for derivation.)

Figure 2.20
Count-rate circuit using diodes.

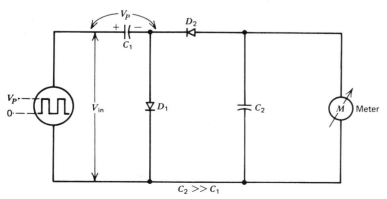

The count-rate meter can be used in such applications as a tachometer, a frequency meter, and a capacitance meter. Notice that with Equation 7 if V_p and C_1 are constants, I_{ave} is directly proportional to F_{req}, or if F_{req} and V_p are held constant, I_{ave} is directly proportional to C_1.

EXAMPLE

Design a count-rate meter to indicate a full-scale current of 1 mA for a frequency of 1 kHz. Assume V_p of the square-wave source is 10 V. Using Equation 7, we have

$$I_{ave} = V_p C_1 F_{req}$$

or

$$1 \times 10^{-3}\ A = 10\ V \times C_1 \times 1 \times 10^3\ Hz$$

$$C_1 = 0.1\ \mu F$$

where

$$C_1 \ll C_2 \quad \text{or} \quad C_1 \times 10 = C_2$$

$$C_2 \geq 1\ \mu F$$

If the frequency were 500 Hz, the 1-mA meter movement would indicate

$$I_{ave} = 10\ V \times 0.1 \times 10^{-6}\ F \times 500\ Hz$$

$$= 0.5\ mA$$

REVIEW QUESTIONS AND PROBLEMS

2-1　What is a diode? List some applications for the diode.

2-2　What is the difference between a real and ideal diode?

2-3　How and why is the barrier potential generated in a PN junction?

2-4　What is the depletion region in a diode and when does it occur?

2-5　When and why is leakage current generated in a diode?

2-6　What is meant by the term forward-biased?

2-7　What is meant by the term reverse-biased?

2-8　Assume that the barrier potential for a diode is 0.63 V at room temperature (25 °C). If the temperature increases to 80 °C, what is the barrier potential at the new temperature?

2-9　If the PIV rating for a diode is exceeded, why is the diode usually destroyed?

2-10　What is a zener diode? List some applications for the zener diode.

2-11　A 5-V power supply is to be designed having a load current from 0 to 150 mA. A 15-volt source is to be used. Calculate the necessary values for R_S and R_z, and the power ratings for the diode and R_s.

2-12　Calculate the effects on the output voltage across the zener diode in Problem 2-11 if the input voltage is changed from 15 to 18 V.

2-13 With the comparison of two zener diodes having the same zener voltage but with different wattages, why does the higher wattage zener have the lower value of zener resistance?

2-14 What is a variable-voltage capacitor? Why does its capacitance change with reverse bias voltage? List some applications for the VVC.

2-15 Can a diode be used as an electronic switch? Explain.

2-16 Why is the barrier potential for a germanium diode (0.3 V) different from the barrier potential for a silicon diode (0.6 V)?

2-17 What is diode bias clamping? Why is the *RC* time constant chosen so that it is much, much greater than the period of the input frequency?

2-18 Assume that an ideal diode is used to suppress the cemf generated by a dc relay. How much time delay does the relay have because of the diode? Assume that the relay has a dc resistance of 1400 Ω, a pull-in current of 10 mA, a drop-out current of 3 mA, and an inductance of 50 mH.

2-19 If a diode and a zener diode were connected as shown, what would the voltage-current curve look like? List some possible applications for this circuit arrangement.

$$A \circ\!\!\!-\!\!\!\!\triangleright\!\!|\!-\!\!\!\overset{\text{10 V}}{|\!\!\triangleleft}\!-\!\!\!\circ B$$

2-20 Assume that a diode has a barrier potential of 0.65 V with a current of 500 mA flowing through the diode. Calculate the power dissipated by the diode if its internal resistance was (a) 25 Ω and (b) 5 Ω.

2-21 Design a count-rate circuit configuration that can be used as a capacitance meter with a range of 0 to 1 μF. The meter used in the device is a 50 μA full-scale instrument. Select the frequency and amplitude for the square-wave generator.

2-22 For a count-rate meter, what is the maximum internal resistance the frequency source can have? Assume a frequency of 1000 Hz with an input capacitor of 0.5 μF. (*Hint*: In five time constants a capacitor is fully charged.)

OBJECTIVES FOR CHAPTER 3,
THE TRANSISTOR AND
CIRCUIT CONFIGURATIONS

At the completion of this chapter you can demonstrate your understanding of the material by answering these questions:

- What is a BJT?
- What are the dc biasing requirements for the BJT?
- What is an amplifier?
- What are the three basic circuit configurations for the BJT?
- How and why is leakage current generated in a BJT?
- What is a decibel?
- Why is it convenient to express the gain of an amplifier in decibels?

3 THE TRANSISTOR AND CIRCUIT CONFIGURATIONS

The bipolar transistor is used in common-emitter, common-base, or common-collector circuit configurations and can function as an amplifier. You should study the transistor and its circuit configurations in order to comprehend why one type of circuit configuration is incorporated instead of another in specific applications.

Chapter 3 introduces the bipolar junction transistor (BJT). The material presented is necessary for the understanding of the NPN or PNP transistor used in a common-emitter, common-base, or common-collector amplifier. Characteristics of the transistor are discussed in both ac and dc circuit configurations.

BIPOLAR JUNCTION TRANSISTOR

The bipolar junction transistor (BJT) is a bipolar, three-terminal, two-junction device. The schematic symbol and semiconductor diagram for the PNP and NPN transistor are shown in Figure 3.1. The PNP transistor is constructed from two-P-type and one N-type blocks of semiconductor material. The NPN transistor is constructed from two N-type and one P-type blocks of semiconductor material. The three terminals of the device are labeled the emitter, the base, and the collector. The two junctions of the device are the emitter-base and base-collector junctions.

The transistors shown in Figure 3.1 are considered bipolar, that is, current through the device is made up of majority current carriers of holes in the P-type material and the majority current carriers of electrons in the N-type material. Such other solid-state devices as the junction field effect transistor (JFET), metal oxide semiconductor field effect transistor (MOS-FET), and unijunction

32

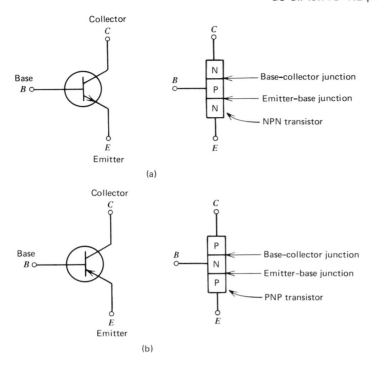

Figure 3.1
Bipolar junction transistors: (a) NPN transistors and (b) PNP transistor.

transistor (UJT) are called unipolar devices. Unipolar implies that the device has only one type of current carrier, holes or electrons. These devices will be discussed later in the text.

THE AMPLIFIER

One of the many applications of the bipolar junction transistor is its use as an amplifier. An amplifier has both an input and an output signal terminal. Figure 3.2 shows the schematic symbol for the amplifier. The input signal is enlarged or amplified in terms of voltage, current, or power and appears at the output terminals of the device. A linear amplifier has an output signal that is the replica of the input signal. A nonlinear amplifier has an output that is not a replica of of the input signal. The next few chapters of this text are concerned with linear amplifiers.

dc BIASING REQUIREMENTS

The bipolar transistor to be used as a linear amplifying device has its emitter-base junction forward-biased and its base-collector junction reverse-biased. These junction biasing requirements are necessary for both the PNP and the

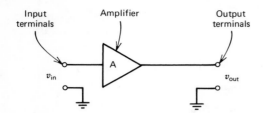

Figure 3.2
Schematic diagram for an amplifier.

NPN transistors and are called "transistor action." The effects of transistor action are shown in Figure 3.3.

With the emitter-base junction forward-biased and the base-collector junction reverse-biased, electrons in the N-type emitter region (NPN transistor) are attracted by the P-type base region because the emitter-base junction is forward-biased. The electrons that cross into the base region are attracted to the collector by the strong electric field caused by the positive potential present at the collector of the transistor. The majority of the electrons that are "emitted" by the emitter region and cross into the base region are "collected" by the collector region of the transistor. Because the electrons are accelerated toward the collector, the electrons do not remain in the base region for a very long period of time; thus the electrons cannot combine with very many holes in the base

Figure 3.3
Current flow: (a) NPN BJT, (b) PNP BJT, and (c) transistor action.

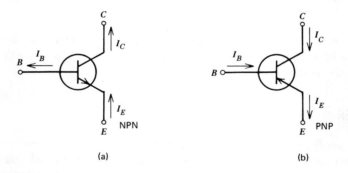

(a)   (b)

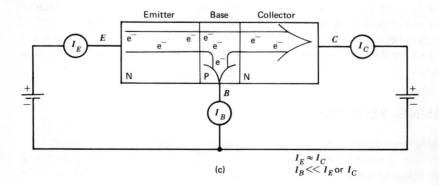

(c) $I_E \approx I_C$
           $I_B \ll I_E \text{ or } I_C$

region. With very few electrons combining with the holes in the base region, the base current is very small compared to either the collector current or emitter current. The dc emitter current is labeled I_E, the dc base current is labeled I_B, and the dc collector current is labeled I_C.

The same reasoning for a small base current can be applied to a PNP (a majority current carrier of holes in the collector and emitter region). There is also a leakage current (I_{CO}) inside the transistor caused by the minority current carriers; leakage current will be discussed at the end of this chapter and in the next few chapters.

The emitter current is equal to the sum of the collector and base currents, or

$$I_E = I_B + I_C \tag{1}$$

where I_E is the dc emitter current, I_B the dc base current, and I_C the dc collector current. With the transistor, the emitter current and the collector current are generally much greater than the base current. As an example, if the collector current is 1.02 mA and the emitter current is 1.0 mA, the base current would be 0.2 mA or 20 μA. The ratio (neglecting leakage current) between the dc base and collector currents is defined as beta (β), or

$$h_{FE} = \beta_{dc} = \frac{I_C}{I_B} \tag{2}$$

The h_{FE} is the hybrid circuit equivalent of beta (β) and will be explained shortly. The ratio (neglecting leakage current) between the dc value of the emitter and the collector current is defined as alpha (α) or

$$h_{FB} = \alpha_{dc} = \frac{I_C}{I_E} \tag{3}$$

The h_{FB} is the hybrid circuit equivalent of alpha (α) and will be explained shortly. Because the base current is much smaller than the emitter or collector currents, the collector current is approximately equal to the emitter current (see Equation 1). The numerical values for h_{FE} and h_{FB} (β_{dc} and α_{dc}) are in the range of 20 to 600 and 0.95 to 0.998, respectively.

The relationships between alpha and beta can be shown with the use of Equations 1, 2, and 3. Substituting Equation 1 into Equation 2, we have

$$h_{FE} = \beta = \frac{I_C}{I_B} = \frac{I_C}{I_E - I_C}$$

or

$$h_{FE} = \beta = \frac{I_C/I_E}{1 - I_C/I_E}$$

but because $h_{FB} = \alpha = I_C/I_E$, it follows that

$$\beta = \frac{\alpha}{1 - \alpha} \tag{4}$$

or

$$h_{FE} = \frac{h_{FB}}{1 - h_{FB}} \tag{5}$$

With like reasoning, substituting Equation 1 into Equation 3, we have

$$h_{FB} = \alpha = \frac{I_C}{I_E} = \frac{I_C}{I_B + I_C}$$

$$= \alpha = \frac{I_C/I_B}{1 + I_C/I_B}$$

Then

$$\alpha = \frac{\beta}{\beta + 1} \tag{6}$$

or

$$h_{FB} = \frac{h_{FE}}{h_{FE} + 1} \tag{7}$$

The relationship between the emitter and base currents is found with the aid of Equations 1 and 2. Because $I_E = I_B + I_C$ and $I_C = h_{FE}I_B$, substituting, we have

$$I_E = I_B + h_{FE}I_B$$

$$= I_B (h_{FE} + 1) \tag{8}$$

or

$$I_E = I_B (\beta + 1) \tag{9}$$

EXAMPLE

Assume that the collector current I_C is equal to 1.5 mA and that h_{FE} (β_{dc}) is equal to 60. Calculate the base current I_B, the emitter current I_E, and the ratio of collector to emitter current h_{FB} (α). Using Equation 2, we can calculate I_B, or

$$I_B = \frac{I_C}{h_{FE}} = \frac{1.5 \text{ mA}}{60}$$

$$= 25 \text{ }\mu\text{A}$$

Using Equation 1, we can calculate I_E, or

$$I_E = I_B + I_C = 0.025 \text{ mA} + 1.5 \text{ mA}$$

$$= 1.525 \text{ mA}$$

Using Equation 7, we can calculate h_{FB}, or

$$h_{FB} = \frac{h_{FE}}{h_{FE} + 1} = \frac{60}{60 + 1}$$

$$= 0.984$$

VOLTAGE AND CURRENT NOMENCLATURE

The method of denoting various currents and voltages has been fairly well standardized. This nomenclature helps denote such variables as dc average values, time-varying values, ac values, dc sources, and rms values.

The ratios of alpha and beta were defined as dc current values. The notation used with these values is an upper-case variable with an upper-case subscript. Thus the variables I_E, I_B, and I_C are all dc steady-state or average values of current.

An ac value of current that varies with time is denoted by a lower-case variable with lower-case subscripts. Thus the variable i_e would be an ac value of emitter current.

If a variable is dc and varies with time, it is denoted by a lower-case variable with an upper-case subscript. Thus the variable i_E is a time-varying dc value of current.

An rms variable is denoted by an upper-case variable with a lower-case subscript. For example, I_e is the rms value of the emitter current. Generally, currents are labeled with only one subscript. Figure 3.4a summarizes the variables discussed here.

A supply voltage source, such as a battery, is labeled by an upper-case variable with two upper-case subscripts. Thus the supply voltage for a transistor would be labeled V_{CC}. Because a voltage is by definition a potential difference, two measurement points are necessary. A voltage variable is generally labeled with two subscripts. The first subscript denotes the point of measurement and the second subscript denotes the reference point. As an example, in Figure 3.4b, the dc voltage across the collector-to-emitter would be labeled V_{CE}. If a voltage variable does not have a second subscript, the reference point for that voltage would be common or ground. In Figure 3.4, the voltage across the resistor R_E would be V_E; the reference point for V_E would be common. As it can be reasoned with the aid of the diagram, $V_C = V_{CE} + V_E$.

The variables in voltages would, as with current variables, be labeled with upper- and lower-case variables. Thus v_{CE} would be a dc, time-varying voltage,

Figure 3.4
Current and voltage nomenclature: (a) current waveforms and (b) voltage measurements of a BJT circuit.

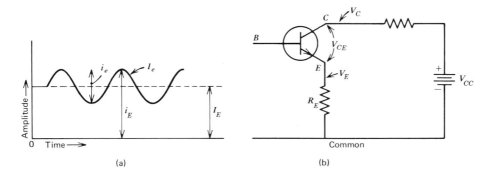

measured from collector-to-emitter of the transistor; v_{ce} would be an ac, time-varying voltage, measured from collector-to-emitter; V_{CE} would be a steady dc voltage, measured from collector-to-emitter of the transistor; and v_C would be a dc time-varying voltage, measured from the collector-to-common or ground. The variable V_c is an rms voltage at the collector of the transistor with respect to common.

CIRCUIT CONFIGURATIONS

An amplifier has an input and output terminal with respect to common. These terminals are shown in Figure 3.2. The bipolar transistor, when used as an amplifier, also has an input and output terminal with respect to common. Because the transistor is a three-terminal device (emitter, base, and collector), one of the terminals is common to the other two terminals. The three possible combinations of circuit configurations for the bipolar transistor used as an linear amplifier are the common-emitter, common-base, or common-collector amplifier. These three circuit configurations are shown in Figure 3.5.

Figure 3.5a shows a common-emitter amplifier (*CE*); its emitter lead is common to both the input terminal (base lead) and the output terminal (collector lead). Figure 3.5b shows a common-base amplifier (*CB*); its base lead is common

Figure 3.5
Amplified configurations: (a) common emitter—CE, (b) common base—CB, and (c) common collector—CC.

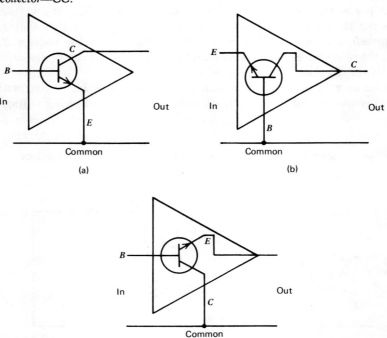

to both the input terminal (emitter lead) and the output terminal (collector lead). Figure 3.5c shows a common-collector amplifier (CC); its collector lead is common to both the input terminal (base lead) and the output terminal (emitter lead).

The three circuit configurations shown in Figure 3.5 are ac circuit configurations. The dc biasing networks are *not* shown in the figure (but are necessary). With each amplifier configuration (CE, CB, and CC), the emitter-base junction is forward-biased and the base-collector junction is reverse-biased. This biasing requirement is necessary for both the NPN and PNP transistor used as a linear amplifier.

Table 3.1 shows the comparison among the three types of circuit configurations just discussed. The table lists the voltage gain (A_v), current gain (A_i), low-frequency input resistance (R_{in}), low-frequency output resistance (R_{out}), and power gain (P_g) for each of the three circuit configurations for the same transistor.

Table 3.1

Comparison of the Common Emitter, Base, and Collector BJT Amplifiers

$$Q = 2N4400, R_L = 4.7 \text{ k}\Omega, I_{CQ} = 1 \text{ mA}, R_g = 100 \text{ k}\Omega$$

	A_i	A_v	R_{in}	R_{out}*	P_g
Common emitter	126	-139	4.2 kΩ	65 kΩ	42.4 dB
Common base	0.992	140	33 Ω	8.5 MΩ	21.4 dB
Common collector	128	0.98	640 kΩ	735 Ω	20.9 dB

* Does not include R_L.

The voltage gain A_v is defined as the ratio of the output voltage to the input voltage, or $A_v = v_{out}/v_{in}$, where v_{out} is the output voltage that is present at the output terminal of the amplifier when an input voltage v_{in} is present at the input terminal of the amplifier. Both of these voltages are measured with respect to common.

The current gain A_i is defined as the ratio of the output current to the input current, of $A_i, = i_{out}/i_{in}$, where i_{out} is the output terminal current with an i_{in} at the input terminal of the amplifier. The low-frequency input resistance is the resistance seen by a source looking into the input terminal of the amplifier with respect to the common. The output resistance is the resistance seen by the load that the amplifier is working into. The load could represent a loud speaker, another amplifier, a load resistor, and so forth.

The power gain P_g for the amplifier is the ratio of the rms output power to the rms input power, or $P_g = P_{out}/P_{in}$. Notice that the voltage gain and current gain are dimensionless quantities; that is, the input and output values have the same units. Having the same units, the input and output voltages and currents could be expressed in peak, peak-to-peak, or rms values.

BIASING TECHNIQUES

Figure 3.6 shows how a single dc voltage source can be used to meet the requirements for transistor action. The figure shows only the dc biasing requirements for the transistor. The NPN transistor is demonstrated. If a PNP transistor were used, the supply voltage would simply be reversed.

Figure 3.6a shows a NPN transistor in the common-emitter circuit configuration. The emitter-base junction is forward-biased because the emitter N-type

Figure 3.6
Biasing requirements for the BJT: (a) CE configuration, (b) CB configuration, and (c) CC configuration.

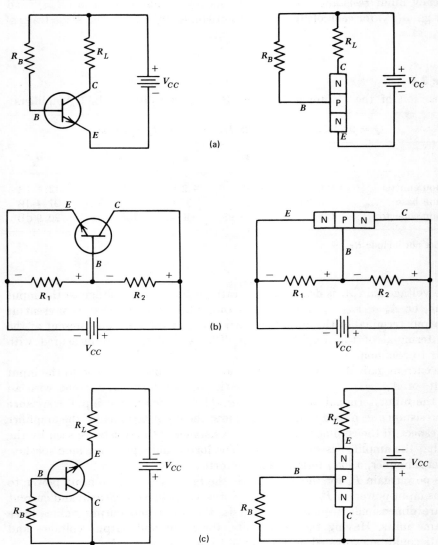

material is at the negative terminal of the voltage source and the base P-type material is connected to the positive terminal of the voltage source through the resistor R_B. The base-collector junction is reverse-biased because the collector N-type material is connected to the positive terminal of the voltage source through the resistor R_L and the base P-type material is connected to the negative terminal of the voltage source through the forward-biased emitter-base junction. The magnitude of the voltage that is reverse-biasing the base-collector junction is approximately $V_{CB} = V_{CE} - V_{BE}$.

Figure 3.6b shows a NPN transistor connected in the common-base circuit configuration. The voltage divider consisting of R_1 and R_2 across the battery voltage supplies the voltage necessary for transistor action. The voltage across R_1 forward-biases the emitter-base junction and the voltage across R_2 reverse-biases the base-collector junction of the transistor. R_2 is much greater than R_1.

Figure 3.6c shows a NPN transistor connected in the common-collector configuration. The emitter-base junction is forward-biased because the emitter N-type material is connected to the negative terminal of the voltage source and the base P-type material is connected to the positive terminal of the voltage source through the resistor R_B. The base-collector junction of the transistor is reverse-biased because the collector N-type material is connected to the positive terminal of the voltage source and the base P-type material is connected to the negative terminal of the voltage source through the forward-biased emitter-base junction of the transistor.

REAL CIRCUIT CONFIGURATIONS

Figure 3.7 shows the *CE*, *CB*, and *CC* circuit configurations, and includes the inputs, outputs, and biasing techniques. The dc biasing technique is only one of the possible techniques that can be used (see Chapters 4, 5, 6, and 7).

Figure 3.7a shows a NPN transistor in the common-emitter configuration. The input and output terminals are measured with respect to the emitter, which is common. The capacitors C_{in} and C_{out} couple the input and output signals of the transistor *without* disturbing the dc biasing points of the junctions, causing the ac signals of the amplifier to be isolated from the dc voltage biasing points necessary for transistor action.

Figure 3.7b shows a NPN transistor in the common-base configuration. The input and output terminals of the amplifier are measured with respect to the base, which is common. The base is common because the capacitor C_1 short circuits the base to the common of the circuit for ac signals. The coupling capacitors C_{in} and C_{out} couple the input and output signals *without* disturbing the dc biasing points of the transistor necessary for transistor action.

Figure 3.7c shows a NPN transistor in the common-collector circuit configuration. The input and output terminals of the amplifier are measured with respect to the collector, which is common. The collector is common because it is connected to the common of the circuit through the capacitor C shown in the circuit. The capacitor C short circuits the collector of the transistor to common for ac values of input and output signals but does not disturb the dc supply voltage V_{CC}.

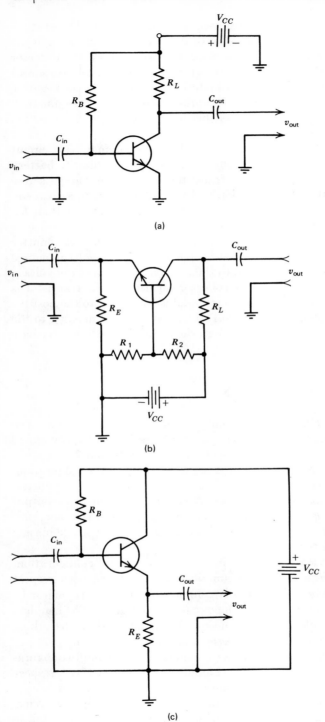

Figure 3.7
Single-stage amplifiers: (a) CE, (b) CB, and (c) CC.

CURRENT RATIOS

As was shown in the common-emitter circuit configuration, the emitter terminal of the transistor was common to the input terminal (base lead) and output terminal (collector lead). With the CE amplifier, the ratio of the dc collector current to the dc base current was given by Equation 2, or

$$h_{FE} = \beta = \frac{I_C}{I_B}$$

As was shown in the common-base circuit configuration, the base terminal of the transistor was common to the input terminal (emitter lead) and output terminal (collector lead). With the CB amplifier, the ratio of the dc collector current to the dc emitter current was given by Equation 3, or

$$h_{FB} = \alpha = \frac{I_C}{I_E}$$

As was seen in the common-collector circuit configuration, the collector terminal of the transistor was common to the input terminal (base lead) and output terminal (emitter lead). With the CC amplifier, the ratio of the dc emitter current to the dc value of the base current can be calculated with the aid of Equation 8, or

$$I_E = I_B(h_{FE} + 1)$$

The value of $h_{FE} + 1$ can be defined to be h_{FC}, or

$$h_{FC} = \frac{I_E}{I_B} = (h_{FE} + 1) \tag{10}$$

The h_{FE}, h_{FB}, and h_{FC} can now be explained. With the different circuit configurations, the input and output currents (dc) were defined with the aid of Equations 2, 3, and 10. The h values are the hybrid values and the subscripts denote the circuit configuration. The h_{FE} is the forward current transfer ratio (subscript F) in the common-emitter circuit configuration (subscript E). The h_{FB} is the forward current transfer ratio in the common-base circuit configuration. The h_{FC} is the forward current transfer ratio in the common-collector circuit configuration. Note that these values are all dc values and are denoted with upper-case subscripts.

As will be demonstrated in Chapters 4, 5, and 8, these hybrid h values can also be ac forward current transfer ratios, denoted as h_{fe}, h_{fb}, and h_{fc}. These ac values will denote the ac input and output currents i_{in} and i_{out}. Thus $h_{fe} = i_c/i_b$, $h_{fb} = i_c/i_e$, and $h_{fc} = i_e/i_b$. The term *hybrid* (h) comes from the circuit model used for the transistor. It is called a hybrid because the circuit model will use a Theveninized voltage source and a Norton current source (see Chapter 8).

TRANSISTOR MANUFACTURING

All bipolar transistors can be thought of as devices that have two junctions, consisting of either PNP- or NPN-structured semiconductor material. The

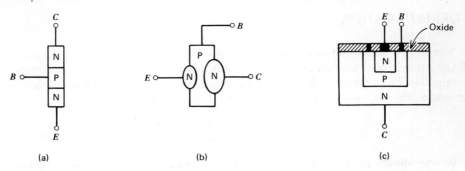

Figure 3.8
Manufacturing BJTs: (a) grown junction, (b) alloy junction, and (c) diffused planar.

two manufacturing techniques that are currently popular are the alloy and
diffused planar processes. The first method used to produce transistors was
the grown-junction method, but because of the limited frequency response and
other drawbacks, this method is not used today. Figure 3.8 illustrates the three
manufacturing methods stated here.

TRANSISTOR SPECIFICATIONS

Some of the following transistor specifications are commonly provided by the
manufacturer.

• BV_{CBO}: This voltage rating is the maximum reverse-biased voltage of the
collector-base junction with the emitter open. With transistor action in the
common-base circuit configuration, this voltage rating is the maximum the
collector-base junction can withstand without breaking down.

• BV_{CEO}: This voltage rating is the maximum reverse-biased voltage of the
collector-emitter junction with the base open. With transistor action in the
common-emitter circuit configuration, this voltage rating is the maximum
voltage between the collector and emitter the device can withstand without
breaking down. The voltage rating BV_{CEO} is less than BV_{CBO} for the same
transistor because the leakage current in the common-emitter mode is much
greater than in the common-base mode. Because there is more leakage current,
avalanche breakdown will occur at a lower voltage across the junction.

• BV_{EBO}: This is the maximum reverse-biased voltage permitted across the
emitter-base junction with the collector open. In using the BJT for a solid-
state switch, BV_{EBO} is a parameter that must not be exceeded.

• P_D: This is the total power that the transistor can dissipate. Generally,
this power rating can be thought of as the power dissipated in the reverse-
biased collector-base junction of the transistor and can be approximated by
$P_D = V_{CE}I_C$ (where $V_{CB} \approx V_{CE}$).

• h_{FE}: This is the current gain or forward transfer current ratio (dc) of the
transistor and is defined as I_C/I_B (neglecting the leakage current I_{CO}).

- $I_{C(\max)}$: This is the maximum current the collector can safely pass without destroying the device. Since the power rating of the device is given as $P_D = V_{CE}I_C$, the maximum collector current would be used with a very low value of V_{CE}. As an example, if P_D equaled 400 mW and $I_{C(\max)}$ equaled 300 mA, the value of V_{CE} for that collector current $I_{C(\max)}$ would be $V_{CE} = P_D/I_C = 400 \text{ mW}/300 \text{ mA} = 1.34 \text{ V}$.

- θ_{JA}: This is the thermal resistance, junction to ambient, of the device and has units of °C/mW or °C/W. The thermal resistance of the transistor is very important when the device is used as a power amplifier. Examples of this parameter will be used and discussed in Chapter 14.

- T_J: This is the maximum junction temperature for the transistor. It is generally given in °C and cannot be exceeded without destroying the device.

Table 3.2 shows the maximum ratings for a 2N4400 NPN silicon transistor and a 2N2141 PNP germanium power transistor. The reader is asked to compare the various ratings listed in the table.

Table 3.2

Comparison of Transistor Characteristics

		2N2141 Ge	2N4400 S$_i$
		Power transistor	General purpose
Collector-base voltage	BV_{CBO}	90 V	60 V
Collector-emitter voltage	BV_{CEO}	65 V	40 V
Emitter-base voltage	BV_{EBO}	45 V	6 V
Collector current (max)	$I_{C(\max)}$	3 A	600 mA
Power dissipation	P_D	62.5 W	310 mW
h_{FE}		25 typical	60 typical
Junction temperature (max)	T_j	100 °C	135 °C
Thermal resistance			
junction-to-ambient	θ_{jA}		0.357 °C/mW
junction-to-case	θ_{jC}	1.2 °C/W	0.137 °C/mW

POWER GAIN

Frequently, for convenience, the gain of an amplifier is expressed in decibels (dB). The decibel is defined as

$$G(\text{dB}) = 10 \log_{10}\left(\frac{P_1}{P_2}\right) \tag{11}$$

where $\log_{10}$ is the logarithm using the base 10 and P_1 and P_2 are powers. If P_1 is defined as the output power of an amplifier and P_2 as the input power of an amplifier, Equation 11 can be written as

$$G(\text{dB}) = 10 \log_{10}\left(\frac{P_{\text{out}}}{P_{\text{in}}}\right) \tag{12}$$

EXAMPLE

If an amplifier has an output power of 10 W and an input power of 10 mW, calculate the dB gain for this amplifier. Using Equation 12, we have

$$G(\text{dB}) = 10 \log_{10}\left(\frac{P_{\text{out}}}{P_{\text{in}}}\right)$$

$$= 10 \log_{10}\left(\frac{10 \text{ W}}{0.01 \text{ W}}\right)$$

$$= 10 \log_{10}(10^3)$$

$$G(\text{dB}) = 30 \text{ dB}$$

It is important to note that the input and output powers that are to be substituted in Equation 12 have the same units (i.e., watts, mWatts, and μWatts).

In order to express the dB gain of an amplifier in terms of voltage (input and output), Equation 12 has to be modified by expressing power in terms of voltage and resistance. Then P_{out} can be expressed in terms of output voltage and output resistance of the amplifier, or

$$P_{\text{out}} = \frac{v_{\text{out}}^2}{R_{\text{out}}} \tag{13}$$

where v_{out} is the ac output voltage of the amplifier, which is across the output resistor R_{out}.

By like reasoning, P_{in} can be expressed as

$$P_{\text{in}} = \frac{v_{\text{in}}^2}{R_{\text{in}}} \tag{14}$$

Substituting P_{out} and P_{in} as expressed in Equations 13 and 14 into Equation 12 we have

$$G(\text{dB}) = 10 \log_{10}\left(\frac{P_{\text{out}}}{P_{\text{in}}}\right)$$

$$= 10 \log_{10}\left(\frac{v_{\text{out}}^2/R_{\text{out}}}{v_{\text{in}}^2/R_{\text{in}}}\right)$$

$$= 10 \log_{10}\left(\frac{v_{\text{out}}}{v_{\text{in}}}\right)^2 + 10 \log_{10}\left(\frac{R_{\text{in}}}{R_{\text{out}}}\right)$$

$$G(\text{dB}) = 20 \log_{10}\left(\frac{v_{\text{out}}}{v_{\text{in}}}\right) + 10 \log_{10}\left(\frac{R_{\text{in}}}{R_{\text{out}}}\right) \tag{15}$$

If the input and output resistances are equal in value ($R_{\text{in}} = R_{\text{out}}$), Equation 15 reduces to

$$G(\text{dB}) = 20 \log_{10}\left(\frac{v_{\text{out}}}{v_{\text{in}}}\right) \tag{16}$$

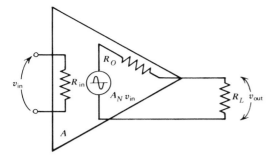

Figure 3.9
A circuit model for a voltage amplifier.

EXAMPLE

Calculate the decibel gain of an amplifier that has an ac input voltage of 10 mV and an output voltage of 10 V. The input resistance in 4 kΩ and the output resistance is 10 Ω (see Figure 3.9). Because the input and output resistances are not the same, Equation 15 will be used.

$$G(\text{dB}) = 20 \log_{10}\left(\frac{v_{\text{out}}}{v_{\text{in}}}\right) + 10 \log_{10}\left(\frac{R_{\text{in}}}{R_{\text{out}}}\right)$$

$$= 20 \log_{10}\left(\frac{10 \, v}{0.01 \, v}\right) + 10 \log_{10}\left(\frac{4000}{10}\right)$$

$$= 20(3) + 10(2.6)$$

$$G(\text{dB}) = 86 \text{ dB}$$

It is important to note that if the gain of an amplifier is expressed in decibels, the power or the voltage and resistance for the amplifier can be calculated.

EXAMPLE

Assume that an amplifier has a gain of 20 dB and that the input and output resistances are equal in value ($R_{\text{in}} = R_{\text{out}}$). Calculate the power and voltage ratio for the amplifier. Using Equation 12 to obtain the power ratio, we have

$$20 \text{ dB} = 10 \log_{10}\left(\frac{P_{\text{out}}}{P_{\text{in}}}\right)$$

$$2 \text{ dB} = \log_{10}\left(\frac{P_{\text{out}}}{P_{\text{in}}}\right)$$

$$10^2 = \frac{P_{\text{out}}}{P_{\text{in}}} \qquad \text{or} \qquad 100 \times P_{\text{in}} = P_{\text{out}}$$

Using Equation 16, we have

$$20 \text{ dB} = 20 \log_{10}\left(\frac{v_{\text{out}}}{v_{\text{in}}}\right)$$

$$1 \text{ dB} = \log_{10}\!\left(\frac{v_{\text{out}}}{v_{\text{in}}}\right)$$

$$10^1 = \frac{v_{\text{out}}}{v_{\text{in}}} \qquad \text{or} \qquad 10 \times v_{\text{in}} = v_{\text{out}}$$

A minus decibel gain $(-\text{dB})$ for a device simply means that the output is less than the input.

EXAMPLE

The output power gain for an amplifier is -30 dB; calculate the power ratio. Using Equation 12 (the $\log_{10}$ has to be equal to or greater than 1),

$$G(\text{dB}) = 10 \log_{10}\!\left(\frac{P_{\text{out}}}{P_{\text{in}}}\right)$$

$$= 10 \log_{10}\!\left(\frac{P_{\text{in}}}{P_{\text{out}}}\right)^{-1}$$

$$= -10 \log_{10}\!\left(\frac{P_{\text{in}}}{P_{\text{out}}}\right)$$

$$-30 \text{ dB} = -10 \log_{10}\!\left(\frac{P_{\text{in}}}{P_{\text{out}}}\right)$$

$$3 \text{ dB} = \log_{10}\!\left(\frac{P_{\text{in}}}{P_{\text{out}}}\right)$$

$$10^3 = \frac{P_{\text{in}}}{P_{\text{out}}} \qquad \text{or} \qquad 1000\, P_{\text{out}} = P_{\text{in}}$$

See Appendix 6 for commonly used decibel values.

REVIEW QUESTIONS AND PROBLEMS

3-1 What is a BJT and how is it constructed?

3-2 What are the biasing requirements for a NPN BJT and a PNP BJT?

3-3 For each of the BJTs shown, list if the base-emitter and base-collector junctions are forward-biased or reverse-biased.

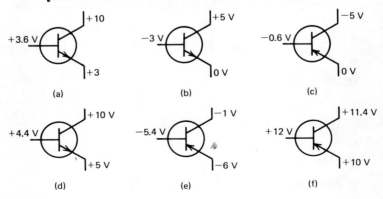

3-4 If a transistor has a collector current of 3 mA and an h_{FE} of 50, calculate the emitter and base currents.

3-5 If a BJT has a beta of 75, calculate its alpha.

3-6 Assume that an amplifier has an input voltage of 30 mV with an input resistance of 4000 Ω. The output voltage is 10 V with an output resistance of 500 Ω. Calculate the gain of this amplifier in decibels.

3-7 Assume that a transmission line has a loss of -2 dB per 100 ft. How many decibels of loss would a mile of this cable have?

3-8 If the cable in Problem 3-7 has an output voltage of 1 V, what is the input voltage? Assume that the input and output resistances are equal.

3-9 What is a common-emitter amplifier? List some of its characteristics.

3-10 What is a common-collector amplifier? List some of its characteristics.

3-11 What is a common-base amplifier? List some of its characteristics.

3-12 If an amplifier has a gain of 30 dB, with an input resistance of 100 kΩ and an output resistance of 50 Ω, what is the ratio of v_{out}/v_{in}?

3-13 What is meant by "A BJT is a current-operated device"?

3-14 For a high beta device, why can the collector current be considered approximately equal to the emitter current?

3-15 Define the current gain for a common-emitter amplifier.

3-16 Define the current gain for a common-base amplifier.

3-17 Define the current gain for a common-collector amplifier.

3-18 For a common-base amplifier, list the input and output currents and voltages.

3-19 Repeat Problem 3-18 for the common-emitter amplifier.

3-20 Repeat Problem 3-18 for the common-collector amplifier.

OBJECTIVES FOR CHAPTER 4, COMMON-EMITTER AMPLIFIER

At the completion of this chapter you can demonstrate your understanding of the material by answering these questions:

- What is a common-emitter amplifier?
- How and why do the dc and ac betas for a BJT differ?
- What is the approximate voltage gain equation for a common-emitter BJT?
- How and why does an emitter-resistor affect the gain and input resistance of a BJT in the common-emitter configuration?
- What is an emitter-bypass capacitor?
- What effect does the dc biasing resistor have on the input resistance of a BJT used in a common-emitter circuit configuration?

4 COMMON-EMITTER AMPLIFIER

The most frequently used circuit configuration for the BJT is the common emitter. You should study the BJT common-emitter amplifier in order to comprehend the specific characteristics of this circuit configuration found in many applications.

The common-emitter configuration is the most frequently used configuration for the bipolar junction transistor BJT. In this chapter approximation equations for voltage and current gain and for input and output resistance are given and dc biasing circuit arrangements for the transistor are discussed. Other methods of dc biasing the BJT may be found in Chapters 6 and 7.

STATIC CHARACTERISTIC CURVES

Figure 4.1 illustrates a transistor in a common-emitter circuit configuration and the plot of its various currents and voltages called static characteristic curves. These curves graphically relate the behavior of the transistor with changes in base current, collector current, collector-to-emitter voltage, and base-to-emitter voltage.

In Figure 4.1a, the base current I_B of the transistor can be varied by changing the ohmic value of R. The emitter-base junction is forward-biased and the base-collector junction is reverse-biased. With fixed values of V_{CE}, the relationship of the base current to the base-to-emitter voltage is shown in Figure 4.1b. These curves are called the input characteristic curves for the transistor in the common-emitter circuit configuration.

Figure 4.1c shows the relationship between the collector-to-emitter voltage and the collector current with fixed increments of base current. These curves are called the common-emitter output characteristic curves. The input and output

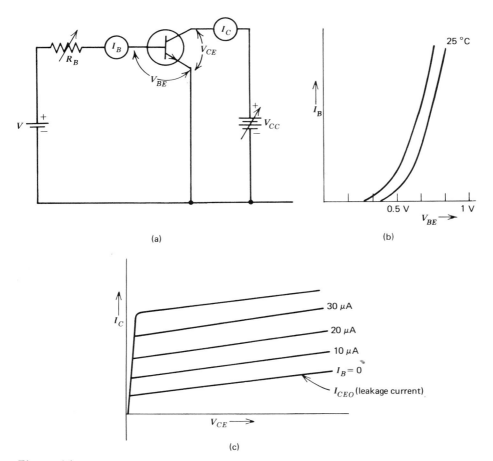

Figure 4.1
Characteristic curves for the CE BJT: (a) circuit to obtain curves, (b) input curves, and (c) output curves.

characteristic curves are concerned with four variables, namely, the input voltage and current (V_{BE} and I_B) and the output voltage and current (V_{CE} and I_C).

dc BIASING

Figure 4.2 shows a circuit configuration for a transistor in the common-emitter mode. The emitter-base junction is forward-biased via V_{CC} (the supply voltage) and R_B (the base resistor). Using Kirchoff's voltage law to obtain the base loop equation in order to calculate the base current I_B, we have

$$0 = V_{CC} - I_B R_B - V_{BE}$$

or

$$I_B = \frac{V_{CC} - V_{BE}}{R_B} \qquad (1)$$

where I_B is the dc base current, V_{CC} is the dc supply voltage, V_{BE} is the dc voltage drop across the base-emitter junction, and R_B is the base resistor. For silicon transistors, V_{BE} is approximately 0.7 V, for germanium transistors V_{BE} is approximately 0.3 V.

Using Kirchoff's voltage law to obtain the collector loop equation in order to calculate the collector-emitter voltage V_{CE}, we have

$$0 = V_{CC} - I_C R_L - V_{CE}$$

or

$$V_{CE} = V_{CC} - I_C R_L \qquad (2)$$

where V_{CC} is the dc supply voltage, V_{CE} the dc voltage drop across the collector-emitter of the transistor, I_C the collector current, R_L the load resistor, and the product $I_C R_L$ the dc voltage drop across the load reistor.

With the aid of Equations 1 and 2, the dc operating or "quiescent" point for the transistor shown in Figure 4.2 can be calculated. To denote the quiescent point, the variables V_{CE} and I_C are labeled V_{CEQ} and I_{CQ}. The upper-case subscript Q denotes that the collector-to-emitter voltage and collector current is the dc quiescent operating point for the transistor without an ac input voltage or current.

For Equations 1 and 2 to be used, a relationship between the base current I_B and the collector current I_C has to be found. The collector current of the transistor is made up of two components, namely, the base current I_B and the leakage current I_{CO}. The leakage current is a product of the minority current carriers in the reverse-biased base-collector junction of the transistor. The collector current of a transistor in the common-emitter configuration is given by

$$I_C = h_{FE} I_B + (h_{FE} + 1) I_{CO} \qquad (3)$$

(Equation 3 will be developed later in this chapter.) The term $(h_{FE} + 1) I_{CO}$ is defined as I_{CEO}. Solving Equation 3 for h_{FE}, we have

$$h_{FE} = \frac{I_C - I_{CO}}{I_B + I_{CO}} \qquad (4)$$

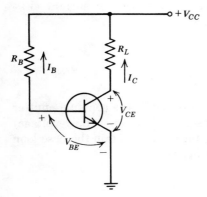

Figure 4.2
Simple biasing method for the CE BJT.

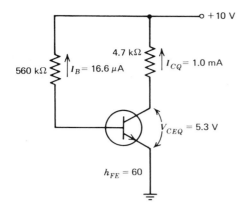

Figure 4.3
Circuit configuration for example problem.

In Equation 4, h_{FE} is the definition for the dc forward current transfer ratio that includes the leakage current I_{CO}. Notice that if I_{CO} is neglected ($I_{CO} = 0\,\mu A$), Equation 4 reduces to $h_{FE} = I_C/I_B$, which was stated in Chapter 3. With Equations 1, 2, and 3 the quiescent point for a transistor in the common-emitter mode shown in Figure 4.2 can be calculated.

EXAMPLE

Assume the following: $h_{FE} = 60$, $R_B = 560$ kΩ, $R_L = 4.7$ kΩ, $I_{CO} = 0.1\ \mu A$, $V_{CC} = 10$ V, and $V_{BE} = 0.7$ V. Calculate I_{CQ} and V_{CEQ} for Figure 4.3.

Using Equation 1 to calculate I_B, we have

$$I_B = \frac{V_{CC} - V_{BE}}{R_B} = \frac{10\ V - 0.7\ V}{560\ k\Omega}$$

$$I_B = 16.6\ \mu A$$

Using Equation 3 to calculate I_{CQ}, we have

$$I_C = h_{FE}I_B + (h_{FE} + 1)\ I_{CO}$$

$$= 60 \times 16.6\ \mu A + (61)(0.1\ \mu A)$$

$$I_C = 1.0\ mA$$

or

$$I_{CQ} = 1.0\ mA$$

Using Equation 2 to calculate V_{CEQ}, we have

$$V_{CE} = V_{CC} - I_C R_L$$

$$= 10\ V - 1\ mA \times 4.7\ k\Omega$$

$$V_{CE} = 5.3\ V$$

or

$$V_{CEQ} = 5.\ 3V$$

GRAPHICAL ANALYSIS

The circuit configuration shown in Figure 4.2 can be joined to the output characteristic curves of the transistor shown in Figure 4.1c. The result of this union is shown in Figure 4.4 and will demonstrate how the transistor can be used as an amplifier.

The loop equation for the collector of the transistor was given by Equation 2, or

$$0 = V_{CC} - I_C R_L - V_{CE}$$

The two variables in the equation are the voltage across the collector-to-emitter voltage (V_{CE}) and the collector current (I_C). If in Equation 2 the collector-to-emitter voltage V_{CE} were equal zero volts, the equation would reduce to

$$I_C = \frac{V_{CC}}{R_L} = I_{C(\text{sat})} \tag{5}$$

where I_C would be the saturation current of the transistor. It is shown in Figure 4.4b at point A.

If the collector current I_C were reduced to zero, Equation 2 would reduce to

$$V_{CE} = V_{CC} \tag{6}$$

where V_{CE} would be the cutoff point of the transistor. It is shown in Figure 4.4b at point B. A straight line connecting points A and B is the dc load line for the circuit configuration illustrated in Figure 4.4a.

As can be reasoned from the dc load line, the variables V_{CE} and I_C for the transistor with a load resistor R_L and supply voltage V_{CC} would always lay on the straight line connecting points A and B. The quiescent point for the transistor would also be located on the dc load line and is labeled (V_{CEQ}, I_{CQ}).

To demonstrate how the circuit configuration illustrated in Figure 4.4a can be used to amplify an input signal, consider the effect of a change in the base

Figure 4.4
dc load: (a) circuit configuration and (b) load line joined to the output curves.

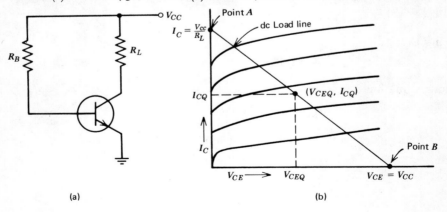

(a) (b)

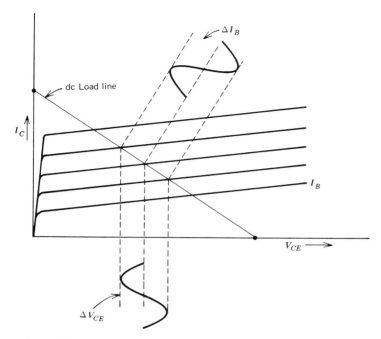

Figure 4.5
dc load line used to represent how voltage gain is achieved with the BJT amplifier.

current for the transistor. If the base current were to change, the collector current of the transistor would change, causing the voltage drop across the load resistor to change in value. If the output signal were taken across the collector-to-emitter of the transistor, a change in the voltage drop across the load resistor would cause a change in the voltage V_{CE} (Equation 2). Because the collector current is equal to the base current times beta, a small change in the base current would cause a relatively large change in the collector current. This relationship is shown in Figure 4.5, where it can be seen that if the base current changed in magnitude (μA), the voltage across the collector-to-emitter would change in magnitude (volts).

Figure 4.6a shows one method of causing the base current of a transistor to change in value. The capacitor C_{in} isolates the ac signal source from the dc biasing network of R_B. The reactance of C_{in} is very small at the frequency of the ac voltage source. The value of R_g is made large in order to convert the voltage source v_s into a current source. The ac current that is pumped into the base of the transistor is labeled i_b and is caused by the voltage source v_s and the resistor R_g. Also at the base of the transistor is the dc base current I_B. The total base current that the transistor sees is the sum of the ac and dc currents, or

$$i_B = i_b + I_B \tag{7}$$

where i_B is the time-varying dc base current, as shown in Figure 4.6b.

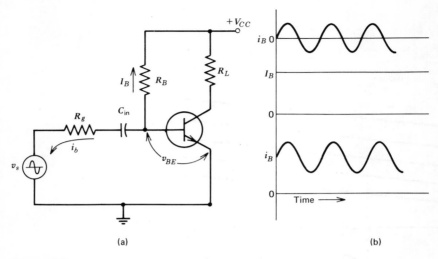

(a)

(b)

Figure 4.6
Voltage source converted into current sources: (a) circuit and (b) waveforms.

EXAMPLE

In the previous example problem, a base current of 16.6 μA produced a quiescent voltage of 5.3 V. Calculate the new value of quiescent voltage V_{CEQ} if the base current were to change to 20 μA. Using Equation 3 to calculate the new collector current, we have

$$I_C = h_{FE}I_B + (h_{FE} + 1)\, I_{CO}$$
$$= 60 \times 20\ \mu\text{A} + (61) \times 0.1\ \mu\text{A}$$
$$I_C = 1.2\ \text{mA}$$

Using Equation 2 to calculate the new value of V_{CEQ}, we have

$$V_{CE} = V_{CC} - I_C R_L$$
$$= 10\ \text{V} - (1.2\ \text{mA}) \times 4.7\ \text{k}\Omega$$
$$V_{CE} = 4.36\ \text{V}$$

Note that in this example problem, a change of 3.4 μA (20 to 16.6 μA) in base current produced a change of 0.94 V (5.3 to 4.36 V) in collector-emitter voltage.

With a change in base current i_B, the base-to-emitter voltage v_{BE} would change. This relationship between i_B and v_{BE} is shown on the input characteristic curves of Figure 4.7. Note that a change in i_B causes a corresponding change in the base-emitter voltage v_{BE}. The magnitude of change in v_{BE} is in millivolts (mV). The voltage gain of the transistor amplifier in a common-emitter circuit configuration is defined as

$$A_v = -\frac{v_{\text{out}}}{v_{\text{in}}} = -\frac{v_{ce}}{v_{be}} \qquad (8)$$

where the negative sign implies the input signal is 180° out of phase with the output signal.

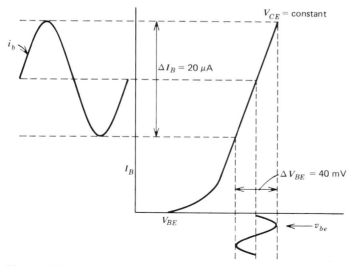

Figure 4.7
Input characteristic curve illustrating how a change in base current produces a change in base-emitter voltage.

Figure 4.8 graphically shows the gain relationship using the input and output characteristic curves for the transistor. The input voltage signal (v_{be}) is 180° out of phase with the output signal (v_{ce}). When the input signal is positive, the base current increases, causing the collector current to increase. An increase in the collector current causes the voltage drop across the load resistor to increase. But an increase in voltage drop across the load resistor causes a decrease in the voltage across the collector-to-emitter of the transistor. Thus the input voltage signal is 180° out of phase with the output voltage signal.

Figure 4.8
Input and output curves used to demonstrate voltage gain for the CE BJT amplifier.

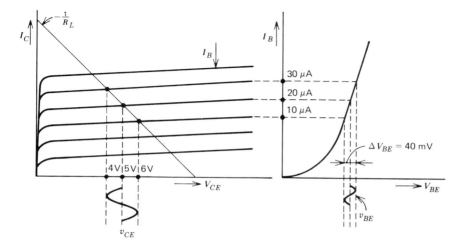

The important concept to retain is that the transistor is a current-operated device but that it can serve or function as a voltage amplifier.

EXAMPLE

Using the data given in Figure 4.8, calculate the voltage gain for this transistor amplifier. In the figure the input voltage v_{be} is equal to 40 mV. The output voltage v_{ce} is equal to 2 V. The voltage gain is then

$$A_v = \frac{-2 \text{ V}}{0.04 \text{ V}} = -50$$

As can be seen in Figure 4.8, a change of 20 μA in base current produces a change of 40 mV in the base-emitter voltage. Because a change of 20 μA is shown in the figure, the value of R_g necessary to produce this change is calculated by

$$R_g \cong \frac{v_s}{i_b}$$

If v_s is assumed to be $2v_{\text{p-p}}$, then R_g would be

$$R_g = \frac{2v_{\text{p-p}}}{20 \ \mu\text{A}} = 100 \text{ k}\Omega$$

ac BETA

The ac forward current transfer ratio for the transistor in the common-emitter configuration is denoted by h_{fe}. The hybrid h has two lower-case subscripts, the first standing for *ac* forward current transfer ratio (f) and the second standing for common-emitter circuit configuration (e). In equation form h_{fe} is defined as

$$h_{fe} = \frac{\Delta I_C}{\Delta I_B} = \frac{i_c}{i_b} \tag{9}$$

The ac h_{fe} is greater in magnitude than the dc h_{fe} because of the leakage current I_{CO}. The leakage current in the dc beta is included with the base current, as indicated in Equation 4. The ac beta is defined as a *change* in currents causing the leakage current not to be included with the base current. (Leakage current is temperature-dependent.)

INPUT RESISTANCE

The ac input resistance of a transistor in the common-emitter circuit configuration is denoted by h_{ie}. The two lower-case subscripts denote the input resistance (i) and the common-emitter circuit configuration (e). The input resistance h_{ie} is defined as the *change* in the input voltage (ΔV_{BE}) over the *change* in the input current (ΔI_B), or

$$h_{ie} = \frac{\Delta V_{BE}}{\Delta I_B} = \frac{v_{be}}{i_b} \tag{10}$$

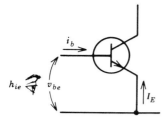

Figure 4.9
Circuit configuration used to calculate h_{ie}.

The input resistance for the transistor facing the base-to-emitter junction of the device can be approximated by knowing the dc emitter current and the h_{fe}. This relationship is shown in Figure 4.9 and is given by

$$h_{ie} = (h_{fe} + 1)\left(\frac{26 \text{ mV}}{I_E \text{ (mA)}}\right) \tag{11}$$

The term $26 \text{ mV}/I_{E(mA)}$ is derived from the diode equation.

EXAMPLE

Assume that the emitter current for a transistor in the common-emitter circuit configuration in 1.0 mA and the h_{fe} is equal to 75. Calculate the input impedance h_{ie}. Using Equation 11, we have

$$h_{ie} = (h_{fe} + 1)\left(\frac{26 \text{ mV}}{I_E \text{ (mA)}}\right)$$

$$= (75 + 1)\left(\frac{26}{1}\right)$$

$$h_{ie} = 1976 \; \Omega$$

APPROXIMATE VOLTAGE GAIN

With an approximate value for the input impedance for the transistor, the ac voltage gain for a transistor with a circuit configuration as shown in Figure 4.10 can be calculated. The ac output voltage (v_{out}) is equal to the voltage across the collector-emitter of the transistor and is also equal to the ac voltage across the load resistor R_L, or

$$v_{\text{out}} = |v_{ce}| = |i_c R_L| \tag{12}$$

where the voltage across the load resistor is given by $i_c R_L$.

The input ac voltage is the voltage across the base-to-emitter junction of the transistor or

$$v_{\text{in}} = v_{be} = i_b h_{ie} \tag{13}$$

where $i_b h_{ie}$ is the input voltage at the base-to-emitter junction of the transistor. With the aid of Equation 8 and with the input and output voltages defined by

Equations 12 and 13, the voltage gain for the common-emitter transistor can be approximated as

$$A_v \approx \frac{-v_{\text{out}}}{v_{\text{in}}} = \frac{-i_c R_L}{i_b h_{ie}} \qquad (14)$$

but $i_c = h_{fe}i_b$, or

$$A_v \approx - \frac{h_{fe}i_b R_L}{i_b h_{ie}}$$

or

$$A_v \approx - \frac{h_{fe} R_L}{h_{ie}} \qquad (15)$$

Equation 15 can be modified by substituting Equation 11 for h_{ie}, or

$$A_v \approx \frac{-h_{fe} R_L}{(h_{fe} + 1) \left[\dfrac{26 \text{ mV}}{I_E(\text{mA})} \right]} \qquad (16)$$

Equation 16 can be compacted by assuming $h_{fe} \approx h_{fe} + 1$, or

$$A_v \approx - \frac{R_L}{26 \text{ mV}/I_E(\text{mA})} \qquad (17)$$

$$A_v \approx - \frac{R_L}{h_{ib}} \qquad (18)$$

where $h_{ib} = 26 \text{ mV}/I_{E(\text{mA})}$.

EXAMPLE

Calculate the voltage gain for a transistor in a common-emitter circuit configuration having an emitter current of 1.0 mA and a load resistor of 4.7 kΩ. Calculating h_{ib}, we have

$$h_{ib} = \frac{26 \text{ mV}}{I_E(\text{mA})}$$

$$= \frac{26}{1}$$

$$h_{ib} = 26 \text{ }\Omega$$

Using Equation 18, we can calculate the voltage gain, or

$$A_v \approx - \frac{4700 \text{ }\Omega}{26 \text{ }\Omega}$$

$$A_v \approx - 181$$

QUIESCENT POINT

With the aid of Equation 18, the ac voltage gain can be approximated for the circuit configuration illustrated in Figure 4.10. The question arises, "Where should the dc operating or quiescent point for the transistor be located on the

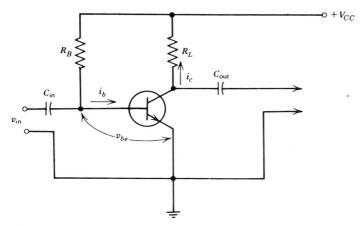

Figure 4.10
Circuit configuration used to calculate A$_v$.

dc load line shown in the output characteristic curves for the common-emitter configuration?" The quiescent point V_{CEQ}, I_{CQ} should be located for a maximum symmetrical voltage swing for the ac output signal at the following points:

$$V_{CEQ} = \frac{V_{CC}}{2} \tag{19a}$$

$$I_{CQ} = \frac{V_{CC}}{2R_L} \tag{19b}$$

for the circuit configuration shown in Figure 4.10. The quiescent point selected with the aid of Equation 19 will permit the output signal (ac) to have a peak value of V_{CEQ} or a peak-to-peak value of $2V_{CEQ}$ or V_{CC}. Thus if the supply voltage is given as 10 V, V_{CEQ} would equal 5 V and I_{CQ} would be calculated with Equation 19.

EXAMPLE

Calculate the value of R_B for a circuit configuration such as the one shown in Figure 4.10 if R_L is equal to 3.3 kΩ, V_{CC} is equal to 15 V, V_{BE} is equal to 0.7 V, and h_{fe} is equal to 50. Using Equation 19 to calculate V_{CEQ} and I_{CQ}, we have

$$V_{CEQ} = \frac{V_{CC}}{2} = \frac{15 \text{ V}}{2} = 7.5 \text{ V}$$

$$I_{CQ} = \frac{V_{CC}}{2R_L} = \frac{15 \text{ V}}{2 \times 3.3 \text{ k}\Omega} = 2.28 \text{ mA}$$

Then

$$I_B = \frac{I_{CQ}}{h_{FE}} = \frac{2.28 \text{ mA}}{50}$$

$$I_B = 46 \text{ μA}$$

Then using Equation 1 and solving for R_B, we have

$$R_B = \frac{V_{CC} - V_{BE}}{I_B} = \frac{15 \text{ V} - 0.7 \text{ V}}{46 \text{ } \mu\text{A}}$$

$$R_B = 311 \text{ k}\Omega$$

CE CONFIGURATION WITH EMITTER-RESISTOR

Figure 4.11 illustrates a circuit configuration for a transistor in the common-emitter circuit configuration with an emitter-resistor. The emitter-resistor R_E is used to obtain better stability against beta and temperature changes for the quiescent point. The base current is given by

$$I_B = \frac{V_{CC} - V_{BE} - I_E R_E}{R_B} \tag{20}$$

But $I_E = (h_{FE} + 1) I_B$, or

$$I_B = \frac{V_{CC} - V_{BE}}{R_B + (h_{FE} + 1) R_E} \tag{21}$$

If the emitter current is assumed to equal the collector current, Kirchhoff's voltage law for the collector loop equation Figure 4.11 is given by

$$V_{CC} = I_{CQ}(R_L + R_E) + V_{CEQ} \tag{22}$$

EXAMPLE

Calculate the quiescent point for the circuit configuration shown in Figure 4.11 if $R_E = 2.7 \text{ k}\Omega$, $R_L = 4.7 \text{ k}\Omega$, $R_B = 750 \text{ k}\Omega$, $V_{CC} = 15 \text{ V}$, $h_{FE} = 75$, $V_{BE} = 0.6 \text{ V}$, and $I_{CO} = 1 \text{ } \mu\text{A}$.

Using Equation 21 to calculate I_B, we have

$$I_B = \frac{15 \text{ V} - 0.6 \text{ V}}{750 \text{ k}\Omega + (76)(2.7 \text{ k}\Omega)}$$

$$I_B = 15 \text{ } \mu\text{A}$$

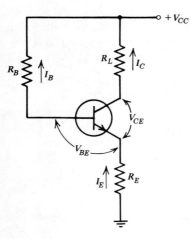

Figure 4.11
A CE BJT amplifier having an emitter resistor R_E.

The collector current is given by Equation 3, or

$$I_C = h_{FE}I_B + (h_{FE} + 1)\, I_{CO}$$
$$= 75 \times 15\ \mu\text{A} + (76)\ 1\ \mu\text{A}$$
$$I_C = 1.2\ \text{mA}$$

Using Equation 22 to calculate V_{CEQ}, we have

$$V_{CEQ} = V_{CC} - I_{CQ}(R_L + R_E)$$
$$= 15\ \text{V} - 1.2\ \text{mA}(4.7\ \text{k}\Omega + 2.7\ \text{k}\Omega)$$
$$V_{CEQ} = 6\ \text{V}$$

INPUT RESISTANCE WITH EMITTER-RESISTOR

Figure 4.12 illustrates the input resistance for a common-emitter transistor with an emitter-resistor R_E. The emitter-resistor R_E is "reflected" into the base of the transistor by a factor of $h_{fe} + 1$. This effect is demonstrated below. The input voltage v_{in} is equal to

$$v_{\text{in}} = i_b h_{ie} + i_e R_E \tag{23}$$

but $i_e = (h_{fe} + 1)\, i_b$, or

$$v_{\text{in}} = i_b h_{ie} + (h_{fe} + 1)\, i_b R_E \tag{24}$$

Then the input resistance is defined as v_{in}/i_b or

$$R_{\text{in}} = \frac{v_{\text{in}}}{i_b} = h_{ie} + (h_{fe} + 1)\, R_E$$

Because $h_{ie} = (h_{fe} + 1)\, h_{ib}$, we have

$$R_{\text{in}} = (h_{fe} + 1)(R_E + h_{ib}) \tag{25}$$

Figure 4.12
The effect of R_E on R_{in}: (a) circuit configuration and (b) equivalent base circuit.

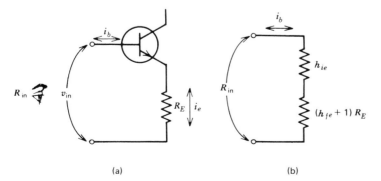

(a)   (b)

EXAMPLE

Assume that $h_{fe} = 75$, $I_E = 1.5$ mA, and $R_E = 2$ kΩ; calculate the input resistance R_{in}. Using Equation 25, we have

$$R_{in} = (h_{fe} + 1)(R_E + h_{ib})$$

$$= (76)\left(2 \text{ k}\Omega + \frac{26}{1.5}\right)$$

$$R_{in} = 153 \text{ k}\Omega$$

ac VOLTAGE GAIN WITH EMITTER-RESISTOR

Figure 4.13 shows an input signal at the base of the transistor with respect to common. The ac input voltage with respect to common is given by Equation 24, or

$$v_{in} = i_b h_{ie} + (h_{fe} + 1) i_b R_E \tag{26}$$

The ac output voltage v_{out} is the dc voltage across the load resistor, or

$$v_{out} = i_c R_L \tag{27}$$

Because the voltage gain is defined as

$$A_v = \frac{-v_{out}}{v_{in}}$$

we have

$$A_v = \frac{-i_c R_L}{i_b h_{ie} + (h_{fe} + 1) i_b R_E}$$

Figure 4.13
A common-emitter amplifier with feedback.

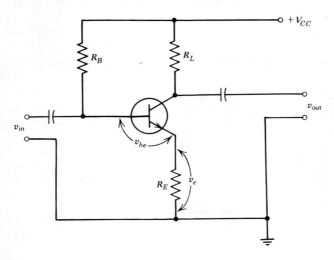

But $i_c = h_{fe}i_b$, or

$$A_v = \frac{-h_{fe}R_L}{h_{ie} + (h_{fe} + 1)\ R_E} \tag{28}$$

If we assume that $h_{ie} = (h_{fe} + 1)$ and $(h_{fe} + 1) \approx h_{fe}$, Equation 28 reduces to

$$A_v = \frac{-R_L}{h_{ib} + R_E} \tag{29}$$

If h_{ib} is much smaller than R_E, Equation 29 reduces to

$$A_v \approx -\frac{R_L}{R_E} \tag{30}$$

EXAMPLE

Calculate the voltage gain for the circuit configuration shown in Figure 4.13. Assume that $R_E = 1\ \mathrm{k\Omega}$, $R_L = 4.7\ \mathrm{k\Omega}$, and $I_{CQ} = 1.0\ \mathrm{mA}$. We can assume that $I_{CQ} \approx I_E$; then

$$h_{ib} = \frac{26\ \mathrm{mV}}{I_E(\mathrm{mA})} = \frac{26\ \mathrm{mV}}{2.0\ \mathrm{mA}}$$

$$h_{ib} = 26\ \Omega$$

Because h_{ib} is 26 Ω and R_E is 1 kΩ, Equation 30 can be used to calculate the approximate voltage gain, or

$$A_v \approx -\frac{R_L}{R_E} = -\frac{4700\ \Omega}{1000\ \Omega}$$

$$A_v \approx -4.7$$

The voltage gain for a circuit configuration that has an emitter-resistor is much less than that for the same circuit without the emitter-resistor (see the example problems). The voltage that is amplified by the transistor is the ac base-emitter voltage v_{be}. As can be seen in Figure 4.13, the input voltage v_{in} is equal to the sum of the base-emitter voltage v_{be} and the ac voltage across the emitter-resistor v_e, or

$$v_{\mathrm{in}} = v_{be} + v_e \tag{31}$$

where $v_e = i_e R_E$. Solving Equation 31 for v_{be}, we have

$$v_{be} = v_{\mathrm{in}} - v_e \tag{32}$$

where v_{be} is the voltage amplified by the transistor.

Notice that the emitter voltage is subtracted from the input voltage. As the input voltage v_{in} increases, the base, and thus the emitter, currents increase. An increase in the emitter current i_e causes the emitter voltage v_e to increase. Thus, as the input voltage increases, the emitter voltage increases also. The voltage gain for this stage is controlled with the selection of R_E and R_L.

LEAKAGE CURRENT

The leakage current for a transistor in the common-emitter configuration is denoted by I_{CEO} and is defined with the base current reduced to zero. If in Equation 4, the base current were equal to zero, the collector current would reduce to

$$I_C = (h_{fe} + 1) \, I_{co} \tag{33}$$

Frequently, the term $(h_{FE} + 1) \, I_{co}$ is defined as I_{CEO}, where the three upper-case subscripts denote the collector-to-emitter current with the base led open (CEO). The leakage current in the reverse-biased base-collector junction is caused by the minority current carriers in the P- and N-type material and is denoted by the variable I_{co}. Because the transistor is in the common-emitter circuit configuration, the leakage current I_{co} must travel through the base-emitter junction and thus is multiplied by the factor $(h_{FE} + 1)$ as illustrated in the following discussion.

The collector current for a transistor in the common-base circuit configuration is given by

$$I_C = \alpha I_E + I_{co} \tag{34}$$

Because of the transistor action, $I_E = I_B + I_C$. Substituting this value of I_E in Equation 34 to obtain variables in I_B and I_C, we have

$$I_C = \alpha(I_B + I_C) + I_{co}$$

or

$$I_C = \frac{\alpha}{1 - \alpha} \, I_B + \frac{1}{1 - \alpha} \, I_{co} \tag{35}$$

Recalling that $\alpha/(1 - \alpha) = \beta$ and $1/(1 - \alpha) = (\beta + 1)$ we find that Equation 35 becomes

$$I_C = \beta I_B + (\beta + 1) \, I_{co} \tag{36}$$

or

$$I_C = h_{FE} I_B + (h_{FE} + 1) \, I_{co} \tag{37}$$

If in Equation 37, the base current were set to zero, I_C would become

$$I_C = (h_{FE} + 1) \, I_{co}$$

which is the same as Equation 33. The collector current can then be defined as leakage current for the transistor in the common-emitter configuration, or

$$I_{CEO} = (h_{FE} + 1) \, I_{co} \tag{38}$$

The term I_{CEO} is noted on the output characteristic curves shown in Figure 4.1c.

EMITTER BYPASS CAPACITOR

The emitter-resistor in a common-emitter circuit configuration was shown to reduce the voltage gain of the stage. The advantage of the emitter-resistor is that it increases the input resistance for the stage and increases temperature stability.

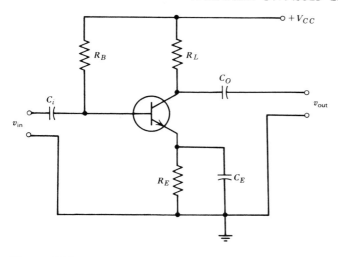

Figure 4.14
A CE circuit configuration with R_E bypassed.

If the emitter-resistor R_E is to be incorporated in a circuit configuration for quiescent point stabilization and if a large voltage gain is needed, the emitter-resistor can be *bypassed* with a capacitor. This arrangement is shown in Figure 4.14.

The capacitor C_E is chosen so that it short-circuits the emitter-resistor R_E for the ac signal (v_e) across the resistor R_E only. As a rule of thumb, the reactance of C_E should be $1/10$ the ohmic value of R_E at the lowest frequency of the input signal v_{in}. As an example, if R_E were 1 kΩ, the reactance of C_E (X_C) should equal 100 Ω at the lowest frequency of the input source. The reactance of a capacitor is given by $X_C = 1/(2\pi fC)$, where f is the frequency in hertz, C is the capacitance in farads, and X_C is in ohms.

The voltage gain for a circuit such as the one illustrated in Figure 4.14 would be given by

$$A_v \approx - \frac{R_L}{h_{ib}}$$

The input resistance for the stage would be given by h_{ie}, or $h_{ie} \approx (h_{fe} + 1)\, h_{ib}$.

PARTIALLY BYPASSED EMITTER-RESISTOR

A compromise between relatively high voltage gain and high input resistance can be accomplished by partially bypassing the emitter-resistor R_E. This arrangement is shown in Figure 4.15.

The voltage gain is given by

$$A_v \approx - \frac{R_L}{h_{ib} + R'_E} \tag{39}$$

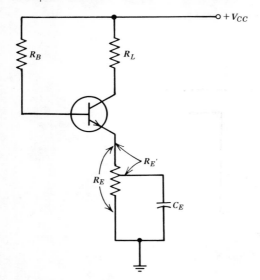

Figure 4.15
A CE circuit configuration with R_E partially bypassed.

where R'_E is the "un-bypassed" portion of the emitter-resistor R_E. The input resistance is given by

$$R_{in} = (h_{fe} + 1)(R'_E + h_{ib}) \tag{40}$$

EXAMPLE

Calculate the voltage gain and input resistance for the circuit shown in Figure 4.16. In the figure, R'_E equals 1 kΩ. Using Equation 39, we can calculate the voltage gain

$$h_{ib} = \frac{26}{1.5} = 17 \ \Omega$$

and

$$A_v = \frac{-4700 \ \Omega}{17 \ \Omega + 1000 \ \Omega} = -4.6$$

Using Equation 40 to calculate the input resistance, we have

$$R_{in} = (h_{fe} + 1)(R'_E + h_{ib})$$

$$= (76)(1017 \ \Omega)$$

$$R_{in} = 77,292 \ \Omega$$

R$_{in}$ INCLUDING R$_B$

The input impedances for the circuit configurations discussed in this chapter have to be modified to include the dc base biasing resistor R_B. The equations given for the input resistances were ac input resistances for the various circuit

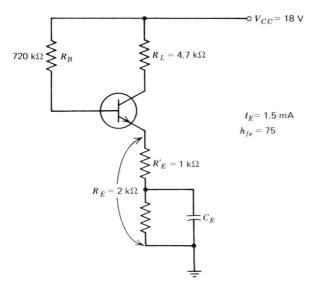

Figure 4.16
Circuit configuration for example problem.

configurations. The ac circuit configuration has to be included with the dc biasing networks discussed. In all the circuitry in this chapter, the base resistor R_B is in parallel with the ac input resistance R_{in}. To denote the parallel combination of R_B and R_{in}, the variable R'_{in} is used and is defined as

$$R'_{in} = R_{in} \| R_B \tag{41}$$

EXAMPLE

Calculate R'_{in} for the circuit shown in Figure 4.16. From the preceding example problem R_{in} is 77,292 and R_B is 720 kΩ. Using Equation 41, we have

$$R'_{in} = R_{in} \| R_B$$
$$= 77.3 \text{ k}\Omega \| 720 \text{ k}\Omega$$
$$R'_{in} = 69.8 \text{ k}\Omega$$

ac LOAD LINE

Frequently, the transistor amplifier is feeding an external load. The external load may be simply the input resistance of another amplifier. Figure 4.17*a* illustrates a BJT amplifier driving an external load resistor of 5 kΩ. The external load is connected to the amplifier via a coupling capacitor C. The capacitor C transfers only the ac signal developed by the amplifier.

The dc load line is calculated as follows (see Figure 4.17*b*).

$$V_{CE} = V_{CC} = 15 \text{ V}$$

$$I_{C(max)} \approx \frac{V_{CC}}{R_L} = \frac{15 \text{ V}}{5 \text{ k}\Omega} = 3 \text{ mA}$$

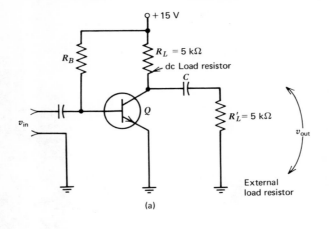

(a)

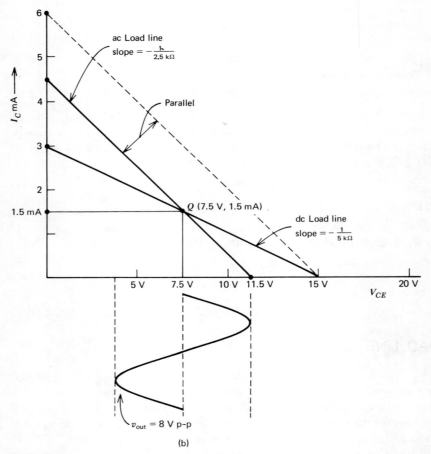

(b)

Figure 4.17
ac load: (a) circuit configuration and (b) load lines.

The base resistor R_B is chosen so that the quiescent point is located at the point $I_{CQ} = 1.5$ mA and $V_{CEQ} = 7.5$ V. The ac load that the transistor "sees" is the parallel combination of the dc load resistor R_L and the external load Resistor R'_L, or

$$R = R_L \| R'_L = 5 \text{ k}\Omega \| 5 \text{ k}\Omega$$

$$R = 2.5 \text{ k}\Omega$$

The transistor would normally have a load line as shown with the dotted line in the figure, with an effective total load resistance of 2.5 kΩ. The two points for calculating this load line would be

$$V_{CE} = 15 \text{ V}$$

$$I_{C(\text{max})} = \frac{15 \text{ V}}{R_L \| R'_L} = \frac{15 \text{ V}}{2.5 \text{ k}\Omega} = 6 \text{ mA}$$

Now because the dc and ac load lines *must pass through the same quiescent point* (7.5 V, 1.5 mA), the ac load line is drawn parallel to the dotted load line and passes through Q (see Figure 4.17b). The slope for the dc load line is $-1/5$ kΩ, and for the ac load line is $-1/2.5$ kΩ. Note in Figure 4.17b that the maximum peak-to-peak ac voltage across the external load resistor is 8 V$_{\text{p}-\text{p}}$.

As can be reasoned from the figure, if the external load resistor is ten times greater than the dc load resistor, the dc and ac load lines are considered the same. Thus if the BJT amplifier is driving an external load, the effective resistance "seen" by the amplifier is the parallel combination of the ac load resistance R'_L and the dc load resistance R_L.

REVIEW QUESTIONS AND PROBLEMS

4-1   Show that h_{fe} is always greater than or equal to h_{FE}.

4-2   Calculate h_{FE} and h_{fe}, assuming an I_B of 15 μA, I_C of 3 mA, I_{CO} of 2 μA, ΔI_C of 1 mA and a ΔI_B of 4 μA.

4-3   For the following, calculate the values for R_L and R_B, assuming an I_{CQ} of 2 mA and a V_{CEQ} of 10 V.

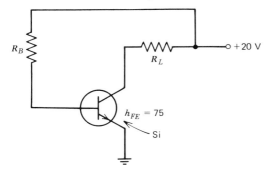

4-4 If the h_{FE} in Problem 4-3 changed to 95, calculate the new values of I_{CQ} and V_{CEQ}.

4-5 Calculate the collector current, collector-to-emitter voltage, and base current for the following circuit configurations. Assume an I_{CO} of 0 μA and a V_{BE} of 0.6 V.

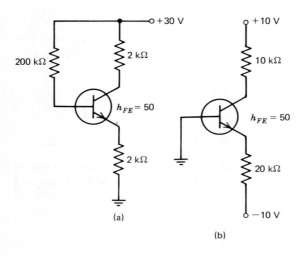

(a)

(b)

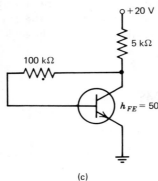

(c)

4-6 For the following circuit configuration, calculate the values for R_g and R_B to obtain the maximum output voltage for the amplifier.

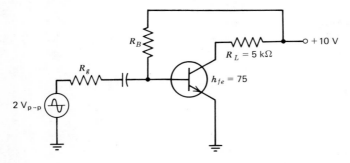

4-7 If an emitter resistor of 2.7 kΩ is to be bypassed with a capacitor, what is the minimum value of capacitance needed if the lowest frequency of the amplifier is 100 Hz?

4-8 For the circuit configuration shown, what is the maximum peak-to-peak output signal? What value of R_g is necessary to obtain this maximum value of output signal? What is the voltage gain of the amplifier?

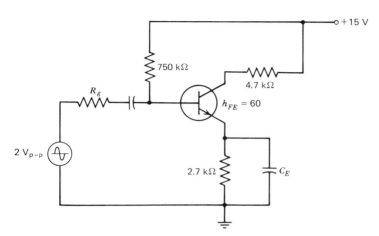

4-9 Calculate the input resistances for the circuits. Assume that all transistors have an h_{fe} of 60.

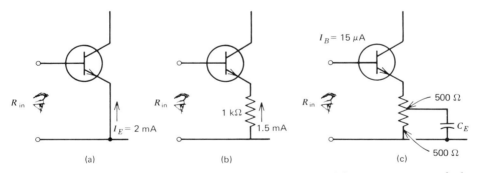

4-10 In Problem 4-5, calculate the new collector and base currents and the collector-emitter voltage if the beta changed to 75.

4-11 Design a common-emitter amplifier with a partially bypassed emitter-resistor having a voltage gain of -30 and an input resistance of 30 kΩ. Assume a supply voltage of 15 V, a quiescent current of 1.0 mA, and an h_{fe} of 99. (*Hint*: Manipulate the voltage gain and input resistance equations for the common-emitter amplifier with a partially bypassed emitter-resistor.)

4-12 If the quiescent collector-emitter voltage is selected to be one-half of the supply voltage for a circuit configuration having only a load and base resistors, why will the output voltage for the amplifier be a maximum?

4-13 Discuss why there is a 180° phase relationship between the input and output signals of a common-emitter amplifier.

4-14 If a BJT is a current-operated device, how and why can it be used as a voltage amplifier?

4-15 Draw the dc and ac load lines for a common-emitter amplifier having an emitter and load resistors R_E and R_L. The R_E is bypassed with a capacitor for ac signals. (*Hint*: the dc and ac load lines must pass through the same quiescent operating point.)

4-16 For the amplifier discussed in Problem 4-15, why is it not a good idea for the operating point to be equal to one-half the supply voltage?

OBJECTIVES FOR CHAPTER 5, COMMON-COLLECTOR AND COMMON-BASE AMPLIFIERS

At the completion of this chapter you can demonstrate your understanding of the material by answering these questions:

- What are the characteristics of the common-collector amplifier and of the common-base amplifier?
- What are the dc biasing requirements for common-collector and common-base amplifiers?
- What are some applications for the common-collector amplifier and for the common-base amplifier?
- What is a BJT current pump?
- What are some applications for the BJT current pump?

5 COMMON-COLLECTOR AND COMMON-BASE AMPLIFIERS

The BJT used in a common-collector or common-base circuit configuration has different circuit characteristics from that used in a common-emitter circuit configuration. You should study common-collector and common-base amplifier to understand how and why these different characteristics can be advantageous in circuit applications.

In this chapter common-collector and common-base BJT circuit configurations are discussed. At the completion of the chapter, the reader is advised to review Table 3.1 in Chapter 3, in which the three types of BJT amplifiers (CE, CB, and CC) are compared.

This chapter also discusses and develops approximate equations for the input and output resistances and voltage gains for common-collector and common-base amplifiers with applications of these amplifiers. The chapter concludes with a discussion of the BJT constant-current pump.

Chapter Eight provides a more detailed discussion of common-collector and common-base amplifiers.

COMMON-COLLECTOR AMPLIFIER

A common-collector amplifier (sometimes referred to as an emitter-follower) is shown in Figure 5.1. The collector of the BJT is common with respect to the base and emitter of the BJT. The input signal is coupled to the base of the BJT via C_{in} and the output signal is taken from the emitter of the BJT. The input and output signals are in phase with each other. As the input signal swings positive, the ac base current increases. As the ac base current increases, the ac emitter current increases. Because the output voltage is the product of i_e and R_L, the input voltage and output voltage are in phase.

78

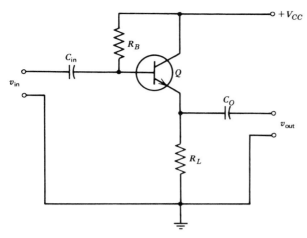

Figure 5.1
Circuit configuration for a common-collector BJT amplifier.

dc BIASING REQUIREMENTS

The biasing requirements for the common-collector amplifier are that the base-emitter junction be forward-biased and that the base-collector junction be reverse-biased. Figure 5.2 illustrates one of the possible dc biasing methods for the common-collector amplifier. The forward bias voltage for the base-emitter junction is the difference between the voltage drops across R_B and R_L. The reverse bias voltage across the base-collector junction is the difference between the supply voltage and the voltage drop across R_B.

As shown in Chapter 3, the relationship between the dc base current and the dc emitter current is given by

$$I_E = (h_{FE} + 1) I_B = h_{FC} I_B \tag{1}$$

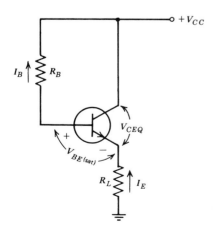

Figure 5.2
Circuit configuration illustrating the biasing requirements for the common-collector amplifier.

The dc collector loop equation, from Figure 5.2, is found to be

$$V_{CC} = V_{CE} + I_E R_L \tag{2}$$

The dc base loop equation is found to be

$$V_{CC} = I_B R_B + V_{BE\text{(sat)}} + I_E R_L \tag{3}$$

Now substituting Equation 1 into Equation 3 and solving for I_B, we have

$$I_B = \frac{V_{CC} - V_{BE\text{(sat)}}}{R_B + (h_{FE} + 1)\, R_L} \tag{4}$$

EXAMPLE

Assume that a circuit configuration such as the one shown in Figure 5.2 has the following components and parameters: $V_{CC} = 10$ V, $R_L = 2.7$ kΩ, $R_B = 150$ kΩ, $V_{BE\text{(sat)}} = 0.6$ V, and $h_{FE} = 75$. Calculate the collector-emitter voltage V_{CE}, the emitter current I_E, and the base current I_B. Using Equation 4 to find I_B, we have

$$I_B = \frac{V_{CC} - V_{BE\text{(sat)}}}{R_B + (h_{FE} + 1)\, R_L}$$

$$= \frac{10\text{ V} - 0.6\text{ V}}{150\text{ k}\Omega + 76 \times 2.7\text{ k}\Omega}$$

$$I_B = 26.5\ \mu\text{A}$$

The emitter current is calculated with Equation 1.

$$I_E = (h_{FE} + 1)\, I_B$$

$$= (76)\ 26.5\ \mu\text{A}$$

$$I_E = 2.01\text{ mA}$$

Using Equation 2 to calculate V_{CE}, we have

$$V_{CC} = V_{CE} + I_E R_L$$

$$10\text{ V} = V_{CE} + 2.01\text{ mA} \times 2.7\text{ k}\Omega$$

or

$$V_{CE} = V_{CEQ} = 4.57\text{ V}$$

If the base resistor R_B and load resistor R_L are to be calculated by our knowing V_{CE} and I_E, Equations 2 and 3 can be modified. The value of R_L is then found by our solving Equation 2 for R_L, or

$$R_L = \frac{V_{CC} - V_{CE}}{I_E} \tag{5}$$

For a maximum ac output voltage, V_{CE} should be equal to one-half of the supply voltage or

$$V_{CE} = V_{CEQ} = \tfrac{1}{2}V_{CC}$$

Substituting this value for V_{CE} into Equation 5, we have

$$R_L = \frac{V_{CC}}{2I_E} \tag{6}$$

Now solving Equation 3 for R_B, we have

$$R_B = \frac{V_{CC} - I_E R_L - V_{BE(\text{sat})}}{I_B} \tag{7}$$

But we assumed that $V_{CE} = \frac{1}{2}V_{CC}$, so that $I_E R_L$ also equals $\frac{1}{2}V_{CC}$. Substituting this value for $I_E R_L$ into Equation 7, we have

$$R_B = \frac{\frac{1}{2}V_{CC} - V_{BE(\text{sat})}}{I_B} \tag{8}$$

EXAMPLE

Assume that a 2N4400 transistor is to be used as a common-collector amplifier with a supply voltage of 15 V and an emitter current of 1.0 mA. The h_{FE} is 75 and $V_{BE(\text{sat})}$ is 0.6 V. Calculate the necessary values for R_B and R_L for this circuit configuration. Assuming that V_{CE} is equal to $\frac{1}{2}V_{CC}$, we can calculate R_L with Equation 6, or

$$R_L = \frac{V_{CC}}{2I_E}$$

$$= \frac{10 \text{ V}}{2 \times 10 \text{ mA}}$$

$$R_L = 7.5 \text{ k}\Omega$$

The base current is calculated with Equation 1, or

$$I_B = \frac{I_E}{(h_{FE} + 1)}$$

$$= \frac{1.0 \text{ mA}}{76}$$

$$I_B = 13.16 \text{ }\mu\text{A}$$

Equation 8 is used to calculate the base resistor R_B, or

$$R_B = \frac{\frac{1}{2}V_{CC} - V_{BE(\text{sat})}}{I_B}$$

$$= \frac{7.5 \text{ V} - 0.6 \text{ V}}{13.16 \text{ }\mu\text{A}}$$

$$R_B = 524 \text{ k}\Omega$$

INPUT RESISTANCE

If we neglect the base resistor R_B, the ac input resistance into the base with respect to common for the common-collector amplifier shown in Figure 5.3 is approximated by

$$R_{\text{in}} = \frac{v_{\text{in}}}{i_{\text{in}}} \tag{9}$$

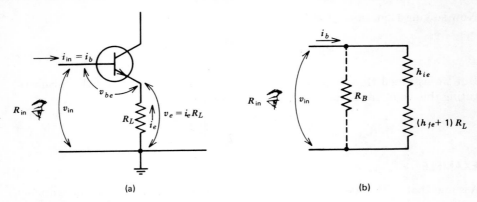

Figure 5.3
Input resistance of the CC amplifier: (a) circuit and (b) equivalent base circuit.

But from the figure we see that $i_{in} = i_b$ and $v_{in} = v_{be} + v_e$. Substituting these relationships into Equation 9, we have

$$R_{in} = \frac{v_{be} + v_e}{i_b} \tag{10}$$

Because $v_e = i_e R_L$, Equation 10 becomes

$$R_{in} = \frac{v_{be} + i_e R_L}{i_b}$$

or

$$R_{in} = \frac{v_{be}}{i_b} + \frac{i_e}{i_b} R_L \tag{11}$$

Recalling that in Chapter 4 the quantity v_{be}/i_b was equal to h_{ie} and that $i_e = i_b(h_{fe} + 1)$ and substituting these relationships into Equation 11, we have

$$R_{in} = h_{ie} + (h_{fe} + 1)\ R_L \tag{12}$$

Equation 12 can be modified by letting h_{ie} be equal to $(h_{fe} + 1)\ h_{ib}$, or

$$R_{in} = (h_{fe} + 1)\ h_{ib} + (h_{fe} + 1)\ R_L$$

or

$$R_{in} = (h_{fe} + 1)(h_{ib} + R_L) \tag{13}$$

where $h_{ib} = (26\ \text{mV})/[I_E(\text{mA})]$

EXAMPLE

From the previous example problem, calculate the ac input resistance for the common-collector amplifier. Assume that $h_{FE} = h_{fe}$. As was calculated, the load resistor and base resistor were $R_L = 7.5\ \text{k}\Omega$ and $R_B = 524\ \text{k}\Omega$. The emitter current I_E equaled 1.0 mA and the beta equaled 75.

In order to calculate the input resistance (neglecting R_B) using Equation 13, we have to find the quantity h_{ib} by

$$h_{ib} = \frac{26 \text{ mV}}{I_E(\text{mA})}$$

$$= \frac{26 \text{ mV}}{1.0 \text{ mA}}$$

$$h_{ib} = 26 \text{ } \Omega$$

Now using Equation 13, we have

$$R_{in} = (h_{fe} + 1)(h_{ib} + R_L)$$

$$= (76)(26 \text{ } \Omega + 7.5 \text{ k}\Omega)$$

$$R_{in} = 572 \text{ k}\Omega$$

As can be seen from Figure 5.3, the total input resistance (R_{in_T} seen by the input voltage) would include the base resistor R_B, or

$$R_{in_T} = R_B \| R_{in} \tag{14}$$

$$= 524 \text{ k}\Omega \| 572 \text{ k}\Omega$$

$$R_{in_T} = 273.5 \text{ k}\Omega$$

VOLTAGE GAIN

The ac voltage gain for the common-collector amplifier can be approximated with the aid of Figure 5.4. The voltage gain of the amplifier can be defined as

$$A_v = \frac{v_0}{v_{in}} \tag{15}$$

The ac output voltage is across R_L, or

$$v_0 = i_e R_L \tag{16}$$

Figure 5.4
Circuit configuration used to calculate the voltage gain for a common-collector amplifier.

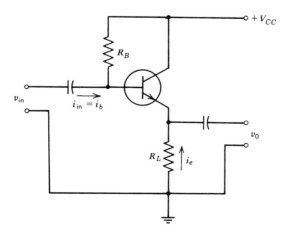

The ac input voltage v_{in} can be calculated in terms of the input current and resistance, or

$$v_{in} = i_{in}R_{in}$$

But i_{in} equals i_b, and R_{in} was defined by Equation 13, or

$$v_{in} = i_b(h_{fe} + 1)(h_{ib} + R_L) \tag{17}$$

Substituting Equations 16 and 17 into Equation 15, we have

$$A_v = \frac{i_e R_L}{i_b(h_{fe} + 1)(h_{ib} + R_L)}$$

$$= \frac{i_b(h_{fe} + 1)\ R_L}{i_b(h_{fe} + 1)(h_{ib} + R_L)}$$

$$A_v = \frac{R_L}{h_{ib} + R_L} \tag{18}$$

Notice that the voltage gain is approximately one if h_{ib} is much less than the load resistor R_L.

OUTPUT RESISTANCE

Figure 5.5*a* shows a common-collector amplifier. A voltage source v_s and its source resistance R_g are feeding the input of the amplifier. The ac output resistance R_0 for the amplifier is the parallel combination of the load resistor R_L and the resistance between the emitter of the transistor and common.

Figure 5.5*b* illustrates a means of calculating the resistance between the emitter and common of the transistor labeled R. The dc base resistor R_B is in parallel with the source resistance R_g. The resistance R is then defined as

$$R = \frac{v_0}{i_e} \tag{19}$$

But $v_0 = i_b(R_g \| R_B)$ and $i_e = i_b(h_{fe} + 1)$. Substituting these relationships into Equation 19, we have

$$R = \frac{v_{be} + i_b(R_g \| R_B)}{i_b(h_{fe} + 1)}$$

or

$$R = \frac{v_{be}}{i_b(h_{fe} + 1)} + \frac{R_g \| R_B}{(h_{fe} + 1)} \tag{20}$$

Recalling from Chapter 4 that the quantity v_{be}/i_b is equal to h_{ie}, we have

$$R = \frac{h_{ie}}{(h_{fe} + 1)} + \frac{R_g \| R_B}{(h_{fe} + 1)} \tag{21}$$

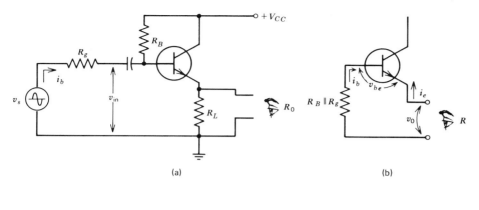

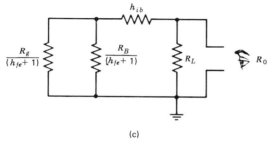

(c)

Figure 5.5
Output resistance of the CC amplifier: (a) circuit configuration, (b) load resistor removed, and (c) complete equivalent circuit.

Now the total resistance R_0 is the parallel combination of R_L and R, or

$$R_0 = R_L \| R$$

$$R_0 = R_L \| \left[\frac{h_{ie}}{(h_{fe} + 1)} + \frac{R_g \| R_B}{(h_{fe} + 1)} \right] \tag{22}$$

Recalling that $h_{ie}/(h_{fe} + 1) = h_{ib}$, we may rewrite Equation 22 as

$$R_0 = R_L \| \left[h_{ib} + \frac{R_g \| R_B}{(h_{fe} + 1)} \right] \tag{23}$$

Figure 5.5c represents the output resistance described with Equation 23. It is not surprising that the resistance on the base side of a transistor appears reduced by a factor of $1/(h_{fe} + 1)$ as viewed from the emitter side of the transistor. Recall that an emitter resistor is magnified by the factor $(h_{fe} + 1)$ as viewed from the base side of the transistor.

As can be reasoned from Equation 23, the value of the source resistance can be a major factor in determining the output resistance for the common-collector amplifier. This relationship between R_g and R_0 is demonstrated in the next example problem.

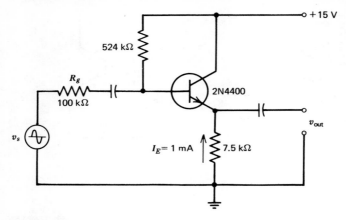

Figure 5.6
Circuit configuration for example problem.

EXAMPLE

The emitter-follower shown in Figure 5.6 has $R_g = 100$ kΩ, $R_L = 7.5$ kΩ, $R_B = 524$ kΩ, $I_E = 1.0$ mA, and $h_{fe} = 75$. Calculate the output resistance R_0. The parallel combination of R_B and R_g is equal to

$$R_B \| R_g = 524 \text{ k}\Omega \| 100 \text{ k}\Omega$$
$$R_g \| R_B = 83.97 \text{ k}\Omega$$

Then h_{ib} is equal to

$$h_{ib} = \frac{26 \text{ mV}}{I_E(\text{mA})} = \frac{26 \text{ mV}}{1.0 \text{ mA}}$$

$$h_{ib} = 26 \ \Omega$$

Using Equation 23 to calculate R_0, we have

$$R_0 = R_L \| \left[h_{ib} + \frac{R_g \| R_B}{(h_{fe} + 1)} \right]$$

$$= 7.5 \text{ k}\Omega \| \left[26 \ \Omega + \frac{83.97 \text{ k}\Omega}{76} \right]$$

$$R_0 = 982 \ \Omega$$

If R_g were changed to 10 kΩ, R_0 would be equal to

$$R_0 = 7.5 \text{ k}\Omega \| \left[26 \ \Omega + \frac{10 \text{ k}\Omega \| 524 \text{ k}\Omega}{76} \right]$$

$$= 158 \ \Omega$$

COMMON-BASE AMPLIFIER

Figure 5.7 illustrates a circuit configuration for a common-base amplifier. The capacitor C provides a low resistance path for ac signals at the base of the

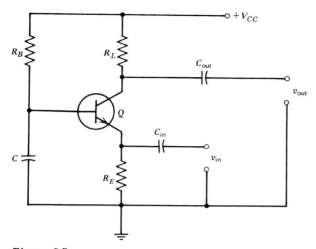

Figure 5.7
Circuit configuration for the common-base amplifier.

transistor to common. The input signal is applied to the emitter of the BJT via C_{in}. The output signal is taken from the collector of the BJT via C_{out}. The input current is i_e and the output current is i_c. As can be reasoned from the figure, if i_e increases, i_c increases, and vice versa. Thus the input and output voltage signals are in phase with each other.

dc BIASING REQUIREMENTS

The common-base circuit configuration has the same biasing requirements as those of the common-collector or common-emitter circuit configuration; that is, the base-emitter junction must be forward-biased and the base-collector junction must be reverse-biased. One such biasing arrangement for the common-base amplifier is shown in Figure 5.8.

The dc loop equation for the collector and emitter of the BJT is given by

$$V_{CC} = I_C R_L + V_{CE} + I_E R_E \tag{24}$$

If h_{FE} (beta) is large, we can assume that $I_C \cong I_E$. Then Equation 24 becomes

$$V_{CC} = I_C(R_L + R_E) + V_{CE} \tag{25}$$

The dc base loop equation can be written so that the base current I_B can be calculated, or

$$I_B = \frac{V_{CC} - V_{BE(sat)}}{R_B + (h_{FE} + 1)\,R_E} \tag{26}$$

The selection for the ratio of R_L to R_E for a maximum ac output voltage swing is discussed in Chapter 6.

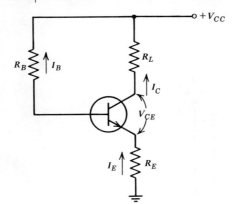

Figure 5.8
Biasing requirements for the CB amplifier.

EXAMPLE

For the circuit configuration shown in Figure 5.7, assume the following: $V_{CC} =$ 1.5 V, $R_L = 4.7$ kΩ, $R_E = 2.7$ kΩ, $R_B = 750$ kΩ, $h_{FE} = 75$, and $V_{BE(\text{sat})} = 0.6$ V. Calculate I_E, I_B, and V_{CE}. Equation 26 is used to calculate I_B

$$I_B = \frac{V_{CC} - V_{BE(\text{sat})}}{R_B + (h_{FE} + 1) R_E}$$

$$= \frac{15 \text{ V} - 0.6 \text{ V}}{750 \text{ k}\Omega + 76 \times 2.7 \text{ k}\Omega}$$

$$I_B = 15 \ \mu\text{A}$$

and I_E is calculated as follows:

$$I_E = I_B(h_{FE} + 1)$$

$$= 15 \ \mu\text{A}(76)$$

$$I_E = 1.14 \text{ mA}$$

Assuming that $I_E = I_C$, we can use Equation 25 to calculate V_{CE}, or

$$V_{CC} = I_C(R_L + R_E) + V_{CE}$$

$$15 \text{ V} = 1.14 \text{ mA}(7.4 \text{ k}\Omega) + V_{CE}$$

$$V_{CE} = 6.56 \text{ V}$$

INPUT AND OUTPUT RESISTANCES

The total input resistance for the common-base amplifier is the parallel combination of the emitter-resistor R_E and the resistance between the emitter and common labeled R (see Figures 5.9a and b). In Figure 5.9b, the value of the resistance R is equal to

$$R = \frac{v_{\text{in}}}{i_e} \tag{27}$$

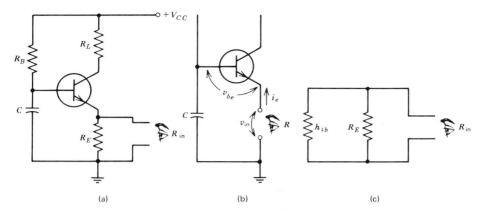

Figure 5.9
Input resistance of the CB amplifier: (a) circuit configuration, (b) load resistor removed, and (c) equivalent circuit.

But because of the bypass capacitor C, $v_{in} = v_{be}$. Recalling that $i_e = i_b(h_{fe} + 1)$, we see that Equation 27 becomes

$$R = \frac{v_{be}}{i_b(h_{fe} + 1)} \tag{28}$$

Because the quantity v_{be}/i_b equals h_{ie}, we have

$$R = \frac{h_{ie}}{h_{fe} + 1}$$

or

$$R = h_{ib}$$

Now the total input resistance R_{in} is equal to the parallel combination of R and R_E, or

$$R_{in} = h_{ib} \| R_E \tag{31}$$

See Figure 5.9c.

For most practical circuit configurations, the value of the emitter-resistor R_E is usually much greater than h_{ib} so that R_{in} can be approximated by

$$R_{in} \approx h_{ib} \tag{32}$$

where $h_{ib} = 26 \text{ mV}/I_E(\text{mA})$. The output resistance for the common-base amplifier can be approximated by R_L ($R_{out} = R_L$).

VOLTAGE GAIN

The voltage gain of the common-base amplifier can be calculated with the aid of Figure 5.10. It can be determined that the output current is equal to the ac collector current i_c and the input current is equal to the emitter current i_e. The voltage of the amplifier is defined as

$$A_v = \frac{v_0}{v_{in}} \tag{33}$$

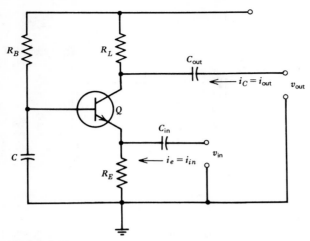

Figure 5.10
Circuit used to calculate A$_v$ for the CB amplifier.

But the output voltage v_0 is equal to $i_c R_L$ and the input voltage v_{in} is equal to $i_e R_{\text{in}}$. Equation 33 defines R_{in} as being equal to h_{ib} so that v_{in} is equal to $i_e h_{ib}$. Substituting these relationships into Equation 33, we have

$$A_v = \frac{v_0}{v_{\text{in}}} = \frac{i_c R_L}{i_e h_{ib}}$$

$$= \frac{i_b h_{fe} R_L}{i_b (h_{fe} + 1)\, h_{ib}}$$

$$A_v \approx \frac{R_L}{h_{ib}}$$

Note that the voltage gain of the common-base amplifier is the same as the voltage gain for a common-emitter amplifier with a bypassed emitter-resistor (see Chapter 4).

EXAMPLE

For the common-base amplifier illustrated in Figure 5.7, assume the following: $R_L = 4.7\ \text{k}\Omega$, $R_E = 2.7\ \text{k}\Omega$, $R_B = 750\ \text{k}\Omega$, and $h_{FE} = 75$. The values of C, C_{in}, and C_{out} are chosen so that their reactance is one-tenth the values of the resistance bypassed at the lowest frequency the amplifier is designed for. Calculate the input resistance and voltage gain for this amplifier.

For the component values given, the emitter current I_E is 1.14 A (see the previous example problem). Equation 32 can be used to calculate R_{in}, or

$$R_{\text{in}} = h_{ib} = \frac{26\ \text{mV}}{I_E(\text{mA})}$$

$$= \frac{26\ \text{mV}}{1.14\ \text{mA}}$$

$$R_{\text{in}} \approx 23\ \Omega$$

The voltage gain is calculated with Equation 34, or

$$A_v \approx \frac{R_L}{h_{ib}}$$

$$= \frac{4700 \; \Omega}{23}$$

$$A_v = 204$$

APPLICATIONS FOR THE COMMON-COLLECTOR AMPLIFIER

As discussed in this chapter, the common-collector amplifier has a high input resistance, a low output resistance, and a voltage gain of approximately one. The power gain of the common-collector amplifier is defined as the product of the voltage gain and current gain. The current gain is defined as the output current over the input current ($A_i = i_0/i_{in}$). As can be reasoned, the maximum theoretical current gain is ($h_{fe} + 1$) or h_{fc}. Thus, although the amplifier has a voltage gain of approximately one, it does have a power gain greater than one.

The common-collector amplifier can serve as a matching device with power gain. Suppose that a resistance of 100 kΩ had to be matched to a resistance of 600 Ω. A transformer could be used to match these two resistances by the proper selection of the turns ratio. However, the transformer does have losses. A common-collector amplifier can be designed to have an input resistance of 100 kΩ and an output resistance of 600 Ω and to provide a power gain. This ability of the common collector makes it an ideal matching device (see Figure 5.11). Another feature of the common-collector amplifier is its frequency response. The frequency response of the amplifier is much greater than that of a transformer.

The frequency response of the common-collector amplifier is much greater than that of the common-emitter amplifier using the same transistor. The higher frequency response is due to the reduction of the Miller effect that dominates at the higher frequencies for the common-emitter amplifier (see Chapter 8). The Miller effect in the common-emitter amplifier is the magnifying of the base-collector capacitance of the transistor by ($1 + A_v$).

APPLICATIONS FOR THE COMMON-BASE AMPLIFIER

As discussed in this chapter, the common-base amplifier has a low input resistance, an output resistance that is equal to the load resistance R_L, and a voltage gain that is much greater than one. Because of its low input resistance, one of the many applications of the common-base amplifier is a microphone amplifier. A low-resistance microphone (carbon or dynamic) can be coupled to the input of the common-base amplifier and the signal amplified (see Figure 5.12).

Another frequent application of the common-base amplifier is a RF amplifier. The common-base amplifier has a voltage gain that is equivalent to a common-emitter stage, but the common-base amplifier has a much higher frequency

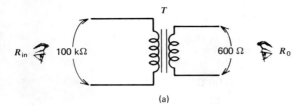

(a)

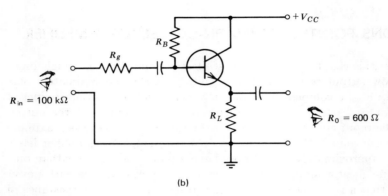

(b)

Figure 5.11
Matching devices: (a) transformer and (b) CC amplifier.

response than the common-emitter amplifier. With the common-base amplifier, the base-collector capacitance is not magnified by $(1 + A_v)$.

CONSTANT CURRENT PUMPS

An interesting circuit configuration is the constant-current generator or pump (BJT). As the name implies, this device generates a constant dc current to an external load. One common circuit configuration for a BJT constant-current pump is illustrated in Figure 5.13. The PNP transistor Q is biased by the voltage across the zener diode V_z. The value of R is chosen so that sufficient current is supplied to the zener diode. The constant current supplied by the BJT is caused by the high internal resistance of the device. A look at the output characteristic curves for a BJT will show that if the base current is held constant, the collector current is a constant over a wide range of collector-to-emitter voltages (V_{CE}).

The dc loop equation for the base emitter of the BJT is

$$V_z - V_{BE(\text{sat})} = I_E R_E \tag{35}$$

Solving Equation 35 for I_E, we have

$$I_E = \frac{V_z - V_{BE(\text{sat})}}{R_E} \tag{36}$$

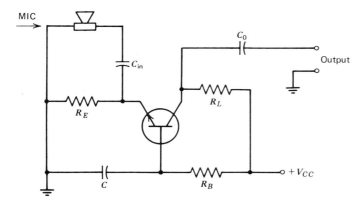

Figure 5.12
Simple application of the CB amplifier.

If the BJT has a large beta, then the emitter current is approximately the collector current (pump current) I_P, or

$$I_P \approx \frac{V_z - V_{BE(\text{sat})}}{R_E} \qquad (37)$$

The BJT current pump shown in Figure 5.13 will pump a constant current I_P until the collector-to-emitter voltage is equal to two-tenths of a volt. Let us assume that this two-tenths of a volt is the saturation voltage of the collector-emitter, or $V_{CE(\text{sat})}$. The maximum voltage across terminals A and B (for I_P) will then be equal to

$$V_{AB(\text{max})} = V_{CC} - (I_P R_E + V_{CE(\text{sat})}) \qquad (38)$$

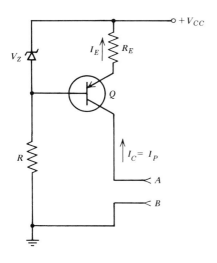

Figure 5.13
BJT constant current pump.

EXAMPLE

Design a 1.0-mA current pump using a 2N4402 BJT. The supply voltage is to be 15 V, and a 3.6 zener is to be used. Assume that $h_{FE} = 75$, $V_{BE(sat)} = 0.6$ V, and $V_{CE(sat)} = 0.4$ V. Calculate the necessary values of R_E, R, and the maximum external resistance that can be placed across terminals A and B for the pump to still generate the 1.0 mA of current.

Solving Equation 37 for R_E, we have

$$R_E = \frac{V_z - V_{BE(sat)}}{I_P}$$

$$= \frac{3.6 \text{ V} - 0.6 \text{ V}}{1.0 \text{ mA}}$$

$$R_E = 3 \text{ k}\Omega$$

With 1 mA of collector current, the base current is equal to

$$I_B = \frac{I_C}{h_{FE}} = \frac{1 \text{ mA}}{75}$$

$$I_B \approx 14 \text{ }\mu\text{A}$$

If I_Z is much greater than I_B, R is calculated by

$$R \approx \frac{V_{CC} - V_Z}{I_z}$$

(see Chapter 2). Choosing I_Z to be 2.5 mA (for convenience), we see that R becomes

$$R = \frac{15 \text{ V} - 3.6 \text{ V}}{2.5 \text{ mA}} = 4.6 \text{ k}\Omega$$

The maximum voltage that can be across terminals A and B is found with Equation 38, or

$$V_{AB(max)} = V_{CC} - (I_P R_E + V_{CE(sat)})$$

$$= 15 - (1 \text{ mA} \times 3 \text{ k}\Omega + 0.4 \text{ V})$$

$$V_{AB(max)} = 11.6 \text{ V}$$

According to Ohm's law, then, the maximum resistance that can be placed across terminals A and B is

$$R_{max} = \frac{V_{AB(max)}}{I_P} = \frac{11.6 \text{ V}}{1 \text{ mA}}$$

$$= 11.6 \text{ k}\Omega$$

Thus any resistance from 0 to 11.6 kΩ placed across terminals A and B will have a current flow through it of 1.0 mA.

If a capacitor is connected across terminals A and B, a constant current is pumped into the capacitor until the voltage across the capacitor reaches $V_{AB(max)}$.

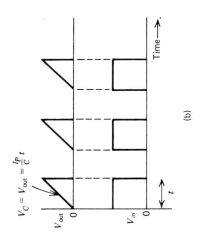

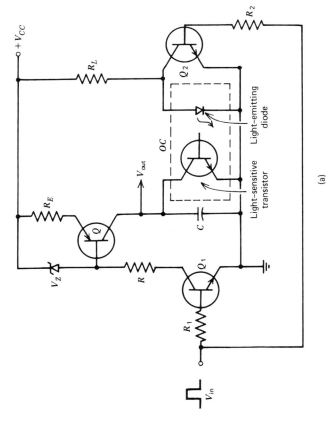

Figure 5.14
Triangular waveform generator: (a) circuit and (b) waveforms.

With a constant current pumped into a capacitor, the voltage across the capacitor rises linearly and is equal to

$$V_C = \frac{I}{C} t \qquad (39)$$

where I is the current in amperes, C the capacitance in farads, and t the time in seconds that the current I is pumped into the capacitor.

APPLICATIONS FOR THE CURRENT PUMP

Some of the many applications of the current pump are in timing circuits (see Chapters 17 and 20), in difference amplifiers (see Chapter 15), and in oscillators (see Chapter 19).

A simple sawtooth oscillator is shown in Figure 5.14. A constant current is pumped into the capacitor C; the current pump can be turned on and off by transitor Q_1, which acts as a switch. If V_{in} is positive, Q_1 is saturated, turning on the current pump. When V_{in} is positive, it also causes Q_2 to be saturated. With Q_2 saturated, the light-emitting diode (LED) inside the optical coupler (see Chapter 18) is off. With the LED off, the light-sensitive transitor is cut off, allowing the voltage across C to increase linearly.

When V_{in} is zero volts, Q_1 and Q_2 are both in cutoff. Transistor Q_1 in cutoff shuts off the current pump, and Q_2 in cutoff causes the LED to be turned on. With the LED turned on, the light from this device causes the light-sensitive transitor to be turned on or saturated. With the light-sensitive transistor saturated, capacitor C discharges quickly (see Figure 5.14b). Thus the output voltage across the capacitor C is a sawtooth waveform that is controlled by the input pulse V_{in}.

REVIEW QUESTIONS AND PROBLEMS

5-1   Calculate the values of R_B and R_L for the following circuit.

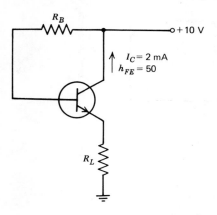

R_B

$+10$ V

$I_C = 2$ mA
$h_{FE} = 50$

R_L

5-2   If the beta of the BJT in Problem 5-1 were to change to 75, calculate the new values for the base and emitter currents and the collector-emitter voltage.

5-3   Discuss why there is a zero-degree phase difference between the input and output signals for the common-collector amplifier.

5-4   Calculate the input and output resistances for the following common-collector amplifier.

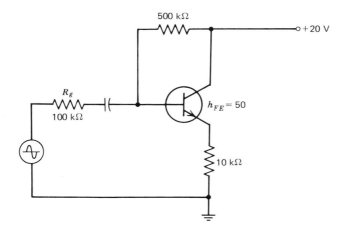

5-5   Explain how there is a power gain associated with the common-collector amplifier even though it has a voltage gain of less than one.

5-6   List some applications for the common-collector amplifier.

5-7   Design a common-collector amplifier having an input resistance of 100 kΩ and an output resistance of 600 Ω. Assume a beta of 75.

5-8   Calculate the input and output resistances and voltage gain for the circuit configuration shown.

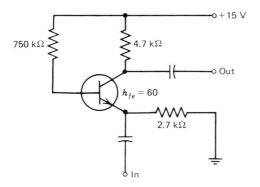

5-9   Compare the characteristics of the common-base amplifier to the common-emitter and collector amplifiers.

5-10 Design a 5-mA BJT current pump using a 5.1-V zener diode, a 20-V supply, and BJT having a beta of 75 and V_{BE} of 0.65 V.

5-11 What is the voltage that will exist across the output terminals of the current pump in Problem 5-10?

5-12 For the circuit shown, assuming that the capacitor has an initial charge of zero, how much time will it take the capacitor voltage to reach 5 V?

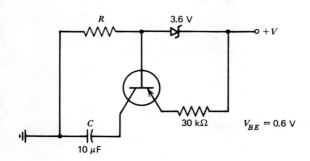

OBJECTIVES FOR CHAPTER 6,
dc BIASING TECHNIQUES

At the completion of this chapter you can demonstrate your understanding of the material by answering these questions:

- What is constant base current biasing?
- What is emitter feedback biasing?
- What is collector feedback biasing?
- What is universal biasing?
- How is the Q point of an amplifier determined?

6 dc BIASING TECHNIQUES

There are many methods that can be employed to bias the BJT. You should study these various biasing techniques because of their frequent use in circuit applications.

In this chapter a variety of biasing techniques commonly used with common-emitter and common-collector circuit configurations are presented. First, the quiescent points of the various circuits are explored and then determined, with the biasing, load, and emitter resistors given. After the above analysis is completed, the selection of biasing resistors when the quiescent point is stated by the designer is discussed.

CONSTANT BASE CURRENT

Figure 6.1 shows a constant base current biasing circuit. The base current is given by

$$I_B = \frac{V_{CC} - V_{BE}}{R_B} \tag{1}$$

If V_{CC} is much greater than V_{BE}, then $I_B \approx V_{CC}/R_B$. The collector current is calculated by

$$I_C = h_{FE}I_B + (h_{FE} + 1)\, I_{CO} \tag{2}$$

and the loop equation around the collector is given by

$$V_{CC} = I_C R_L + V_{CE} \tag{3}$$

With the use of Equations 1, 2, and 3, we can calculate the values of base and collector currents.

100

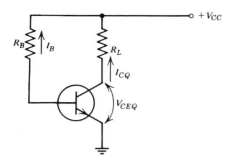

Figure 6.1
Constant base current biasing.

EXAMPLE

From Figure 6.2, calculate the quiescent voltage and current; $R_L = 4.7$ kΩ, $R_B = 680$kΩ, $h_{FE} = 75$, $I_{CO} = 1$ μA, and $V_{BE} = 0.6$ V. Using Equation 1, we have

$$I_B = \frac{V_{CC} - V_{BE}}{R_B} = \frac{(10 - 0.6)\ \text{V}}{680\ \text{k}\Omega}$$

$$I_B = 13.8\ \mu\text{A}$$

Using Equation 2 to calculate the collector current, we have

$$I_{CQ} = h_{FE}I_B + (h_{FE} + 1)\ I_{CO}$$

$$= 75 \times 13.8\ \mu\text{A} = 76 \times 1\ \mu\text{A}$$

$$I_{CQ} = 1.11\ \text{mA}$$

and using Equation 3 to calculate the quiescent voltage V_{CEQ}, we have

$$V_{CEQ} = V_{CC} - I_{CQ}R_L$$

$$= 10\ \text{V} - 1.11\ \text{mA} \times 4.7\ \text{k}\Omega$$

$$V_{CEQ} = 4.78\ \text{V}$$

If the temperature were to increase to a point that caused I_{CO} to be equal to 8 μA, the new quiescent point would be

$$I_{CQ} = 75 \times 13.8\ \mu\text{A} + 76 \times 8\ \mu\text{A}$$

$$I_{CQ} = 1.64\ \text{mA}$$

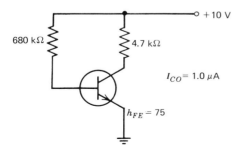

Figure 6.2
Circuit configuration for example problem.

and

$$V_{CEQ} = 10 \text{ V} - 1.64 \text{ mA} \times 4.7 \text{ k}\Omega$$
$$V_{CEQ} = 2.3 \text{ V}$$

EMITTER FEEDBACK

Another type of biasing arrangement is shown in Figure 6.3. This configuration is called emitter-feedback biasing. From the figure, we have

$$V_{CC} = I_{CQ}R_L + V_{CEQ} + I_E R_E \tag{4}$$

$$V_{CC} = I_B R_B + V_{BE} + I_E R_E \tag{5}$$

Letting $I_E = (h_{FE} + 1)I_B$, we have

$$I_B = \frac{V_{CC} - V_{BE}}{R_B + (h_{FE} + 1)\, R_E} \tag{6}$$

EXAMPLE

In Figure 6.3, we have $R_L = 4.7$ kΩ, $R_B = 750$ kΩ, $V_{CC} = 15$ V, $R_E = 2.7$ kΩ, $h_{FE} = 75$, and $I_{CO} = 1$ μA. Calculate the quiescent point. First, I_B is calculated with equation 6.

$$I_B = \frac{V_{CC} - V_{BE}}{R_B + (h_{FE} + 1)\, R_E} = \frac{(15 - 0.6) \text{ V}}{750 \text{ k}\Omega + 76 \times 2.7 \text{ k}\Omega}$$

$$I_B = 15 \text{ } \mu\text{A}$$

Then I_{CQ} and V_{CEQ} are calculated with the aid of Equations 2 and 4.

$$I_{CQ} = h_{FB}I_B + (h_{FE} + 1)\, I_{CO} = 75 \times 15 \text{ } \mu\text{A} + 76 \times 1 \text{ } \mu\text{A}$$

$$I_{CQ} = 1.2 \text{ mA}$$

$$V_{CEQ} = V_{CC} - I_C R_L - I_E R_E =$$
$$= 15 - (1.2 \text{ mA} \times 4.7 \text{ k}\Omega) - (1.215 \text{ mA} \times 2.7 \text{ k}\Omega)$$

$$V_{CEQ} = 6.08 \text{ V}$$

Figure 6.3
Emitter feedback biasing.

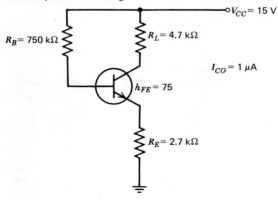

UNIVERSAL BIASING

Figure 6.4 shows a universal biasing circuit configuration. Figure 6.4a shows the placement of the load, emitter, and voltage-dividing resistors. The voltage-dividing resistors R_1 and R_2 can be Theveninized, as shown in Figure 6.4b. We have

$$R_B = R_1 \| R_2 = \frac{R_1 R_2}{R_1 + R_2} \tag{7}$$

$$V_{BB} = V_{CC} \frac{R_2}{R_1 + R_2} \tag{8}$$

Substituting Equation 8 into Equation 7 for $R_1 + R_2$, we have

$$R_1 = \frac{V_{CC}}{V_{BB}} R_B \tag{9}$$

And solving Equation 7 for R_2, we have

$$R_2 = \frac{R_1 R_B}{R_1 - R_B} \tag{10}$$

The base and collector loop equations are given by

$$V_{BB} = I_B R_B + V_{BE} + I_E R_E$$
$$V_{BB} = I_B[R_B + (h_{FE} + 1)R_E] \tag{11}$$
$$V_{CC} = I_{CQ}(R_L + R_E) + V_{CEQ} \tag{12}$$

EXAMPLE

Figure 6.5 shows a universal biasing configuration. Looking into the base of the transistor ($R_{\text{in-dc}}$), calculate I_{CQ}, V_{CEQ}, and the dc resistance; $R_L = 4.7$ kΩ, $R_E = 2.7$ kΩ, $R_1 = 39$ kΩ, $R_2 = 15$ kΩ, $I_{CO} = 1$ μA, and $h_{FE} = 75$.

Figure 6.4
Universal biasing: (a) circuit configuration and (b) equivalent circuit.

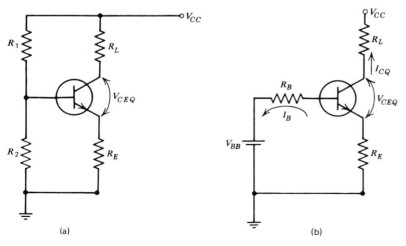

(a)   (b)

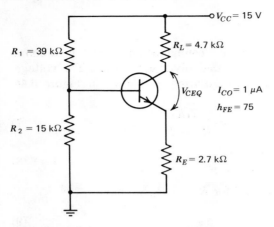

Figure 6.5
Universal biasing example problem.

With the aid of Equations 7 and 8, R_B and V_{BB} are calculated

$$R_B = \frac{R_1 R_2}{R_1 + R_2} = \frac{15 \text{ k}\Omega \times 39 \text{ k}\Omega}{54 \text{ k}\Omega} = 10.8 \text{ k}\Omega$$

$$V_{BB} = \frac{V_{CC} \times R_2}{R_1 + R_2} = \frac{15 \text{ V} \times 15 \text{ k}\Omega}{54 \text{ k}\Omega} = 4.17 \text{ V}$$

and I_B is calculated using Equation 11.

$$I_B = \frac{V_{BB} - V_{BE}}{R_B + (h_{FE} + 1) R_E} = \frac{(4.17 - 0.6) \text{ V}}{10.8 \text{ k}\Omega + 76 \times 2.7 \text{ k}\Omega}$$

$$I_B = 16.5 \text{ } \mu\text{A}$$

Then I_{CQ} and V_{CEQ} are calculated with the aid of Equations 2 and 12.

$$I_{CQ} = h_{FE} I_B + (h_{FE} + 1) I_{CO} = 75 \times 16.5 \text{ } \mu\text{A} + 76 \times 1 \text{ } \mu\text{A}$$

$$I_{CQ} = 1.31 \text{ mA}$$

$$V_{CEQ} = V_{CC} - I_C R_L - I_E R_E = 15\text{V} - 1.31 \text{ mA}$$
$$\times 4.7 \text{ k}\Omega - 1.32 \text{ mA} \times 2.7 \text{ k}\Omega$$

$$V_{CEQ} = 5.28 \text{ V}$$

Now $R_{\text{in-dc}}$ is calculated from Figure 6.5.

$$R_{\text{in-dc}} = R_1 \| R_2 \| (h_{FE} + 1) R_E$$

$$= 15 \text{ k}\Omega \| 39 \text{ k}\Omega \| 205 \text{ k}\Omega$$

$$R_{\text{in-dc}} \approx 10.8 \text{ k}\Omega$$

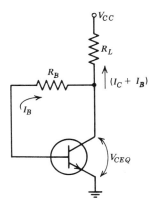

Figure 6.6
Collector feedback biasing.

COLLECTOR FEEDBACK

Another type of biasing arrangement is shown in Figure 6.6. With this method the base resistor is connected to the collector of the transistor rather than to the supply voltage. The total current through the load resistor is the emitter current. From Figure 6.6 we have

$$V_{CC} = (I_C + I_B)\,R + V_{CE} \tag{13}$$

$$I_B = \frac{V_{CE} - V_{BE}}{R_B} \tag{14}$$

where $I_E = (h_{FE} + 1)\,I_B$.

EXAMPLE

Calculate the quiescent point of the circuit shown in Figure 6.7. $R_L = 4.7$ kΩ, $R_B = 330$ kΩ, $h_{FE} = 75$. Using Equations 13 and 14, we have

$$V_{CC} = I_E R_L + V_{CE}$$

or

$$10 = 76 \times I_B \times 4.7 \text{ k}\Omega + V_{CE} \tag{A}$$

$$I_B = \frac{V_{CE} - V_{BE}}{R_B}$$

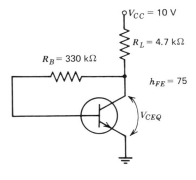

Figure 6.7
Circuit for example problem.

$$I_B = \frac{(V_{CE} - 0.6) \text{ V}}{330 \text{ k}\Omega} \qquad \text{(B)}$$

Substituting Equation B into A and solving for V_{CEQ}, we have

$$10 = 76 \times 4.7 \text{ k}\Omega\left(\frac{V_{CE} - 0.6 \text{ V}}{330 \text{ k}\Omega}\right) + V_{CE}$$

$$V_{CE} = V_{CEQ} = 5.12 \text{ V}$$

Using Equation B to calculate I_B, we have

$$I_B = \frac{V_{CE} - V_{BE}}{R_B} = \frac{(5.12 - 0.6) \text{ V}}{330 \text{ k}\Omega}$$

$$I_B = 13.7 \ \mu\text{A}$$

$$I_E = (h_{FE} + 1) \ I_B = 76(13.7 \ \mu\text{A})$$

$$I_E = 1.04 \text{ mA}$$

COMMON COLLECTOR

Figure 6.8 shows a common-collector amplifier using an R_E (emitter and load resistor) and two voltage-divider resistors R_1 and R_2. We have

$$V_{CC} = V_{CEQ} + I_E R_E \qquad \text{(15)}$$

$$V_{BB} = I_B R_B + V_{BE} + I_E R_E$$

$$I_B = \frac{V_{BB} - V_{BE}}{R_B + (h_{FE} + 1) \ R_E} \qquad \text{(16)}$$

Figure 6.8
CC biasing: (a) circuit configuration and (b) equivalent circuit.

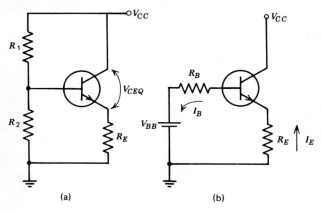

(a)   (b)

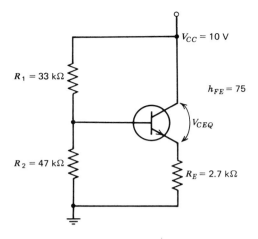

Figure 6.9
Circuit for example problem.

EXAMPLE

Figure 6.9 shows a common-collector amplifier with $R_E = 2.7$ kΩ, $R_1 = 33$ kΩ, $R_2 = 47$ kΩ, $h_{FE} = 75$, and $V_{CC} = 10$ V. Calculate I_{CQ} and V_{CEQ} for the circuit. Using Equation 7 to find R_B, we have

$$R_B = R_1 \| R_2 = 33 \text{ k}\Omega \| 47 \text{ k}\Omega$$

$$R_B = 19.4 \text{ k}\Omega$$

Then V_{BB} is found by the use of Equation 8,

$$V_{BB} = V_{CC} \frac{R_2}{R_1 + R_2} = \frac{10 \text{ V} \times 47 \text{ k}\Omega}{80 \text{ k}\Omega}$$

$$V_{BB} = 5.87 \text{ V}$$

and I_B with the aid of Equation 16.

$$I_B = \frac{V_{BB} - V_{BE}}{R_B + (h_{FE} + 1) R_E} = \frac{(5.87 - 0.6) \text{ V}}{19.4 \text{ k}\Omega + 76 \times 2.7 \text{ k}\Omega}$$

$$I_B = 23.5 \text{ } \mu\text{A}$$

Finally, the emitter current and the collector-emitter voltage are found.

$$I_E = (h_{FE} + 1) I_B = 76 \times 23.5 \text{ } \mu\text{A}$$

$$I_E = 1.79 \text{ mA}$$

$$V_{CEQ} = V_{CC} - I_E R_E = 10 - 1.79 \text{ mA} \times 2.7 \text{ k}\Omega$$

$$V_{CEQ} = 5.17 \text{ V}$$

SELECTION OF THE VOLTAGE-DIVIDER RESISTORS

In the previous section, the values of R_1 and R_2 were given and the quiescent point was calculated. Let us reverse that process. Now I_{CQ} and V_{CEQ} are selected and the values of the voltage-divider resistors are calculated.

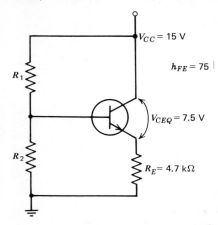

Figure 6.10
Circuit configuration used to calculate the voltage divider resistors R₁ and R₂.

Figure 6.10 shows a common-collector amplifier with a load resistor of 4.7 kΩ. If the supply voltage is 15 V, the quiescent voltage point should be one-half the supply voltage. From the figure $V_{CEQ} = 7.5 \text{ V} = V_E$. The emitter current, then, is calculated by

$$I_E = \frac{V_E}{R_E} = \frac{7.5 \text{ V}}{4.7 \text{ k}\Omega}$$

$$I_E = 1.6 \text{ mA}$$

The base current is

$$I_B = \frac{I_E}{(h_{FE} + 1)} = \frac{1.6 \text{ mA}}{76}$$

$$I_B = 21 \text{ μA}$$

Then V_{BB} is calculated with Equation 16.

$$V_{BB} = I_B R_B + V_{BE} + V_E$$

$$V_{BB} = 21 \text{ μA} \times R_B + 0.6 \text{ V} + 7.5 \text{ V}$$

The relationship between R_B and R_E can be stated as a rule of thumb:

$$R_B \leq \frac{(h_{FE} + 1) \, R_E}{10}$$

or

$$R_B \leq \frac{76 \times 4.7 \text{ k}\Omega}{10} = 35.72 \text{ k}\Omega$$

(we shall use 33 kΩ). Then

$$V_{BB} = 21 \text{ μA} \times 33 \text{ k}\Omega + 0.6 + 7.5$$

$$V_{BB} = 8.8 \text{ V}$$

(see Chapter 7). To determine the values of R_1 and R_2, we use Equations 9 and 10.

$$R_1 = \frac{V_{CC}}{V_{BB}} R_B = \frac{15 \text{ V}}{8.8 \text{ V}} \times 33 \text{ k}\Omega$$

$$R_1 = 56 \text{ k}\Omega$$

$$R_2 = \frac{R_1 R_B}{R_1 - R_B} = \frac{56 \text{ k}\Omega \times 33 \text{ k}\Omega}{56 \text{ k}\Omega - 33 \text{ k}\Omega}$$

$$R_2 = 80.3 \text{ k}\Omega$$

SELECTION OF THE Q POINT

Figure 6.11 shows a universal biasing, common-emitter configuration. As can be seen from the figure, R_E is bypassed. The bypassing of R_E permits a large voltage gain. The circuit configuration shown has a dc and an ac load line. These load lines are shown in Figure 6.11b. The intersection of these two load lines should be the ideal quiescent point for the circuit configuration shown. If the Q point were selected at $V_{CC}/2$, the ac signal would *not* be the maximum peak-to-peak value permitted by the circuit.

To determine the ideal Q point, we use the dc load line and the ac load line equations from the figure.

$$\text{dc: } V_{CC} \approx I_C(R_L + R_E) + V_{CE} \tag{17}$$

$$\text{ac: } v_{ce} = i_c R_L \tag{18}$$

Figure 6.11
Selecting the ratio of R_L to R_E: (a) circuit configuration and (b) dc and ac load lines.

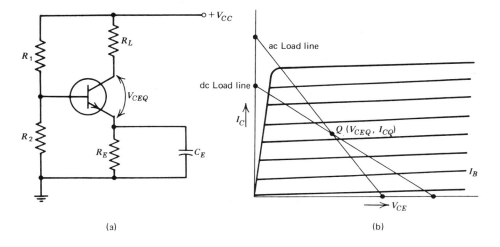

The ac and dc load line equations must both pass through the quiescent point. This condition is shown in Equations 19 and 20.

$$V_{CC} = I_{CQ}(R_L + R_E) + V_{CEQ} \qquad (19)$$

$$0 = I_{CQ}R_L - V_{CEQ} \qquad (20)$$

Solving Equations 19 and 20 simultaneously, we have

$$I_{CQ} = \frac{V_{CC}}{2R_L + R_E} \qquad (21)$$

$$V_{CEQ} = \frac{V_{CC}}{2 + \dfrac{R_E}{R_L}} \qquad (22)$$

Equation 21 reduces to $I_{CQ} = V_{CC}/2R_L$ and Equation 22 reduces to $V_{CEQ} = V_{CC}/2$ if $R_E = 0$.

EXAMPLE

For Figure 6.11, assume that $R_L = 4.7\ k\Omega$, $R_E = 2.7\ k\Omega$, $V_{CC} = 15\ V$, $h_{FE} = 75$, and $I_{CO} = 1\ \mu A$. Calculate R_1 and R_2. Now I_{CQ} and V_{CEQ} are found by the use of Equations 21 and 22.

$$I_{CQ} = \frac{V_{CC}}{2R_L + R_E} = \frac{15\ V}{2 \times 4.7\ k\Omega + 2.7\ k\Omega}$$

$$I_{CQ} = 1.24\ mA$$

$$V_{CEQ} = \frac{V_{CC}}{2 + R_E/R_L} = \frac{15\ V}{2 + (2.7\ k\Omega/4.7\ k\Omega)}$$

$$V_{CEQ} = 5.84\ V$$

The base current I_B is calculated with the aid of Equation 2,

$$I_B = \frac{I_C - (h_{FE} + 1)\ I_{CQ}}{h_{FE}} = \frac{1.24\ mA - 76 \times 1\ \mu A}{75}$$

$$I_B = 15.5\ \mu A$$

and V_{BB} is calculated with the use of Equation 11.

$$R_B \leq \frac{(h_{FE} + 1)\ R_E}{10} = \frac{76 \times 2.7\ k\Omega}{10}$$

$$R_B \leq 20.52\ k\Omega$$

(We shall use 20 kΩ.)

$$V_{BB} = I_B[R_B + (h_{FE} + 1)\ R_E] + V_{BE}$$

$$= 15.5\ \mu A(20\ k\Omega + 76 \times 2.7\ k\Omega) + 0.6\ V$$

$$V_{BB} = 4.09\ V$$

To calculate the values of R_1 and R_2, we use Equations 9 and 10.

$$R_1 = \frac{V_{CC}}{V_{BB}} R_B = \frac{15}{4.09} \times 20 \text{ k}\Omega$$

$$R_1 = 73.35 \text{ k}\Omega$$

$$R_2 = \frac{R_1 R_B}{R_1 - R_B} = \frac{73.35 \text{ k}\Omega \times 20 \text{ k}\Omega}{73.35 \text{ k}\Omega - 20 \text{ k}\Omega}$$

$$R_2 = 27.5 \text{ k}\Omega$$

The ac, low-frequency input resistance of the stage is calculated by

$$R_{\text{in-ac}} \approx R_B \| h_{ie} = R_B \| (h_{FE} + 1) h_{ib}$$

but

$$h_{ib} \approx \frac{26 \text{ mV}}{I_E(\text{mA})} = \frac{26 \text{ mV}}{1.24 \text{ mA}}$$

$$h_{ib} = 21\Omega$$

Then

$$R_{\text{in-ac}} = 20 \text{ k}\Omega \| 76 \times 21\Omega$$

$$R_{\text{in-ac}} \approx 1.48 \text{ k}\Omega$$

REVIEW QUESTIONS AND PROBLEMS

6-1 (a) Calculate I_{CQ}, V_{CEQ}, and I_B for the circuit shown. Assume a $I_{CO} = 1$ μA.

(b) If h_{FE} is 100, what are the new values for I_{CQ} and V_{CEQ}?

(c) If I_{CO} changes to 10 μA($h_{FE} = 50$), what would the new values of I_{CQ} and V_{CEQ} be?

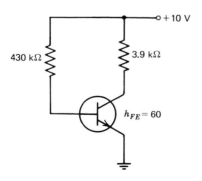

6-2 Repeat Question 6.1 for the following circuit.

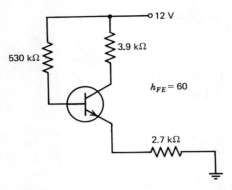

6-3 Repeat Question 6.1 for the following circuit.

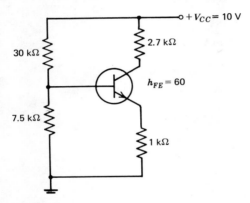

6-4 Calculate the necessary values for V_{CEQ}, I_{CQ}, R_1, R_2, and R_B for the circuit shown. Assume that $R_B = [(h_{FE} + 1) R_E]/10$.

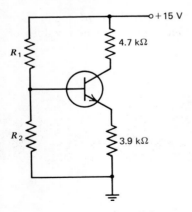

6-5 Calculate I_{CQ}, V_{CEQ}, R_1, R_2, and R_B for the circuit shown. Assume that $R_B = [(h_{FE} + 1) R_E]/15$.

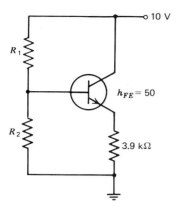

6-6 Calculate V_{CEQ} for the following circuit.

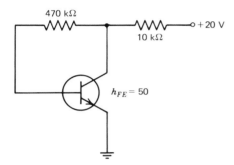

6-7 In the circuit shown, what is the maximum peak-to-peak voltage of the undistorted output signal that is possible. Assume that the transistor is ideal.

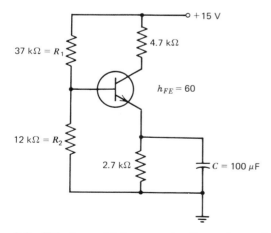

6-8 What would happen to the values of the quiescent voltage and current if the supply voltage V_{CC} were to change? Calculate this change in Q for the circuit in Review Question 6-4 if the supply voltage were to (a) increase 3 V and (b) decrease 3 V.

OBJECTIVES FOR CHAPTER 7,
QUIESCENT POINT STABILIZATION

At the completion of this chapter you can demonstrate your understanding of the material by answering these questions:

- Why is it important to consider the effects of changes in temperature and beta on the Q point of an amplifier?
- How do changes in beta affect the Q point?
- How do changes in temperature affect the Q point?
- What are stability factors?

7 QUIESCENT POINT STABILIZATION

The magnitude of the effect for changes in temperature and beta on the operating or Q point of an amplifier is determined by the type of biasing method employed. You should study the effects of temperature and beta changes on the Q point in order to understand why one type of biasing method is superior to another in specific circuit applications.

In order to maintain the linearity of class A amplifier with large output waveforms, the quiescent point (V_{CEQ}, I_{CQ}) should remain stationary with changes in temperature and h_{FE} (beta). The selection of a biasing technique, as will be shown in this chapter, will determine how much the quiescent point changes with temperature and beta.

EFFECTS OF BETA CHANGES ON THE Q POINT

As an example of change in I_{CQ} and V_{CEQ}, consider the circuit shown in Figure 7.1. The base current is given by

$$I_B = \frac{V_{CC} - V_{BE}}{R_B} \tag{1}$$

In equation 1 V_{BE} is approximately constant for silicon transistors ($V_{BE} = 0.65$ V and $T = 25$ °C). Transistor manufacturers list h_{FE} from a maximum to a minimum value. This range of values may have a 3:1 ratio. The substituting of another transistor in the circuit arrangement shown in Figure 7.1 will change the collector current by Δh_{FE} because the base current (Equation 1) is considered constant.

116

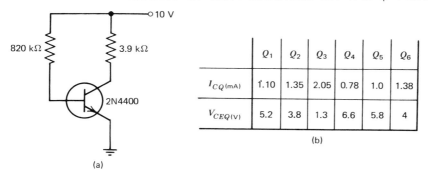

(b)

(a)

Figure 7.1
Effects of beta changes: (a) circuit and (b) table showing results.

	Q_1	Q_2	Q_3	Q_4	Q_5	Q_6
$I_{CQ\,(mA)}$	1.10	1.35	2.05	0.78	1.0	1.38
$V_{CEQ\,(V)}$	5.2	3.8	1.3	6.6	5.8	4

To demonstrate the above effect for a change in beta for the circuit arrangement shown in Figure 7.1a, six transistors were randomly selected (2N4400) and substituted in the circuit shown in Figure 1. The values of V_{CEQ} and I_{CQ} are recorded in Figure 7.1b. Note the wide variations of the quiescent point because of the different beta values of each transistor.

Figure 7.2a shows a second biasing technique. An emitter-resistor is placed is series with the emitter lead of the transistor. The base current of the transistor is given by Equation 2

$$I_B = \frac{V_{CC} - (V_{BE} + V_E)}{R_B} \tag{2}$$

where $V_E = I_E R_E$.

In Equation 2 the base current of the transistor is dependent on V_{CC}, V_{BE}, R_B, and V_E (V_E can be thought of as a feedback voltage that is a function of the emitter current). In Figure 7.2b the results of substituting six randomly selected 2N4400 transistors in the circuit are listed. Note that the variations

Figure 7.2
Effects of beta changes: (a) circuit and (b) table showing results.

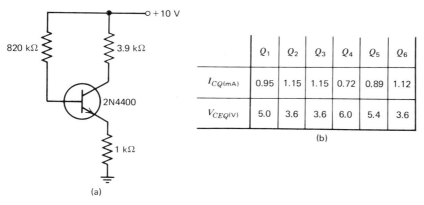

	Q_1	Q_2	Q_3	Q_4	Q_5	Q_6
$I_{CQ\,(mA)}$	0.95	1.15	1.15	0.72	0.89	1.12
$V_{CEQ\,(V)}$	5.0	3.6	3.6	6.0	5.4	3.6

(b)

(a)

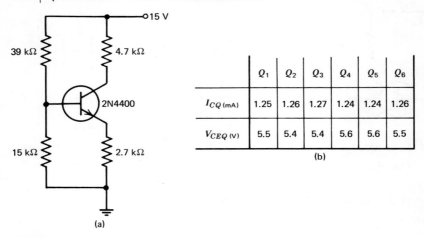

	Q_1	Q_2	Q_3	Q_4	Q_5	Q_6
I_{CQ} (mA)	1.25	1.26	1.27	1.24	1.24	1.26
V_{CEQ} (V)	5.5	5.4	5.4	5.6	5.6	5.5

(b)

(a)

Figure 7.3
Effects of beta changes: (a) circuit and (b) table showing results.

in the quiescent point of the transistors are not as great as the variations shown in Figure 7.1b.

The results in Figure 7.2b can be accounted for because of the feedback voltage across the emitter-resistor V_E. If a transistor with a higher beta than the original one is substituted in the circuit, a greater emitter current is generated [$I_B(h_{FE} + 1) = I_E$]. This greater emitter current causes a greater voltage drop across $R_E(V_E)$ to be generated. In Equation 2, this greater V_E will cause I_B to be less; hence the collector current will be closer to the designed quiescent point. As will be shown, the values of R_E and R_B will determine the variations in the quiescent point.

Figure 7.3a shows another technique in biasing. This circuit configuration is generally referred to as a universal biasing circuit. Two biasing resistors (R_1 and R_2), an emitter-resistor (R_E), and a load resistor (R_L) are shown in the figure. Figure 7.3b shows the variations in the quiescent point for various beta values for a 2N4400.

EFFECTS OF TEMPERATURE CHANGES ON THE Q POINT

A change of temperature in a transistor amplifier changes the collector current of the amplifier. Because collector current is a function of V_{BE}, h_{FE}, and I_{CO}, and because these values are a function of temperature, temperature will change the collector current. Figure 7.4 shows the effects of temperature on the quiescent point for the three biasing arrangements thus far discussed. Figure 7.4a shows the greatest change in Q and Figure 7.4c shows the smallest change in Q.

The temperature change affected V_{BE}, I_{CO}, and h_{FE}. In order to analyze the effects of changes in temperature and beta on the Q point, a general equation will now be developed. This equation will relate the variables of I_{CO}, V_{BE}, and h_{FE} to the collector current I_C. Stability factors will be determined to show the effects of the various biasing techniques demonstrated thus far.

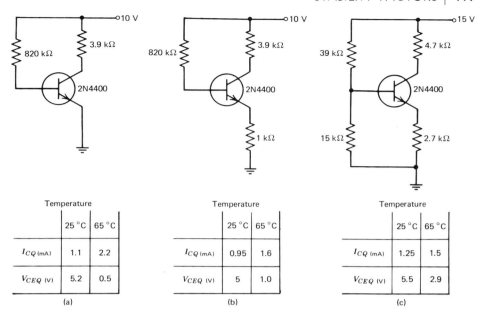

	Temperature			Temperature			Temperature	
	25 °C	65 °C		25 °C	65 °C		25 °C	65 °C
I_{CQ} (mA)	1.1	2.2	I_{CQ} (mA)	0.95	1.6	I_{CQ} (mA)	1.25	1.5
V_{CEQ} (V)	5.2	0.5	V_{CEQ} (V)	5	1.0	V_{CEQ} (V)	5.5	2.9
	(a)			(b)			(c)	

Figure 7.4
Effects of temperature changes: (a) constant base current biasing, (b) emitter feedback biasing, and (c) universal biasing.

STABILITY FACTORS

To understand why one biasing method seems better for a stabilized Q than another, let us examine the universal biasing circuit.

Figure 7.5 shows the universal biasing circuit.

$$V_{BB} = V_{CC} \frac{R_2}{R_1 + R_2} \tag{3}$$

Figure 7.5
Universal biasing: (a) circuit configuration and (b) equivalent circuit.

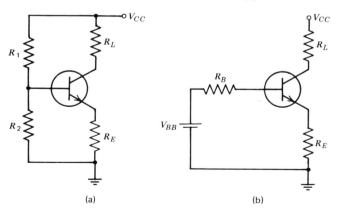

and

$$R_B = R_1 \mid\mid R_2 = \frac{R_1 R_2}{R_1 + R_2} \tag{4}$$

Here V_{BB} and R_B are the Thevininized open-circuit voltage and short-circuit resistance.

Applying Kirchhoff's voltage equations to the base loop, we have

$$V_{BB} = I_B R_B + V_{BE} + I_E R_E \tag{5}$$

The effect of I_{CO} with temperature on the collector and emitter current is given by

$$I_C = h_{FE} I_B + (h_{FE} + 1) I_{CO} \tag{6}$$

and

$$I_C = \alpha I_E + I_{CO} \tag{7}$$

When we substitute Equations 6 and 7 into Equation 5, the collector current is given as a function of several variables, or

$$I_C = \frac{h_{FE}(V_{BB} - V_{BE}) + I_{CO}(h_{FE} + 1) (R_B + R_E)}{R_B + (h_{FE} + 1) R_E} \tag{8}$$

Equation 8 could be used to directly calculate I_C for various changes in h_{FE}, I_{CO}, and V_{BE}; but for convenience, stability factors (S) are defined and used to indicate the changes in collector current for changes in the various parameters, V_{BE}, and so forth. Equation 8 is differentiated to calculate the stability factors. The stability factors are defined in Equations 9, 10, and 11. It is important to note that the change in collector current from the designed Q point is given from the various stability factors.

$$S_{I_{CO}} \cong \frac{\Delta I_C}{\Delta I_{CO}}\bigg|_{\substack{V_{BE},\, h_{FE} \\ V_{BB}\, \text{const}}} = \frac{(h_{FE} + 1) (R_B + R_E)}{R_B + (h_{FE} + 1) R_E} \tag{9}$$

$$S_{V_{BE}} \cong \frac{\Delta I_C}{\Delta V_{BE}}\bigg|_{\substack{I_{CO},\, h_{FE} \\ V_{BB}\, \text{const}}} = \frac{h_{FE}}{R_B + (h_{FE} + 1) R_E} \tag{10}$$

$$S_{h_{FE}} \cong \frac{\Delta I_C}{\Delta h_{FE}}\bigg|_{\substack{I_{CO},\, V_{BE} \\ V_{BB}\, \text{const}}} = \frac{(I_C - I_{CO})}{h_{FE}}\left[\frac{R_B + R_E}{R_B + (h_{FE} + 1) R_E}\right] \tag{11}$$

The results of the change in collector current (from the designed Q point) because of temperature and h_{FE} changes are given by Equation 12.

$$\Delta I_C = S_{I_{CO}}\Delta I_{CO} + S_{V_{BE}}\Delta V_{BE} + S_{h_{FE}}\Delta h_{FE} \tag{12}$$

A fourth stability factor could have been defined, $S_{V(CC)}$. This stability factor would relate the change in collector current due to changes in the supply voltage. However, we shall assume that a regulated power supply is used with the amplifier and thus eliminate the need for this stability factor.

In order to understand why the constant base current circuit (Figure 7.1) is the poorest in stability, let us calculate the possible collector current changes with temperature and beta changes. Let $R_E = 0\ \Omega$, $\Delta h_{FE} = 100$, $R_B = 560\ \text{k}\Omega$, $h_{FE} = 60$, $I_C = 1\ \text{mA}$, $I_{CO} = 1\ \mu\text{A}$, $\Delta I_{CO} = 16\ \mu\text{A}$ for a ΔT of 40 °C. Calculating the effects of beta change only ($T = 25$ °C), we have

$$S_{h_{FE}} = \frac{10^{-3}}{60} = 16.7\ \mu\text{A}$$

Using Equation 12, we have

$$\Delta I_C = S_{hFE}\ \Delta h_{FE} = 16.7\ \mu\text{A} \times 100 = 1.67\ \text{mA}$$

Calculating the effects of a temperature change only (assume beta does not change), we have

$$S_{ICO} = (h_{FE} + 1) = 61$$

$$S_{VBE} = \frac{h_{FE}}{R_B} = -10.7 \times 10^{-4}\ \Omega^{-1}$$

$$\Delta V_{BE} = -2.5\ \text{mV/°C} \times 40\ \text{°C} = -100\ \text{mV}$$

and

$$\Delta I_C = S_{ICO}\ \Delta I_{CO} + S_{VBE}\ \Delta V_{BE}$$
$$= 61 \times 0.016\ \text{mA} + (-10.7 \times 10^{-4}\ \Omega^{-1} \times -10^{-1}\ \text{V})$$
$$\Delta I_C = 1.083\ \text{mA}$$

The effects of only a beta change ($T = 25$ °C) could produce a possible change in collector current of 1.67 mA with a maximum change of beta of 100. If the temperature is changed from 25 to 65 °C, the collector current change is then 1.08 mA, assuming that beta does not change with temperature. To calculate the total current change for this constant base current circuit, all the stability terms would be used to describe the effects of temperature and beta changes.

In Figure 7.3, the universal biasing method, the stability factors are $R_E = 1\ \text{k}\Omega$, $R_B = 6.1\ \text{k}\Omega$, $I_C = 1\ \text{mA}$, $I_{CO} = 1\mu\text{A}$ ($T = 25$ °C), $h_{FE} = 60$, $\Delta h_{FE} = 100$, and $\Delta I_{CO} = 16\ \mu\text{A}$.

Calculating the effects of a change only in beta, we find that

$$S_{hFE} = \frac{(10^{-3})}{60}\left[\frac{7.1\ \text{k}\Omega}{6.1\ \text{k}\Omega + 61 \times 1\ \text{k}\Omega}\right] = 0.0018\ \text{mA}$$

$$\Delta I_C = S_{hFE}\ \Delta h_{FE} = 0.0018\ \text{mA} \times 100 = 0.18\ \text{mA}$$

Calculating the effects of a change only in temperature, we have

$$\Delta V_{BE} = -2.5\ \frac{\text{mV}}{\text{°C}} \times 40\ \text{°C} = -100\ \text{mV}$$

$$S_{VBE} = \frac{-60}{6.1\ \text{k}\Omega + 61\ \text{k}\Omega} = -0.89 \times 10^{-3}\ \Omega^{-1}$$

$$S_{ICO} = \frac{(61)\ (7.1 \times 10^3)}{6.1\ \text{k}\Omega + 61\ \text{k}\Omega} = 6.45$$

$$\Delta I_C = S_{ICO} \, \Delta I_{CO} + S_{VBE} \, \Delta V_{BE}$$

$$= 6.45 \times .016 \text{ mA} + (-0.89 \times 10^{-3} \, \Omega^{-1} \times -10^{-1} \text{ V})$$

$$= 0.103 \text{ mA} + 0.09 \text{ mA}$$

$$\Delta I_C = 0.193 \text{ mA}$$

A comparison of the two previous example problems shows that the constant base current circuit could have a change in collector current of 1.67 mA for a beta change and a change in collector current of 1.083 mA for a temperature change. The universal biasing circuit, for the same beta and temperature changes, could have a collector current change of 0.18 mA for a beta change and of 0.193 mA for a temperature change. The universal biasing circuit is a better circuit arrangement for a stable Q point than the constant base current circuit.

If we assume that $h_{FE} >> 1$ and that $(h_{FE} + 1) \, R_E >> R_B$, the stability equations (Equations 9, 10, and 11) reduce to

$$S_{ICO} \approx 1 + \frac{R_B}{R_E} \tag{13}$$

$$S_{VBE} \approx \frac{-1}{R_E} \tag{14}$$

$$S_{hFE} \approx \frac{(I_C - I_{CO})}{h_{FE}{}^2} \left[1 + \frac{R_B}{R_E} \right]$$

$$S_{hFE} \approx \frac{(I_C - I_{CO})}{h_{FE}{}^2} S_{ICO} \tag{15}$$

With our assumption, $R_E(h_{FE} + 1) >> R_B$, the selection of R_1 and R_2 can now be demonstrated. Figure 7.5 shows a universal biasing arrangement with its Thevininized equivalent. Kirchhoff's voltage law for the base loop is

$$V_{BB} = I_B R_B + V_{BE} + I_E R_E$$

$$V_{BB} = I_B R_B + V_{BE} + I_B(h_{FE} + 1) \, R_E \tag{16}$$

But if $R_E(h_{FE} + 1) >> R_B$, then Equation 16 reduces to

$$V_{BB} \approx V_{BE} + V_E \tag{17}$$

where

$$V_E = I_E R_E$$

The calculation of R_1 and R_2 can be made from Equations 3 and 4.

$$V_{BB} = V_{CC} \frac{R_2}{R_1 + R_2} \tag{18}$$

but

$$R_1 + R_2 = \frac{R_1 R_2}{R_B} \tag{19}$$

Substituting Equation 19 into Equation 18, we have

$$V_{BB} = V_{CC} \frac{R_B}{R_1}$$

or

$$R_1 = \frac{V_{CC}}{V_{BB}} R_B \qquad (20)$$

And solving Equation 4 for R_2, we have

$$R_2 = \frac{R_1 R_B}{R_1 - R_B} \qquad (21)$$

EXAMPLE

Calculate R_B, R_1, R_2, and V_{BB} for a universal biasing circuit, where $I_{CQ} = 1$ mA, $V_{CC} = 10$ V, $V_{BE} = 0.65$ V, $h_{FE} = 60$, $R_E = 1$ kΩ, $I_{CO} = 1$ μA, and $\Delta I_{CO} = 16$ μA

From Equation 17, $V_{BB} = V_{BE} + V_E$; $V_E = I_E R_E \approx I_C R_E$

$$V_{BB} = 0.65 \text{ V} + 1 \text{ mA} \times 1 \text{ kΩ}$$

then

$$V_{BB} = 1.65 \text{ V}$$

Let $(h_{FE} + 1) R_E >> R_B$ or

$$R_B = \frac{(h_{FE} + 1) R_E}{10} = \frac{61 \times 1 \text{ kΩ}}{10}$$

$$R_B = 6.1 \text{ kΩ}$$

Using Equation 20 we have

$$R_1 = \frac{V_{CC}}{V_{BB}} R_B = \frac{10 \text{ V}}{1.65 \text{ V}} \times 6.1 \text{ kΩ}$$

$$R_1 = 37 \text{ kΩ}$$

And using Equation 21, we have

$$R_2 = \frac{R_1 R_B}{R_1 - R_B} = \frac{37 \text{ kΩ} \times 6.1 \text{ kΩ}}{37 \text{ kΩ} - 6.1 \text{ kΩ}}$$

$$R_2 = 7.3 \text{ kΩ}$$

As can be noted, R_1 and R_2 are not standard values for resistors. If R_1 is chosen to be 33 kΩ, according to Equation 20, R_B is 5.44 kΩ. With 5.44 kΩ $R_B = [(h_{FE} + 1)R_E,]/11.2$ which is better than our requirements of $[(h_{FE} + 1)R_E]/10$. With the use of Equation 21, R_2 is calculated to be 6.5 kΩ. We shall use $R_1 = 33$ kΩ and $R_2 = 6.8$ kΩ.

The stability factors for the example problem can be calculated with Equations 13, 14, and 15.

$$S_{ICO} \approx 1 + \frac{R_B}{R_E} = 1 + \frac{5.44 \text{ k}\Omega}{1 \text{ k}\Omega}$$

$$S_{ICO} = 6.44$$

$$S_{VBE} \approx -\frac{1}{R_E} = -\frac{1}{1 \text{ k}\Omega}$$

$$S_{VBE} = -0.001 \ \Omega^{-1}$$

$$S_{hFE} \approx \frac{(I_C - I_{CO})}{h_{FE}{}^2} S_{ICO} = \frac{10^{-3}}{60^2} \times 6.44$$

$$S_{hFE} = 0.0019 \text{ mA}$$

Combining the stability factors, we have

$$\Delta I_C = S_{ICO} \ \Delta I_{CO} + S_{VBE} \ \Delta V_{BE} + S_{hFE} \ \Delta h_{FE}$$

$$= 6.44 \times 15 \ \mu\text{A} + (-10^{-3} \ \Omega^{-1} \times -10^{-1} \ \text{V}) + 0.0019 \text{ mA} \times 100$$

$$= 0.103 \text{ mA} + 0.1 \text{ mA} + 0.19 \text{ mA}$$

$$\Delta I_C = 0.393 \text{ mA}$$

For a temperature range from 25 °C to 40 °C and an h_{FE} change of 100, we would expect the I_{CQ} to change 0.393 mA under the worst possible conditions. Figure 7.6 shows this biasing arrangement with the various parameters.

DIODE COMPENSATION BIASING TECHNIQUES

Figure 7.7 shows the diode compensation biasing technique. The diode D is manufactured to have the identical temperature characteristics as those of the transistor.

$$I_D + I_B = I_T \tag{22}$$

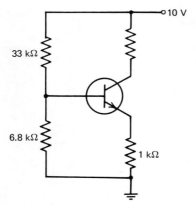

Figure 7.6
Circuit configuration for example problem.

and

$$I_E = I_B + I_C \tag{23}$$

Substituting Equation 23 into Equation 22, we have

$$I_D - I_C = I_T - I_E \tag{24}$$

Now if the diode is matched to the transistor, $I_E = I_D$, then Equation 24 reduces to $I_C = I_D$. Because I_D can be calculated, we have

$$I_D \approx \frac{V_{CC}}{R} = I_{CQ} \tag{25}$$

Then if $R/2 = R_L$, we have

$$V_{CEQ} = V_{R(L)} = I_{CQ}R_L$$

or

$$V_{CEQ} = \frac{V_{CC}}{R}R_L = \frac{V_{CC}}{R}\frac{R}{2}$$

or

$$V_{CEQ} = \frac{V_{CC}}{2} \tag{26}$$

The quiescent point V_{CEQ} is equal to one-half the supply voltage. Thus changes in the supply voltage or the temperature will not affect V_{CEQ}'s being equal to one-half the supply voltage. The requirement is that the diode be matched to the transistor ($I_E = I_D$). To accomplish this matching, an identical transistor is used as the diode in Figure 7.7. This transistor-used-as-a-diode is usually connected with its base and collector tied together. The arrangement is shown in Figure 7.8, where R_L is made equal to one-half the resistance of R. This identity permits the quiescent voltage of the transistor (V_{CEQ}) to be equal to $V_{CC}/2$.

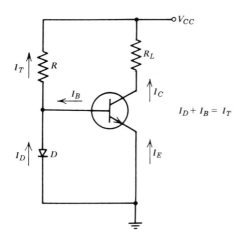

Figure 7.7
Diode compensation biasing.

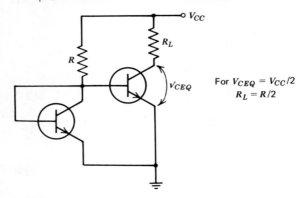

For $V_{CEQ} = V_{CC}/2$
$R_L = R/2$

Figure 7.8
Transistor used as a diode.

With monolithic integrated circuits, this diode compensation technique is used extensively in linear, IC amplifiers. This technique eliminates the need of an emitter-resistor and a bypass capacitor for a stable biasing circuit.

SHUNT FEEDBACK BIASING TECHNIQUES

Another biasing technique commonly found in solid-state devices is shown in Figure 7.9. This curcuit arrangement has the biasing resistor connected to the collector of the transistor rather than to the supply voltage. If the quiescent current I_{CQ} increases because of temperature, the base current is decreased because the V_{CEQ} voltage decreases. Thus the quiescent point of the transistor is very near its designed Q point. The above is demonstrated by Equation 27, which gives the base current.

$$I_B = \frac{V_{CE} - V_{BE}}{R_B} \tag{27}$$

From Figure 7.9

$$V_{CC} = I_E R_L + V_{CE} \tag{28}$$

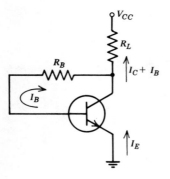

Figure 7.9
Collector feedback biasing.

Now, to find the collector current as a function of I_{CO}, h_{FE}, R_L, and R_B, we have

$$I_C = h_{FE} I_B + (h_{FE} + 1) I_{CO}$$

and

$$I_E = \frac{I_C - I_{CO}}{\alpha}$$

Substituting the above two equations into Equations 27 and 28, we have

$$I_C = \frac{h_{FE}(V_{CC} - V_{BE}) + I_{CO}(h_{FE} + 1)(R_B + R_L)}{R_B + (h_{FE} + 1) R_L} \tag{29}$$

Finding $S_{I_{CO}}$, we have

$$S_{ICO} = \frac{\Delta I_C}{\Delta I_{CO}} = \frac{(h_{FE} + 1)(R_B + R_L)}{R_B + (h_{FE} + 1) R_L} \tag{30}$$

$$S_{VBE} = \frac{\Delta I_C}{\Delta V_{BE}} = \frac{-h_{FE}}{R_B + (h_{FE} + 1) R_L} \tag{31}$$

$$S_{HFE} = \frac{\Delta I_C}{\Delta h_{FE}} = \frac{(I_C - I_{CO})}{h_{FE}} \left[\frac{R_B + R_L}{R_B + (h_{FE} + 1) R_L} \right] \tag{32}$$

For very low stability factors, R_L should be greater than R_B ($R_L > R_B$). The total current change in the collector current can be calculated using Equation 12.

REVIEW QUESTIONS AND PROBLEMS

7-1 Calculate $S_{I_{CO}}$, $S_{V_{BE}}$, and $S_{h_{FE}}$ for the circuits shown in the following figures in Chapter 6: (a) 6.2, (b) 6.3, (c) 6.5, (d) 6.7, (e) 6.9.

7-2 Calculate $S_{I_{CO}}$, $S_{V_{BE}}$, and $S_{h_{FE}}$ for the circuits shown in the following review questions from Chapter 6: (a) 6.1, (b) 6.2, (c) 6.3, (d) 6.4, (e) 6.5, and (f) 6.6.

7-3 Assuming that $\Delta h_{FE} = 100$, $\Delta I_{CO} = 10\mu A$, and $\Delta T = 40\ °C$, calculate the total possible change in the collector currents for the problems in Question 7-1.

7-4 Repeat Question 7-3 for the problems in Question 7-2.

7-5 Using Equation 8, calculate the stability factor $S_{V_{CC}}$. Assume that

$$V_{BB} = V_{CC} \frac{R_2}{R_1 + R_2} \quad \text{and} \quad R_B = R_1 \| R_2$$

OBJECTIVES FOR CHAPTER 8, CIRCUIT MODELS

At the completion of this chapter you can demonstrate your understanding of the material by answering these questions:

- What is a circuit model and why is it used?
- What are hybrid parameters?
- What circuit model is used for an ideal transistor?
- What is the effect of frequency on a circuit model?
- How do independent and dependent variables determine the configuration of a circuit model?
- How and why can a single device have more than one circuit model?

8 CIRCUIT MODELS

The behavior of a solid-state device can be explained through the use of a circuit model. Circuit models are used for devices because it is impossible to get inside the device to measure various parameters. You should study the various circuit models because of this increased understanding of the workings of the devices.

In this chapter the various circuit models commonly used to describe the behavior of solid-state devices are discussed. By the use of a *circuit model* the behavior of active or passive devices in terms of their input and output voltages and currents can be described. Some circuit models are applied to particular devices in lieu of other models because of the ease of measurements for the circuit parameters. As an example, if a BJT is used as a current amplifier, the current amplifier circuit model is applied because the model uses input and output currents in the measurement of circuit parameters. It is important to realize that a circuit model may be applied to single- or multistage amplifiers. Because of the commonality of the hybrid parameters describing BJTs, the hybrid circuit model is discussed first in this chapter.

LOW-FREQUENCY HYBRID MODEL

The behavior of an electronic device can be described by its input and output voltages and currents. Figure 8.1 illustrates an electronic device inside a black box with the input voltage and current labeled V_{in} and I_{in} and the output voltage and current labeled V_{out} and I_{out}. The four variables V_{in}, I_{in}, V_{out}, and I_{out} can be measured and mathematically defined as functions of each other. If the input voltage were defined as a function of I_{in} and V_{out} and the output current I_{out} were defined as a function of I_{in} and V_{out}, we would have a circuit

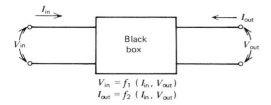

$$V_{in} = f_1 \, (I_{in}, \, V_{out})$$
$$I_{out} = f_2 \, (I_{in}, \, V_{out})$$

Figure 8.1
Black box.

model commonly called the *hydrid circuit model*. These relationships are expressed as

$$V_{in} = f_1(I_{in}, V_{out}) \tag{1}$$

$$I_{out} = f_2(I_{in}, V_{out}) \tag{2}$$

In order to understand the ac behavior of the device inside the black box, Equations 1 and 2 are differentiated, or

$$dV_{in} = \frac{\partial f_1}{\partial I_{in}} \, dI_{in} + \frac{\partial f_1}{\partial V_{out}} \, dV_{out} \tag{3}$$

$$dI_{out} = \frac{\partial f_2}{\partial I_{in}} \, dI_{in} + \frac{\partial f_2}{\partial V_{out}} \, dV_{out} \tag{4}$$

If we assume that $dV_{in} = \Delta V_{in} = v_{in}$ and $dI_{out} = \Delta I_{out} = i_{out}$, Equations 3 and 4 can be expressed as

$$v_{in} = \frac{\partial f_1}{\partial I_{in}} \, i_{in} + \frac{\partial f_1}{\partial V_{out}} \, v_{out} \tag{5}$$

$$i_{out} = \frac{\partial f_2}{\partial I_{in}} \, i_{in} + \frac{\partial f_2}{\partial V_{out}} \, v_{out} \tag{6}$$

Equation 5 describes the input voltage v_{in} in terms of the input current i_{in} and the output voltage v_{out}. Equation 6 describes the output current i_{out} in terms of the input current i_{in} and the output voltage v_{out}. The only task left is to define the partial derivatives $\partial f_1/\partial I_{in}$, $\partial f_1/\partial V_{out}$, $\partial f_2/\partial I_{in}$, and $\partial f_2/\partial V_{out}$. These partial derivatives are defined with the aid of Equations 1 and 2. Thus

$$\frac{\partial f_1}{\partial I_{in}} = \frac{\partial V_{in}}{\partial I_{in}} \, (\text{unit—}\Omega) \tag{7}$$

$$\frac{\partial f_1}{\partial V_{out}} = \frac{\partial V_{in}}{\partial V_{out}} \, (\text{no units}) \tag{8}$$

$$\frac{\partial f_2}{\partial I_{in}} = \frac{\partial I_{out}}{\partial I_{in}} \, (\text{no units}) \tag{9}$$

$$\frac{\partial f_2}{\partial V_{out}} = \frac{\partial I_{out}}{\partial V_{out}} \, (\text{units—}\mho) \tag{10}$$

PARAMETERS

In equation 7 we have a voltage over a current with the resulting units of ohms. In Equation 8 we have a voltage over a voltage with the resulting units being dimensionless. In Equation 9 we have a current over a current with the resulting units being dimensionless. In Equation 10 we have a current over a voltage with the resulting units being mhos. The circuit model for this configuration is called hybrid because of the mixture of units.

The partial derivatives described with Equations 7 through 10 can be represented by the letter h, with two subscripts used to identify the particular derivative of interest. Thus

$$\frac{\partial V_{in}}{\partial I_{in}} = h_{11} = h_i \tag{11}$$

$$\frac{\partial V_{in}}{\partial V_{out}} = h_{12} = h_r \tag{12}$$

$$\frac{\partial I_{out}}{\partial I_{in}} = h_{21} = h_f \tag{13}$$

$$\frac{\partial I_{out}}{\partial V_{out}} = h_{22} = h_o \tag{14}$$

Substituting Equations 11 through 14 into Equations 5 and 6, we have

$$v_{in} = h_{11}i_{in} + h_{12}v_{out} \tag{15}$$

$$i_{out} = h_{21}i_{in} + h_{22}v_{out} \tag{16}$$

Equations 15 and 16 describe the behavior of the device inside the black box.

Figure 8.2 illustrates how Equations 15 and 16 can be interpreted as a circuit model. The input side of the circuit model is described as Equation 15. The input resistance is labeled h_{11}. The feedback voltage that is in the input side of the model is h_{12}, multiplied by the output voltage v_{out}. The output side of the circuit model is described by Equation 16. A current pump of magnitude $h_{21}i_{in}$ supplies current to the output conductance h_{22} and whatever load is placed across the output terminals of the device. Note that the input side of the circuit model is a Thevininized voltage source and resistance and that the output side of the circuit model is a Norton current source and conductance.

The hybrid parameters h_{11}, h_{12}, h_{21}, and h_{22} can be measured or calculated with the aid of Equations 15 and 16. If the ac output voltage v_{out} were zero, Equations 15 and 16 would reduce to

$$h_{11} = \frac{v_{in}}{i_{in}} \text{ (units—ohms)} \tag{17}$$

$$h_{21} = \frac{i_{out}}{i_{in}} \text{ (no units)} \tag{18}$$

The input resistance is h_{11}, and h_{21} is the forward current gain for the device inside the black box with the output voltage v_{out} reduced to zero volts.

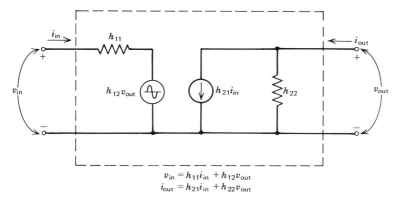

$$v_{in} = h_{11}i_{in} + h_{12}v_{out}$$
$$i_{out} = h_{21}i_{in} + h_{22}v_{out}$$

Figure 8.2
Hybrid circuit model.

If the ac input current i_{in} were reduced to zero amperes, Equations 15 and 16 would reduce to

$$h_{12} = \frac{v_{in}}{v_{out}} \text{ (no units)} \tag{19}$$

$$h_{22} = \frac{i_{out}}{v_{out}} \text{ (units—mhos)} \tag{20}$$

The reverse voltage transfer ratio is h_{12}, and h_{22} is defined as the output conductance of the device inside the black box with the input current i_{in} reduced to zero amperes. It is important to realize that the hybrid circuit model is an ac model only. If the device inside the box were a transistor that had dc biasing requirements, the hybrid model would express only the behavior of the device with ac voltages and currents.

EXAMPLE

Figure 8.3a illustrates a resistor network inside a black box. The "T" resistor network consists of three 100-Ω resistors. Calculate the hybrid parameters for this network.

Figure 8.3b illustrates the method used to calculate the parameters h_{11} and h_{21} with the output voltage v_{out} reduced to zero volts. Because h_{11} is the input resistance, we have a 100-Ω resistor in series with a 50-Ω resistance (100‖100), or

$$h_{11} = 150 \text{ Ω} \tag{A}$$

The ratio of the output current to input current is defined at h_{21} (Equation 18). If we assume an input current of 1 mA, from Figure 8.3b we see that the output current would be 0.5 mA. Note that the output current in this example is flowing in the same direction as the input current. Thus the output current is negative (flowing in the opposite direction), compared to the current of the hybrid model illustrated in Figure 8.3a. Thus

$$h_{21} = \frac{-0.5 \text{ mA}}{1.0 \text{ mA}}$$
$$h_{21} = -\tfrac{1}{2} \tag{B}$$

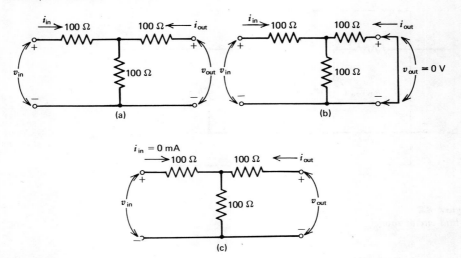

Figure 8.3
Example problem: (a) circuit, (b) v_{out} reduced to zero, and (c) l_{in} reduced to zero.

Figure 8.3c illustrates the method used to calculate theparameters h_{12} and h_{22}, with the input current reduced to zero. The ratio of input voltage to output voltage is h_{12} (Equation 19). If we assume that the output voltage is 1 V, voltage division would dictate that the input voltage would be 0.5 V. Because the input and output voltages are of the same polarity as in the hybrid circuit model, h_{12} is calculated to be

$$h_{12} = \frac{0.5 \text{ V}}{1.0 \text{ V}}$$

$$h_{12} = \frac{1}{2} \tag{C}$$

Equation 20 defines the output conductance h_{22}. Looking into the output side of the network shown in Figure 8.3c with the input current reduced to zero, we see the conductance as

$$h_{22} = \frac{1}{200 \text{ }\Omega}$$

$$h_{22} = 0.005 \text{ } \mho \tag{D}$$

Now, substituting Equations A through D into Equations 15 and 16, we have the hybrid circuit equations for the resistor network shown, or

$$v_{in} = 150 \text{ } i_{in} + 0.5 \text{ } v_{out} \tag{E}$$

$$i_{out} = -0.5 \text{ } i_{in} + 0.005 \text{ } v_{out} \tag{F}$$

We shall be using these two equations shortly.

CIRCUIT MODEL WITH LOAD

The circuit model shown in Figure 8.2, along with the hybrid parameters h_{11}, h_{12}, h_{21}, and h_{22}, describe the behavior of a device. For a complete model, the device is usually driven by a signal source and is working into a load.

Figure 8.4 illustrates the hybrid circuit model connected to a load resistance R_L and having a source v_s with an internal resistance R_g. With this circuit arrangement the input and output voltages can be calculated by

$$v_{in} = v_s - i_{in}R_g \tag{21}$$

$$v_{out} = -i_{out}R_L \tag{22}$$

where $-i_{out}R_L$ implies that the output voltage across R_L is of opposite polarity, as shown in Figure 8.4, because of the direction of the output current. The four equations that now describe the behavior of the device inside the black box connected to a load resistor and driven by a signal source are

$$v_{in} = h_{11}i_{in} + h_{12}v_{out} \tag{15}$$

$$i_{out} = h_{21}i_{in} + h_{22}v_{out} \tag{16}$$

$$v_{in} = v_s - i_{in}R_g \tag{21}$$

$$v_{out} = -i_{out}R_L \tag{22}$$

With the four equations listed above, five general equations can be derived. The five general equations define R_{in} (input resistance), R_{out} (output resistance), A_i (current gain), A_v (voltage gain), and P_g (power gain) for the circuit configuration shown in Figure 8.4.

As an example of the algebra involved, let us derive the general equation for the current gain A_i. Solving Equation 22 for i_{out} and substituting this value into Equation 16, we have

$$-\frac{v_{out}}{R_L} = h_{21}i_{in} + h_{22}v_{out}$$

$$v_{out} = \frac{-i_{in}h_{21}R_L}{h_{22}R_L + 1}$$

Figure 8.4
Hybrid circuit with load and source.

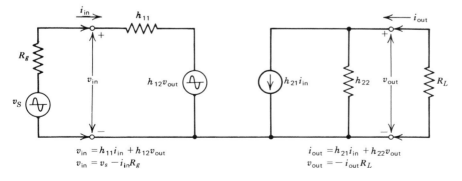

$$v_{in} = h_{11}i_{in} + h_{12}v_{out}$$
$$v_{in} = v_s - i_{in}R_g$$

$$i_{out} = h_{21}i_{in} + h_{22}v_{out}$$
$$v_{out} = -i_{out}R_L$$

But Equation 22 relates v_{out} as $-i_{out}R_L$, or

$$-i_{out}R_L = \frac{-i_{in}h_{21}R_L}{h_{22}R_L + 1}$$

$$A_i = \frac{i_{out}}{i_{in}} = \frac{h_{21}}{h_{22}R_L + 1} \tag{23}$$

With like reasoning and manipulation, we have

$$R_{in} = \frac{v_{in}}{i_{in}} = \frac{\Delta^h R_L + h_{11}}{h_{22}R_L + 1} \tag{24}$$

$$R_{out} = \frac{v_{out}}{i_{out}} = \frac{R_g + h_{11}}{\Delta^h + h_{22}R_g} \tag{25}$$

$$A_v = \frac{v_{out}}{v_{in}} = \frac{-h_{21}R_L}{\Delta^h R_L + h_{11}} \tag{26}$$

$$P_g = |A_v A_i| \tag{27}$$

where $\Delta^h = h_{11}h_{22} - h_{21}h_{22}$ (see the Review Questions).

Equations 23 through 27 can be used to calculate the voltage and current gains, input and output resistances, and power gain for a device inside the black box with specified source and load resistances. It is important to note that the hybrid parameters do not change for the device inside the box if the source and load resistors change in value. Changes in the source and load resistances would change the gains and resistances expressed by Equations 23 through 27 but would not change the numerical values of the h parameters.

EXAMPLE

Figure 8.5 illustrates a "T" resistor network coupled to a load resistor of 100 Ω and a source that has zero internal resistance ($R_g = 0$). Calculate the input and output resistances and the voltage and current gains for this circuit using the hybrid equations.

The h parameters for this "T" network were found in the previous example: $h_{11} = 150\,\Omega$, $h_{12} = 0.5$, $h_{21} = -0.5$, $h_{22} = 0.005\,\mho$. Using Equations 23 through 27, we have

$$\Delta^h = h_{11}h_{22} - h_{21}h_{12}$$

$$= (150)(0.005) - (0.5)(-0.5)$$

$$\Delta^h = 1$$

$$A_i = \frac{h_{21}}{h_{22}R_L + 1}$$

$$= \frac{-0.5}{(0.005)(100) + 1}$$

$$A_i = -0.33$$

$$R_{in} = \frac{\Delta^h R_L + h_{11}}{h_{22}R_L + 1}$$

$$= \frac{(1)(100) + 150}{(0.005)(100) + 1}$$

$$R_{\text{in}} = 167 \ \Omega$$

$$A_v = \frac{-h_{21}R_L}{\Delta^h R_L + h_{11}}$$

$$= \frac{-(-0.5)(100)}{1(100) + 150}$$

$$A_v = 0.2$$

$$R_{\text{out}} = \frac{R_g + h_{11}}{\Delta^h + h_{22}R_g}$$

$$= \frac{0 + 150}{1 + (0.005)(0)}$$

$$R_{\text{out}} = 150 \ \Omega$$

The reader is asked to verify the results in this example using conventional network theorems.

CE, CB, AND CC HYBRID PARAMETERS

With BJT devices the hybrid parameters (h_{11}, h_{12}, h_{21}, h_{22}) are modified to reflect the transistor's use in a common-emitter (CE), common-base (CB), or common-collector (CC) circuit configuration. The first subscript of the h denotes input resistance (i), output conductance (o), reverse voltage ratio (r), or forward current gain (f). The second subscript for the h denotes the BJT's use in a common-emitter (e), common-base (b), or common-collector (c) circuit configuration. Thus for a BJT, we have the following parameters listed in Table 8.1.

Figure 8.5
Example problem with load resistor.

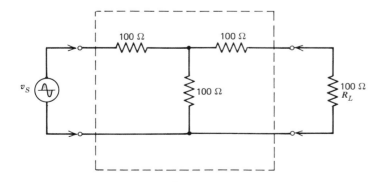

Table 8.1

			CE	CB	CC
Input resistance	h_{11}	=	h_{ie}	h_{ib}	h_{ic}
Reserve voltage gain	h_{12}	=	h_{re}	h_{rb}	h_{rc}
Forward current gain	h_{21}	=	h_{fe}	h_{fb}	h_{fc}
Output conductance	h_{22}	=	h_{oe}	h_{ob}	h_{oc}

Figure 8.6 illustrates the hybrid circuit configuration with the appropriate hybrid equations for the common-emitter, common-base, and common-collector circuit configuration. The general hybrid equations expressed by Equations 23 through 27 are valid for all three circuit configurations (CE, CB, and CC). As an

Figure 8.6
Hybrid models: (a) CB BJT, (b) CE BJT, and (c) CC BJT.

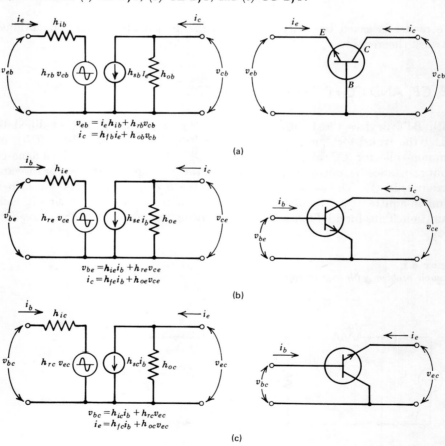

$$v_{eb} = i_e h_{ib} + h_{rb} v_{cb}$$
$$i_c = h_{fb} i_e + h_{ob} v_{cb}$$

(a)

$$v_{be} = h_{ie} i_b + h_{re} v_{ce}$$
$$i_c = h_{fe} i_b + h_{oe} v_{ce}$$

(b)

$$v_{bc} = h_{ic} i_b + h_{rc} v_{ec}$$
$$i_e = h_{fc} i_b + h_{oc} v_{ec}$$

(c)

example, to calculate the voltage gain for a BJT in the common-emitter circuit configuration, Equation 26 would become

$$A_v = \frac{-h_{fe}R_L}{\Delta^{he}R_L + h_{ie}}$$

where $\Delta^{he} = h_{oe}h_{ie} - h_{fe}h_{re}$. The manufacturer usually lists the hybrid parameters for a BJT for the common-emitter circuit configuration.

EXAMPLE

The 2N4400 BJT has the following parameters: $h_{ie} = 4$ kΩ, $h_{re} = 4 \times 10^{-4}$, $h_{fe} = 135$, $h_{oe} = 15 \times 10^{-6}$ mhos. If a load resistor of 4.7 kΩ and a source resistor of 100 kΩ are used with the BJT in the CE circuit configuration, calculate the following: A_i, A_v, R_{in}, R_{out}, and P_g.

$$\Delta^{he} = h_{oe}h_{ie} - h_{fe}h_{re}$$
$$= 15 \times 10^{-6} \times 4 \times 10^3 - 135 \times 4 \times 10^{-4}$$
$$\Delta^{he} = 0.114$$

$$A_v = \frac{-h_{fe}R_L}{\Delta^{he}R_L + h_{ie}}$$
$$= \frac{-135 \times 4700}{0.114 \times 4700 + 4000}$$
$$A_v = -139$$

$$A_i = \frac{h_{fe}}{h_{oe}R_L + 1}$$
$$= \frac{135}{15 \times 10^{-6} \times 4700 + 1}$$
$$A_i = 126$$

$$R_{in} = \frac{\Delta^{he}R_L + h_{ie}}{h_{oe}R_L + 1}$$
$$= \frac{0.114 \times 4700 + 4000}{15 \times 10^{-6} \times 4700 + 1}$$
$$R_{in} = 4240 \ \Omega$$

$$R_{out} = \frac{R_g + h_{ie}}{\Delta^{he} + h_{oe}R_g}$$
$$= \frac{100 \ \text{k}\Omega + 4 \ \text{k}\Omega}{0.114 + 15 \times 10^{-6} \times 100 \ \text{k}\Omega}$$
$$R_{out} = 65 \ \text{k}\Omega$$

$$P_g = |A_v A_i|$$
$$= |139 \times 126|$$
$$P_g = 17,514 = 42.4 \ \text{dB}$$

	General	Common Base (CB)	Common Emitter (CE)	Common collector (CC)
A_i	$\dfrac{h_{21}}{h_{22}R_L + 1}$	$\dfrac{h_{fb}}{h_{ob}R_L + 1}$	$\dfrac{h_{fe}}{h_{oe}R_L + 1}$	$\dfrac{h_{fc}}{h_{oc}R_L + 1}$
A_v	$\dfrac{-h_{21}R_L}{\Delta^h R_L + h_{11}}$	$\dfrac{-h_{fb}R_L}{\Delta^{hb}R_L + h_{ib}}$	$\dfrac{-h_{fe}R_L}{\Delta^{he}R_L + h_{ie}}$	$\dfrac{-h_{fc}R_L}{\Delta^{hc}R_L + h_{ic}}$
R_{in}	$\dfrac{\Delta^h R_L + h_{11}}{h_{22}R_L + 1}$	$\dfrac{\Delta^{hb}R_L + h_{ib}}{h_{ob}R_L + 1}$	$\dfrac{\Delta^{he}R_L + h_{ie}}{h_{oe}R_L + 1}$	$\dfrac{\Delta^{hc}R_L + h_{ic}}{h_{oc}R_L + 1}$
R_{out}	$\dfrac{h_{11} + R_g}{\Delta^h + h_{22}R_g}$	$\dfrac{h_{ib} + R_g}{\Delta^{hb} + h_{ob}R_g}$	$\dfrac{h_{ie} + R_g}{\Delta^{he} + h_{oe}R_g}$	$\dfrac{h_{ic} + R_g}{\Delta^{hc} + h_{oc}R_g}$

$$\Delta^h = h_{22}h_{11} - h_{12}h_{21} \qquad \Delta^{hb} = h_{ob}h_{ib} - h_{rb}h_{fb}$$

$$\Delta^{he} = h_{oe}h_{ie} - h_{re}h_{fe} \qquad \Delta^{hc} = h_{oc}h_{ic} - h_{rc}h_{fc}$$

Figure 8.7
General hybrid equations.

Figure 8.7 relates the voltage and current gains and the input and output resistances for the same transistor used in a CE, CB, or CC circuit configuration.

CONVERSION OF HYBRID PARAMETERS

In the previous example the common-emitter h parameters were used for the 2N4400 BJT. If this BJT were to be used in a common-base circuit configuration, the common-emitter h parameters would have to be converted to the common-base h parameters in order to calculate the voltage gain, and so forth. The conversion process is now outlined.

Figure 8.8 illustrates the hybrid equations for the common-base and common-emitter circuit configurations. If we assume that $v_{be} \approx v_{eb}$ and the $v_{cb} \gg v_{be}$ or v_{eb}, then it follows that $v_{cb} \approx v_{ec}$. From the figure

$$i_c = h_{fe}i_b + h_{oe}v_{ce}$$

but $i_b = i_e - i_c$ or

$$i_c = \frac{h_{fe}}{h_{fe} + 1}(-i_e) + \frac{h_{oe}}{h_{fe} + 1}(v_{ce}) \tag{28}$$

where i_e is negative $(-i_e)$ because it flows in the opposite direction for the common-base circuit configuration. The output current for the common-base circuit configuration is given by

$$i_c = h_{fb}i_e + h_{ob}v_{cb} \tag{29}$$

Comparing Equations 28 and 29 term by term, we have

$$h_{fb} = -\frac{h_{fe}}{h_{fe} + 1}$$

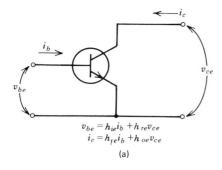

$$v_{be} = h_{ie}i_b + h_{re}v_{ce}$$
$$i_c = h_{fe}i_b + h_{oe}v_{ce}$$

(a)

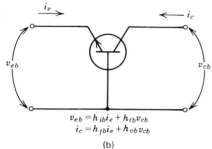

$$v_{eb} = h_{ib}i_e + h_{rb}v_{cb}$$
$$i_c = h_{fb}i_e + h_{ob}v_{cb}$$

(b)

Figure 8.8
*Hybrid equations: (a) CE circuit and
(b) CB circuit.*

$$h_{ob} = \frac{h_{oe}}{h_{fe} + 1}$$

Now, using the input voltage equation for the common-emitter circuit configuration and substituting $(i_e - i_c)$ for i_b and $(h_{fe}i_e + h_{ob}v_{cb})$ for i_c, we have

$$v_{be} = (h_{ie} - h_{ie}h_{fe})\, i_e + (h_{re} - h_{ie}h_{ob})\, v_{ce} \qquad (30)$$

Substituting h_{fb} and h_{ob} into Equation 30, we have

$$v_{be} = \frac{h_{ie}}{h_{fe} + 1}\, i_e + \left(\frac{\Delta^{he} - h_{re}}{h_{fe} + 1}\right) v_{ce} \qquad (31)$$

The equation for the input voltage for the common-base circuit configuration is

$$v_{eb} = h_{ib}i_e + h_{rb}v_{ce} \qquad (32)$$

Comparing Equations 31 and 32 term by term, we have

$$h_{ib} = \frac{h_{ie}}{h_{fe} + 1}$$

$$h_{rb} = \frac{\Delta^{he} - h_{re}}{h_{fe} + 1}$$

With like reasoning, we can equate the h parameters in the common-base, common-emitter, or common-collector circuit configuration to the common-base, common-emitter, or common-collector circuit configuration A summary for h conversions is given in Figure 8.9.

CB from CE and CC

$$h_{ib} = \frac{h_{ie}}{h_{fe} + 1} = \frac{-h_{ic}}{h_{fc}} \qquad\qquad h_{ob} = \frac{h_{oe}}{h_{fe} + 1} = \frac{-h_{oc}}{h_{fc}}$$

$$h_{fb} = \frac{-h_{fe}}{h_{fe} + 1} = \frac{-(h_{fc} + 1)}{h_{fc}} \qquad\qquad h_{rb} = \frac{\Delta^{he} - h_{re}}{h_{fe} + 1} = \frac{-(\Delta^{hc} + h_{fc})}{h_{fc}}$$

CE from CB and CC

$$h_{ie} = \frac{h_{ib}}{h_{fb} + 1} = h_{ic} \qquad\qquad h_{oe} = \frac{h_{ob}}{h_{fb} + 1} = h_{oc}$$

$$h_{fe} = \frac{-h_{fb}}{h_{fb} + 1} = -(h_{fc} + 1) \qquad\qquad h_{re} = \frac{\Delta^{hb} - h_{rb}}{h_{fb} + 1} = (1 - h_{rc})$$

CC from CE and CB

$$h_{ic} = h_{ie} = \frac{h_{ib}}{h_{fb} + 1} \qquad\qquad h_{oc} = h_{oe} = \frac{h_{ob}}{h_{fb} + 1}$$

$$h_{fc} = -(h_{fe} + 1) = \frac{1}{(h_{fb} + 1)} \qquad\qquad h_{rc} = (1 - h_{re}) = \frac{(1 - h_{ib}h_{ob})}{(h_{fb} + 1)}$$

Figure 8.9
Conversion table for the hybrid parameters.

EXAMPLE

Calculate the A_i, A_v, R_{in}, and R_{out} for a common-base amplifier using the 2N4400 BJT. Assume a load resistor of 4.7 kΩ and a source resistance R_g of 100 kΩ. Because the common-emitter h parameters are known for the 2N4400 BJT, Figure 8.9 can be used to convert these parameters into the common-base parameters. Thus $h_{ib} = 29.4$ Ω, $h_{fb} = -0.9926$, $h_{rb} = 8.4 \times 10^{-4}$, and $h_{ob} = 1.1 \times 10^{-7}$ ℧. Substituting these parameters into Equations 23 through 26, we have

$$\Delta^{hb} = h_{ob}h_{ib} - h_{fb}h_{rb}$$

$$= 1.1 \times 10^{-7} \times 29.4 - (-0.9926)(8.4 \times 10^{-4})$$

$$\Delta^{hb} = 8.37 \times 10^{-4}$$

$$A_i = \frac{h_{fb}}{h_{ob}R_L + 1}$$

$$A_i = \frac{(-0.9926)}{1.1 \times 10^{-7} \times 4700 + 1}$$

$$A_i = -0.992$$

$$A_v = \frac{-h_{fb}R_L}{\Delta^{hb}R_L + h_{ib}}$$

$$= \frac{-(-0.9926)(4700)}{8.30 \times 10^{-4} \times 4700 + 29.4 \ \Omega}$$

$$A_v = 140$$

$$R_{\text{in}} = \frac{\Delta^{hb}R_L + h_{ib}}{h_{ob}R_L + 1}$$

$$= \frac{8.37 \times 10^{-4} \times 4700 + 29.4 \ \Omega}{1.1 \times 10^{-7} \times 4700 + 1}$$

$$R_{\text{in}} = 33 \ \Omega$$

$$R_{\text{out}} = \frac{R_g + h_{ib}}{\Delta^{hb} + h_{ob}R_g}$$

$$= \frac{100 \ \text{k}\Omega + 29.4 \ \Omega}{8.37 \times 10^{-4} + 1.1 \times 10^{-7} \times 100 \ \text{k}\Omega}$$

$$R_{\text{out}} = 8.5 \ \text{M}\Omega$$

$$P_g = |A_i A_v|$$

$$= |140 \times 0.992|$$

$$P_g = 138.88 = 21.4 \ \text{dB}$$

THE IDEAL TRANSISTOR

For many practical approximations, a BJT in the common-emitter circuit configuration can be considered as having a very high output resistance ($h_{oe} \approx 0$) and a very small value of feedback voltage ($h_{re} \approx 0$). With these conditions for a transistor in the common-emitter circuit configuration, the BJT may be considered ideal. Figure 8.10 illustrates this concept of an ideal transistor in the common-emitter circuit configuration.

If we assume that $h_{oe} = 0$ and $h_{re} = 0$, Equations 23 through 26 for the common-emitter amplifier would reduce to

$$A_i \approx h_{fe} \tag{33}$$

$$A_v \approx \frac{-h_{fe}R_L}{h_{ie}} \tag{34}$$

$$R_{\text{in}} \approx h_{ie} \tag{35}$$

$$R_{\text{out}} \approx \infty \ \Omega \tag{36}$$

Note that Equations 33 through 36 are the same equations derived in Chapter 4.

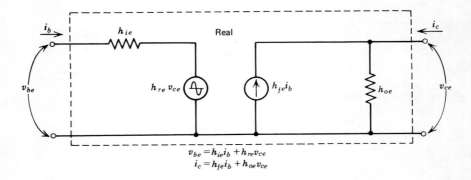

$$v_{be} = h_{ie}i_b + h_{re}v_{ce}$$
$$i_c = h_{fe}i_b + h_{oe}v_{ce}$$

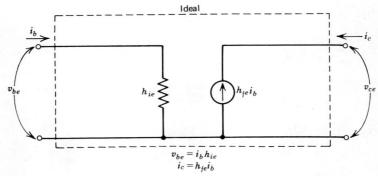

$$v_{be} = i_b h_{ie}$$
$$i_c = h_{fe}i_b$$

Figure 8.10
Hybrid circuit model for the ideal and real transistors.

For a common-base circuit configuration, the ideal BJT would have an h_{ob} equal to zero and an h_{rb} equal to zero. Thus the gain and resistance equations would reduce to

$$A_i \approx h_{fb} = \frac{-h_{fe}}{h_{fe}+1} \tag{37}$$

$$A_v \approx \frac{h_{fb}R_L}{h_{ib}} = \frac{h_{fe}R_L}{h_{ie}} \tag{38}$$

$$R_{in} \approx h_{ib} = \frac{h_{ie}}{h_{fe}+1} \tag{39}$$

$$R_{out} \approx \infty \ \Omega \tag{40}$$

For a BJT used in a common-collector circuit configuration, h_{oc} can be considered zero and h_{rc} equal to one. Thus the gain and resistance equations would reduce to

$$A_i \approx h_{fc} = -(h_{fe}+1) \tag{41}$$

$$A_v \approx \frac{-h_{fc}R_L}{-h_{fc}R_L + h_{ic}} = 1 \tag{42}$$

$$R_{in} \approx -h_{fc}R_L + h_{ic} \approx h_{ie} + (h_{fe}+1)\,R_L \tag{43}$$

$$R_{\text{out}} \approx \frac{h_{ic} + R_g}{-h_{fc}} = \frac{h_{ie} + R_g}{h_{fe} + 1} \tag{44}$$

Note that Equations 37 through 43 are the same equations derived in Chapter 5.

HYBRID-PI MODEL

The hybrid model (CE) just discussed for the BJT is valid only for the low frequencies (fewer than a few hundred kilohertz). For a valid circuit model for operating frequencies above a few hundred kilohertz, the hybrid model is modified and called the hybrid-pi model. The hybrid-pi model is illustrated in Figure 8.11 for a common emitter circuit configuration. This model assumes that the feedback voltage ratio h_{re} is zero.

Comparing Figure 8.11 with the low-frequency hybrid model, note that h_{ie} is replaced with r_{be}, $1/h_{oe}$ is replaced with r_{ce}, and the current generator $h_{fe}i_b$ is replaced with $g_m v_{be}$. The input resistance r_{be} is shunted with the capacitance of the base-to-emitter junction C_{be} and a feedback capacitance C is placed between the input and output of the device. For the common-emitter circuit configuration, this capacitance C is equal to $C_{bc}(1 + A_v)$ and is sometimes referred to as the Miller capacitance.

The capacitance of the forward-biased junction (C_{be}) is commonly called diffusion capacitance and is caused by the delay time of the current carriers in the forward-biased junction. The capacitance of the reverse-biased junction (C_{be}) is commonly called depletion capacitance and is caused by the depletion area that exists in a reverse-biased junction. As can be reasoned from this circuit model, the internal capacitance of the BJT is taken into account when the device is operating at high frequencies.

The input impedance Z_{in} is the parallel combination of r_{be} and $(C_{be} + C)$, where C is equal to $C_{bc}(1 + A_v)$. Thus

$$Z_{\text{in}} = r_{be} \| (C_{be} + C)$$

or

$$Z_{\text{in}} = \frac{r_{be}}{1 + j\omega r_{be}(C_{be} + C)} \tag{45}$$

Figure 8.11
High-frequency pi (hybrid) circuit model.

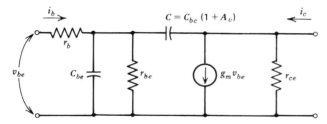

Because the input impedance consists of resistance and capacitance, the input current i_{in} is divided: Part goes through the resistance and part through the capacitance. The input current is thus bypassed to common through the total capacitance $(C_{be} + C)$, causing the output voltage to be reduced for the amplifier. The point at which this input current is divided equally between the resistance and the reactance of the total capacitance is called the -3 dB point. The frequency at which the -3 dB point occurs is called the beta cutoff frequency and is found by letting the resistance equal the reactance, or

$$r_{be} = \omega(C_{be} + C)$$

or

$$f_{\beta(\text{cutoff})} = \frac{1}{2\pi r_{be}(C_{be} + C)} \tag{46}$$

where $C = C_{bc}(1 + A_v)$.

Manufacturers frequently list a frequency called the gain bandwidth product labeled f_T. This f_T is the frequency at which the current gain of the transistor in the common-emitter configuration is equal to one. The relationship between the beta cutoff frequency and the gain bandwidth product frequency is

$$f_T = h_{fe} f_{\beta(\text{cutoff})} \tag{47}$$

EXAMPLE

The manufacturer for a 2N4400 lists the gain bandwidth product frequency f_T as equal to 200 Mhz. If the h_{fe} is assumed to be 135, calculate the beta cutoff frequency f_β. Using Equation 46, we have

$$200 \text{ Mhz} = 135 f_\beta$$

or

$$f_\beta = 1.48 \text{ Mhz}$$

Using Equation 46, we can calculate the total capacitance as

$$(C_{be} + C) = \frac{1}{f_\beta 2\pi r_{be}} = \frac{1}{1.48 \times 10^6 \times 2 \times \pi \times 4000}$$

$$(C_{be} + C) = 27 \text{ pF}$$

COMMON-BASE AND COMMON-COLLECTOR FREQUENCY RESPONSE

Common-base and common-collector amplifiers have a higher frequency response than a common-emitter amplifier. Because of the Miller effect in the common-emitter amplifier, the capacitance between base-collector is magnified by approximately the voltage gain of the stage. The common-base and common-collector amplifiers, because of their in-phase output voltage and the elements in common to both the input and output signals, do not have this increase in effective capacitance. Thus the frequency response for a common-base and common-collector amplifier is much higher for the same BJT that is used in a

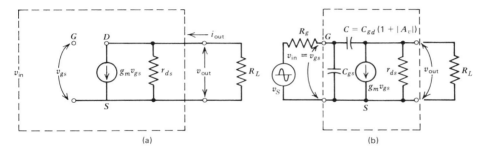

Figure 8.12
Circuit models for the common-source FET: (a) low-frequency and (b) high frequency.

common-emitter circuit configuration. In fact, a well-designed amplifier in the common-base or common-collector circuit configuration will have an upper frequency limit approaching that of the gain bandwidth product f_T.

LOW-FREQUENCY

As will be discussed in Chapter 10, junction and metal oxide semiconductor field effect transistors (JFET, MOSFET) are voltage-operated devices in which the BJT is a current-operated device. The voltage-operated device functions by controlling the current flow with electostatic fields. With this type of control (assuming no input current) the input of the device is isolated from the output. Because the field effect transistor has a high output resistance, the circuit model for the FET has a current generator coupled to an output resistance (see Chapter 10). Figure 8.12a illustrates such a model (common-source).

Notice that the input side of the model is isolated from the output side of the model. The input voltage that controls the output current of the device is labeled v_{gs} (gate-to-source). The transconductance of the device g_m is approximately equal to $\Delta I_{ds}/\Delta V_{gs}$. The manufacturer generally lists this parameter in micromhos (μmhos). The output resistance of the device is given by r_{ds} (drain-to-source). With a load resistor connected to the device, the voltage gain can be calculated by

$$A_v = -\frac{v_{\text{out}}}{v_{gs}} = \frac{-i_o R_i}{v_{gs}}$$

but

$$i_0 = \frac{g_m v_{gs} r_{ds}}{r_{ds} + R_L}$$

or

$$A_v = \frac{-g_m v_{gs} r_{ds} R_L}{(r_{ds} + R_L) v_{gs}}$$

dividing by r_{ds}, we have

$$A_v = \frac{-g_m R_L}{1 + (R_L/r_{ds})}$$

if $R_L \ll r_{ds}$, then it follows that

$$A_v \approx -g_m R_L$$

COMMON-SOURCE HIGH-FREQUENCY MODEL FOR THE FET

Figure 8.12*b* illustrates the high-frequency model for the field effect transistor. The capacitance from gate-to-source (C_{gs}) is shunting the input of the device. The Miller effect that was prevalent in the circuit model for the common-emitter BJT also exists in a FET common-source circuit configuration. The Miller effect causes the capacitance from gate-to-drain (C_{gd}) to be magnified by approximately the voltage gain for the state $(g_m R_L)$. The internal resistance of the source driving the input of the FET amplifier is labeled R_g.

As can be seen from the circuit model, C_{gs} is paralleled with C, and this capacitance combination is in series with R_g. The -3 dB point for this circuit is calculated by letting the resistance R_g equal the reactance of the capacitance combination. Thus

$$f_{UL} \cong \frac{1}{2\pi(C + C_{gs})}$$

where f is the upper frequency limit for the device, $|A_v| \approx g_m R_L$ and $C \approx (1 + A_v) C_{gd} \approx g_m R_L C_{gd}$.

THE *z* AND *y* CIRCUIT MODELS

Referring to Figure 8.1, we recall that the input voltage and output current for the device inside the black box were defined as functions of the output voltage (V_{out}) and the input current (I_{in}). Defining these functions generated the hybrid, parameter circuit model. Because we are dealing with four measurable quantities, other functions can be defined.

If the input and output currents were defined as functions of the input and output voltages (V_{out}, V_{in}), we would have

$$I_{in} = f_1(V_{in}, V_{out}) \tag{48}$$

$$I_{out} = f_2(V_{in}, V_{out}) \tag{49}$$

Equations 48 and 49 define what is commonly called the *y* parameter circuit model or admittance circuit model. Figure 8.13*a* illustrates this model. Notice that the model is constructed of two current generators.

If the input and output voltage were defined in terms of the input and output currents (I_{in}, I_{out}), the circuit model generated would be called the *z* or impedance parameter circuit model. Figure 8.13*b* illustrates this circuit model. Note that the model consists of two voltage generators described by

$$V_{in} = f_1(I_{in}, I_{out}) \tag{50}$$

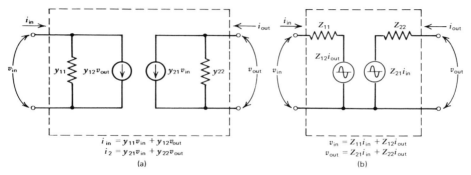

Figure 8.13
Circuit models: (a) y parameters and (b) z parameters.

$$V_{out} = f_2(I_{in}, I_{out}) \tag{51}$$

The procedures used with Equations 1 and 2 would define the various parameters depicted in the circuit models illustrated in Figures 8.13a and b (see the Review Questions).

VOLTAGE AMPLIFIER CIRCUIT MODEL

Figure 8.14a illustrates a circuit model for a voltage amplifier. The gain of the amplifier is expressed in terms of its output voltage over its input voltage, or

$$A_v = \frac{v_o}{v_{in}} \tag{52}$$

Note that with Equation 52 it is assumed that R_L is much greater than R_0 ($R_L \gg R_0$).

Figure 8.14
Circuit model for the (a) voltage amplifier and (b) current amplifier.

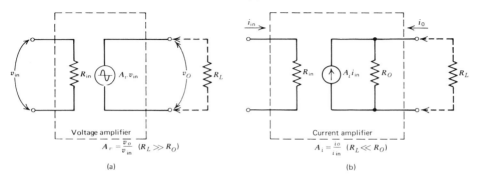

CURRENT AMPLIFIER CIRCUIT MODEL

Figure 8.14*b* illustrates a circuit model for a current amplifier. The gain of the amplifier is expressed in terms of its output current over its input current, or

$$A_i = \frac{i_o}{i_{in}} \tag{53}$$

Note that with Equation 53 it is assumed that R_L is much less than R_0 $(R_L \ll R_0)$.

TRANSRESISTANCE AMPLIFIER CIRCUIT MODEL

Sometimes it is convenient to express the gain of an amplifier in terms of its output voltage and input current. Figure 8.15*a* illustrates a circuit model for a transresistance amplifier. The gain of the amplifier is its output voltage over its input current, or

$$R_m = \frac{v_o}{i_{in}} \tag{54}$$

Note that with Equation 54 it is assumed that R_L is much greater than R_0 $(R_L \gg R_0)$.

TRANSCONDUCTANCE AMPLIFIER CIRCUIT MODEL

Sometimes it is convenient to express the gain of an amplifier in terms of its output current over its input voltage. Figure 8.15*b* illustrates a circuit model for a transconductance amplifier. The gain of the amplifier is

$$G_m = i_o/v_{in} \tag{55}$$

Note that in Equation 55 it is assumed that R_L is much less than R_0 $(R_L \ll R_0)$.

Figure 8.15
Circuit models for the (a) transresistance amplifier and (b) transconductive amplifier.

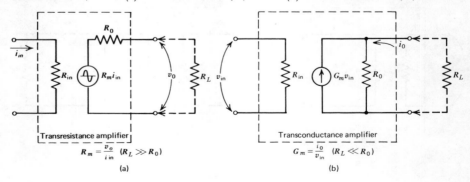

REVIEW QUESTIONS AND PROBLEMS

8-1 Discuss the significance of a circuit model.

8-2 Is it possible for a single device to have several circuit models? Explain.

8-3 What are hybrid parameters?

8-4 Calculate the hybrid parameters for the following resistor network and draw the circuit model.

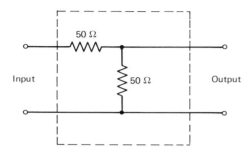

8-5 If a 50-Ω load resistor R_L is placed across the output terminals of the circuit shown in Problem 8-4, calculate the input resistance R_{in}, the voltage gain A_v, the current gain A_i, and the output resistance R_0.

8-6 Using Ohm's law, check the answers found in Problem 8-5.

8-7 A BJT used in a common-emitter circuit configuration has the following parameters: $h_{fe} = 75$, $h_{ie} = 3$ kΩ, $h_{oe} = 15$ μmhos, and $h_{re} = 4 \times 10^{-4}$. Assume that the load resistor is 5 kΩ and the source resistance is 50 kΩ. Calculate A_i, A_v, R_{in}, R_0, and the power gain for this circuit.

8-8 Convert the parameters listed in Problem 8-7 into the parameters for the BJT used in (a) a common-base circuit configuration and (b) a common-collector circuit configuration.

8-9 Assuming an ideal transistor ($h_{oe} = h_{re} = 0$), calculate the input and output resistances and voltage and current gains for this transistor used in the common-emitter circuit configuration. Compare your results with the equations found in Chapter 4.

8-10 Repeat Problem 8-9 for the transistor used in (a) a common-base circuit configuration and (b) a common-collector circuit configuraiton. Compare your results with the equations found in Chapter 5.

8-11 Assume that a common-emitter amplifier has a load resistor of 4700 Ω and an h_{ib} of 26 Ω with a C_{be} of 12 pF and a C_{bc} of 4 pF. Because of the Miller effect, calculate the effective input capacitance of this amplifier.

8-12 What limits the upper frequency response of a BJT common-emitter amplifier?

8-13 Discuss (a) a transresistance amplifier, (b) a transconductance amplifier, (c) a voltage amplifier, and (d) a current amplifier.

OBJECTIVES FOR CHAPTER 9, FEEDBACK PRINCIPLES

At the completion of this chapter you can demonstrate your understanding of the material by answering these questions:

- What are positive feedback and negative feedback?
- How does negative feedback affect the input resistance, the output resistance, and the gain of an amplifier?
- What are the four types of negative feedback?
- What are some of the circuit configurations using feedback principles?
- Why are feedback principles used?

9 FEEDBACK PRINCIPLES

Feedback principles are employed in both oscillators and amplifiers. Positive feedback is used with oscillators, and negative feedback is used with amplifiers. You should study feedback principles in order to comprehend its effect on stabilizing the input resistances, voltage and current gains, and frequency response and distortion associated with amplifiers.

As will be demonstrated in this chapter, negative feedback can be incorporated with an amplifier to stabilize its gain (A_v, A_i, G_m, or R_m), increase its frequency response, increase or decrease its input resistance, increase or decrease its output resistance, and decrease its distortion. The changes in the input and output resistance of the amplifier are related to the type of feedback used. As will be demonstrated, the feedback voltage can be a function of the output voltage or current, or the feedback current can be a function of the output voltage or current.

If the feedback voltage is a function of output voltage, the circuit model used for the amplifier is the *voltage amplifier*. If the feedback current is a function of the output current, the circuit model used for the amplifier is the *current amplifier*. If the feedback voltage is a function of the output current, the circuit model used for the amplifier is the *transconductance amplifier*. If the feedback current is a function of output voltage, the circuit model used for the amplifier is the *transresistance amplifier*.

It is important for the reader to realize that negative feedback techniques can be applied to an amplifier having one or many active devices such as BJTs or FETs. The feedback techniques discussed in this chapter are also used with the operational amplifier (see Chapter 15).

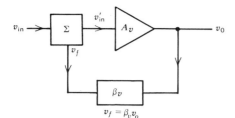

Figure 9.1
Block diagram of a voltage amplifier
using feedback.

VOLTAGE AMPLIFIER WITH FEEDBACK

Figure 9.1 illustrates a circuit arrangement for a voltage amplifier using feedback. The amplifier has an open loop gain of A_v, which is defined as

$$A_v = \frac{v_o}{v'_{in}} \qquad (1)$$

With feedback the overall amplifier has a closed loop gain, which is defined as

$$A_{vf} = \frac{v_o}{v_{in}} \qquad (2)$$

The feedback voltage v_f is a function of the output voltage v_o and is defined as

$$v_f = \beta_v v_o \qquad (3)$$

where β is the feedback factor. The block labeled β_v in the figure indicates the components necessary to sample a portion of the output voltage and return this sample to the input of the amplifier. The summing point labeled sigma in the figure is where the input voltage v_{in} is algebraically added to the feedback voltage v_f. With negative or degenerative feedback, the difference between v_{in} and v_f equals the input voltage at the amplifier v'_{in}.

The closed loop gain for the amplifier is defined as A_{vf} and is equal to

$$A_{vf} = \frac{v_o}{v_{in}} \qquad (2)$$

but $v_{in} - v_f = v'_{in}$. Substituting this relationship in the above equation, we have

$$A_{vf} = \frac{v_o}{v'_{in} + v_f}$$

$$A_{vf} = \frac{v_o}{v'_{in} + \beta_v v_o}$$

Dividing by v'_{in}, we have

$$A_{vf} = \frac{v_o/v'_{in}}{1 + \beta_v(v_o/v'_{in})}$$

but v_o/v'_{in} is defined as the open loop gain A_v, or

$$A_{vf}{}^1 = \frac{A_v}{1 + \beta_v A_v} \qquad (4)$$

If A_v is much greater than one, Equation 4 becomes

$$A_{vf} \approx \frac{1}{\beta_v} \qquad (5)$$

EXAMPLE

Assume that a voltage amplifier has an open loop gain of 100. Using degenerative feedback, 10 percent of the output voltage is fed back to the input. Calculate the closed loop voltage gain of this amplifier. Using Equation 3 to calculate β_v, we have

$$v_f = 0.1 v_o$$

or

$$\beta_v = 0.1$$

Using Equation 4 to calculate the closed loop gain, we have

$$A_{vf} = \frac{A_v}{1 + \beta_v A_v}$$

$$= \frac{100}{1 + 0.1 \times 100}$$

$$A_{vf} \approx 9.10$$

Note that the closed loop gain is 9, whereas the open loop gain is 100. At first one would think that the reduction in gain is not beneficial; but look at what happens if the open loop gain of the amplifier were to change to 150. Using the same percentage of feedback (10 percent), the closed loop gain is

$$A_{vf} = \frac{150}{1 + 0.1 \times 150}$$

$$A_{vf} = 9.38$$

Thus with a change of 50 in the open loop gain, the closed loop gain changed only 0.38. When negative feedback is used with a voltage amplifier, the closed loop gain for the amplifier is stabilized. The change in the open loop voltage gain for an amplifier can be the result of changing the BJT, aging components, and so forth.

INPUT AND OUTPUT RESISTANCES

The circuit model for a voltage amplifier is illustrated in Figure 9.2. With the aid of the circuit model, the input and output resistances can be calculated.

---

[1] If positive or regenerative feedback were used (v_f would be added or in phase with v_{in}), Equation 4 would become $A_{vf} = A_v/(1 - \beta_v A_v)$. If certain criteria are meant for the denominator, the circuit would become an oscillator (see Chapter 19).

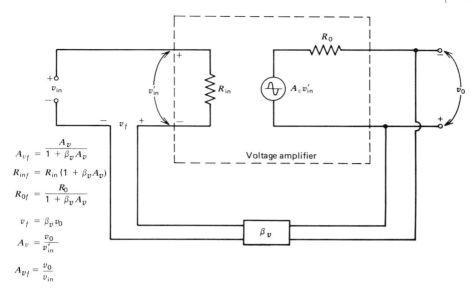

$$A_{vf} = \frac{A_v}{1 + \beta_v A_v}$$

$$R_{inf} = R_{in}(1 + \beta_v A_v)$$

$$R_{0f} = \frac{R_0}{1 + \beta_v A_v}$$

$$v_f = \beta_v v_0$$

$$A_v = \frac{v_0}{v'_{in}}$$

$$A_{vf} = \frac{v_0}{v_{in}}$$

Figure 9.2
Circuit model for a voltage amplifier using negative feedback.

We shall assume that the load resistor R_L is much greater than the internal resistance of the amplifier R_0 and that the β_v feedback network does not load the amplifier.

The input resistance with feedback is defined as

$$R_{inf} = \frac{v_{in}}{i_{in}}$$

but $v_{in} = v_f + v'_{in}$. Substituting, we have

$$R_{inf} = \frac{v'_{in}}{i_{in}} + \frac{v_f}{i_{in}}$$

Now $v_f = \beta_v v_0$, or

$$R_{inf} = \frac{v'_{in}}{i_{in}} + \frac{\beta_v v_0}{i_{in}}$$

and $v_0 = A_v v'_{in}$, or

$$R_{inf} = \frac{v'_{in}}{i_{in}} + \frac{\beta_v A_v v'_{in}}{i_{in}}$$

$$R_{inf} = \frac{v'_{in}}{i_{in}} (1 + \beta_v A_v)$$

but v'_{in}/i_{in} is the input resistance without feedback, or

$$R_{inf} = R_{in}(1 + \beta_v A_v) \tag{6}$$

Equation 6 tells us that the input resistance for a voltage amplifier using negative feedback is increased by the factor of $(1 + \beta_v A_v)$.

The output resistance for the voltage amplifier using negative feedback is defined as

$$R_{of} = \frac{v_o}{i_o}$$

The output current i_o is equal to

$$i_o = \frac{A_v v'_{in} + v_o}{R_o}$$

Setting the input voltage v_{in} equal to zero, the feedback voltage v_f is

$$v_f = v'_{in} = \beta_v v_o$$

Substituting this relationship into the equation for i_o, we have

$$i_o = \frac{B_v A_v v_o + v_o}{R_o}$$

$$R_{of} = \frac{R_o}{(1 + \beta_v A_v)} \tag{7}$$

Equation 7 tells us that the output resistance for a voltage amplifier using negative feedback is decreased by the factor of $(1 + \beta_v A_v)$.

EXAMPLE

Assume that a voltage amplifier has an open loop gain of 100, an input resistance R_{in} of 4kΩ, and an output resistance of 20 kΩ. The percentage of feedback is 15. Calculate A_{vf}, R_{inf}, and R_{of}. Using Equation 4, we have

$$A_{vf} = \frac{A_v}{1 + \beta_v A_v}$$

$$= \frac{100}{1 + 0.15 \times 100}$$

$$A_{vf} = 6.25$$

Using Equations 6 and 7 we find R_{inf} and R_{of},

$$R_{inf} = R_{in}(1 + \beta_v A_v)$$

$$= 4 \text{ k}\Omega(16)$$

$$R_{inf} = 64 \text{ k}\Omega$$

$$R_{of} = \frac{R_o}{1 + \beta_v A_v}$$

$$= \frac{20 \text{ k}\Omega}{16}$$

$$R_{of} = 1.25 \text{ k}\Omega$$

A means of calculating the effects of a change in the open loop gain of a voltage amplifier on its closed loop gain can be obtained by modifying Equation

4. If Equation 4 is differentiated with respect to the open loop gain, we would have

$$\Delta A_{vf} = \frac{\Delta A_v}{(1 + \beta_v A_v)^2}$$

Dividing by A_{vf}, we have

$$\frac{\Delta A_{vf}}{A_{vf}} = \frac{1}{(1 + \beta_v A_v)} \frac{\Delta A_v}{A_v} \tag{8}$$

With the use of Equation 8, the magnitude of the feedback factor and the open loop gain necessary for an amplifier can be calculated when the change in the closed loop gain is specified.

EXAMPLE

Assume that a voltage amplifier is to have a closed loop gain of 50 with no more than a 1 percent change in A_{vf}. The open loop gain of the amplifier can change as much as 50 percent. Calculate the necessary open loop gain and feedback factor for this amplifier. Using Equation 8, we have

$$0.01 = \frac{1}{(1 + \beta_v A_v)} \times 0.5$$

$$(1 + \beta_v A_v) = 50$$

In order to calculate the open loop gain, we must use Equation 4.

$$A_{vf} = \frac{A_v}{(1 + \beta_v A_v)}$$

$$50 = \frac{A_v}{50}$$

$$A_v = 2500$$

Substituting A_v into the quantity $(1 + \beta_v A_v)$, we have

$$1 + \beta_v A_v = 50$$

$$\beta_v = \frac{49}{2500}$$

$$\beta_v = 0.0196 \qquad \text{or} \qquad 1.96 \text{ percent}$$

CIRCUIT CONFIGURATION

Figure 9.3a illustrates a circuit configuration for a multistage BJT amplifier with feedback; Q_1 and Q_2 are identical voltage amplifiers with open loop gains of A_1 and A_2. The total open loop gain for the amplifier is the product $A_1 A_2$. The capacitors labeled C isolate the dc from the ac signals in the amplifier. With the two stages of amplifiers connected as common-emitter amplifiers, the input voltage v_{in} is in phase with the output voltage v_{out}. With these phase

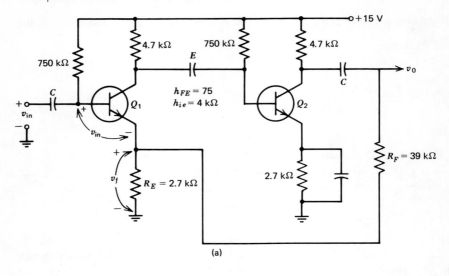

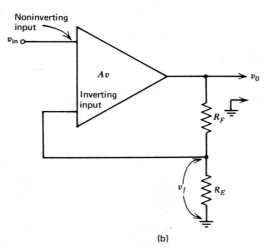

Figure 9.3
Voltage amplifier using negative feedback: (a) discrete BJT amplifier and (b) operational amplifier.

relationships, the two inputs for the total amplifier can be labeled inverting and noninverting (see Figure 9.3b). The voltage that Q_1 amplifies is the difference between v_{in} and v_f; thus the amplifier uses negative feedback.

The feedback voltage is given by Equation 3, or

$$v_f = \beta_v v_o$$

$$v_f = \frac{R_E}{R_E + R_F} v_o$$

Then β_v is equal to

$$\beta_v = \frac{R_E}{R_E + R_F} \tag{9}$$

Because $A_1 A_2$ is the open loop gain and is much greater than 1, Equation 5 can be used to find the closed loop gain of the total amplifier, or

$$A_{vf} \approx \frac{1}{\beta_v} = \frac{R_F}{R_E} + 1 \tag{10}$$

EXAMPLE

With the circuit components given in Figure 9.3a, calculate the closed loop voltage gain and input resistance for the amplifier in the figure. With the methods discussed in Chapter 4, the collector current for Q_1 or Q_2 is calculated to be 1.2 mA. The closed loop gain is found with Equation 10, or

$$A_{vf} = \frac{1}{\beta_v} = \frac{R_F}{R_E} + 1$$

$$A_{vf} = \frac{39 \text{ k}\Omega}{2.7 \text{ k}\Omega} + 1$$

$$A_{vf} = 15.4$$

The open loop gain of the total amplifier is $A_1 A_2$, or

$$A_1 \cong \frac{R_L}{h_{ib}} = \frac{4700\Omega}{30 \text{ mV}/1.2 \text{ mA}} = 188$$

$$A_1 A_2 \cong 188^2 \approx 35{,}000$$

$$A_v = 35{,}000$$

Then β_v is calculated with Equation 9, or

$$\beta_v = \frac{R_E}{R_E + R_F} = \frac{2.7 \text{ k}\Omega}{2.7 \text{ k}\Omega + 39 \text{ k}\Omega}$$

$$\beta_v = 0.065$$

and R_{inf} is calculated to be

$$R_{inf} = h_{ie}(1 + \beta_v A_v)$$

$$= 4 \text{ k}\Omega(1 + 0.065 \times 35{,}000)$$

$$R_{inf} = 9 \text{ M}\Omega$$

With the voltage amplifier using feedback, the voltage gain A_{vf} is stabilized, the input resistance is increased by the factor $(1 + \beta_v A_v)$, and the output resistance is decreased by a factor of $(1 + \beta_v A_v)$.

TRANSCONDUCTANCE AMPLIFIER WITH FEEDBACK

With the voltage amplifier just discussed, the feedback voltage v_f was a function of the output voltage v_o. If the feedback voltage v_f were a function of the output current i_o, the amplifier would be considered a transconductance amplifier. Figure 9.4 illustrates a transconductance amplifier with feedback. The feedback

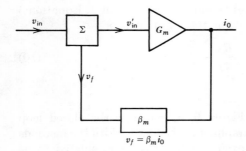

Figure 9.4
Block diagram of a transconductance amplifier using feedback.

voltage v_f is a function of the output current i_o. The amplifier has an open loop transconductance gain of G_m, which is defined as

$$G_m = \frac{i_o}{v'_{in}} \qquad (11)$$

With feedback, the overall amplifier has a closed loop transconductance gain G_{mf}, which is defined as

$$G_{mf} = \frac{i_o}{v_{in}} \qquad (12)$$

The feedback voltage v_f is a function of the output current i_o and is defined as

$$v_f = \beta_m i_o \qquad (13)$$

where β_m is the feedback factor. The block labeled β_m in the figure is the arrangement of circuit components necessary to sample a portion of the output current and return it as a feedback voltage. The input voltage of the amplifier is labeled v'_{in} and is equal to

$$v'_{in} = v_{in} - v_f \qquad (14)$$

Substituting Equation 14 into Equation 12 in order to calculate G_{mf}, we have

$$G_{mf} = \frac{i_o}{v_{in}} = \frac{i_o}{v'_{in} + v_f}$$

but $v_f = \beta_m i_o$, or

$$G_{mf} = \frac{i_o}{v'_{in} + \beta_m i_o}$$

Now dividing by v'_{in}, we have

$$G_{mf} = \frac{i_o/v'_{in}}{1 + \beta_m(i_o/v'_{in})}$$

With the use of Equation 11, we have

$$G_{mf} = \frac{G_m}{1 + \beta_m G_m} \qquad (15)$$

INPUT AND OUTPUT RESISTANCES

The circuit model for a transconductance amplifier is illustrated in Figure 9.5. With the aid of the circuit model, the input and output resistances can be calculated with feedback. We shall assume that the load resistor R_L is much smaller than the internal resistance R_o and that the feedback sampling network produces a small voltage drop.

The input resistance with feedback is defined as

$$R_{inf} = \frac{v_{in}}{i_{in}}$$

but $v_{in} = v_f + v'_{in}$, or

$$R_{inf} = \frac{v'_{in}}{i_{in}} + \frac{v_f}{i_{in}}$$

Because $v_f = \beta_m i_o$, we have

$$R_{inf} = \frac{v'_{in}}{i_{in}} + \frac{\beta i_o}{i_{in}}$$

With Equation 11, $i_o = G_m v'_{in}$, or

$$R_{inf} = \frac{v'_{in}}{i_{in}} + \frac{\beta_m G_m v'_{in}}{i_{in}}$$

$$R_{inf} = R_{in}(1 + \beta_m G_m) \tag{16}$$

The output resistance for the transconductance amplifier using feedback is defined as

$$R_{of} = \frac{v_o}{i_o}$$

Figure 9.5
Circuit model for a transconductance amplifier using negative feedback.

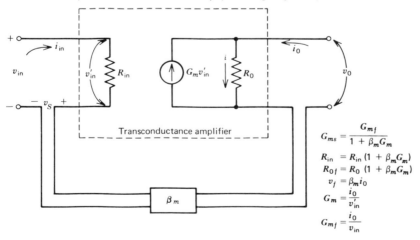

From the figure we can determine that

$$v_o = iR_o$$

but

$$i = G_m v'_{in} + i_o$$

or

$$v_o = (G_m v'_{in} + i_o) R_o$$

Setting the input voltage v_{in} equal to zero, the feedback voltage v_f is

$$v_f = v'_{in} = \beta_m i_o$$

Substituting this relationship into the equation for v_o, we have

$$v_o = (\beta_m G_m i_o + i_o) R_o$$

or

$$R_{of} = R_o(1 + \beta_m G_m) \qquad (17)$$

Notice that the input and output resistances for the transconductance amplifier are increased by a factor of $(1 + \beta_m G_m)$. Thus a transconductance amplifier that has a feedback voltage v_f that is a function of the output current i_o has higher input and output resistances than the transconductance amplifier without feedback.

CIRCUIT CONFIGURATION

Figure 9.6 illustrates a BJT amplifier with an unbypassed emitter-resistor R_E. As can be reasoned from the figure, the feedback voltage is across the emitter-resistor R_E, and if we assume $i_c \approx i_e$, the feedback voltage is a function of the output current i_c. Using Equation 13, we have

$$v_f = \beta_m i_o$$

Figure 9.6
Discrete BJT amplifier using negative feedback.

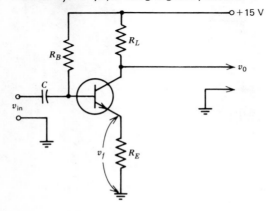

but if $i_c \approx i_e = i_o$, then $v_f = i_o R_E$, or

$$\beta_m = R_E \qquad (18)$$

The transconductance open loop gain is found by

$$G_m = \frac{i_o}{v'_{in}} = \frac{i_c}{i_b h_{ie}}$$

but $i_c \approx h_{fe} i_b$, or

$$G_m = \frac{h_{fe}}{h_{ie}} \qquad (19)$$

Substituting Equations 18 and 19 into Equation 16, we have

$$R_{inf} = R_{in}(1 + \beta_m G_m)$$

$$= h_{ie}\left(1 + R_E \frac{h_{fe}}{h_{ie}}\right)$$

or

$$R_{inf} = h_{ie} + h_{fe} R_E \qquad (20)$$

Equation 20 is similar to the input resistance developed in Chapter 4.

The output resistance with feedback, Equation 17, then becomes

$$R_{of} = \frac{1}{h_{oe}}\left(1 + R_E \frac{h_{fe}}{h_{ie}}\right) \qquad (21)$$

The voltage gain for the circuit shown in Figure 9.7 can be calculated. If, in Equation 15, we assume that $\beta_m \gg 1/G_m$, we have

$$G_{mf} \approx \frac{1}{\beta_m} \qquad (22)$$

but we know that

$$G_{mf} = \frac{i_o}{v_{in}}$$

or

$$i_o = v_{in} G_{mf}$$

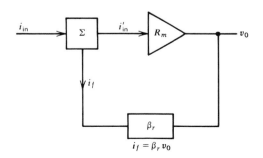

Figure 9.7
Block diagram of a transresistance amplifier using feedback.

Multiplying both sides of the above equation by R_L, we have

$$v_{in}G_{mf}R_L = i_oR_L$$

but $i_oR_L = v_o$ and $G_{mf} = 1/\beta_m = 1/R_E$, so that

$$v_{in}\frac{R_L}{R_E} = v_o$$

or

$$A_{vf} = \frac{R_L}{R_E} \tag{23}$$

Note the similarity between Equation 23 and the voltage gain equation of an amplifier with an unbypassed emitter resistor discussed in Chapter 4.

EXAMPLE

Assume that an amplifier has the circuit configuration shown in Figure 9.6, with the following parameters: $h_{ie} = 4$ kΩ, $h_{fe} = 100$, $R_E = 2.7$ kΩ, and $h_{oe} = 5 \times 10^{-5}$ mhos. Calculate the input and output resistances and the voltage gain of the amplifier. Using Equation 18, we have

$$\beta_m = R_E$$

$$\beta_m = 2.7 \text{ k}\Omega$$

Equation 19 defines G_m, or

$$G_m = \frac{h_{fe}}{h_{ie}} = \frac{100}{4000}$$

$$G_m = 0.025 \ \Omega^{-1}$$

Using Equation 20, we have

$$R_{inf} = h_{ie} + h_{fe}R_E$$

$$= 4 \text{ k}\Omega + 100 \times 2.7 \text{ k}\Omega$$

$$R_{inf} = 274 \text{ k}\Omega \text{ (neglecting } R_B)$$

Then R_{of} is calculated with Equation 21, or

$$R_{of} = \frac{1}{h_{oe}}\left(1 + R_E\frac{h_{fe}}{h_{ie}}\right)$$

$$= \frac{1}{5 \times 10^{-5}}\left(1 + 2.7 \times 10^3 \times \frac{100}{4 \times 10^3}\right)$$

$$R_{of} = 1.37 \text{ M}\Omega$$

With the transconductance amplifier using feedback, both the input and output resistances are greater than the transconductance amplifier without feedback.

TRANSRESISTANCE AMPLIFIER WITH FEEDBACK

With the voltage and transconductance amplifiers using feedback, the feedback voltage v_f was a function of the output voltage or current. Feedback can also be current. With current feedback, i_f is fed back to the input of the amplifier to combine with the input current of the amplifier i_{in}. If the feedback current is a function of the output voltage, the amplifier is called a transresistance amplifier. Figure 9.7 illustrates a block diagram of a transresistance amplifier using feedback. The summing point at the input of the amplifier results in

$$i'_{in} = i_{in} - i_f \tag{24}$$

assuming that the amplifier uses negative feedback. This type of amplifier stabilizes the transresistance gain by using negative feedback.

The transresistance closed loop gain is defined as

$$R_{mf} = \frac{v_o}{i_{in}} \tag{25}$$

The feedback current i_f is defined as

$$i_f = \beta_r v_o \tag{26}$$

Substituting Equations 24 and 26 into Equation 25 and recalling that the transresistance without feedback is $R_m = v_o/i'_{in}$, we have

$$R_{mf} = \frac{R_m}{1 + \beta_r R_m} \tag{27}$$

INPUT AND OUTPUT RESISTANCES

To aid in the calculation of the input and output resistances of the amplifier, we shall use Figure 9.8. From the figure it can be seen that the input resistance with feedback is

$$R_{inf} = \frac{v_{in}}{i_{in}} \tag{28}$$

Substituting Equations 24 and 26 into Equation 28 and noting that $v_o = R_m i'_{in}$, we have

$$R_{inf} = \frac{R_{in}}{(1 + \beta_r R_m)} \tag{29}$$

The output resistance with feedback is defined as

$$R_{of} = \frac{v_o}{i_o} \tag{30}$$

From Figure 9.8, it can be determined that i_o is equal to

$$i_o = \frac{v_o + R_m i'_{in}}{R_o} \tag{31}$$

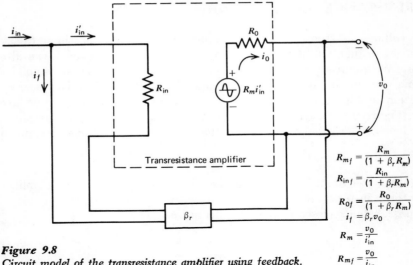

Figure 9.8
Circuit model of the transresistance amplifier using feedback.

Alongside the figure:

$$R_{mf} = \frac{R_m}{(1 + \beta_r R_m)}$$

$$R_{inf} = \frac{R_{in}}{(1 + \beta_r R_m)}$$

$$R_{of} = \frac{R_o}{(1 + \beta_r R_m)}$$

$$i_f = \beta_r v_o$$

$$R_m = \frac{v_o}{i'_{in}}$$

$$R_{mf} = \frac{v_o}{i_{in}}$$

Setting the input current equal to zero, we have $i'_{in} = i_f = \beta_r v_o$, or

$$i_o = \frac{v_o + \beta_r R_m v_o}{R_o}$$

or

$$R_{of} = \frac{R_o}{(1 + \beta_r R_m)} \tag{32}$$

With the transresistance amplifier using feedback, the input and output resistances are much lower than the transresistance amplifier without feedback. Feedback with this amplifier also stabilizes the transresistance gain.

CIRCUIT CONFIGURATION

Figure 9.9a illustrates a BJT circuit configuration for a transresistance amplifier using feedback. The feedback current i_f is a function of the output voltage v_o. The resistor R_F provides the feedback. The feedback is considered negative because the amplifier has a 180° phase reversal. The amplifier shown in Figure 9.9a is redrawn in Figure 9.9b. The open loop transresistance is calculated by assuming that $R_o \ll R_L$, or

$$R_m = \frac{v_o}{i'_{in}} = \frac{i_c R_L}{i'_{in}} = \frac{i_c R_L}{i_b}$$

but $i_b = i_c/h_{fe}$, or

$$R_m = \frac{i_b h_{fe} R_L}{i_b} = h_{fe} R_L \tag{33}$$

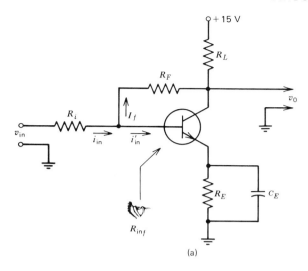

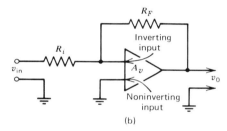

Figure 9.9
Transresistance amplifier using negative feedback: (a) discrete BJT amplifier and (b) operational amplifier.

With the input resistance low, the feedback current i_f is equal to

$$i_f \approx \frac{v_o}{R_f} = \beta_r v_o$$

or

$$\beta_r = \frac{1}{R_f} \tag{34}$$

The input and output resistances are then

$$R_{inf} = \frac{R_{in}}{(1 + \beta_r R_m)}$$

$$R_{inf} = \frac{h_{ie}}{[1 + (1/R_F)(h_{fe}R_L)]} \tag{35}$$

$$R_{of} = \frac{R_o}{\left(1 + \dfrac{h_{fe}R_L}{R_F}\right)} \tag{36}$$

If we assume that β_r is much greater than $1/R_m$, Equation 27 reduces to

$$R_{mf} \approx \frac{1}{\beta_r} \tag{37}$$

but $R_{mf} = v_o/i_{in}$, or

$$\frac{v_o}{i_{in}} = \frac{1}{\beta_r}$$

Solving for i_{in}, we have

$$i_{in} = \beta_r v_o$$

Because the input resistance is low, i_{in} is also equal to

$$i_{in} \approx \frac{v_{in}}{R_i}$$

Thus $\beta_r v_o = v_{in}/R_i$.

Solving for v_o/v_{in} (which is the closed loop voltage gain) and letting $\beta_r = 1/R_F$ (Equation 34), we have

$$Av_f = \frac{R_F}{R_i} \tag{38}$$

Equation 38 agrees with the gain equation discussed in Chapter 15.

EXAMPLE

Figure 9.9 illustrates a BJT transresistance amplifier with feedback. Resistor R_F is the feedback resistor causing i_f to be a function of v_o. Calculate R_{inf}, R_{of}, and the closed loop voltage gain if $R_L = 4.7$ kΩ, $R_E = 2.7$ kΩ, $R_i = 10$ kΩ $R_F = 68$ kΩ, $h_{fe} = 75$, $h_{ie} = 2.5$ kΩ, and $h_{oe} = 5 \times 10^{-5}$ $\mho$.

Using Equation 35, we have

$$R_{inf} = \frac{h_{ie}}{[1 + (1/R_F)(h_{fe}R_L)]}$$

$$= \frac{2500}{[1 + (1/68 \text{ k}\Omega)(75 \times 4.7 \text{ k}\Omega)]}$$

$$R_{inf} = 403 \ \Omega$$

Using Equation 36, we have

$$R_{of} = \frac{1/h_{oe}}{(1 + \beta_r R_m)}$$

$$= \frac{20{,}000}{6.2}$$

$$R_{of} = 3.2 \text{ k}\Omega$$

Using Equation 38, we have

$$A_{vf} = \frac{R_F}{R_i} = \frac{68 \text{ k}\Omega}{10 \text{ k}\Omega}$$

$$A_{vf} = 6.8$$

Note that becuase of the low input resistance R_{inf} the resistance seen by the input source v_{in} is approximately R_i.

CURRENT AMPLIFIER WITH FEEDBACK

With the transresistance amplifier, the feedback current i_f was a function of the output voltage v_o. If the feedback current is a function of the output current, the amplifier is considered a current amplifier. Figure 9.10 illustrates a block diagram for the current amplifier with feedback. With this amplifier the feedback current is equal to

$$i_f = \beta_i i_o \tag{38}$$

The closed loop current gain is defined as

$$A_{if} = \frac{i_o}{i_{in}} \tag{39}$$

and the open loop current gain is defined as

$$A_i = \frac{i_o}{i'_{in}} \tag{40}$$

Using Equations 38, 39, and 40, we find that the closed loop current gain is

$$A_{if} = \frac{A_i}{1 + \beta_i A_i} \tag{41}$$

INPUT AND OUTPUT RESISTANCES

Figure 9.11 illustrates the circuit model for the current amplifier incorporating feedback. The input and output resistances with feedback are found to be

$$R_{inf} = \frac{v_{in}}{i_{in}}$$

$$R_{inf} = \frac{R_{in}}{1 + \beta_i A_i} \tag{42}$$

Figure 9.10
Block diagram for a current amplifier using feedback.

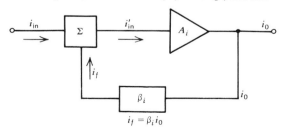

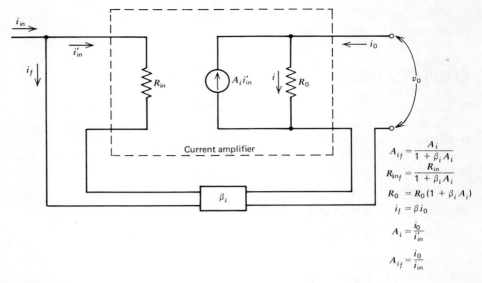

$$A_{if} = \frac{A_i}{1 + \beta_i A_i}$$

$$R_{inf} = \frac{R_{in}}{1 + \beta_i A_i}$$

$$R_0 = R_0 (1 + \beta_i A_i)$$

$$i_f = \beta i_0$$

$$A_i = \frac{i_0}{i'_{in}}$$

$$A_{if} = \frac{i_0}{i_{in}}$$

Figure 9.11
Circuit model for a current amplifier using negative feedback.

$$R_{of} = \frac{v_o}{i_o}$$

$$R_{of} = R_o(1 + \beta_i A_i) \qquad (43)$$

CIRCUIT CONFIGURATION

Figure 9.12 illustrates two BJT amplifiers connected. The output current i_o is sampled by the emitter-resistor of Q_2. The feedback current i_f is shunted with the input current i_{in} via the feedback resistor R_F. Because of the phase relationships in this amplifier, the feedback current i_f is out of phase with the input current i_{in}. Thus the amplifier has negative feedback. Because the amplifier has low input resistance with feedback, the feedback current i_f is approximately equal to

$$i_f \approx \frac{v_{RE}}{R_F} \qquad (44)$$

But the current through the emitter-resistor R_E is the difference between i_o and i_f, or

$$i_e = i_o - i_f \qquad (45)$$

Substituting Equation 45 into Equation 44 and solving for i_f/i_o, we have

$$\frac{i_f}{i_o} = \frac{R_E}{R_E + R_F} \qquad (46)$$

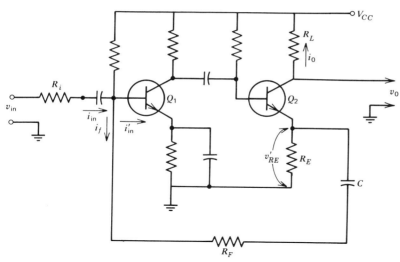

Figure 9.12
Discrete BJT current amplifier using feedback.

But Equation 46 defines β_i, or

$$\beta_i = \frac{R_E}{R_E + R_F} \qquad (47)$$

If we assume that R_L is much smaller than R_o for the BJTs, then the open loop current gain for both transistors is

$$A_i \approx h_{fe_1} h_{fe_2} \qquad (48)$$

The closed loop voltage gain for Figure 9.12 is defined as

$$A_{vf} = \frac{v_o}{v_{in}} \qquad (49)$$

but because R_{inf} is low in value, $v_{in} \approx i_{in} R_i$, or

$$A_{vf} = \frac{v_o}{i_{in} R_i} = \frac{i_o R_L}{i_{in} R_i} \qquad (50)$$

If in Equation 41, A_i is much greater than one, A_{if} becomes

$$A_{if} = \frac{1}{\beta_i} = \frac{i_o}{i_{in}} \qquad (51)$$

Substituting Equation 51 into Equation 50 and defining β_i with Equation 47, we have

$$A_{vf} = \frac{1}{\beta_i} \frac{R_L}{R_i}$$

or

$$A_{vf} = \frac{(R_E + R_F)}{R_E} \times \frac{R_L}{R_i} \qquad (52)$$

EXAMPLE

Calculate R_{inf}, R_{of}, and A_{inf} for the circuit configuration shown in Figure 9.12 if $R_L = 4.7$ kΩ, $R_E = 470$ Ω, $R_F = 10$ kΩ, $R_i = 10$ kΩ, $A_i = 500$, $h_{ie} = 1.5$ kΩ, and $R_o = 30$ kΩ.

Using Equation 47, we have

$$\beta_i = \frac{R_E}{R_E + R_F} = \frac{470}{470 + 10,000}$$

$$\beta_i = 0.045$$

The quantity $(1 + \beta_i A_i)$ then equals

$$(1 + \beta_i A_i) = (1 + 0.045 \times 500)$$

$$(1 + \beta_i A_i) = 23.5$$

The input and output resistances are found with Equations 42 and 43.

$$R_{inf} = \frac{R_{in}}{1 + \beta_i A_i} = \frac{1500}{23.5}$$

$$R_{inf} = 64 \ \Omega$$

$$R_{of} = R_o(1 + \beta_i A_i) = 30 \text{ k}\Omega(23.5)$$

$$R_{of} = 705 \text{ k}\Omega$$

Equation 52 is used to calculate the closed loop voltage gain.

$$A_{vf} = \frac{(R_E + R_F)}{R_E} \times \frac{R_L}{R_i} = 22.3 \times \frac{4700 \ \Omega}{10,000 \ \Omega}$$

$$A_{vf} \approx 10.5$$

Table 9.1

Comparison of the Four Basic Types of Negative Feedback

Circuit model	Feedback	Gain	R_{inf}	R_{of}
Voltage amplifier	$v_f = \beta_v v_0$	$A_{vf} = \dfrac{A_v}{(1 + \beta_v A_v)}$	$R_{in}(1 + \beta_v A_v)$	$\dfrac{R_0}{(1 + \beta_v A_v)}$
Transconductance amplifier	$v_f = \beta_m i_0$	$G_{mf}{}^1 = \dfrac{G_m}{(1 + \beta_m G_m)}$	$R_{in}(1 + \beta_m G_m)$	$R_0(1 + \beta_m G_m)$
Transresistance amplifier	$i_f = \beta_r v_0$	$R_{mf}{}^2 = \dfrac{R_m}{(1 + \beta_r R_m)}$	$\dfrac{R_{in}}{(1 + \beta_r R_m)}$	$\dfrac{R_0}{(1 + \beta_r R_m)}$
Current amplifier	$i_f = \beta_i i_0$	$A_{if} = \dfrac{A_i}{(1 + \beta_i A_i)}$	$\dfrac{R_{in}}{(1 + \beta_r R_m)}$	$R_0(1 + \beta_r R_m)$

1. $G_{mf} = i_0/v_{in}$ and $G_m = i_0/v'_{in}$
2. $R_{mf} = v_0/i_{in}$ and $R_m = i_0/i'_{in}$

Table 9.1 lists the various equations developed in this chapter. Note that if the amplifier uses feedback voltage v_f, the input resistance of the amplifier with feedback is increased. If the amplifier uses feedback current i_f, the input resistance of the amplifier with feedback is decreased. If the sampling network for the feedback is in parallel or shunted across the output of the amplifier, the output resistance of the amplifier with feedback is decreased. If the sampling network for the feedback is in series with the output of the amplifier, the output resistance of the amplifier with feedback is increased.

The example problems worked in this chapter dealt with BJT amplifiers. Feedback techniques are also used with FET and operational amplifiers and are discussed in Chapters 10 and 15.

FREQUENCY RESPONSE AND DISTORTION WITH NEGATIVE FEEDBACK

The frequency response of an amplifier, whether single or multiple stage, can be graphically displayed by simply plotting the gain of the amplifier versus the frequency of the input voltage. Such a plot is illustrated in Figure 9.13. The lower-frequency limit of an amplifier is labeled f_L and is defined at the point at which the open loop voltage gain decreases to $0.707 A_v$; A_v is considered the mid-frequency gain for the amplifier.

The upper-frequency limit of the amplifier is labeled f_H and is defined at the point at which the open loop voltage gain of the amplifier decreases to $0.707 A_v$. These two frequency limits are shown in Figure 9.13. It is sometimes convenient to express the gain of an amplifier in decibels. If the gain is expressed in decibels, the 0.707 or half-power points are labeled -3 dB points (see Chapter 3).

If the same amplifier discussed above uses negative feedback, its frequency response increases. This increase in frequency is also illustrated in Figure 9.13. The lower-frequency limit of the amplifier with feedback is labeled f_{Lf}, and the upper-frequency limit of the amplifier is labeled f_{Hf}. The lower-frequency limit with feedback is calculated by

$$f_{Lf} = \frac{f_L}{(1 + \beta_v A_v)} \tag{53}$$

and the upper-frequency limit is calculated by

$$f_{Hf} = f_H(1 + \beta_v A_v) \tag{54}$$

If the lower-frequency limit of an amplifier is subtracted from the upper-frequency limit of an amplifier, the result if called the bandwidth, or

$$BW_f = f_{Hf} - f_{Lf} \tag{55}$$

EXAMPLE

Assume that an amplifier has an open loop gain of 800 with an f_H of 0.5 MHz and f_L of 300 Hz. Calculate the new lower- and upper-frequency limits of the amplifier if 1.0 percent feedback is applied.

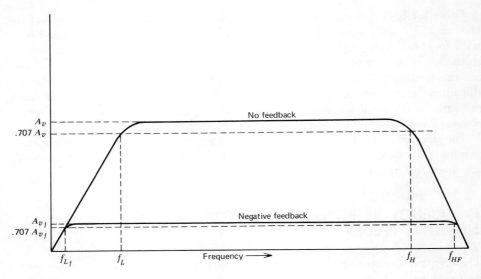

Figure 9.13
Frequency response for an amplifier: with and without feedback.

Calculating the quantity $(1 + \beta_v A_v)$, we have

$$(1 + \beta_v A_v) = (1 + 0.01 \times 800) = 9$$

Using Equation 53, we have

$$f_{Lf} = \frac{f_L}{(1 + \beta_v A_v)} = \frac{300 \text{ Hz}}{9}$$

$$f_{Lf} = 33 \text{ Hz}$$

Using Equation 54, we have

$$f_{Hf} = f_H(1 + \beta_v A_v) = 0.5 \text{ MHz}(9)$$

$$f_{Hf} = 4.5 \text{ MHz}$$

Distortion in an amplifier with negative feedback is reduced by the factor of $(1 + \beta_v A_v)$. The feedback loop feeds back not only a portion of the output voltage, but also a portion of the distortion D. The reduction in distortion for a voltage amplifier with negative feedback is given by

$$D_f = \frac{D}{(1 + \beta_v A_v)} \tag{56}$$

where D is the distortion without feedback. *It is important to realize that the distortion D is generated by the amplifier.*

REVIEW QUESTIONS AND PROBLEMS

9-1   What is meant by the term *feedback*?

9-2   Discuss negative feedback.

9-3   Discuss positive feedback.

9-4   List and discuss the four types of negative feedback.

9-5 List the effects of feedback on the input and output resistances of an amplifier.

9-6 What is the meaning of the terms *open loop voltage gain* and *closed loop voltage gain*?

9-7 If the feedback voltage is a function of the output voltage, why is it convenient to use a voltage amplifier circuit model?

9-8 If the feedback voltage is a function of the output current, why is it convenient to use a transconductance amplifier circuit model?

9-9 If the feedback current is a function of the output voltage, why is it convenient to use a transresistance amplifier circuit model?

9-10 If the feedback current is a function of the output current, why is it convenient to use a current amplifier circuit model?

9-11 If a voltage amplifier has an open loop voltage gain of 2500 and 10 percent feedback is applied, what is the closed loop voltage gain for this amplifier?

9-12 A voltage amplifier is to have a closed loop gain of 25 with no more than 2 percent change. Calculate the necessary percentage of feedback and the open loop gain required. Assume that the open loop gain will vary as much as 40 percent.

9-13 For the circuit shown, calculate A_{vf}, R_{inf}, and R_{of}.

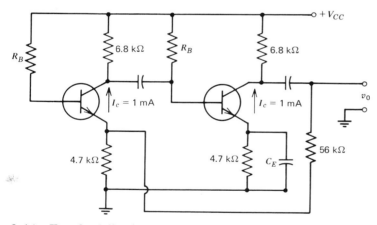

9-14 For the following circuit, calculate A_{vf}, R_{inf}, and R_{of}.

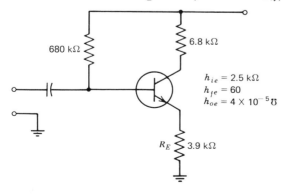

9-15 For the circuit shown, calculate A_{vf}, R_{inf}, and R_{of}.

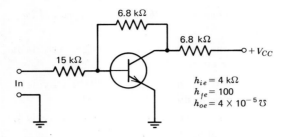

9-16 For the following, calculate A_{vf}, R_{inf}, and R_{of}.

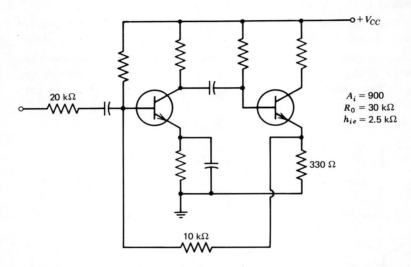

9-17 Derive Equation 55.

9-18 Assume that an amplifier has an open loop voltage gain of 60 dB with a upper frequency limit of 0.5 MHz. How much feedback is needed to extend the upper frequency limit of this amplifier to 5.0 MHz?

OBJECTIVES FOR CHAPTER 10, FIELD EFFECT TRANSISTORS

At the completion of this chapter you can demonstrate your understanding of the material by answering these questions:

- What is a JFET and what are its characteristics?
- What are the two types of MOSFETs?
- What are the common-circuit configurations for the JFET and the MOSFET?
- What is a JFET current pump?
- How and why can a JFET be operated as a variable resistor?
- Why is a FET considered a voltage-operated device?

10 THE FIELD EFFECT TRANSISTOR

The FET is becoming increasingly popular in linear and digital circuit applications. You should study the various field effect transistors and their characteristics because of the many circuit applications that take advantage of the special characteristics of the JFET and the MOSFET.

Another type of solid-state device that has become quite important in circuit applications is the field effect transistor (FET). The current flow through a FET is controlled by an electrostatic field (voltage). The current flow through a BJT is controlled by a current. The BJT is called a *bipolar* device because the current through the device is made up of both holes and electrons. The FET is called a *unipolar* device because the current through the device is made up of only electrons or holes.

There are two major categories for the FET: the junction field effect transistor (JFET) and the metal oxide semiconductor field effect transistor (MOSFET). As will be discussed in this chapter, the JFET operates in the depletion mode, whereas the MOSFET operates in the depletion and/or enhancement mode.

In comparison to the BJT, the FET has a much higher input resistance, no offset voltage (base-to-emitter voltage to overcome), and less effect on operating parameters due to certain radiation. It can be made to draw very low power in digital circuitry.

CONSTRUCTION OF JFETs

Figure 10.1*a* illustrates a cross-sectional view of an N-type JFET. The channel is constructed of N-type silicone semiconductor material with *ohmic contacts* on both ends. The ohmic contacts simply allow conductors to be connected to the channel. One end of the channel is labeled source (*S*) and the other, drain

(D). On both sides of the channel, P-type material is formed by diffusion techniques. The parallel connection of the two P-type materials forms the gate (G) of the device. The P-gate and the N-channel form PN junctions—thus the name JFET.

In the operation of the JFET, the PN junctions are seldom forward-biased, and thus have no offset voltage. In fact, the PN junctions are reverse-biased. Because the junctions are reverse-biased, the input resistance of the device is very high. The means of controlling current flow through the channel of the device is illustrated in Figure 10.1b.

The potential V_{DD} is placed across the channel of the device at the terminals D and S. Electron current enters at the source terminal and exits from the drain terminal of the device. If the P-type material were not present, the current through the channel would be simply a function of the voltage across the drain and source terminals V_{DS} and the resistance of the channel. However, with the P-type material forming a junction with the channel and being reverse-biased, a *depletion* region exists in the channel. If the gate-to-source voltage V_{GS} were to increase to such a magnitude that the two depletion regions would touch, the drain current I_D would be zero (see Figure 10.1c). The magnitude of V_{GS} that causes the drain current to be reduced to zero is called the *pinchoff voltage* V_P (or $V_{GS(\text{cutoff})}$).

The extent of the depletion regions in the channel is determined by V_{GS}. Thus the current flow through the channel is controlled by the magnitude of the gate-to-source voltage V_{GS}. As can be reasoned, the JFET functions in the depletion mode.

Figure 10.1d shows the popular method for the construction of a JFET. Different varieties of P- and N-type materials are diffused only on one side. This type of construction eliminates the difficulties encountered in maunfacturing the type of JFET illustrated in Figure 10.1a. The JFET illustrated in Figure 10.1d functions in the same manner as the one shown in Figure 10.1a.

Figure 10.1e shows the schematic symbol used for the N-type JFET. A P-type JFET is also possible. With a P-type JFET, the channel is constructed of P-type semiconductor material with the gate material constructed from N-type material. The operating voltages for the P-type JFET would be the reverse of those of the N-type JFET. The schematic symbol for the P-type JFET is illustrated in Figure 10.1f. As a point of interest, because of the construction of a JFET, the source and drain connections can be interchanged in circuit configuration.

CHARACTERISTIC CURVES FOR THE JFET

Figure 10.2a illustrates a circuit diagram for obtaining the output characteristic curve of an N-type JFET shown in Figure 10.2b. The output characteristic curves illustrate graphically the relationships between the output current I_D, the gate-to-source voltage V_{GS}, and drain-to-source voltage V_{DS}. The curves illustrated in Figure 10.2b are divided into two parts, separated by a dashed line. The region to the left of the dashed line is labeled the linear or ohmic region and to the right, the saturated or constant-current region.

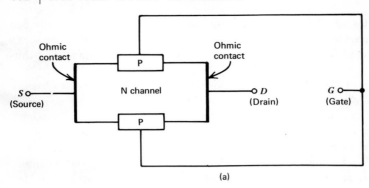

(a)

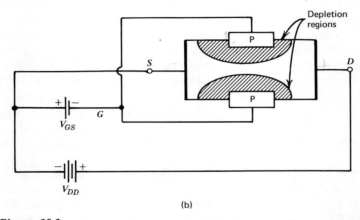

(b)

Figure 10.1
JFET: (a) construction, (b) depletion regions, (c) channel cutoff, (d) usual construction of JFET, (e) symbol for N-channel JFET, and (f) symbol for P-channel JFET.

In the linear region, the drain-to-source voltage V_{DS} is equal to the difference between the gate-to-source voltage V_{GS} and the pinchoff voltage V_P. With this magnitude of V_{DS}, the drain current increases linearly because the channel resistance is constant. Once V_{DS} increases beyond ($V_{GS} - V_P$), the channel resistance becomes very high, causing a constant current I_D to flow for particular values of V_{GS}.

If the V_{GS} voltage is equal to zero, the point at which I_D becomes a constant is called the saturation current and is labeled I_{DSS}. In the saturation region of the JFET, the increase in V_{DS} simply changes the geometry of the depletion regions without the depletion regions touching. The changing in the geometry of the depletion regions with increasing V_{DS} causes the channel resistance to change, keeping the drain current a constant. With any reverse-biased junction, if the voltage across the junction is large enough, the junction breaks down (see Chapter 2). The breakdown or zener effect for the JFET is labeled in Figure 10.2b.

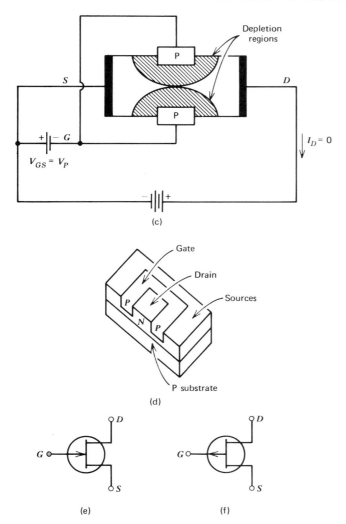

Figure 10-1 (continued)

TRANSFER CHARACTERISTIC CURVE

Figure 10.3 shows the plot of the drain current versus the gate-to-source voltage for the N-type transfer characteristic curve (JFET) in the saturated region. This curve is commonly called the transfer curve. The drain current I_D can be approximated by

$$I_D \approx I_{DSS}\left(1 - \frac{V_{GS}}{V_P}\right)^2 \tag{1}$$

assuming that the V_{GS} voltage is equal to or less than the pinchoff voltage V_P.

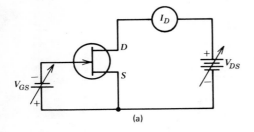

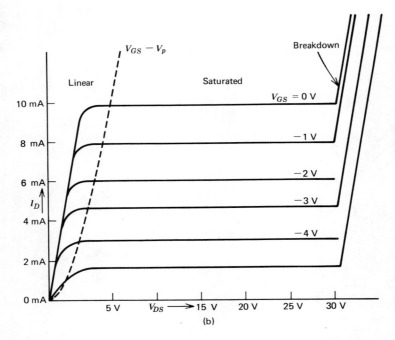

Figure 10.2
Output curves for the CS JFET: (a) circuit and (b) curves.

The transconductance g_m of the JFET can be obtained by taking the slope of the transfer curve, or

$$g_m = \frac{\Delta I_D}{\Delta V_{GS}} \bigg|_{V_{Df}=\text{const}} \tag{2}$$

It is important to realize that the transfer curve illustrated in Figure 10.3 is obtained by holding the drain-to-source voltage constant.

Figure 10.4 shows the coupling of the output characteristic and transfer curves for an N-type JFET. The figure demonstrated how a change in V_{GS} can produce a change in the drain current I_D for the JFET. This relationship will be used to demonstrate voltage amplifications for the JFET.

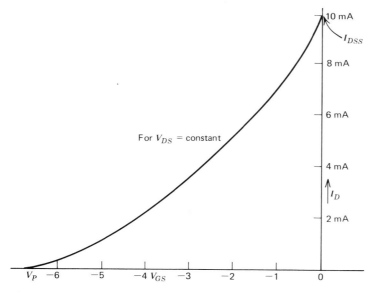

Figure 10.3
Transfer curve for the JFET.

JFET VOLTAGE AMPLIFIER (GRAPHIC)

Figure 10.5a illustrates a JFET connected as a common-source voltage amplifier. The load resistor R_L is connected between the drain and the supply voltage V_{DD}. The gate is returned to common through the gate resistor R_g. The source resistor R_S biases the JFET for linear voltage amplification. At a specified quiescent drain current, a voltage drop exists across R_S that is equal to

$$V_{R_S} = I_{DQ}R_S \qquad (3)$$

Figure 10.4
Relationship between the output and transfer curves.

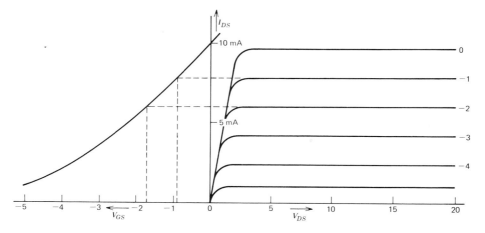

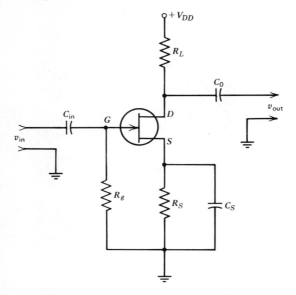

Figure 10.5
Circuit configuration for the common source JFET.

where I_{DQ} is the quiescent drain current and V_{RS} is the voltage drop across R_S. The voltage across R_S is used to bias the JFET for linear operation.

The capacitor C_S short-circuits the ac voltage across R_S to common. The value of the reactance of C_S is chosen to be one-tenth the ohmic value of R_S at the lowest frequency of the input signal. The capacitors C_{in} and C_{out} simply isolate the ac signal from the dc operating voltages of the JFET. The voltage gain of the device is defined as

$$A_v = \frac{v_o}{v_{\text{in}}} = \frac{-v_{ds}}{v_{gs}} \tag{4}$$

where the minus sign $(-)$ indicates a 180° phase difference between input and output voltages.

Figure 10.6 illustrates how voltage gain is accomplished with the JFET. The transfer and output characteristic curves are shown in the figure. The dc load line for the JFET shown in Figure 10.5*a* given by

$$V_{DD} = I_D(R_L + R_S) + V_{DS} \tag{5}$$

The ac load line for this circuit configuration is given by

$$i_d = -\frac{1}{R_L} v_{ds} \tag{6}$$

If an input voltage is applied between the gate and common, this signal voltage is additive with the bias voltage across R_S. In Figure 10.6, if the biasing voltage is 2 V with an input signal of 2 V peak to peak, the total voltage across gate to source would change from -3 to -1 V. The change in V_{GS} would cause

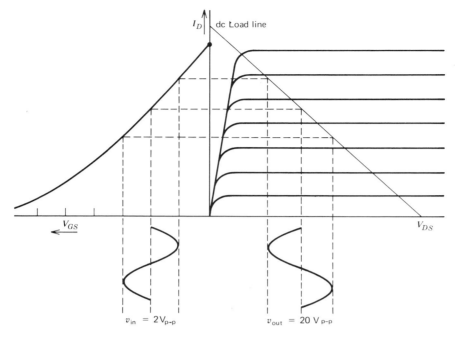

Figure 10.6
Illustration of voltage gain using the JFET.

a change in the drain current which is illustrated in Figure 10.6. Because of the change in I_D, the drain-to-source voltage would change as shown in Equation 6. Because the output voltage is equal to v_{ds} and also to the voltage across the load resistor R_L (i_dR_L), voltage amplification is accomplished. In the common-source JFET amplifier, the input voltage is 180° out of phase with respect to the output voltage.

The input resistance of the JFET at low frequencies is equal to the value of the gate resistor R_g. The high input resistance of the JFET is equal to the reverse-biased resistance of the gate-channel junction. Because this junction resistance is very high, the parallel combination of R_g and the reverse-biased junction resistance is approximately equal to R_g.

JFET CURRENT PUMP

Because of the high value of r_{ds}, the JFET can be made to function as a constant current generator or current pump. Figure 10.7a illustrates the schematic symbol for a current pump; Figure 10.7b shows the circuit configuration of a JFET current pump. The voltage across the drain source is assumed to be greater than the difference between ($V_{GS} - V_P$) in order to keep the device in its constant-current or saturated region.

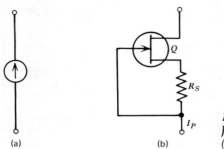

Figure 10.7
JFET current pump: (a) symbol and
(b) circuit.

(a)   (b)

The magnitude of the pump current is given by

$$I_P = I_D = \frac{|V_{GS}|}{R_S} \tag{7}$$

Equation 7 can be solved for the value of R_S necessary to pump a specific current I_P. The transfer curve gives the relationship between V_{GS} and I_D (which is I_P), from which the value of R_S can be determined. Another means of obtaining the V_{GS} necessary for a specific pump current is solving Equation 1 for V_{GS}, or

$$V_{GS} = V_P \left(1 - \sqrt{\frac{I_P}{I_{DSS}}} \right) \tag{8}$$

EXAMPLE

Calculate the value of R_S in order to pump a constant current of 500 μA, using a JFET with a $I_{DSS} = 6$ mA and $V_P = -4$ V. Using Equation 8, we can calculate the necessary value of V_{GS}.

$$V_{GS} = V_P \left(1 - \sqrt{\frac{I_P}{I_{DSS}}} \right)$$

$$= -4 \left(1 - \sqrt{\frac{0.5 \text{ mA}}{6 \text{ mA}}} \right)$$

$$V_{GS} = -2.85 \text{ V}$$

Substituting this value of V_{GS} into Equation 7, we have

$$R_S = \frac{|V_{GS}|}{I_P}$$

$$= \frac{2.85 \text{ V}}{500 \times 10^{-6} \text{ A}}$$

$$R_S = 5700 \ \Omega$$

MOSFET (ENHANCEMENT-DEPLETION)

Figure 10.8a illustrates the construction of a metal oxide semiconductor field effect transistor (MOSFET) along with its schematic symbol. Note in the figure that there are no junctions as in the JFET. The small channel of N-type

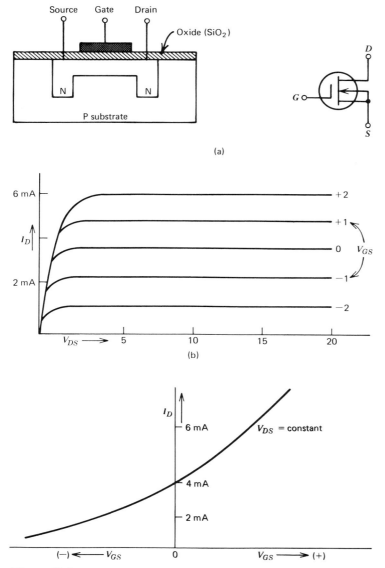

Figure 10.8
Enhancement-depletion MOSFET: (a) construction, (b) output curves, and (c) transfer curves.

material for current conduction exists between two wells. Contacts are brought out from the two wells and are labeled the source and drain. The gate is simply a metal conductor isolated from the channel by an oxide coating of silicone (SiO_2). The MOSFET has *very high* input resistance because the gate is isolated from the channel by the oxide coating. The gate together with the channel form a capacitor.

If a positive potential is applied at the gate terminal with respect to the source, because of capacitor action a negative charge is induced into the N-type channel region. A negative charge induced into the channel decreases the channel's resistance. Because of more negative current carriers in the channel, I_D increases (see Figure 10.8b).

If a negative potential is applied at the gate terminal with respect to the source, again because of capacitor action, a positive charge is induced into the channel, increasing the channel's resistance. Because of fewer current carriers in the channel, I_D decreases. Thus the conductivity of the channel is changed according to the polarity and magnitude of the potentail at the gate terminal of the device. A positive V_{GS} *enhances* the channel with additional current carriers, and a negative V_{GS} *depletes* the channel, causing fewer current carriers.

Because the input resistance is very high, static charge can collect on the gate and puncture the oxide coating. So that the device will not be damaged when it is not in a circuit, the leads are frequently connected together with foil to prevent any charge from building up at the gate terminal. Manufacturers recommend that when working with MOSFETs, one should not remove the foil until the device is placed in the circuit and that tools (soldering irons, pliers, etc.) should be grounded.

Some manufacturers produce the MOSFET with an internal zener diode across the gate-source terminals of the device. The zener diode conducts if the gate-to-source voltage exceeds the zener voltage rating, thus preventing damage to the MOSFET.

MOSFET (ENHANCEMENT)

Figure 10.9a illustrates the construction of an enhancement MOSFET. The enhancement MOSFET is constructed with two N-type wells and no N-channel. The gate terminal is metallic. The gate and P-substrate form a capacitor. A positive potential at the gate terminal with respect to the source causes a negative charge to be induced into the area underneath the gate. If the gate voltage is large enough, a negative channel is induced between the two N-type wells. With a channel of negative charge induced between the two wells, current can flow between the source and drain. Thus the channel is enhanced by a positive potential at the gate terminal of the device (see Figure 10.9b). Notice that the schematic symbol for the enhancement MOSFET is different from that for the enhancement-depletion MOSFET.

An important point to realize is that the MOSFET takes much less physical space to construct than does the BJT. Physical space is a very important consideration in the number of circuits that can be "fitted" on an IC chip (see Chapter 11).

DUAL-GATE MOSFET

Figure 10.10 illustrates the construction of a dual-gate enhancement-depletion MOSFET together with its schematic symbol. The current through the channel is controlled by the potential of both the gates. In practice, gate one is operated at a negative potential and gate two is operated as a positive potential with

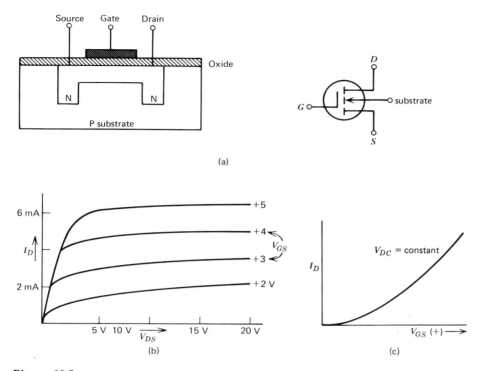

Figure 10.9
Enhancement MOSFET: (a) construction, (b) output curves, and (c) transfer curve.

respect to the source. With this type of operation, the gate-to-drain capacitance is reduced, permitting the device to be operated at a higher frequency in the common-cource circuit configuration.

Another frequent use for the dual-gate MOSFET is in a mixer circuit. A mixer used in RF circuitry has two input frequencies and produces sum and difference frequencies. If the two input frequencies were labeled f_1 and f_2, the output of the mixer stage would include $(f_1 + f_2)$ and $(f_1 - f_2)$. The input signal f_1 is at gate 1, and the input signal f_2 is at gate 2.

Figure 10.10
Construction and symbol for the dual gate MOSFET.

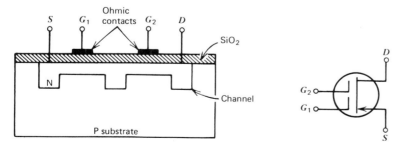

BIASING THE FET

For linear operation, the JFET is operated with a negative gate voltage with respect to the source. Figure 10.11a illustrates a circuit configuration that is sometimes called a "self-biased" JFET. The voltage drop across the source resistor R_S develops the necessary bias voltage for linear operation. In practice, the source resistor is bypassed with a capacitor. The gate terminal has to have a direct current returned to common for the circuit to function correctly. Because the gate current is approximately zero, there is no voltage drop across R_g. With no voltage drop across R_g, the gate-to-source voltage is equal to the voltage drop across R_S.

Figure 10.11b illustrates a simple circuit configuration of an enhancement-depletion MOSFET. Because the enhancement-depletion MOSFET works with both a positive and negative gate-to-source voltage, for small input signals no special biasing network is needed.

Figure 10.11c illustrates one type of biasing arrangement for the enhancement MOSFET. The enhancement MOSFET does not conduct until a positive gate-to-source voltage is present. Thus, for linear operation fo the device, a

Figure 10.11
Biasing: (a) JFET, (b) enhancement-depletion MOSFET, (c) enhancement MOSFET, (d) dual gate MOSFET, and (e) JFET.

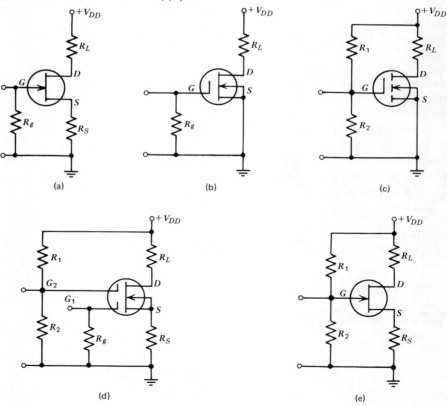

positive bias voltage is needed for the N-type MOSFET. The voltage drop across R_2 is the bias voltage and is equal to $V_{DD}(R_2/(R_1 + R_2))$.

Figure 10.11d illustrates a simple biasing method for the dual-gate depletion MOSFET. The negative voltage at gate 1 with respect to the source is the potential developed across the source resistor R_S. The positive voltage at gate 2 is the voltage drop across R_2 minus the voltage drop across the source resistor R_S.

Figure 10.11e shows a biasing method that incorporates the voltage drop across the source resistor R_S and the voltage drop across resistor R_2. This biasing method, in appearance, is similar to the universal biasing method discussed with BJTs. The biasing network should stabilize the operating or quiescent (Q) point of the circuit for changes that occur in the parameters of the device. The manufacturer lists the parameters for the FET (V_P, I_{DSS}, g_m) between some maximum and minimum values. The biasing method illustrated in Figure 10.11e can be used to minimize a change in the Q point of the circuit with changes in device parameters.

In Figure 10.11e we can see that the voltage drop across R_2 is equal to

$$V_{R_2} = I_D R_S + V_{GS} \tag{9}$$

The drain current for the JFET is given by

$$I_D = I_{DSS}\left(1 - \frac{V_{GS}}{V_P}\right)^2 \tag{10}$$

For the maximum values of V_P and I_{DSS}, if we solve for V_{GS}, Equations 9 and 10 become

$$V_{GS(max)} = V_{(max)}\left(1 - \sqrt{\frac{I_{D(max)}}{I_{DSS(max)}}}\right) \tag{11}$$

$$V_{R_2} = I_{D(max)} R_S + V_{GS(max)} \tag{12}$$

For the minimum values of V_P and I_{DSS}, Equations 9 and 10 become

$$V_{GS(min)} = V_{(min)}\left(1 - \sqrt{\frac{I_{D(min)}}{I_{DSS(min)}}}\right) \tag{13}$$

$$V_{R_2} = I_{D(min)} R_S + V_{GS(min)} \tag{14}$$

Now, because the voltage drop across R_2 will be a constant (assuming no gate current), we can subtract Equation 12 from Equation 14,

$$0 = R_S(I_{D(max)} - I_{D(min)}) + (V_{GS(max)} - V_{GS(min)})$$

or

$$R_S = \frac{V_{GS(min)} - V_{GS(max)}}{I_{D(max)} - I_{(min)}} = \frac{\Delta V_{GS}}{\Delta I_D} \tag{15}$$

where $V_{GS(max)}$ and $V_{GS(min)}$ are calculated with Equations 11 and 13. Equation 15 gives us the value of the source resistor necessary to stabilize the Q point of the circuit if we assume a constant voltage source of V_{R_2}.

EXAMPLE

A biasing method such as the one shown in Figure 10.11e is to be used for a JFET having the following parameters:

$$I_{DSS(\text{max})} = 10 \text{ mA} \qquad V_{P(\text{max})} = -6 \text{ V}$$

$$I_{DSS(\text{min})} = 4 \text{ mA} \qquad V_{P(\text{min})} = -2 \text{ V}$$

Assume that the drain current (I_D) is to vary no more than 1.0 mA between the values of 2.0 mA and 3.0 mA. Calculate the values of R_S and $V_{R(2)}$. (Note: The selection of R_L and the supply voltage will be discussed shortly.) Equation 11 is used to find $V_{GS(\text{max})}$.

$$V_{GS(\text{max})} = V_{P(\text{max})} \left(1 - \sqrt{\frac{I_{D(\text{max})}}{I_{DSS(\text{max})}}} \right)$$

$$= -6\text{v} \left(1 - \sqrt{\frac{3 \text{ mA}}{10 \text{ mA}}} \right)$$

$$V_{GS(\text{max})} = -2.7 \text{ V}$$

and $V_{GS(\text{min})}$ is found with Equation 13.

$$V_{GS(\text{min})} = V_{P(\text{min})} \left(1 - \sqrt{\frac{I_{D(\text{min})}}{I_{DSS(\text{min})}}} \right)$$

$$= -2\text{v} \left(1 - \sqrt{\frac{2 \text{ mA}}{4 \text{ mA}}} \right)$$

$$V_{GS(\text{min})} = -0.59 \text{ V}$$

Equation 15 is used to find R_S

$$R_S = \frac{\Delta V_{GS}}{\Delta I_D}$$

$$= \frac{-0.56 \text{ V} + 2.7 \text{ V}}{3 \text{ mA} - 2 \text{ mA}}$$

$$R_S = 2.11 \text{ k}\Omega$$

Substituting the values of R_S into Equation 12 (or 14), we have

$$V_{R(2)} = I_{D(\text{max})} R_S + V_{GS(\text{max})}$$

$$= 3 \text{ mA} \times 2.11 \text{ k}\Omega - 2.7 \text{ V}$$

$$V_{R(2)} = 3.63 \text{ V}$$

CIRCUIT CONFIGURATIONS

With FETs, the three possible circuit configurations are common-source (CS), common-gate (CG), and common-drain (CD). The two most common circuit configurations are the common-source and the common-drain. For the circuits discussed below, it is assumed that the FET is operated in its saturation region and that the necessary biasing voltage permits linear operation of the device.

LOW-FREQUENCY COMMON-SOURCE CIRCUIT

Figure 10.12*a* shows a JFET used in a common-source circuit configuration. The bias voltage for linear operation is developed across R_S and R_2. Capacitor C_S bypasses the ac signal across R_S to common. The reactance of C_S is chosen so that it is one-tenth the ohmic value of R_S. Figure 10.12*b* illustrates the ac circuit model for the CS JFET amplifier. The current generator is labeled $g_m v_{gs}$, and the internal output resistance for the JFET is labeled r_{ds} (see Chapter 8).

As can be reasoned from Figure 10.12, the input resistance of the amplifier is simply $R_1 \,||\, R_2$, or

$$R_{in} = R_1 \,||\, R_2 \tag{16}$$

The output resistance R_o is the parallel combination of the load resistor R_L and r_{ds}, or

$$R_o = R_L \,||\, r_{ds} \tag{17}$$

The voltage gain of the amplifier is defined as

$$A_v = \frac{v_o}{v_{in}}$$

where $v_o = i_o R_L$ and $v_{in} = v_{gs}$.

The output current i_o is found by using current division, or

$$i_o = \frac{-g_m v_{gs} v_{ds}}{r_{ds} + R_L}$$

but $v_o = i_o R_L$ or

$$v_o = \frac{-g_m v_{gs} r_{ds} R_L}{r_{ds} + R_L}$$

Figure 10.12
Common-source amplifier: (a) circuit configuration and (b) circuit model.

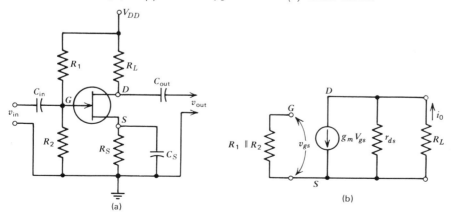

Because $v_{gs} = v_{in}$ and $A_v = v_o/v_{in}$, we have

$$A_v = \frac{-g_m r_{ds} R_L}{r_{ds} + R_L}$$

Now, dividing by r_{ds}, we have

$$A_v = \frac{-g_m R_L}{1 + R_L/r_{ds}} \tag{18}$$

If the drain resistance r_{ds} is much greater than the load resistor R_L, Equation 18 reduces to

$$A_v \approx -g_m R_L \tag{19}$$

where the minus sign implies that the input and output signal voltages are 180° out of phase with each other.

EXAMPLE

Design a common-source amplifier as illustrated in Figure 10.12 having an input resistance of 100 kΩ or more and a voltage gain of −20. Use the parameters discussed in the previous example problem. Assume that the JFET has a g_m of 4250 μmhos. Calculate the necessary values of R_L, R_1, R_2, and V_{DD}.

The results of the previous example problem were an R_S value of 2.11 kΩ, a $V_{R(2)}$ of 3.63 V, and an average drain current of 2.5 mA. The necessary value of load resistance for R_L can be calculated with Equation 19.

$$A_v = -g_m R_L$$

or

$$R_L = \frac{-20}{-4.25 \times 10^{-3}} = 4.7 \text{ k}\Omega$$

Because the source resistor R_S is bypassed, the JFET amplifier has an ac load line different from its dc load line (Equation 6). With the same reasoning that was discussed in Chapter 6 for maximum output voltage swing for a BJT, the drain quiescent current is given by

$$I_{DQ} = \frac{V_{DD}}{2R_L + R_S}$$

Substituting the values of I_{DQ}, R_L, and R_S into the above equation and solving for the supply voltage, we have

$$V_{DD} = I_{DQ}(2R_L + R_S)$$

$$= 2.5 \text{ mA}(2 \times 4.7 \text{ k}\Omega + 2.11 \text{ k}\Omega)$$

$$V_{DD} \approx 30 \text{ V}$$

The quiescent drain-to-source voltage V_{DSQ} is calculated with Equation 5.

$$V_{DSQ} = V_{DD} - I_{DQ}(R_L + R_S)$$

$$= 30 \text{ V} - 2.5 \text{ mA}(4.7 \text{ k}\Omega + 2.11 \text{ k}\Omega)$$

$$V_{DSQ} = 13 \text{ V}$$

Thus, for the ideal JFET used in this example, the maximum output voltage swing is twice V_{DSQ}, or 26 V_{p-p}.

The necessary value of C_S is obtained by assuming that the reactance of C_S at the lowest input frequency is one-tenth R_S, or $X_{C_S} = 210 \ \Omega$. The values for R_1 and R_2 are calculated from knowing the values for R_{in} and V_{R2}, or

$$R_{in} = 100 \text{ k}\Omega = \frac{R_1 R_2}{R_1 + R_2} \tag{A}$$

$$V_{R_2} = 3.63 \text{ V} = 30 \text{ V} \left(\frac{R_2}{R_1 + R_2} \right) \tag{B}$$

Solving Equations A and B for R_1 and R_2, we have

$$R_1 = 826 \text{ k}\Omega$$

$$R_2 = 113.7 \text{ k}\Omega$$

LOW-FREQUENCY COMMON-DRAIN CIRCUIT

Figure 10.13a illustrates a circuit configuration for a JFET in the common-drain arrangement. The output voltage is taken across the source resistor R_S. Because of the isolation between the input and output terminals of the JFET, the input resistance is simply R_g, or

$$R_{in} = R_g \tag{20}$$

Figure 10.13b illustrates the ac circuit model for the CD JFET circuit configuration. The output resistance and voltage gain can be calculated using the feedbackequations discussed in Chapter 9. For the CD configuration, note that 100 percent of the output voltage is fed back to the input, or β_v equals one. The output resistance without feedback is simply R_S, assuming that R_S is much less than r_{ds}.

The voltage gain without feedback is approximated by $g_m R_L$. Using these relationships in Equation 7, from Chapter 9, for the output resistance of a voltage amplifier with feedback, we have

$$R_{of} = \frac{R_o}{(1 + \beta_v A_v)}$$

but $R_o = R_S$, $\beta_v = 1$, and $A_v \approx g_m R_S$, or

$$R_{of} = \frac{R_S}{1 + g_m R_S} \tag{21}$$

The voltage gain for the CD JFET can be found with Equation 4, from Chapter 9, for the voltage gain of an amplifier with feedback, or

$$A_{vf} = \frac{A_v}{1 + \beta_v A_v}$$

or

$$A_{vf} = \frac{g_m R_S}{1 + g_m R_S} \tag{22}$$

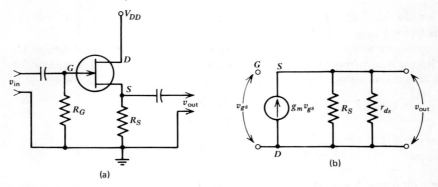

Figure 10.13
Common-source amplifier (a) circuit configuration and (b) circuit model.

EXAMPLE

Figure 10.14 illustrates a common-drain circuit configuration that is to be used as a matching device. The input resistance R_{in} is to be 10 kΩ, and the output resistance R_o is to be 300 Ω. Assume that the JFET has a $g_m = 3000$ μmhos, $V_P = -5$ V, and $I_{DSS} = 9$ mA. Calculate the values necessary for R_S, R_1, and R_2.

To calculate the value of R_S, we solve Equation 21 for R_S.

$$R_S = \frac{R_o}{1 - g_m R_o}$$

$$= \frac{300}{1 - 3 \times 10^{-3} \times 300}$$

$$R_S = 3 \text{ k}\Omega$$

For maximum output voltage swing, $V_{DSQ} = \frac{1}{2}V_{DD}$, or

$$V_{RS} = 10 \text{ V} = I_{DQ}R_S$$

Figure 10.14
Circuit configuration used to calculate input and output resistances.

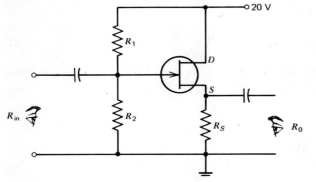

then

$$I_{DQ} = \frac{10 \text{ V}}{3 \text{ k}\Omega} = 3.33 \text{ mA}$$

The bias voltage necessary for the JFET is calculated by solving Equation 1 for V_{GS}.

$$V_{GS} = V_P \left(1 - \sqrt{\frac{I_D}{I_{DSS}}}\right)$$

$$= -5 \text{ V} \left(1 - \sqrt{\frac{3.33 \text{ mA}}{9 \text{ mA}}}\right)$$

$$V_{GS} = 1.96 \text{ V}$$

The necessary voltage drop across R_2 is found with

$$V_{R2} = I_{DQ} R_S + V_{GS}$$

$$= 3.33 \text{ mA} \times 3 \text{ k}\Omega + (-1.96 \text{ V})$$

$$V_{R2} = 8.04 \text{ V}$$

Because R_{in} is equal to $R_1 \| R_2$ and V_{R2} is known, the necessary values for R_1 and R_2 are found to be

$$R_1 \approx 50 \text{ k}\Omega$$

$$R_2 \approx 33 \text{ k}\Omega$$

Figure 10.15
Output curves for a JFET expanded around small values of V$_{DS}$.

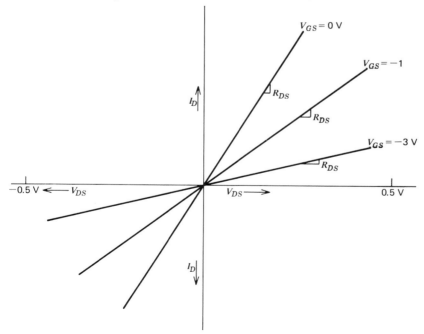

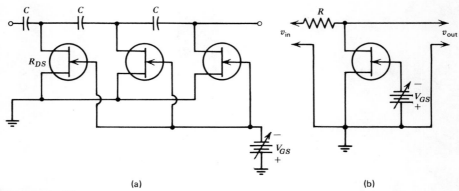

(a)   (b)

Figure 10.16
JFET used as a variable resistance: (a) RC phase shift network and (b) simple voltage divider.

OHMIC OPERATION OF THE JFET

As was mentioned early in the chapter, if the drain-to-source voltage V_{DS} is less than $(V_{GS} - V_P)$, the JFET is in its linear or ohmic region. If the FFET is operated in its ohmic region, the resistance between drain and source (R_{DS}) is a function of the gate-to-source voltage V_{GS}. This relationship is shown in Figure 10.15. In the figure the region for small values of V_{DS} is enlarged. Note that the resistance R_{DS} increases for small values of V_{GS}.

Also, because of the symmetry of the drain and source, the resistance is linear when the V_{DS} voltage is negative for the N-channel JFET. The resistance R_{DS} is sometimes referred to as a voltage variable resistance (VVR). It is important to realize that the VVR is only linear with small values of V_{DS}.

One of the many applications of the VVR of the JFET is in a phase shifting network. Figure 10.16a illustrates a phase shifting network consisting of three RC combinations used for an oscillator (see Chapter 19). The value of the resistance R can be varied by simply changing the gate supply voltage.

Figure 10.16b illustrates a simple volume control. The voltage divider consists of R and R_{DS} of the JFET. The magnitude of the output voltage is determined by the value of the gate supply voltage.

REVIEW QUESTIONS AND PROBLEMS

10-1   What are the differences between the JFET and MOSFET?

10-2   What does the term *pinchoff voltage* mean?

10-3   What is the meaning of the term *drain-source saturation current*?

10-4   What are the biasing requirements for the N-channel JFET for linear voltage operation?

10-5   What is the meaning of "A FET is a voltage-operated device."

10-6   What are the biasing requirements for the enhancement-type, N-channel MOSFET for linear voltage operation?

10-7 What are the biasing requirements for the enhancement-depletion-type N-channel MOSFET for linear voltage operation?

10-8 A JFET has the following parameters: $g_m = 4200$ μmhos, $V_P = -5$ V and $I_{DSS} = 8$ mA. For a common-source amplifier, what are the values for R_L and R_S so that the amplifier has a voltage gain of -12 and a biasing voltage of -2 V?

10-9 Discuss why there is a 180° phase difference between the input and output voltages for a common-source amplifier.

10-10 Discuss why there is a zero-degree phase difference between the input and output voltages for a common-drain amplifier.

10-11 Compare the parameters for a BJT and a JFET.

10-12 Design a JFET current pump having a pump current of 0.2 mA. Assume that the JFET has the same parameters as the one in Problem 10-8.

10-13 For the following common-source amplifier, calculate the voltage gain using the feedback techniques discussed in Chapter 9.

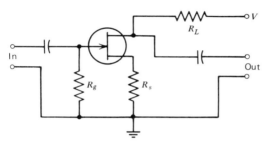

10-14 For the common-drain amplifier shown, calculate the input resistance, output resistance, and voltage gain, using the feedback techniques discussed in Chapter 9.

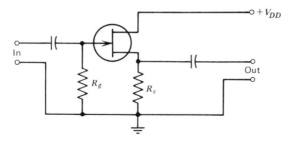

10-15 Explain how the enhancement type of MOSFET can be used as an electronic switch.

10-16 List some applications for the JFET used as a voltage-variable resistance (VVR).

OBJECTIVES FOR CHAPTER 11, IC APPLICATIONS

At the completion of this chapter you can demonstrate your understanding of the material by answering these questions:

- What does the term *monolithic* mean?
- What is a substrate?
- How are resistors and capacitors formed with ICs?
- What is the diffused planar process?
- What are some chips that are available and what are their applications?

11 INTEGRATED CIRCUIT APPLICATIONS

Integrated circuits have revolutionized electronic circuitry. You should study IC applications because of the increased use of chips in circuit applications.

Integrated circuits (ICs) have revolutionized electronics. With the variety of ICs available, the technician can become an applications person. With the studying of the parameters for a chip, the technician can be concerned with the different applications of that chip. In this chapter some of the common or "off the shelf" ICs that are readily available are introduced. In the Appendix the reader will find the manufacturer's specification sheets for the ICs discussed in this chapter.

Because of the complexity in the manufacturing of integrated circuits, a very cursory treatment is given here. There are many excellent texts devoted to the

Figure 11.1
Example of chip manufacturing (Texas Instruments Inc.): (a) circuit configuration and (b) integrated circuit process steps.

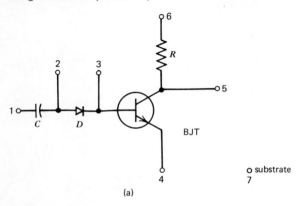

(a)

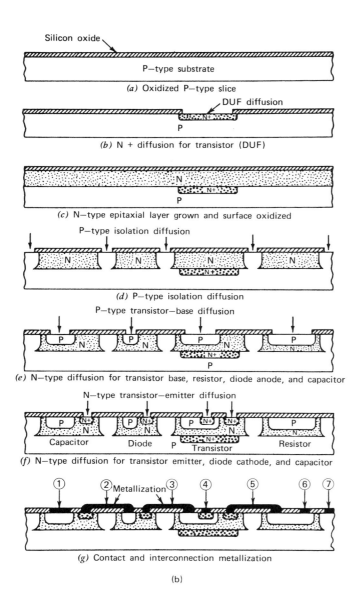

Silicon oxide

P-type substrate

(a) Oxidized P-type slice

DUF diffusion

N+

P

(b) N + diffusion for transistor (DUF)

N

N+

P

(c) N-type epitaxial layer grown and surface oxidized

P-type isolation diffusion

N   N   N   N

N+

(d) P-type isolation diffusion

P-type transistor-base diffusion

P   N   P   N   P   N   P   N

N+

P

(e) N-type diffusion for transistor base, resistor, diode anode, and capacitor

N-type transistor-emitter diffusion

P N+   P N+   P N+ N+   P

N   N   N   N

Capacitor   Diode   P   Transistor   Resistor

N+

(f) N-type diffusion for transistor emitter, diode cathode, and capacitor

①   ②Metallization③   ④   ⑤   ⑥ ⑦

(g) Contact and interconnection metallization

(b)

Figure 11.1 (continued)

entire processing of ICs. It is recommended that the reader review the literature on the manufacturing of ICs.

Integrated circuits associated with digital and optical electronics are discussed in Chapters 16 and 18, respectively.

FABRICATION

Resistors, capacitors (passive elements), BJTs, FETs, and diodes (active elements) can be connected to form an *integrated circuit* (IC). If the passive

Figure 11.2
Specifications for a power supply (Datel Systems Inc.).

and active elements are processed at the same time to form a complete circuit, the IC is called *monolithic*. The material on which the IC is fabricated is called the *substrate*.

Resistors can be constructed for ICs by selecting the length, width, and depth of a P- (or N-) type strip of semiconductor material. Because $R = \rho L/A$, the resistivity (ρ) is determined by the amount of dopen in the P- (or N-) type material.

Capacitors can be constructed for the IC by two methods. The first method is the formation of a reverse-biased PN junction. The width of the depletion area and the physical size of the P- and N-type material determine the capaci-

Figure 11.2 (continued)

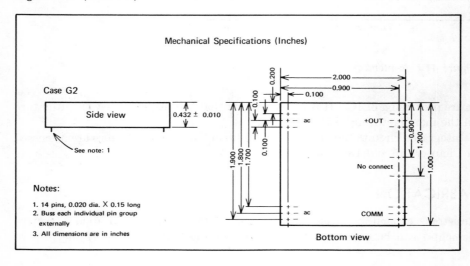

Single Output

Model Number			UPM-5/350*
Outputs	Output voltage		5V dc
	Output voltage accuracy		±0.5%
	Output current		350 mA
	Current limiting		Yes
	Output capacitor (internal)		20 μF
Inputs	Input voltage		115 Vac
	Input voltage range		±10 Vac
	Frequency range		50 to 420 Hz
	Maximum input current		NA
	Isolation →	Resistive coupling	100 MΩ
		Capacitive coupling	100 pF
		Breakdown voltage	300 Vdc
Regulation	Line reguation		0.05%
	Load regulation		0.05%
	Temperature coefficient		0.005%/°C
	Noise & ripple (rms)		2 mV
	Output impedance @ 10 kHz		150 mΩ
	Transient response		50 μsec max
Thermal	Operating temperature		0 to 70 °C
	Storage temperature		−55 to 85 °C
	Heat sinking requirements		NA
Physical	Case material		Diallyl phthalate/epoxy
	Case/pin configuration		G2
	Weight		1.5 oz

Figure 11.2 (*continued*)

tance (which ranges in Picofarads). The second method is the construction of two conductors separated by a dielectric. The various circuit elements in the IC are isolated from each other by either an oxide coating or a reverse-baised junction.

One process of forming the various elements needed for a circuit in an IC is called the diffused planar process. Figure 11.1 illustrates a simple circuit consisting of a capacitor, diode, BJT, and a resistor. Figure 11.1*b* shows the process of making the circuit shown in Figure 11.1*a*.

POWER SUPPLIES

The power supply is the heart of all electronic systems (see Chapter 12). The power supply is capable of delivering a specified output voltage up to a maximum load current.

Let us assume that we need a power supply for a system that is constructed of BJTs, FETs, and so forth, that must supply 5 Vdc at a maximum load current of 350 mA. For our system we have available 115 Vac input voltage. A look through catalogs reveals that Datel Systems, Inc., offers a power supply that requires an input voltage of 115 Vac, 60 Hz, with an output capability of 5 Vdc at 360 mA. Some of the features of this device are (a) overvoltage and overload protection, (b) stability, (c) high input and output isolation, and (d) low cost.

The output voltage is within ±0.5 percent (±0.025 Vdc), and the input voltage can change within a range of ±10 Vac (UPM-5/350). Figure 11.2 illustrates the specifications and physical dimensions for the UPM-5/350 power supply. The physical dimensions are only 2 × 2 × 0.432 in. This power supply has current limiting, which means that if the output is short-circuited, the output current is limited to 140 percent (490 mA) of its rated value. With the output current limiting feature, the power supply module is not destroyed with accidental overloads.

With the power supply modules supplied by Datel Systems the output voltage may be doubled by simply connecting the UPM-5/350 in series. Let us assume that we need a 5 Vdc and a 10 Vdc supply capable of delivering a total current of 350 mA. Figure 11.3 illustrates a simple means of obtaining the voltage requirements needed with the UPM-5/350 modules.

If a complete power supply is not needed for our 5 Vdc, 350 mA power requirements, possibly Signetics LM109H voltage regulator could be used. The LM109H is a complete 5 Vdc voltage regulator constructed on a single silicon chip. The regulator has a built-in current-limiting circuit. It requires an input voltage between 7 and 25 Vdc to deliver 5 Vdc at a maximum output current of 500 mA in the TO-5 package. Figure 11.4a illustrates the physical package, and Figure 11.4b shows the equivalent electrical circuit for the regulator. Some of the features of this device are (a) internal thermal overload protection, (b) internal current limiting, and (c) no external components required.

Figure 11.5a illustrates the electrical characteristics, and Figure 11.5b shows some possible circuit applications for the LM109H voltage regulator. The 5 Vdc regulator is also available in the TO-5 package, which is capable of delivering an output current of 1.5 A. Note that the thermal resistances for the TO-3 and TO-5 packages are listed as 15 and 3°C/W, respectively. For maximum

Figure 11.3
Doubling the output voltage of a power supply using two UPM 5/350s.

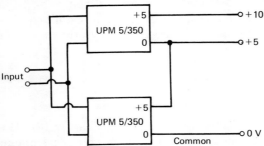

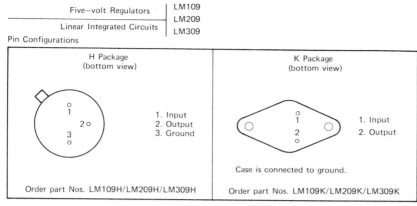

(a)

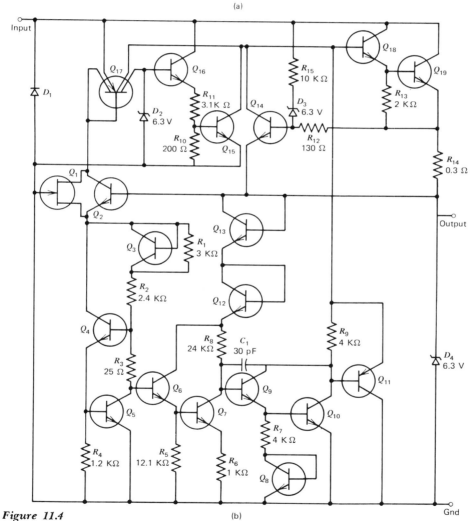

Figure 11.4

(b)

Gnd

Five volt regulator (Signetics Corp.): (a) pin configuration and (b) circuit diagram.

output current, a heat sink should be used with the device. The input voltage for this regulator is simply a rectifier circuit with a capacitor filter that can supply the required dc input voltage for a maximum output current (see Chapter 12).

Figure 11.5
LM109 voltage regulator: (a) electrical characteristics and (b) application.

LINEAR INTEGRATED CIRCUITS: LM 109

Electrical Characteristics (Note 1)

Parameter	Conditions	LM109			LM309			Unit
		Min	Typ	Max	Min	Typ	Max	
Output voltage	$T_j = 25°C$	4.7	5.05	5.3	4.8	5.05	5.2	V
Line regulation	$T_j = 25°C$ $7\ V \leqslant V_{in} \leqslant 25\ V$		4	50		4	50	mV
Load regulation	$T_j = 25°C$							
TO-5	$5\ mA \leqslant I_{out} \leqslant 0.5\ A$		20	50		20	50	mV
TO-3	$5\ mA \leqslant I_{out} \leqslant 1.5\ A$		50	100		50	100	mV
Output voltage	$7\ V \leqslant V_{in} \leqslant 25\ V$ $5\ mA \leqslant I_{out} \leqslant I_{max}$ $P \leqslant P_{max}$	4.6		5.4	4.75		5.25	V
Quiescent current	$7\ V \leqslant V_{in} \leqslant 25\ V$		5.2	10		5.2	10	mA
Quiescent current change	$7\ V \leqslant V_{in} \leqslant 25\ V$ $5\ mA \leqslant I_{out} \leqslant I_{max}$			0.5 0.8			0.5 0.8	mA mA
Output noise voltage	$T_A = 25°C$ $10\ Hz \leqslant f \leqslant 100\ kHz$		40			40		μV
Long term stability				10			20	mV
Thermal resistance Junction to case (note 2)								
TO-5			15			15		°C/W
TO-3			3			3		°C/W

Notes:

1. Unless otherwise specified, these specifications apply for $-55\ °C \leqslant T_j \leqslant 150\ °C$ for the 109 or $0\ °C \leqslant T_j \leqslant 125\ °C$ for the 309, $V_{in} = 10\ V$ and $i_{out} = 0.1\ A$ for the TO-5 package or $I_{out} = 0.5\ A$ for the TO-3 package. For the TO-5 package, $I_{max} = 0.2\ A$ and $P_{max} = 2.0\ W.$, For the TO-3 package, $I_{max} = 1.0\ A$ and $P_{max} = 20\ W.$

2. Without a heat sink, the thermal resistance of the TO-5 package is about 150 °C/W while that of the TO-3 package is approximately 35 °C/W. With a heat sink, the effective thermal resistance can only approach the values specified, depending on the efficiency of the sink.

Typical Applications

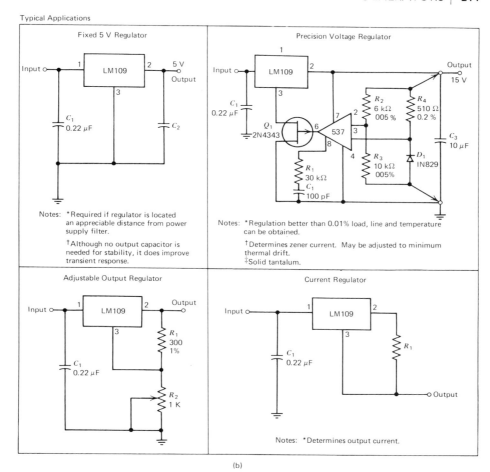

(b)

Figure 11.5 *(continued)*

A review of the literature will reveal that many companies offer very good and reliable power supply modules and voltage regulators. One simply selects the device knowing the requirements of the system.

SIGNAL GENERATORS

A *signal generator* is a circuit configuration that produces various waveforms at a particular frequency (the frequency can, in some cases, be varied). Let us suppose that a square and triangular signal generator was needed for service applications. The frequency range of this generator needs to cover a range of 5 Hz to 100 kHz. Such a generator could be built from discrete components, but a simpler means of constructing such a device would use a Signetics 566 function generator. This function generator has a frequency range from 3 Hz to 100 kHz and generate both a square and triangular waveform.

Figure 11.6a illustrates the physical package; Figure 11.6b shows the block diagram and electrical equivalent circuit for the 566 function generator. The function generator has a wide range of dc input voltages (10 to 24 V); frequency programming by means of an external resistor, capacitor, and voltage; and a frequency range that is adjustable over a 10 : 1 range with the same timing capacitor.

Figure 11.7a shows some typical performance characteristics for the function generator, and Figure 11.7b illustrates a simple circuit application for the function generator. The bias voltage V_C at pin 5 must have a voltage that meets the requirement of $(3/4 \, V \leq V_C \leq V)$. The frequency fo the function generator is given by $f_o = 2(V - V_C)/R_1 C_1 \, V$, where R_1 should be in the range of 2 kΩ 20 kΩ. Because V_C also controls the frequency of the generator, V_C can be used to frequency modulate the function generator. One simple application of this frequency modulation is an electronic siren.

Figure 11.6
566 function generator (Signetics Corp.): (a) pin configuration and (b) block diagram and equivalent circuit.

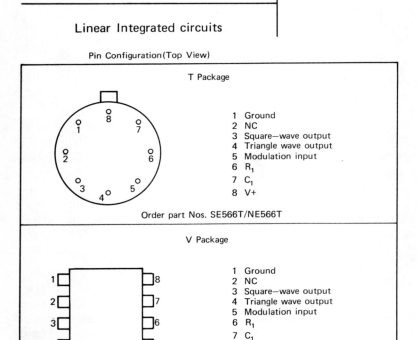

(a)

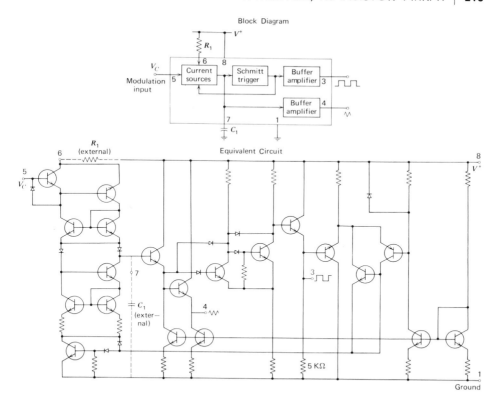

Figure 11.6 (continued)

This function generator can be used in the construction of a digital read-out temperature indicator (see Chapter 20). There are several companies that offer function generators that produce not only square and triangular waveforms but also sine waves.

THYRISTOR/TRANSISTOR ARRAY

An interesting IC produced by RCA is the monolithic silicon CA3097E chip. This chip is called a thyristor/transistor array. The IC chip is a 16-pin dual-in-line package (DIP) and is shown in Figure 11.8a. This array contains (see Figure 11.8b) a NPN transistor, a sensitive-gate silicon-controlled rectifier (SCR—see Chapter 13), a programmable unijunction transistor (PUT—see Chapter 13), a PNP/NPN transistor pair (see Chapter 14), and a zener diode all in a single DIP package.

Some of the features of this array are (a) complete isolation between elements, (b) a transistor pair having a beta of 8000 or individual NPN, PNP transistor operation, (c) a zener voltage of 8 V with an r_z of 15 Ω, and (d) a SCR with a

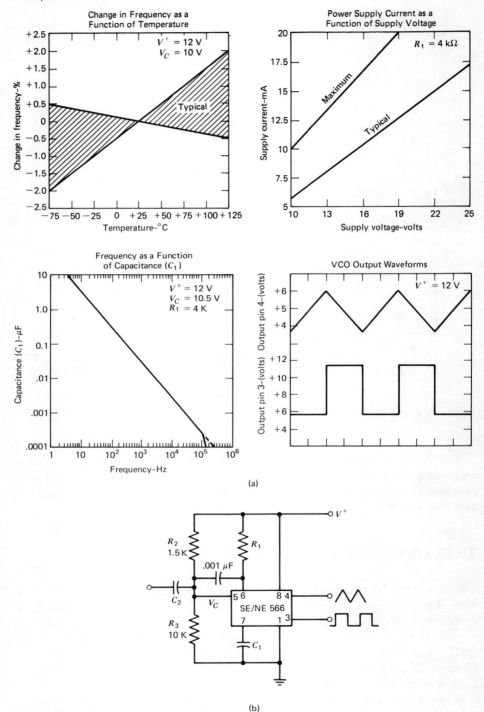

Figure 11.7
566 function generator: (a) performance characteristics and (b) circuit connections.

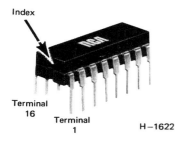

Index

Terminal 16

Terminal 1

H−1622

16−Lead dual−in−line− plastic package

(a)

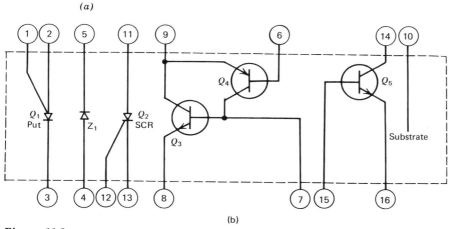

(b)

Figure 11.8

Thyristor/transistor array (RCA): (*a*) *package and* (*b*) *pin configuration for the CA 3097E.*

maximum forward current of 150 mA. The maximum operating parameters are listed by the manufacturer for each of the solid-state devices.

Some simple applications of the array are illustrated in Figure 11.9. A pulse generator is shown in Figure 11.9*a*. Along with eight "outboard" resistors and two timing capacitors, the array can be used in the construction of a pulse generator with a pulse rate adjusted by varying R_T or C_T and a pulse width adjusted by varying the time constant of R_1C_1.

Figure 11.9*b* illustrates a simple series voltage regulator constructed using the array with four external resistors and one capacitor (see Chapter 12). As can be seen from the figure, the output voltage can be adjusted between 9.5 and 15 V with an input dc voltage (unregulated) of 20 V. With this array, many other circuits can be developed.

THE ANALOG SWITCH

We have all had experiences with electromechanical relays. The major advantages of the relay are its low contact resistance and its ability to switch large

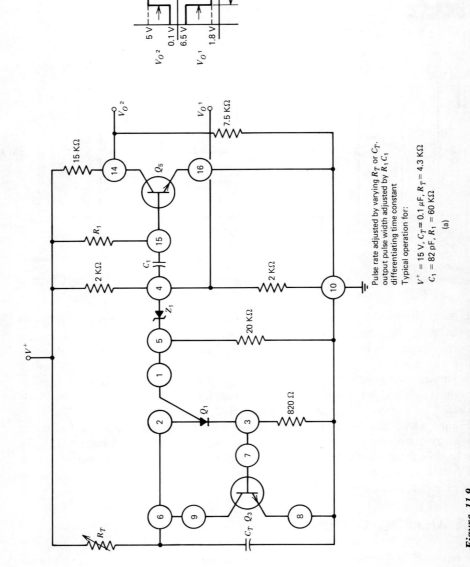

Figure 11.9
Application for the CA 3097E: (a) pulse generator and (b) voltage regulator.

Pulse rate adjusted by varying R_T or C_T.
output pulse width adjusted by $R_1 C_1$
differentiating time constant
Typical operation for:

$V^+ = 15\ V,\ C_T = 0.1\ \mu F,\ R_T = 4.3\ K\Omega$
$C_1 = 82\ pF,\ R_1 = 60\ K\Omega$

(a)

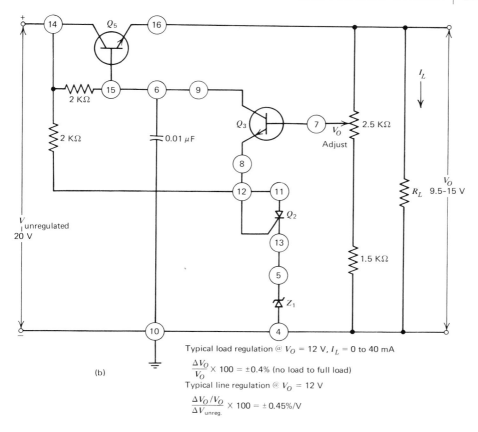

Typical load regulation @ V_O = 12 V, I_L = 0 to 40 mA

$$\frac{\Delta V_O}{V_O} \times 100 = \pm 0.4\% \text{ (no load to full load)}$$

Typical line regulation @ V_O = 12 V

$$\frac{\Delta V_O / V_O}{\Delta V_{unreg.}} \times 100 = \pm 0.45\%/V$$

(b)

Figure 11.9 (*continued*)

values of voltage and current. However, in solid-state applications the relay frequently has some disadvantages. A few disadvantages are contact bounce (contacts open and close because of mechanical construction when the contact is closed), relatively slow closure, and large physical size. The *analog switch* or solid-state relay does not suffer contact bounce, is much faster, and requires much less space than its electromechanical relay equivalent.

In many solid-state circuits, high voltage and large currents are not required for switching purposes. A mechanical relay used for these circuit applications can be outperformed by an all solid-state analog switch such as the ones produced by Intersil, Inc. Figure 11.10 shows some of the solid-state equivalent switches with single- and multicontacts. Note that an anlaog switch requires a specific input voltage to change the contact closers. The contacts themselves are constructed with monolithic complementary MOSFETs. The driver of the MOSFET can be either of the MOSFET or BJT family. The contacts constructed with MOSFETs are operated in the cutoff or saturated mode.

Some of the features of the INTERSIL analog switches are (a) the ability to switch analog voltages up to ±14 V with ±15 V supplies, (b) overvoltage

FORM A

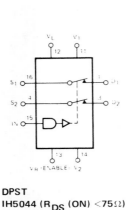

SPST
IH5040 (R_{DS} (ON) <75Ω)

DPST
IH5044 (R_{DS} (ON) <75Ω)

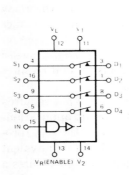

4PST
IH5047 (R_{DS} (ON) <75Ω)

DUAL A

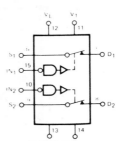

DUAL SPST
IH5041 (R_{DS} (ON) <75Ω)

DUAL DPST
IH5049 (R_{DS} (ON) <30Ω)

IH5045 (R_{DS} (ON) <75Ω)

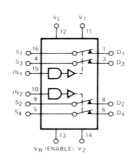

DUAL B

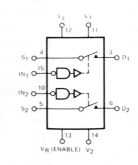

DUAL SPST
IH5048 (R_{DS} (ON) <30Ω)

FORM C

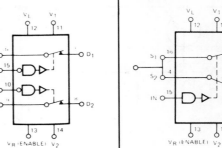

SPDT
IH5050 (R_{DS}(ON) < 30Ω)

IH5042 (R_{DS}(ON) < 75Ω)

DPDT
IH5046 (R_{DS} (ON) <75Ω)

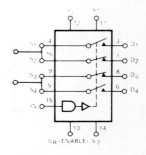

DUAL C

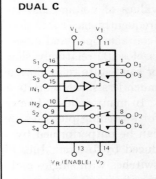

DUAL SPDT
IH5051 (R_{DS} (ON) <30Ω)

IH5043 (R_{DS} (ON) <75Ω)

Figure 11.10
Analog switches (Intersil Corp.)

218

protection, (c) a switching speed of $t_{on} = 500$ nsec and $t_{off} = 250$ nsec, and (d) on-contact resistance $R_{DS(on)}$ of less than 75.

Some of the applications of these analog switches are for adjusting the gain of OP amperes (Chapter 15) and selecting various input voltages (i.e., transducers) to be fed to an amplifier (multiplexing).

TONE DECODER

The generation of tones for control purposes has become very popular. Thus, by generating a specific frequency, a control function can be performed. The SIGNETICS 567 IC can be used to *decode* (detect) various tones. This 567 tone decoder delivers an output at one input frequency by using what is called a phase-locked loop (PLL). A PLL is similar to an automatic frequency control (AFC).

Figure 11.11a shows the package configuration; Figure 11.11b illustrates the block diagram for the 567 tone decoder. The IC chip has a current-controlled oscillator, which has a free running frequency (with no input signal) that is determined by the external components R_1 and C_1. Let us assume that values of R_1 and C_1 are chosen so that the oscillator has a frequency of 1000 Hz. The input signal, which may vary in frequency, is applied to pin 3 of the 567. If the input frequency is 1000 Hz, the output voltage level at pin 8 is approximately zero volts. For all other input frequencies (excluding 1000 Hz), the output voltage at pin 8 is approximately the supply voltage V. Thus this decoder acts as a switch, having a low output at 1000 Hz and a high output at all other frequencies.

Some of the features of the 567 are (a) a wide frequency range (0.01 Hz to 500 kHz), (b), an adjustable frequency over a $20 : 1$ range with an external resistor, and (c) a controlled bandwidth (range of input frequencies) of 14 percent.

One of the many interesting applications of the 567 is its use to decode the Touch-Tone dialing used by the telephone company. The Touch-Tone pad used by the telephone company uses seven discrete tones that range from 697 Hz to 1477 Hz. Each time a number is depressed (0 through 9), two tones are generated that are algebraically added and sent over the telephone lines. Seven 567 tone decoders can be used to decode these telephone tones. Such a scheme is illustrated in Figure 11.12

Because each digit consists of the combination of two tones, only seven 567 tone decoders are needed. The logic gates detect which of the possible two 567s are at a low output (see Chapter 16). The outputs of the various logic gates are labeled 0 through 9 with two digits labeled * and #. These various outputs can be used to drive relays, transistors, and so on.

OTHER DEVICES

Other IC devices include audio-frequency amplifiers, preamplifiers, video amplifiers, balanced modulators, complete radio receivers, just to mention a few of the possible chips available. The reader is asked to review one of the many IC catalogs published by manufacturers in order to gain a feel for the many types of ICs that are available.

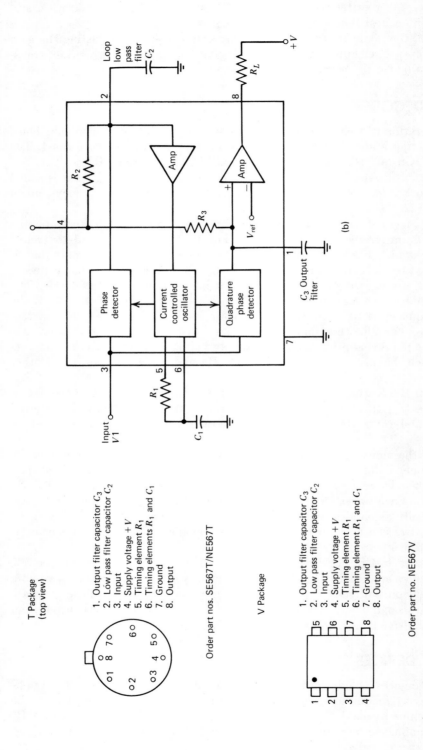

Figure 11.11
567 tone generator (Signetics Corp.): (a) pin configuration and (b) block diagram.

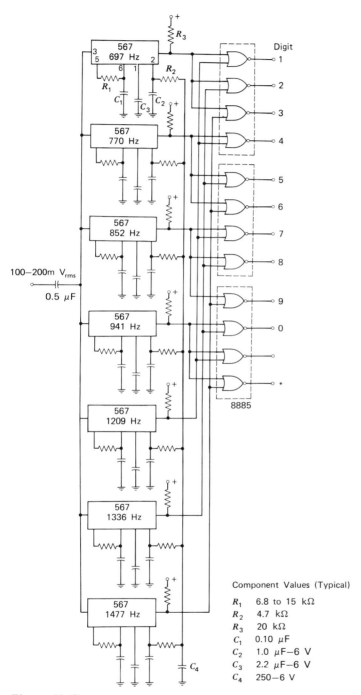

Figure 11.12
Touch-tone decoder using 567 PLL.

OBJECTIVES FOR CHAPTER 12, POWER SUPPLY REGULATION

At the completion of this chapter you can demonstrate your understanding of the material by answering these questions:

- What are the important parameters associated with power supplies?
- What are half-wave, full-wave, and bridge rectifiers?
- What is the difference between a regulated and a nonregulated power supply?
- How is voltage regulation accomplished in a power supply?
- What is a current limiter and how does it function?
- What are the differences between a series and a switching regulator?
- What is a voltage doubler?

12 POWER SUPPLY REGULATION

The power supply is the heart of all electronic systems. You should study the various types of power supplies along with the many types of regulators incorporated because of their frequent use in circuitry.

A circuit used to change alternating current (ac) to direct current (dc) is called a rectifier. The rectifying circuit along with its filter and regulator is called a power supply. The filter smooths out the pulsating direct current. The regulator, if needed, insures a constant output voltage with changing load currents. The power supply in a system generates the required voltage(s) at varying load current(s) for the proper operation of amplifiers, oscillators, and so forth, which make up a system. This chapter discusses the operating parameters of a power supply and the circuitry that is commonly used in the construction of a power supply.

PERCENT REGULATION

An important term associated with power supplies is its *percent voltage regulation*. This percentage indicates the amount of change in output voltage with load current changes and is given by Equation 1.

$$\text{percent regulation} = \frac{V_{NL} - V_L}{V_L} \times 100 \tag{1}$$

where V_{NL} is the output (no load) voltage of the supply and V_L is the output (load) voltage of the supply. A well-designed power supply will have a very small percent regulation.

EXAMPLE

The output voltage of a power supply is 10 V with no load current and drops to 9 V with the load connected to the power supply. Calculate the percentage of voltage regulation for this power supply.

The voltage V_{NL} is equal to 10 V, and V_L is equal to 9 V. Using Equation 1, we have

$$\text{percent regulation} = \frac{10 \text{ V} - 9 \text{ V}}{9\text{V}} \times 100$$

or

$$\text{percent regulation} = 11.1 \text{ percent}$$

RIPPLE VOLTAGE

Another important term associated with power supplies is *ripple voltage*. An ideal power supply will have zero ripple voltage. The magnitude of ripple voltage is determined by the type of filtering techniques incorportaed in the power supply and is a result of "smoothing-out" the pulsating dc voltage (as will be discussed shortly). Ripple voltage is the magnitude of change in the dc voltage of the power supply under load conditions and is expressed in either its peak-to-peak or rms value.

Ripple voltage is minimized via filtering networks and regulating devices. With solid-state systems, power requirements are generally made up of low voltage and high current. Filtering is generally accomplished with large values of capacitance and with regulating devices including zener diodes, difference amplifiers, and OP AMPs.

As a point of interest, power requirements for vacuum tubes are made up of high voltage and low current. With the requirement of high voltage and low current, filtering and regulation are accomplished with capacitors and inductors (pi networks, choke input filters, etc.). Because inductors have a dc resistance, the voltage drop (Ohm's law) across the inductor because of the low-current requirement of vacuum tubes presents no problem when inductors are used in filtering networks. However, with solid-state systems requiring low voltage and high current, the voltage drop across an inductor does present serious problems in series regulators. Therefore, inductors are infrequently used in power supplies in solid-state systems (see Switching Voltage Regulator in this chapter).

FIXED AND VARIABLE OUTPUT VOLTAGES

A power supply may have a fixed and/or variable output voltage(s). A fixed output voltage for a power supply simply means that the output voltage is constant and cannot be changed. If a power supply has a variable output voltage, its output voltage can be changed. The output voltage may be changed in discrete steps or continuously called its *resolution*. Fixed and/or variable output voltage depends on the design of the power supply.

CURRENT LIMITERS

As was stated earlier, a power supply is designed to deliver a certain voltage at a specified current. If this specified current is exceeded, the devices incorporated in the power supply may be damaged. Special circuitry can be used within the power supply to make sure that the power supply will not be damaged if this specified current is exceeded. Such circuitry is called a *current limiter*.

A power supply may have variable or fixed current liniting capability. It is important to note that the current limiter of a power supply is not the same as a current pump. Some supplies are labeled short-circuit proof. If the terminals of a supply so labeled are short-circuited, the supply will deliver a fixed amount of current to the load without damage to the supply. The amount of current depends on the setting of the current limiter. Another type of current protection device incorporated in a power supply is called a *foldback current limiter*. If the terminals of a supply with a foldback current limiter are shorted, the current delivered to the load by the supply is less than the normal load current.

If a supply is shorted momentarily, the output voltage is reduced. Once the short is removed, the amount of time the output voltage takes to come up to the output voltage before the short occurred is generally referred to as its *recovery time*.

INPUT VOLTAGE

A power supply requires a specific input voltage and frequency to function correctly. The input voltage is given at a certain voltage level and frequency (i.e., 120 Vac, 60 Hz). The output voltage change with respect to input voltage change of the power supply is called its *line regulation*.

TEMPERATURE

Temperature does affect the electronic devices incorporated in a power supply. The ambient temperature range over which the power supply can be operated under load conditions is given by the manufacturer. This *operating temperature range* should not be confused with the storage temperature range of the power supply. The *storage temperature range* for a power supply is the temperature over which the power supply can be stored in a nonoperating condition. The manufacturer also gives the effects of temperature on the output voltage in the operating parameters for the power supply. This characteristic is sometimes called its *temperature coefficient*.

Tabel 12.1 lists the operating parameters given by Tektronix for their PS501 power supply. The PS501 has both a fixed and a variable output supply voltage. The variable output voltage (0 to 20 Vdc) has an adjustable current limiter (0 to 400 mA).

Table 12.1

Tektronix Operating Parameters for the PS501 Power Supply

20-V Floating Supplies

Primary power input: Determined by power module (TM 501 or TM 503).

Output: Continuously variable from 0 to at least 20 Vdc. Insulated for 350 Vdc + peak ac above ground.

Current limit: 0 mA to 400 mA.

Stability: (0.1% + 5 mV) or less drift in 8 hours at constant line, load, and temperature.

Minimum resolution: 10 mV.

Line regulation: Within 5 mV for a ±10% line voltage change.

Load regulation: Within 1 mV with a 400-mA load change.

Ripple and noise: 0.5 mV P-P or less, 0.1 mV rms or less.

Temperature coefficient: 0.01%/°C or less.

Transient recovery time: 20 μsec or less for a constant voltage to recover within 20 mV of nominal output voltage after a 400-mA change in current output.

Indicator Lights: Voltage variation and current limit.

+5 V Ground Referenced Supply

Output: (+20° C to + 30 °C): 4.8 Vdc to 5.2 Vdc at 1 A.

Line regulation: (+20 to +30 °C): within 50 mV for a ±10% line voltage change.

Load regulation: Within 100 mV with a 1-A load change.

Ripple and noise: (1 A): 5 mV P-P or less. 100 μV rms or less.

Overload protection: Automatic current limiting and over-temperature shutdown.

General

Temperature: Operating, 0 °C to +50 °C.
Nonoperating, −40 °C to + 75 °C.

Altitude: Operating, to 15,000 feet.
Nonoperating, to 50,000 feet.

Shipping weight: 4 pounds.

Dimensions: 2.6 inches wide, 5 inches high, 12.2 inches deep.
(6.6 x 12.7 × 31 cm).

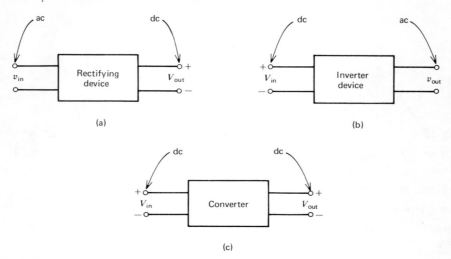

Figure 12.1
Black box of: (a) a rectifier, (b) an inverter, and (c) a converter.

INVERTERS AND CONVERTERS

A rectifying circuit is a circuit that has an ac input voltage and a dc output voltage (i.e., 120 Vac, 12 Vdc). Two other types of supplies are the inverter and the converter. The *inverter* has a dc input voltage and changes this voltage to an ac output voltage at a certain frequency. A *converter* is a circuit arrangement that has a dc input voltage and changes this voltage to a dc output voltage. Examples of these devices are the inverters used to power ac appliances from the car (12-Vdc input, 120-Vac, 60-Hz output) and part of a capacitor discharge ignition system (12-Vdc input, 400-Vdc output). These systems are illustrated in Figure 12.1. The rest of this chapter is concerned with rectifiers and regulators.

RECTIFYING CIRCUIT CONFIGURATIONS

With the use of diodes, an ac voltage (or current) can be converted into a dc voltage (or current). This circuit arrangement is called a *rectifier*. The manner in which the diodes are connected determines the name given to the rectifying circuit configuration. The three most common circuit configurations are called half-wave, full-wave, and bridge rectifiers.

HALF-WAVE RECTIFIER

Figure 12.2 illustrates the circuit configuration for a *half-wave rectifier* with its input and output waveforms. The diode D is connected in series with the load resistor R_L and the secondary of the transformer T. The diode D functions as a switch, turning "on" or "off" depending on the polarity of the voltage across the diode.

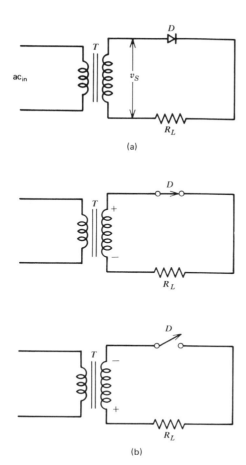

(a)

(b)

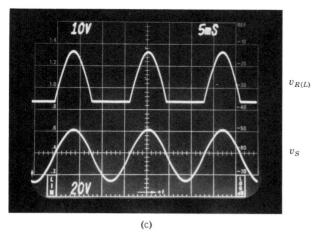

$v_{R(L)}$

v_S

(c)

Figure 12.2
Half-wave rectifier: (a) circuit, (b) diode being switched, and (c) waveforms.

When the anode of diode D is positive with respect to its cathode, the diode is forward-biased. The low resistance of the forward-biased diode allows the positive alternation of the secondary voltage v_s to be placed across the load resistor R_L. When the anode of diode D is negative with respect to its cathode, the diode is reverse-biased. The high resistance of the reverse-biased diode blocks the secondary voltage v_s from reaching the load resistor R_L. The diode D is in effect switched "on" for the positive alternation and "off" for the negative alternation of v_s. Because only one-half of the secondary voltage is across the load, the circuit is called a half-wave rectifier. The frequency of the pulsating direct current across R_L is the line frequency (60 pps).

The average or dc voltage across the load resistor *without filtering* is calculated by

$$V_{dc} = \frac{1}{\pi} \, v_{s(peak)}$$

$$V_{dc} = 0.318 \, v_{s(peak)} \tag{2}$$

With the aid of Equation 2, the average or dc voltage across the load resistor R_L can be calculated. The dc voltage calculated would be the value indicated on a dc voltmeter if it were placed across the load resistor R_L. If an ac voltmeter were placed across the load resistor, the value of voltage indicated could be calculated by

$$V_{ac} = 0.385 \, v_{s(peak)} \tag{3}$$

The ac value of voltage calculated with Equation 3 is referred to as the ripple voltage. For comparison of various rectifying circuit configurations, a term called *ripple factor* is defined as the ac voltage divided by the dc voltage, or

$$\text{R.F.} = \frac{V_{ac}}{V_{dc}} \tag{4}$$

When the diode is reverse-biased in a half-wave rectifier circuit configuration, the magnitude of the reverse-biased voltage is the peak value of the secondary voltage v_s. Therefore, the diode used in the circuit configuration would have to have a PIV (peak inverse voltage) rating equal to or greater than the peak value of the secondary voltage v_s.

EXAMPLE

A transformer has a secondary voltage of 12 V and is to be used in a half-wave rectifier. Calculate the dc and ac voltages across a load resistor R_L, the necessary PIV rating for the diode, and the ripple factor. Using Equation 2 to calculate V_{dc}, we have

$$V_{dc} = 0.318 \, v_{s(peak)}$$
$$= 0.318 \times 12 \times \sqrt{2}$$
$$V_{dc} = 5.4 \text{ V}$$

Using Equation 3 to calculate V_{ac}, we have

$$V_{ac} = 0.385 \, v_{s(peak)}$$
$$= 0.385 \times 12 \times \sqrt{2}$$
$$V_{ac} = 6.5 \text{ V}$$

The PIV rating for the diode is equal to or greater than the peak value of the secondary voltage, or

$$PIV \geq v_{s(peak)}$$

$$\geq 12 \times \sqrt{2}$$

$$PIV \geq 17 \text{ V}$$

Using Equation 4 to calculate the ripple factor, we have

$$R.F. = \frac{V_{ac}}{V_{dc}}$$

$$= \frac{6.5 \text{ V}}{5.4 \text{ V}}$$

$$R.F. = 1.21 \qquad \text{or} \qquad 121 \text{ percent}$$

As can be reasoned from the circuit illustrated in Figure 12.2, the magnitude of the current through R_L is determined by the values of R_L and the secondary voltage v_s. With the use of Ohm's law, the peak value of the current through R_L is given by

$$I_{(peak)} = \frac{v_{s(peak)}}{R_L} \tag{5}$$

In order to calculate the dc and ac current values through the load resistor, Equations 6 and 7 can be used.

$$I_{dc} = \frac{1}{\pi} I_{(peak)}$$

$$I_{dc} = 0.318 I_{(peak)} \tag{6}$$

$$I_{ac} = 0.385 I_{(peak)} \tag{7}$$

FULL-WAVE RECTIFIER

Figure 12.3 illustrates the circuit arrangement for a *full-wave rectifier*. As can be reasoned from the figure, the full-wave rectifier circuit configuration can be thought of as two connected half-wave rectifiers. The secondary winding of the transformer T is center-tapped, with one end of the load resistor connected to the center tap. The total secondary voltage is split in half, with $v_{s(1)}$ and $v_{s(2)}$ being equal in magnitude ($|v_{s(1)}| = |v_{s(2)}|$), being additive ($v_s = v_{s(1)} + v_{s(2)}$), and having a 180° phase difference ($v_{s(1)} = -v_{s(2)}$).

With reference to Figure 12.3, when the anode of diode D_1 is positive with respect to the center tap of the transformer, the anode of diode D_2 is negative with respect to the center tap of the transformer. This voltage polarity causes diode D_1 to be forward-biased and diode D_2 to be reverse-biased. Thus the positive alternation of $v_{s(1)}$ is placed across the load resistor R_L.

When the anode of diode D_1 is negative with respect to the center tap of the transformer, the anode of diode D_2 is positive with respect to the center tap of

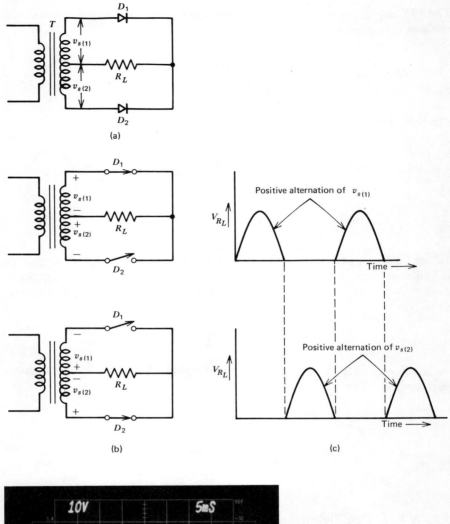

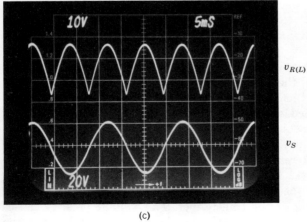

(c)

Figure 12.3
Full-wave rectifier: (a) circuit, (b) diodes being switched, and (c) waveforms.

the transformer. This voltage polarity causes diode D_1 to be reverse-biased and diode D_2 to be forward-biased. Thus the positive alternation of v_{s2} is placed across the load resistor R_L. These relationships are illustrated in Figure 12.3a and 12.3b. The frequency of the pulsating dc voltage across R_L is twice the line frequency (120 pps).

The dc or average voltage without filtering across the load resistor R_L is given by

$$V_{dc} = \frac{2}{\pi} v_{s(1)(peak)} = \frac{2}{\pi} v_{s(2)(peak)}$$

$$V_{dc} = 0.636\ v_{s(1)(peak)} = 0.636\ v_{s(2)(peak)} \tag{8}$$

where $v_{s(1)}$ or $v_{s(2)}$ is one-half of the total secondary voltage v_s. The ac or ripple voltage across R_L is given by

$$V_{ac} = 0.308\ v_{s(1)(peak)} = 0.308\ v_{s(2)(peak)} \tag{9}$$

The maximum reverse-biased voltage for either diode (D_1 or D_2) is the peak voltage of the total secondary voltage v_s. The PIV rating for each diode then should be equal to or greater than the peak value of the total secondary voltage.

$$PIV \geq 2\ v_{s(1)(peak)} = v_{s(peak)} \tag{10}$$

As can be reasoned from the illustration in Figure 12.3, the dc and ac currents through the load resistor R_L are given by

$$I_{peak} = \frac{v_{s(1)\ (peak)}}{R_L} \tag{11}$$

$$I_{dc} = 0.636\ I_{(peak)} \tag{12}$$

$$I_{ac} = 0.308\ I_{(peak)} \tag{13}$$

EXAMPLE

A 24-V center-tapped transformer (24VCT) is to be used with a full-wave rectifier circuit configuration. Calculate the dc and ac voltages across a load resistor of 10 Ω, the ripple factor, and the dc and ac currents. Because the total secondary voltage is 24 V, $v_{s(1)} = v_{s(2)} = \frac{1}{2}v_s$, or

$$v_{s(1)} = \frac{1}{2} \times 24\ V = 12\ V$$

Using Equation 8 to calculate the dc voltage, we have

$$V_{dc} = 0.636\ v_{s(1)(peak)}$$

$$= 0.636 \times 12\ V \times \sqrt{2}$$

$$V_{dc} = 10.7\ V$$

Using Equation 9 to calculate the ac or ripple voltage, we have

$$V_{ac} = 0.308\ v_{s(1)(peak)}$$

$$= 0.308 \times 12\ V \times \sqrt{2}$$

$$V_{ac} = 5.2\ V$$

Using Equation 4 to calculate the ripple factor, we have

$$\text{R.F.} = \frac{V_{ac}}{V_{dc}} = \frac{5.2}{10.7}$$

$$\text{R.F.} = 0.483 \quad \text{or} \quad 48.3 \text{ percent}$$

Using Equation 11 to calculate the peak value of the current, we have

$$I_{(peak)} = \frac{v_{s(1)(peak)}}{R_L} = \frac{17 \text{ V}}{10 \text{ }\Omega}$$

$$I_{(peak)} = 1.7 \text{ A}$$

Using Equations 12 and 13 to calculate the dc and ac load currents, we have

$$I_{dc} = 0.636 \ I_{(peak)} = 0.636 \times 1.7 \text{ A}$$

$$I_{dc} = 1.08 \text{ A}$$

$$I_{ac} = 0.308 \ I_{(peak)} = 0.308 \times 1.7 \text{ A}$$

$$I_{ac} = 0.52 \text{ A}$$

It is important to note that the ripple factor for the full-wave rectifier is much less than that for the half-wave rectifier. It is easier to filter a power supply with a low ripple factor than one with a high ripple factor.

BRIDGE RECTIFIER

Figure 12.4 illustrates the circuit configuration for a bridge rectifier. The bridge rectifier uses four diodes and does not require a transformer with a center-tapped secondary winding. The diodes function as switches, enabling the load resistor R_L to be switched or alternated across the secondary of the transformer. This switching action of the diodes permits the current to flow in only one direction (dc) through the load resistor.

When the top of the secondary winding is positive, diodes D_1 and D_3 are forward-biased and diodes D_2 and D_4 are reverse-biased (see Figure 12.4b). The low resistance of the forward-biased diodes D_1 and D_3 places the load resistor so that current flows from point B to point A. When the top of the secondary winding is negative, diodes D_2 and D_4 are forward-biased and the diodes D_1 and D_3 are reverse-biased. The low resistance of the forward-biased diodes D_2 and D_4 places the load resistor so that the current flows from point B to point A. The switching actions of the diodes always place the load resistor across the secondary of the transformer so that the current through the load resistor is always in one and only one direction. The voltage waveform across R_L is shown in Figure 12.4c. The frequency of the pulsating dc voltage across R_L is twice the line frequency (120 pps).

The dc and ac voltages across the load resistor are calculated by

$$V_{dc} = \frac{2}{\pi} \ v_{s(peak)}$$

$$V_{dc} = 0.636 \ v_{s(peak)} \tag{14}$$

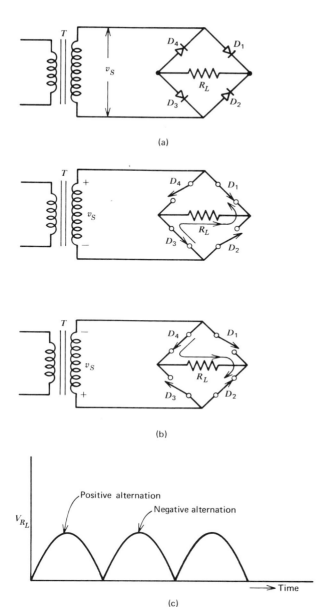

Figure 12.4
Bridge rectifier: (a) circuit, (b) diodes being switched, and (c) waveforms.

$$V_{ac} = 0.308 \, v_{s(peak)} \qquad (15)$$

The reverse-biased voltage across the diodes, as can be reasoned from Figure 12.4b is the peak value of the secondary voltage v_s. Thus the PIV rating for each of the diodes in a bridge circuit should be equal to or greater than the peak value of the secondary voltage v_s. As with the other rectifier configurations discussed so far, the peak value of load current through R_L is related to the ohmic value of R_L and the peak value of the secondary voltage, or

$$I_{(\text{peak})} = \frac{v_{s(\text{peak})}}{R_L} \qquad (16)$$

The dc and ac values of current through the load resistor can be calculated by

$$I_{\text{dc}} = \frac{2}{\pi} I_{(\text{peak})}$$

$$I_{\text{dc}} = 0.636 \, I_{(\text{peak})} \qquad (17)$$

$$I_{\text{ac}} = 0.308 \, I_{(\text{peak})} \qquad (18)$$

EXAMPLE

A power supply is to be constructed incorporating full-wave and bridge rectifying circuit configurations. A 24 VCT transformer is to be used in the power supply. A switch will select the appropriate circuit configuration (see Figure 12.5). With the switch in position A, calculate the dc and ac values of voltage and current for a 10 Ω load resistor. Repeat the previous calculations with the switch in position B. With the switch in position A, the power supply is a full-wave rectifier. The previous example (see p. 233) illustrates the calculations for the various currents and voltages. With the switch in position B, the power supply is a bridge rectifier.

The dc and ac voltages output are calculated with Equations 14 and 15, or

$$V_{\text{dc}} = 0.636 \, v_{s(\text{peak})}$$

$$= 0.636 \times 24 \times \sqrt{2}$$

$$V_{\text{dc}} = 21.5 \text{ V}$$

$$V_{\text{ac}} = 0.308 \, v_{s(\text{peak})}$$

$$= 0.308 \times 24 \times \sqrt{2}$$

$$V_{\text{ac}} = 10.5 \text{ V}$$

Figure 12.5
Diagram used in example problem.

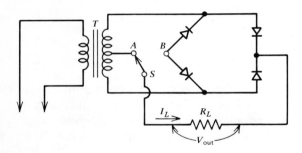

Using Equation 16, we have

$$I_{(\text{peak})} = \frac{v_{s(\text{peak})}}{R_L}$$

$$= \frac{24 \times \sqrt{2}}{10 \ \Omega} \text{ V}$$

$$I_{(\text{peak})} = 3.4 \text{ A}$$

To calculate I_{dc} and I_{ac} using Equations 17 and 18, we have

$$I_{\text{dc}} = 0.636 \ I_{(\text{peak})}$$

$$= 0.636 \times 3.4 \text{ A}$$

$$I_{\text{dc}} = 2.15 \text{ A}$$

$$I_{\text{ac}} = 0.308 \ I_{(\text{peak})}$$

$$= 0.308 \times 3.4 \text{ A}$$

$$I_{\text{ac}} = 1.05 \text{ A}$$

COMPARISON OF RECTIFIERS

The comparison of the half-wave, full-wave, and bridge rectifiers is summarized in Table 12.2. The output-voltage and current equations for the full-wave and bridge rectifiers are the same. The full-wave rectifier requires a center-tapped transformer and uses two diodes. The bridge rectifier does not require a secondary winding that is tapped but does require four diodes. A difference exists in the current waveforms through the secondary of the transformer for the half- and full-wave rectifiers.

Table 12.2
Comparison of Half-wave, Full-wave, and Bridge Rectifiers (Without Filtering)

	Half-wave	Full-wave[1]	Bridge
V_{dc}	$\dfrac{1}{\pi} v_{s(\text{peak})}$	$\dfrac{2}{\pi}\left(\dfrac{v_{s(\text{peak})}}{2}\right)$	$\dfrac{2}{\pi} v_{s(\text{peak})}$
I_{dc}	$\dfrac{1}{\pi} I_{(\text{peak})}$	$\dfrac{2}{\pi} I_{(\text{peak})}$	$\dfrac{2}{\pi} I_{(\text{peak})}$
I_{ac}	$0.385 \ I_{(\text{peak})}$	$0.308 \ I_{(\text{peak})}$	$0.308 \ I_{(\text{peak})}$
PIV	$v_{s(\text{peak})}$	$v_{s(\text{peak})}$	$v_{s(\text{peak})}$
R.F.	1.21	0.48	0.48
no. diodes	1	2	4
freq. (pps)[2]	60	120	120

[1] Secondary winding is center-tapped.
[2] Line frequency assumed to be 60 Hz.

With the full-wave rectifier, the current flow through the top half of the winding and through the bottom half of the secondary is for only one alternation. With the bridge rectifier, because the diodes "switch" the load resistance across the secondary, the current flow through the secondary winding is continuous for both alternations of the output current. The continuous flow of current through the secondary of the transformer in the bridge in comparison with the full-wave rectifier permits less loss in the transformer. The lower loss in the transformer with the bridge rectifier permits the transformer to deliver approximately 12 percent more power to the load. Thus the bridge rectifier is approximately 12 percent more efficient than the full-wave rectifier if both rectifiers use equivalent VA-rated transformers. However, because the bridge rectifier uses four diodes instead of two, the bridge has twice as much power loss due to diode dissipation than does the full-wave rectifier.

FILTERING

A filtering network reduces the ripple voltage (V_{ac}) to a minimum value (ideally zero). Capacitors can be used to filter out the ripple voltage of a power supply because they oppose any voltage change. Figure 12.6 illustrates the waveform across a load resistor R_L for a full-wave or bridge rectifier configuration. Note that if the valleys could be "filled up," the output voltage across R_L would be a smooth, dc value. This filling up can be accomplished simply by placing a capacitor across R_L.

Figure 12.7a shows a capacitor C placed across the load resistor, coupled to a bridge rectifier circuit configuration. The capacitor C charges up to the peak value of the secondary voltage v_s. Between the peaks of the secondary voltage, the capacitor discharges into the load resistor. The time between the peaks is approximately $1/120$ s or 8.33 ms.

For all practical purposes, the time constant of R_L and C is much longer than the 8.33 ms between peaks. Because of the long time constant of R_L and C, the voltage across C decreases linearly (see Figure 12.7b). If we assume that the

Figure 12.6
Output waveforms for a fullwave or bridge rectifier.

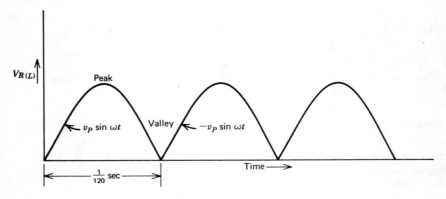

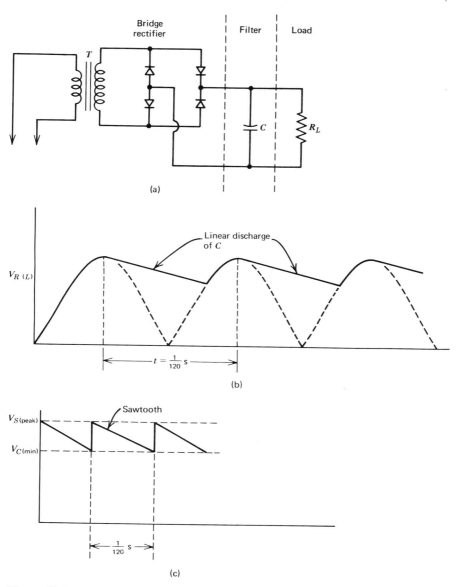

Figure 12.7
Filtering: (a) circuit, (b) waveform across filter capacitor, and (c) approximation of waveform.

capacitor C charges to the peak value of v_s instantaneously, the voltage wave-form across C and the load resistor is a sawtooth with a period of 8.33 ms (see Figure 12.7c).

The minimum voltage across the capacitor C is given by

$$V_{C(\min)} = V_{s(\text{peak})} - \frac{I_{\text{dc}}}{C}t \qquad (19)$$

where I_{dc} is the load current through R_L, C is the value of the capacitor in farads, and t is the period of the sawtooth waveform. Because the voltage across the capacitor is also the output voltage of the power supply, the ripple voltage is equal to the difference between $V_{s(peak)}$ and $V_{c(min)}$, or

$$V_{ripple(p-p)} = V_{s(peak)} - V_{c(min)} \qquad (20)$$

Solving Equation 19 for this difference and letting t equal $\frac{1}{120}$ sec, the ripple voltage (peak-to-peak) is

$$V_{ripple(p-p)} = \frac{I_{dc}}{120\,C} \qquad \text{full-wave or bridge} \qquad (21)$$

Equation 21 is an approximation for the ripple voltage of a full-wave or bridge rectifier using a simple capacitance filter network. The ripple voltage for a half-wave rectifier can be obtained by letting t in Equation 19 be equal to $\frac{1}{60}$ sec, or

$$V_{ripple(p-p)} = \frac{I_{dc}}{60\,C} \qquad \text{half-wave} \qquad (22)$$

The dc output voltage for a rectifying circuit using a simple capacitor filter can be calculated by

$$V_{dc} = V_{s(peak)} - \tfrac{1}{2}V_{ripple(p-p)} \qquad (23)$$

where $\frac{1}{2} \times V_{ripple(p-p)}$ is the dc voltage for a sawtooth waveform. It is important to note the differences between the dc output voltage of a rectifier with and one without a filtering network (Equations 8, 14, and 23).

EXAMPLE

A bridge rectifier is to be used with a 12-V transformer. A capacitor filtering network is to be used with a C of 2000 μF and a dc load current of 0.5 A. Calculate the ripple and dc output voltages under (a) no load and (b) load conditions. For no load conditions $I_{dc} = 0$ mA.

Using Equations 21 and 13, we have

$$V_{ripple(p-p)} = \frac{I_{dc}}{120\,C} = \frac{0\ \text{A}}{120 \times 2 \times 10^{-3}\ \text{F}}$$

$$V_{ripple(p-p)} = 0\ \text{V}$$

$$V_{dc} = V_{s(peak)} - \tfrac{1}{2}V_{ripple(p-p)}$$

$$= 12 \times \sqrt{2} - \tfrac{1}{2}(0)$$

$$V_{dc} = 17\ \text{V}$$

Under full load conditions $I_{dc} = 500$ mA. Using the same equations, we have

$$V_{ripple(p-p)} = \frac{I_{dc}}{120\,C} = \frac{0.5\ \text{A}}{120 \times 2 \times 10^{-3}\ \text{F}}$$

$$V_{ripple(p-p)} = 2.1\ \text{V}$$

$$V_{dc} = V_{s(peak)} - \tfrac{1}{2}V_{ripple(p-p)}$$

$$= 17\ \text{V} - \tfrac{1}{2}(2.1\ \text{V})$$

$$V_{dc} = 15.9\ \text{V}$$

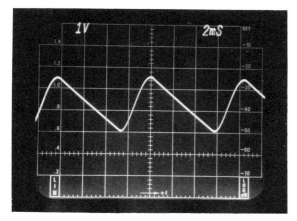

Figure 12.8
Photo of ripple voltage calculated in example problem.

Figure 12.8 illustrates the ripple voltage for the example problem under full load conditions. The ripple voltage is 2.6 $V_{\text{(p-p)}}$, which approximates the ripple value calculated with Equation 21. In the derivation of Equation 21, we assumed that the ripple voltage was an ideal sawtooth wave. But as can be seen in Figure 12.8, the ripple voltage approximates a sawtooth wave.

VOLTAGE DOUBLERS

A voltage doubler circuit configuration has a dc output voltage that is twice the magnitude of the peak value of the ac input voltage. The voltage doubling process is accomplished with diodes and capacitors. Thus, without transformers, the output voltage can be "stepped up" in comparison to the line voltage. Generally, voltage doubling circuits are used for small load currents.

HALF-WAVE DOUBLER

Figure 12.9a illustrates a circuit arrangement for a half-wave voltage doubler. When the input voltage v is positive (Figure 12.9b), diode D_1 is forward-biased and diode D_2 is reverse-biased. With D_1 forward-biased, capacitor C_1 is charged up to the peak value of the input voltage. When the input voltage is negative (see Figure 12.9c), it is additive and in series with the voltage across C_1. With this polarity, D_1 is reverse-biased and D_2 is forward-biased; thus c_2 charges up to twice the peak value of the input voltage. The dc output voltage is

$$V_{\text{dc}} = 2V_{\text{(peak)}} \tag{24}$$

The term *half-wave* for the circuit configuration comes from the fact that the current pulses through C_2 are the line frequency (60 pps).

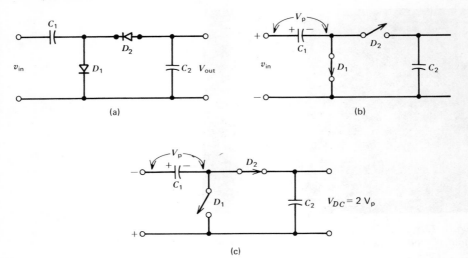

Figure 12.9
Half-wave voltage doubler:·(a) circuit (b) v_{in} positive and v_{in} negative.

FULL-WAVE DOUBLER

Figure 12.10*a* illustrates a circuit arrangement for a full-wave voltage doubler. When the input voltage v is positive (Figure 12.10*b*), diode D_1 is forward-biased and D_2 is reverse-biased. With D_1 forward-biased, capacitor C_1 charges up to the peak value of the input voltage. When the input voltage is negative (Figure 12.10*c*), diode D_2 is forward-biased and D_1 is reverse-biased. With D_2 forward-biased, capacitor C_2 charges up to the peak value of the input voltage. These two voltages across C_1 and C_2 are in series and additive. The output voltage across the load resistor is twice the peak input voltage or

$$V_{dc} = 2V_{(peak)} \tag{25}$$

The term *full-wave* for this circuit configuration comes from the fact that current pulses through C_1 and C_2 are twice the line frequency (120 pps).

VOLTAGE QUADRUPLER

A voltage quadrupler circuit configuration is shown in Figure 12.11. The output voltage is four times the peak value of the input voltage. As can be reasoned from the figure, the quadrupler is two, half-wave voltage doublers cascaded.

SERIES VOLTAGE REGULATORS

A series voltage regulator makes sure that the output voltage of a power supply remains constant with varying load currents (see Equation 1). The voltage

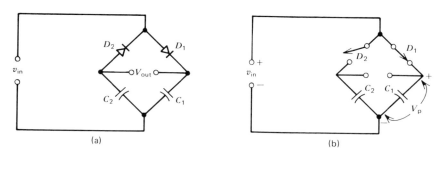

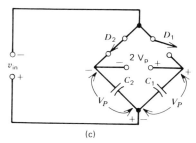

Figure 12.10
Full-wave voltage doubler: (a) circuit, (b) v_{in} positive, and (c) v_{in} negative.

regulator circuit follows the filter network in a power supply. This arrangement is shown in Figure 12.12. The devices used for voltage regulation in a power supply may include a zener diode, operational amplifier, or difference amplifier. As will be demonstrated, the ripple voltage that is across the output terminals of a power supply can be greatly reduced in magnitude by the use of a voltage regulator.

Figure 12.11
Circuit configuration for a voltage quadrupler.

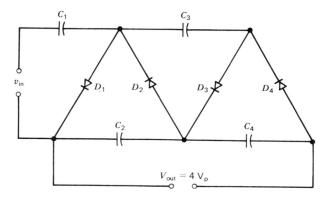

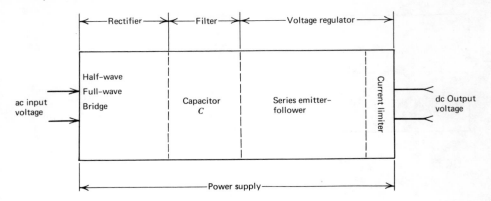

Figure 12.12
Block diagram of a complete power supply.

SERIES EMITTER-FOLLOWER REGULATOR

Figure 12.13*a* illustrates how a zener diode can be coupled with a BJT device to achieve good voltage regulation with a minimum number of components. The transistor Q is connected as an emitter-follower (common collector), its collector connected to the output of the rectifier and its emitter connected to the load resistor. This configuration places Q in series between the rectifier and the load resistor. The zener diode causes a constant voltage to appear between the base of Q and common. The load current for the zener diode is the base current of Q. The base current of Q is approximately the dc load current divided by the h_{FE} of the transistor Q. Resistor R suppleis a constant current to the node point of I_Z and I_B. Resistor R_1 simply provides a path for the emitter of Q if the load is removed from the terminals of the power supply.

As can be determined from the figure, the dc loop equation for the output voltage is

$$V_{\text{out}} = V_z - V_{BE} \tag{26}$$

The power dissipated by Q can be approximated by

$$P_{CE} \cong V_{CE}I_E$$

or

$$P_{CE} = (V_{\text{in}} - V_{\text{out}})I_{\text{load}} \tag{27}$$

Because the regulating transistor Q is connected as an emitter-follower ($A_v \approx 1$), the ripple voltage across the load resistor is approximately the same magnitude as the ripple voltage across the zener diode. The ripple voltage across the zener diode is considerably less than the ripple voltage across the filter network because of the ohmic values of R and R_z, which serve as a voltage divider.

Looking into the base of Q, we can approximate R_T as

$$R_T \approx (h_{FE} + 1)(h_{ib} + R_L) \tag{28}$$

where $h_{ib} \approx 26 \text{ mV}/I_E(\text{mA})$ and R_T is in parallel with the zener resistance R_z. This combination fo $R_T \| R_z$ is in series with R (see Figure 12.13*b*). As can be

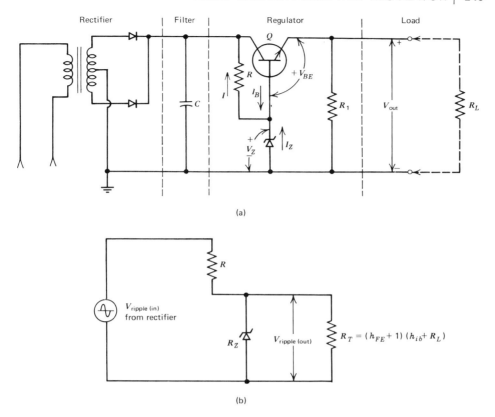

Figure 12.13
Emitter follower voltage regulator (a) circuit configuration, (b) equivalent circuit.

reasoned from the figure, $R_T \| R_z$ and R form a voltage divider. The input voltage for this voltage divider is the ripple voltage across the filtering capacitor C, and the output voltage for the voltage divider is the ripple voltage across the zener diode and the load resistor for the power supply. Using the voltage divider theorem, we have

$$V_{\text{ripple(out)}} = V_{\text{ripple(in)}} \left[\frac{R_z \| R_T}{R + R_z \| R_T} \right] \qquad (29)$$

where $V_{\text{ripple(in)}}$ is the ripple voltage across the filter capacitor C and $V_{\text{ripple(out)}}$ is the ripple voltage across the load resistor for the power supply.

EXAMPLE

Design a power supply for a dc output voltage of 5 V that is capable of delivering 500 mA. The transformer to be used is a 26.8 VCT unit rated at 1.0 A. The regulating transistor Q is a silicon unit with a h_{FE} of 85 and a power rating of 25 W(HEP 703). Assume a V_{BE} of 0.6 V, a full-wave rectifier configuration, and a filter C of 2000 μF.

The voltage rating for the zener diode can be calculated with Equation 26, or

$$V_{\text{out}} = V_z - V_{BE}$$

$$5 \text{ V} = V_z - 0.6 \text{ V}$$

$$V_z = 5.6 \text{ V}$$

A HEP z0212 5.6-V zener diode can be used (see the Appendix for specification sheets). This zener has an R_z of 15 Ω and V_z of 5.6 V at a zener current of 20 mA. The maximum base current I_B for Q is found by

$$I_B = \frac{I_{\text{load}}}{h_{FE}} = \frac{500 \text{ mA}}{85}$$

$$I_B = 5.9 \text{ mA}$$

The current I through R is the sum of I_B and I_z, or

$$I = I_B + I_z$$

$$= 5.9 \text{ mA} + 20 \text{ mA}$$

$$I = 25.9 \text{ mA}$$

The output voltage of the rectifier with the filter capacitor C is found with Equations 21 and 23, or

$$V_{\text{in}} = V_{\text{dc}} = V_{s(\text{peak})} - \tfrac{1}{2}V_{\text{ripple}}$$

$$= 26.8 \text{ V} \times \tfrac{1}{2} \times \sqrt{2} - \tfrac{1}{2} \times 2.1 \text{ V}$$

$$V_{\text{in}} = V_{\text{dc}} = 18 \text{ V}$$

The value of R is calculated by

$$R = \frac{V_{\text{in}} - V_z}{I}$$

$$= \frac{18 \text{ V} - 5.6 \text{ V}}{25.9 \text{ mA}}$$

$$R = 479 \ \Omega \qquad (\text{Use } 470 \ \Omega)$$

The value of R_1 is not critical. If the load is disconnected from the power supply, the emitter current of Q should be a couple milliamperes, or

$$R_1 = \frac{V_{\text{out}}}{2 \text{ mA}} = \frac{5 \text{ V}}{2 \text{ mA}}$$

$$R_1 = 2.5 \text{ k}\Omega \qquad (\text{Use } 2.7 \text{ k}\Omega)$$

The power dissipated by Q is found with Equation 27, or

$$P_{CE} = (V_{\text{in}} - V_{\text{out}})I_{\text{load}}$$

$$= (18 \text{ V} - 5 \text{ V}) \times 0.5 \text{ A}$$

$$P_{CE} = 6.5 \text{ W}$$

The maximum power dissipated by the zener diode occurs when I_{load} is zero, or $I_{z(\text{max})} = 25.9$ mA. Using Equation 3 in Chapter 2, we have

$$P_{z(\text{max})} = I^2_{z(\text{max})}R_z + V_z I_{z(\text{max})}$$

$$= (25.9 \text{ mA})^2 \times 15 \ \Omega + 5.6 \text{ V} \times 25.9 \text{ mA}$$

$$P_{z(\text{max})} = 160 \text{ mW}$$

The HEP z0212 zener had a wattage rating of 500 mW, which is more than adequate for this power supply.

The ripple voltage across the load resistor R_L can be approximated with the use of Equations 28 and 29. The value of R_L is 5V/0.5 A or 10 Ω. Using Equation 28, we have

$$R_T = (h_{FE} + 1) \left(\frac{26 \text{ mV}}{I_E(\text{mA})} + R_L \right)$$

$$= 86 \times \left(\frac{26}{500} \ \Omega + 10 \ \Omega \right)$$

$$R_T = 850 \ \Omega$$

Using Equation 29, we have

$$V_{\text{ripple(out)}} = V_{\text{ripple(in)}} \left(\frac{R_z \| R_T}{R + R_z \| R_T} \right)$$

$$= 2.1 \text{ V} \left(\frac{15 \ \Omega \| 850 \ \Omega}{470 \ \Omega + 15 \ \Omega \| 850 \ \Omega} \right)$$

$$V_{\text{ripple(out)}} = 63 \text{ mV}_{(\text{p-p})}$$

The example circuit was constructed in the laboratory as shown in Figure 12.14. Figure 12.14b illustrates the changes in output voltage with load current changes. Note that the ripple voltage was measured to be 31 mV. Apparently, the zener diode had an R_z of less than 15 Ω. The 15 Ω given by the manufacturer was a maximum value. The regulating transistor Q was used with a heat sink (see Chapter 14).

OP AMP VOLTAGE REGULATOR

An operational amplifier (OP AMP) can be used as a voltage regulator. The OP AMP is a high, open loop gain amplifier with inverting and noninverting inputs (see Chapter 15). Figure 12.15 illustrates a circuit arrangement using an OP AMP to control the series regulating transistor Q. The one input to the OP AMP is the reference voltage of the zener diode; the other input to the OP AMP is the feedback voltage across resistor R_2. This feedback voltage monitors the output voltage for the supply. The closed loop gain for the OP AMP with this type of feedback is given by

$$A_{vf} = \left(1 + \frac{R_1}{R_2} \right)$$

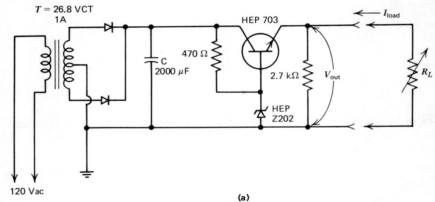

(a)

$I_{L(mA)}$	$V_{out\ (volts)}$	$V_{ripple\ (out)}$
500	5.08	31 mV
400	5.10	—
300	5.13	—
200	5.15	—
100	5.17	—
0	5.30	0 mV

(b)

Figure 12.14
Example problem: (a) circuit and (b) table showing regulation.

where $A_{vf} \approx V_{out}/V_z$ (see Chapter 15). Substituting A_{vf} into the above equation, we have

$$V_{out} = V_z \left(1 + \frac{R_1}{R_2}\right) \tag{30}$$

The two major requirements for the OP AMP are that it can supply the base current needed by Q and that the output voltage is less than the supply voltage of the OP AMP.

EXAMPLE

Design an OP AMP regulator that will supply 9 V dc with a load current of 500 mA. The transformer to be used is a 26.8 VCT unit. The regulating transistor is a HEP 703. The rectifier configuration is full-wave and the filter capacitor C is a 2000-μF unit. The OP AMP to be used is a 741C. A 5.6-V zener is to be used [$V_{z(measured)} = 5.67V$]. Thus Equation 30 can be solved for the quantity $(1 + R_1/R_2)$, or

$$V_{out} = V_z \left(1 + \frac{R_1}{R_2}\right)$$

$$9\text{ V} = 5.67 \left(1 + \frac{R_1}{R_2}\right)$$

$$0.59 R_2 = R_1$$

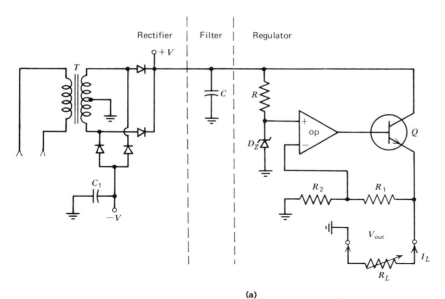

(a)

Components	$I_{L(\text{mA})}$	$V_{\text{out (volts)}}$
$R = 4.7 \text{ k}\Omega$	500	8.96
$R_1 = 2.7 \text{ k}\Omega$	400	8.97
$R_2 = 4.7 \text{ k}\Omega$	300	8.98
$C = 1000 \ \mu\text{F}$	200	8.98
$C_1 = 100 \ \mu\text{F}$	100	8.99
$op = 741 \text{ C}$	0	9.00
$T = 26.8 \text{ VCT, 1 A}$		

(b)

Figure 12.15
OP AMP *used as regulator:* (a) *circuit and* (b) *table showing regulation.*

If R_2 is 4700 Ω, then R_1 is equal to

$$0.59 \times 4700 = R_1$$

$$R_1 = 2770 \ \Omega \qquad (\text{Use } 2700 \ \Omega)$$

Figure 12.15b lists the components for the OP AMP regulator and the output voltage versus load current. Note the very good regulation.

DISCRETE DIFFERENCE AMP REGULATOR

Figure 12.16 shows a circuit configuration for a discrete difference voltage regulator. Transistor Q_1 functions as a difference amplifier. The base voltage of Q_1 is the feedback voltage across resistor R_2, or

$$V_{R2} = V_{\text{out}} \left(\frac{R_2}{R_1 + R_2} \right) \qquad (31)$$

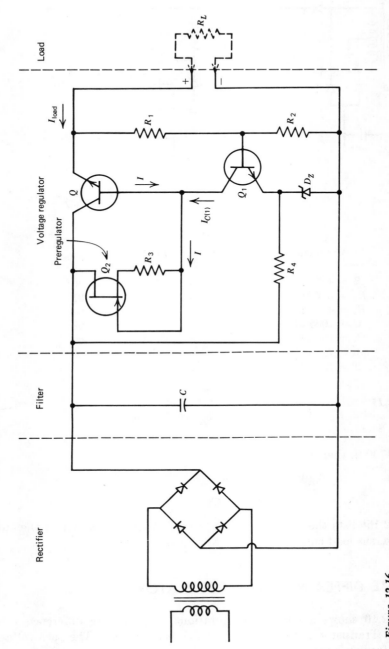

Figure 12.16
Difference voltage regulator.

The emitter voltage of Q_1 is the voltage of the zener diode. Transistor Q_1 amplifies the difference of these two voltages. The dc loop equation for the emitter-to-base loop of Q_1 is

$$V_z + V_{BE} = V_{R2}$$

$$V_z + V_{BE} = V_{out}\left(\frac{R_2}{R_1 + R_2}\right) \tag{32}$$

Solving Equation 32 for the output voltage, we have

$$V_{out} = (V_z + V_{BE})\left(1 + \frac{R_1}{R_2}\right) \tag{33}$$

The JFET pump Q_2 supplies a constant current to the base of Q and collector of Q_1 (see Chapter 10). The R_4 supplies current to the zener diode to make sure that it will remain in its zener region.

If the output voltage were to momentarily increase (if the load were removed), V_{R2} would increase. With V_{R2} increased, Q_1 conducts more current. With Q_1 conducting more current, less current is supplied to the base of Q because of the constant current supplied by the JFET. Less base current supplied to Q causes it to conduct less, bringing the output voltage back to normal. A momentary decrease in output voltage (a load applied) causes the base current of Q to increase. This increase in base current causes Q to conduct more, causing the output voltage to be brought up to normal. A current pump circuit configuration in a regulator circuit is called a "preregulator."

CURRENT LIMITERS

Figure 12.17a illustrates a simple circuit configuration for a current limiter. Resistor R_5 is in series with the load resistor of the power supply. The voltage drop across R_5 is related to the magnitude of load current that the supply is delivering ($V = IR$). With normal load current, the voltage drop across R_5 is not sufficient to cause transistor Q_3 to become forward-biased. If the load current were to increase beyond its normal value, the voltage across R_5 would increase to cause Q_3 to become forward-biased. The forward-biased Q_3 pulls current away from the base of the output transistor regulator Q. This action causes the output current delivered by the supply to be limited. If R_5 were made variable, the current limit of the supply could be varied (see the Review Questions).

Figure 12.17b illustrates how diodes can be used as a current limiter for a power supply. When the voltage drop across R_5 is sufficient to cause the diodes to become forward-biased, the regulator Q is limited to the magnitude of output current it can supply (see the Review Questions). Figure 12.17c shows the relationship between the output voltage and the output current for a power supply with a current limiter.

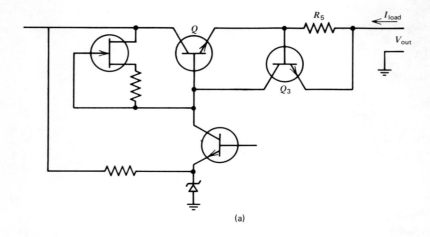

(a)

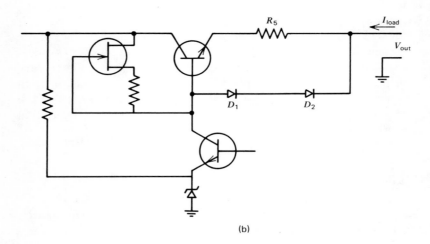

(b)

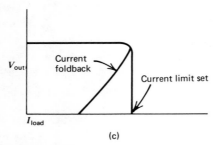

(c)

Figure 12.17
Current limiters: (a) BJT, (b) diodes, and (c) characteristics.

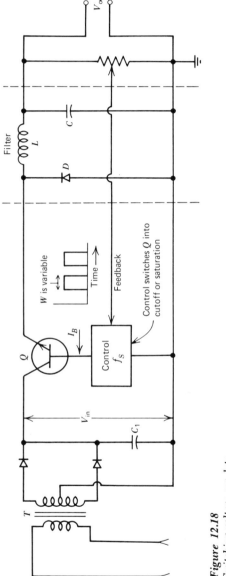

Figure 12.18
Switching voltage regulator.

SWITCHING VOLTAGE REGULATOR

Figure 12.18 illustrates the functional block diagram for a switching voltage regulator. Unlike the series voltage regulators discussed so far, the regulating transistor Q is switched between the saturation state and cutoff state. Because the transistor is switched between cutoff and saturation, the power dissipated by the transistor is very low (ideally zero). Thus the switching regulator is much more efficient than the conventional series regulator whose power dissipation, which is of no benefit, is approximately $(V_{in} - V_{out})I_{load}$.

The output voltage for the switching regulator is found by calculating the average or dc value for the waveform at the emitter of the regulating transistor Q, or

$$V_{out}(t_{on} + t_{off}) = V_{in}t_{on}$$

or

$$V_{out} = V_{in}\left(\frac{t_{on}}{t_{on} + t_{off}}\right) \qquad (34)$$

where $t_{on} + t_{off} = 1/f_s$.

The switching frequency f_s is generally between 10 kHz and 50 kHz. This high switching frequency as compared to the line frequency (60 Hz) enables the filtering network to be built using small values of inductors and capacitors. The filtering network smooths out the pulsating direct current. The inductor L helps to prevent any sudden changes in current, while the capacitor helps to prevent any sudden changes in voltage. The diode D suppresses the voltage spikes generated by the inductor when the transistor Q is being switched from saturation to cutoff ($e = -Ldi/dt$).

Voltage regulation is obtained by varying the duty cycle of the switching frequency, which in turn varies the output voltage (Equation 34). The output voltage is sampled and fed back to the control network. Changes in this feedback voltage cause the duty cycle (pulse width) to change accordingly. The duty cycle generally can be changed from 10 to 90 percent. It is important to note that the duty cycle changes, not the frequency f_s.

REVIEW QUESTIONS AND PROBLEMS

12-1 Discusses the differences between the half-wave and full-wave rectifier circuits.

12-2 Why is the full-wave rectifier easier to filter than the half-wave rectifier circuit?

12-3 What relationship exists between the value of the filtering capacitor C and the magnitude of the ripple voltage?

12-4 With solid-state devices, why should the power supply be well regulated and have a low value of ripple voltage?

12-5 What is the difference between the current limiter and the foldback current limiter?

12-6 Demonstrate that a switching regulator is more efficient than a series regulator.

12-7 What value of filter capacitor is required to insure a maximum ripple voltage of $0.2\ V_{\text{p-p}}$ if the dc voltage is 12 V with a load current of 500 mA?

12-8 Design a power supply having an output voltage of 9 V with a maximum output load current of 750 mA. Calculate the value and power rating of all components. This power supply is shown in the following diagram.

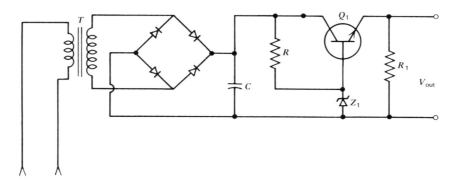

12-9 Repeat Problem 12-8 for the following power supply.

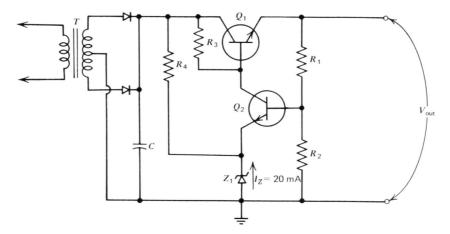

12-10 Derive the value of the limiter resistor used with the diode current limiter shown in Figure 12.17b in terms of the diode barrier voltage and limiting current.

12-11 How can two regulated power supplies be connected to supply both a positive and negative output voltage (example: ±15 V)?

12-12 Design an OP AMP voltage regulator that will deliver 9 V using a reference voltage of 3.6 V.

OBJECTIVES FOR CHAPTER 13, THYRISTORS

At the completion of this chapter you can demonstrate your understanding of the material by answering these questions:

- What is a thyristor?
- What is a SCR and how does it function?
- What is a PUT and how does it function?
- What is a TRIAC and how does it function?
- What is a PNPN switch?
- What are some dc and some ac applications of the thyristor?

13 THYRISTORS

Solid-state devices that belong to the thyristor family are gaining popularity in circuit applications because of their reliability, size, and efficiency. You should study thyristors in order to comprehend such circuit applications as oscillators, power controls, alarm systems, and switching circuitry.

Thyristor is a name given to a group of solid-state devices that includes silicon controlled rectifiers (SCR), silicon controlled switches (SCS), programmable unijunction transistors (PUT), diacs, triacs, and PNPN switches. Thyristors in circuit applications are used as switches, being "on" in the saturated state or "off" in the cutoff state. Because the SCR is more commonly known than a PNPN switch, the SCR is discussed first in this chapter.

SILICON CONTROLLED RECTIFIER

The SCR, as its name implies, is a rectifier that can be controlled or turned on by a gate pulse. The SCR is a unidirectional (passes current in one direction), a three terminal (consists of an anode, gate, and cathode), and a reverse blocking (can be made not to conduct with a positive anode voltage) thyristor. If the SCR is turned on, the gate loses its control over the device. In the on-state the device behaves as a conventional rectifier.

Figure 13.1 shows the schematic symbol for the SCR and its characteristic curves. From the figure it can be seen that the SCR is a current-controlled device. In order for the device to conduct, the anode has to have a positive voltage with respect to the cathode. The magnitude of the anode-cathode voltage (V_{AK}) determines the amount of gate current that will turn on the device. Once the device is turned on (≈ 1 V across the device), the SCR has very low internal resistance and the gate has no further effect on the device. When the V_{AK} voltage

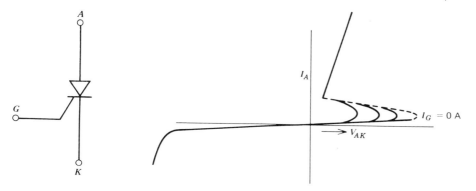

Figure 13.1
Symbol and characteristic curve for the SCR.

is negative, the device is turned off and has a very high internal resistance. As can be seen from the curves, if the SCR has a very large reverse voltage across it, the device will break down.

CIRCUIT MODEL

To understand the various methods of turning the SCR "on" and "off", a circuit model will be developed. The SCR is a PNPN device with one gate, as shown in Figure 13.2. The PNPN device can be thought of as two bipolar transistors, one a PNP and the other a NPN. The anode current of the SCR can now be developed. Referring to Figure 13.2, we have

$$I_A = I_{C(1)} + I_{C(2)}$$

and

$$I_{B(1)} = I_{C(2)}, \quad I_{B(2)} = I_{C(1)}$$

Figure 13.2
Equivalent circuit for the SCR.

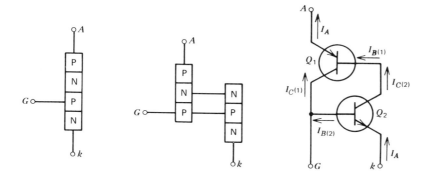

Calculating $I_{C(1)}$ and $I_{C(2)}$, we have

$$I_{C(1)} = h_{FE(1)}I_{B(1)} + (h_{FE(1)} + 1)I_{CO(1)}$$

or

$$I_{C(1)} = h_{FE(1)}I_{C(2)} + (h_{FE(1)} + 1)I_{CO(1)} \qquad (1)$$

and

$$I_{C(2)} = h_{FE(2)}I_{B(2)} + (h_{FE(2)} + 1)I_{CO(2)}$$

or

$$I_{C(2)} = h_{FE(2)}I_{C(1)} + (h_{FE(2)} + 1)I_{CO(2)} \qquad (2)$$

Equations 1 and 2 are two simultaneous equations relating $I_{C(1)}$ and $I_{C(2)}$, or

$$I_{C(1)} - h_{FE(1)}I_{C(2)} = (h_{FE(1)} + 1)I_{CO(1)}$$

$$-h_{FE(2)}I_{C(1)} + I_{C(2)} = (h_{FE(2)} + 1)I_{CO(2)}$$

Solving the above two equations for $I_{C(1)}$ and $I_{C(2)}$, we have

$$I_{C(1)} + I_{C(2)} = I_A$$

or

$$I_A = \frac{(h_{FE(1)} + 1)(h_{FE(2)} + 1)(I_{CO(1)} + I_{CO(2)})}{1 - h_{FE(1)}h_{FE(2)}} \qquad (3)$$

TURNING ON THE SCR

Equation 3 was developed assuming that there was no gate current. From Equation 3, I_A is maximum if the denominator is zero, or $h_{FE(1)}h_{FE(2)} = 1$. From Figure 13.1 it can be seen that with $I_G = 0$ A and V_{AK} large in magnitude, the SCR will conduct. If V_{AK} is made large, the h_{FE} of the individual transistors will vary depending on $I_{CO(1)}$ and $I_{CO(2)}$ whose magnitude is small for silicon. A point in V_{AK} is reached ($I_G = 0$ A) that will cause the product of $h_{FE(1)}h_{FE(2)}$ to be equal to 1. This condition will cause the SCR to be turned on. It is important to note that the I_A current, which is the external load current (I_L), is limited only by the external resistance in the circuit.

The second method of turning on the SCR with the correct polarity of V_{AK} is by supplying gate current. In Figure 13.2, if an extenal positive gate current is supplied to the base of Q_2, $I_{C(2)}$ will increase. If $I_{C(2)}$ increases, $I_{B(1)}$ will increase. But $I_{B(1)}$ increases $I_{C(1)}$ and will cause $I_{B(2)}$ to further increase. This regeneration feedback will cause Q_1 and Q_2 to become saturated quickly, causing the device to be in the on-state. The gate current can then be removed.

The third method of turning on the SCR is with a sudden change in V_{AK}. The junctions of the SCR have capacitance. Because the capacitance has the ability to store a charge, a sudden change in the voltage across the capacitance will generate a current. This current is given by $i_c = C\,\Delta V_{AK}/\Delta t = C\,dv/dt$

The current i_c is generated in the bases of the transistors Q_1 and Q_2. The i_c current will cause the regeneration to occur, and the SCR will be turned on. Because the capacitance (C) is small in the junction regions, a large dv/dt is required to turn on the SCR. The dv/dt is generally listed in the manufacturers' specification sheets.

Once the SCR is turned on, a minimum load current $(I_L = I_A)$ is needed to keep the device in the saturated state. This minimum current is called the holding current or I_H. Once the SCR is turned on, the load current should be greater than the holding current to keep the device in the on-state $(I_L > I_H)$.

TURNING OFF THE SCR

One method commonly used to turn off the SCR (once it has been turned on) is to reduce the voltage across the device (V_{AK}) to zero volts. This zero voltage causes the device to go into cutoff.

A second method (very similar to the first) to turn off a SCR is to reverse the polarity of the voltage across the device (V_{AK}). This reverse voltage sould not exceed the reverse breakdown voltage of the device (see Figure 13.1). This method is commonly used to turn off the device when the device is used in ac circuit applications. If the source voltage is an ac power source (60 Hz), the device is turned off on every negative alternation.

The third method of turning off the SCR is to reduce the load current to a value less than the holding current.

MANUFACTURER'S LISTING FOR SCR

From Table 13.1 commonly used voltage, current, and gate definitions can be found. It is suggested that the reader carefully look at each definition to acquaint himself with the symbols and the definitions of these symbols.

Table 13.1

Definitions of Terms and Symbols for Silicon Controlled Rectifiers

---

Principal Voltage Definitions

Repetitive peak reverse voltage—V_{RROM} [Formerly v_{RM} (rep)]—The maximum instantaneous value of reverse voltage that occurs across a thyristor, including all repetitive transient voltages, but excluding all nonrepetitive transient voltages with the gate open.

Repetitive peak off-state voltage—V_{DROM} [Formerly V_{FBOM} (rep)]—The maximum instantaneous value of off-state voltage that occurs across a thyristor, including all repetitive transient voltages, but excluding all nonrepetitive transient voltares which will not cause switching from the off-state to the on-state with the gate open.

Breakover voltage—$V_{(BO)O}$ (Formerly v_{BOO})—The value of positive principal voltage at the breakover point with the gate open and at specified conditions of junction temperature.

Forward off-state voltage—V_{DO} (Formerly V_{FBO})—The value of positive off-state voltage applied between anode and cathode with the gate open.

Reverse voltage—V_{RO} (Formerly V_{RBO})—The value of negative voltage applied between anode and cathode with the gate open.

Instantaneous on-state voltage—v_T (Formerly v_F)—The instantaneous value of positive principal voltage when the thyristor is in the on-state at a given instantaneous current.

(Continued)

Table 13.1

(Continued)

Critical rate of rise of off-state voltage—Critical dv/dt—The maximum value of the rate of the rise of positive principal voltage that will not cause switching from the off-state to the on-state under specified conditions.

Principal Current Definitions

Average on-state current—$I_{T(AV)}$ (Formerly I_{FAV})—The average value of the principal current when the thyristor is in the on-state.

RMS on-state current—$I_{T(RMS)}$ (Formerly I_{FRMS})—The rms value of the principal current when the thyristor is in the on-state.

Peak surge (nonrepetitive) on-state current—I_{TSM} [Formerly $I_{FM(surge)}$]—An overload on-state current of specific time duration and peak value that may be conducted through the thyristor for one half-cycle from a 60-Hz supply in a single-phase circuit with a resistive load. The thyristor shall be operating within its specified operating voltage, average on-state current, gate power, and temperature ratings prior to the surge current. The surge current may be repeated after sufficient time has elapsed for the device to return to presurge thermal equilibrium conditions.

Critical rate of rise of on-state current—Critical di/dt—The maximum value of the rate of rise of on-state current that a thyristor can withstand under specified conditions.

Peak forward off-state current—I_{DOM} (Formerly I_{FBOM})—The peak value of the forward principal current when the thyristor is in the off-state with the gate open.

Peak reverse blocking current—I_{RROM} (Formerly I_{RBOM})—The peak value of the principal current when the thyristor is in the reverse blocking state with the gate open.

Gate Definitions

dc gate-trigger voltage—V_{GT}—The value of gate voltage required to produce the gate trigger current under specified conditions.

dc gate-trigger current—I_{GT}—The minimum value of gate current required to switch a thyristor from the off-state to the on-state under specified conditions.

Peak gate power dissipation—P_{GM}—The maximum instantaneous value of gate power that may be dissipated between the gate and cathode for a specified time duration.

Miscellaneous Definitions

Gate-controlled turn-on time—t_{gt} (Formerly t_{on})—The time interval between the 10 percent point at the beginning of the gate-trigger voltage pulse and the instant when the principal current has risen to the 90 percent point of its peak value during switching of the thyristor from the off-state to the on-state by a gate pulse.

Circuit-commutated turn-off time—t_q (Formerly t_{off})—The time interval between the instant when the principal current has decreased to zero after external switching of the principal voltage circuit, and the instant when the thyristor is capable of supporting a given principal voltage without turning on under specified conditions.

Note: These terms and symbols follow the latest recommended standards of JEDEC. "JEDEC Suggested Standard No. 7 on Thyristors, April 1967." This standard may be purchased from EIA, Engineering Department, 2001 Eye St., N.W., Washington, D.C. 20006. For convenience, formerly used symbols have been cross-referenced to the new standards.

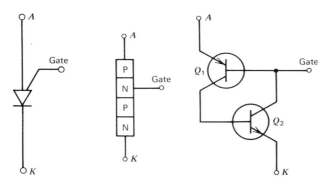

Figure 13.3
Equivalent circuit for the PUT.

PROGRAMMABLE UNIJUNCTION TRANSISTOR

The programmable unijunction transistor (PUT) is another PNPN device with the gate at the N junction (the SCR has its gate at the P junction). Figure 13.3 shows the schematic symbol for the PUT with its circuit model. Like the SCR, once the PUT is in the on-state, its gate loses control. To turn off the PUT, the voltage across the device has to be reduced to zero volts or the current through the device has to be made some minimum value. The PUT, unlike the SCR, is generally used as a timing device.

Figure 13.4 shows a common-circuit configuration for the PUT. The voltage at the gate (V_G) is determined by the voltage dividers R_1 and R_2. When the emitter voltage (V_E) is approximately equal to the voltage across R_2 (V_G is negative with respect to V_E), the PUT will go from the off-state to the on-state. Once the PUT is turned on, the gate loses its control. To turn off the PUT, the voltage across the device has to be reduced to zero volts.

Figure 13.5 shows a diagram of a PUT oscillator. The voltage across C builds up exponentially or $v_c = V(1 - e^{-t/RC})$. If V_E is less than the voltage at the gate (V_G), the PUT is in the cutoff state. When V_E is approximately equal to V_G, the PUT is turned on (low resistance) and causes C to be discharged through the PUT. The discharging of C is very fast and v_c is reduced to approximately zero volts. Zero volts across the PUT will cause it to go into the cutoff state, in which the PUT has very high resistance and v_c starts to build up exponentially again. The timing waveform V_E is shown in Figure 13.5.

Figure 13.4
Triggering the PUT.

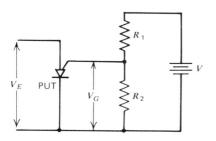

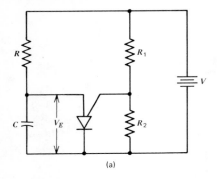
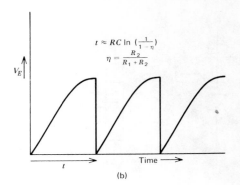

Figure 13.5
A PUT oscillator.

The voltage across R_2 is given by

$$V_G = V \frac{R_2}{R_1 + R_2}$$

When the V_E voltage reaches the V_G voltage, the PUT fires (turns on). This condition is given by

$$V_E \approx V \frac{R_2}{R_1 + R_2} + V_{\text{diode}} \approx V \frac{R_2}{R_1 + R_2} \tag{4}$$

where V_{diode} is the V_{BE} of Q_1, but

$$V_E = v_c = V(1 - e^{-t/RC}) \tag{5}$$

Substituting Equation 5 into Equation 4, we have

$$V(1 - e^{-t/RC}) = V \frac{R_2}{R_1 + R_2}$$

or

$$t = RC \ln \left(\frac{1}{1 - R_2/(R_1 + R_2)} \right)$$

or

$$t = RC \ln \left(\frac{1}{1 - \eta} \right)$$

where

$$\eta = \frac{R_2}{R_1 + R_2} \tag{6}$$

Equation 6 gives the approximate time t for V_E in Figure 13.5. Because η is defined as $R_2/(R_1 + R_2)$, the time (t) for the oscillator can be controlled by the voltage dividers R_1 and R_2 or by the RC combination.

The name programmable unijunction transistor (PUT) is derived from the fact that η (eta) is determined by the external resistors R_1 and R_2. The timing equation for a unijunction transistor (UJT) oscillator is also given by Equation 6, but the η in the UJT is fixed by the manufacturer (see Chapter 17).

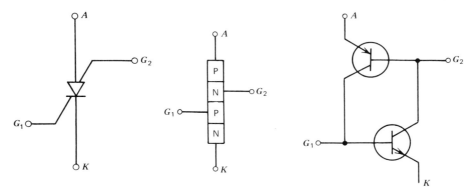

Figure 13.6
Equivalent circuit for the SCS.

SILICON CONTROLLED SWITCH

The silicon controlled switch SCS is a PNPN device with two gates (gate$_1$ and gate$_2$). Figure 13.6 shows the schematic symbol and circuit model for the device. As is true for the SCR and the PUT, if V_{AK} is positive, a plus voltage at gate $_1$ (G_1), with respect to terminal K, will cause the device to be turned on. A negative voltage at gate$_2$ (G_2), with respect to terminal A, will cause the device to be turned on. As with SCR and the PUT, the SCS, having been turned on, can be turned off by the V_{AK} voltage being reduced to zero volts.

PNPN SWITCH

The PNPN switch has no gates. The device breaks down according to Equation 3. The PNPN switch is used primarily as a triggering device. Figure 13.7 shows the schematic diagram and characteristic curve for the device.

Through manufacturing techniques, the PNPN switch can be made to break down (turn on) with a specified voltage V_{AK}. Once the device is turned on, the V_{AK} voltage has to be small to turn off the device. Note that the device in Figure 13.7 is unidirectional (passes current in one direction only).

DIAC

A diac is a bidirectional PNPN switch. The diac passes current in both directions. The device can be thought of as two PNPN switches connected in reverse parallel. Figure 13.8 shows the schematic symbol and the characteristic curves for the diac. If T_1 is positive, conduction takes place by junctions $P_1N_2P_2N_3$. If T_1 is negative, conduction takes place by junctions $N_1P_1N_2P_2$.

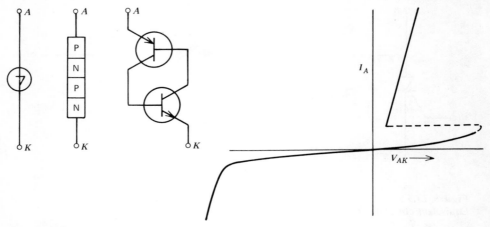

Figure 13.7
Symbol, equivalent circuit and characteristic curve for the PNPN switch.

TRIAC

The triac can be thought of as two SCRs connected in reverse parallel, with one control gate to trigger the two SCRs. Figure 13.9 shows the schematic symbol and characteristic curves for the triac. The triac can be triggered with either positive or negative gate voltages with either positive or negative voltage across the terminals T_1 and T_2. This triggering ability allows the triac to be used in ac application when the load current has to be alternating. As shown in the figure, if T_2 is positive, the junctions for conduction are $P_1N_2P_2N_4$ with a positive gate at P_2 or a negative gate at N_3. If T_2 is negative, the junctions for conduction are $N_1P_1N_2P_2$ with a positive gate at P_2 or a negative gate at N_3.

Figure 13.8
Symbol, construction and characteristic curve for the DIAC.

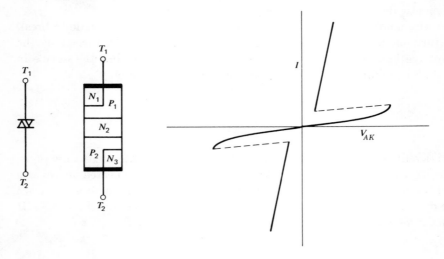

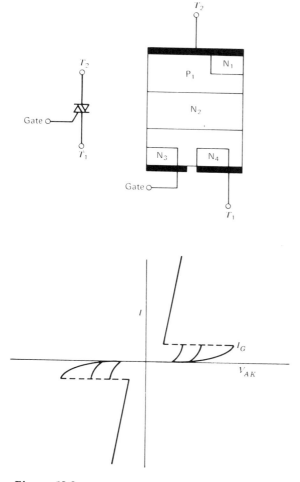

Figure 13.9
Symbol, construction, and characteristic curves for the TRIAC.

dc APPLICATIONS

In Figure 13.10, which shows a SCR used to turn on a lamp, S_1 supplies current through R to the gate of the SCR. Once the SCR is turned on, the device will remain on as long as the current through the lamp is greater than the holding current of the SCR. Thus if S_1 is momentarily depressed, the lamp is turned on. As a point of interest, a bipolar transistor could be used to turn on the lamp, but a constant base current would have to be supplied to keep the lamp on. With the SCR, the gate current is supplied only long enough to cause the SCR to go into conduction.

To turn off the lamp, S_2 is momentarily depressed. This S_2 reduces the V_{AK} of the SCR to zero volt. A bipolar transistor could replace S_2 (Figure 13.10b) as a turnoff mechanism. The capacitor C in the figure is used to suppress any sudden

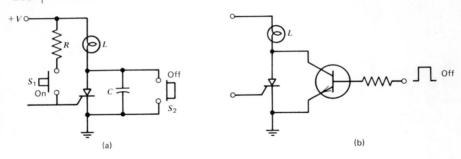

Figure 13.10
Turning the SCR on and off: (a) mechanical switches and (b) with a transistor switch.

changes in the supply voltage (V) that would prematurely turn on the lamp (dv/dt).

Figure 13.11 shows a SCR that has a relay for its load. The diode across the relay is used to suppress the counter electromotive force generated by the relay when the relay is going from its pulled-in state to its dropout state.

The figure shows a simple security system. The loop is constructed from wire. If the loop were broken, the voltage that would be present across R_2 would supply gate current to the SCR and turn it on. The relay contacts could be used to supply current to a bell or some other warning device. The circuit shown would be shut down by removing battery voltage V.

Figure 13.12 shows a dc circuit breaker constructed with SCRs. The "on" switch is momentarily depressed to supply current to the load. With SCR_1 turned on, the load current causes a voltage drop across R_3. The capacitor C is called a commutating capacitor; C charges up to the polarity shown in Figure 13.12. If an overload is generated, the current through R_3 is increased and the voltage drop across R_3 is increased. If the voltage increase across R_3 is large

Figure 13.11
Simple alarm using the SCR

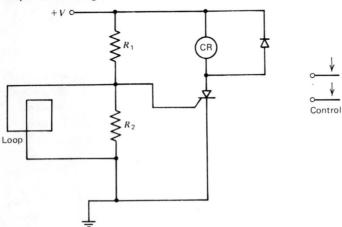

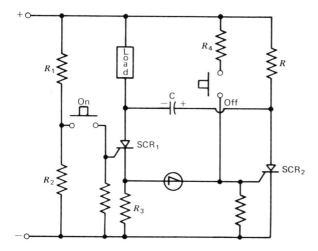

Figure 13.12
Circuit breaker using SCRs.

enough to cause the PNPN switch to trigger, current is supplied to the gate of the second SCR (SCR₂). This current causes SCR₂ to turn on. With SCR₂ turned on, capacitor C is placed across SCR₁ to cause it to be turned off. This cycling action causes the load to be removed from the main dc lines. If the load is to be normally disconnected from the dc line (load current turned off), momentarily depressing the "off" switch will cause the commutating capacitor C to turn off SCR₁.

Capacitor C is selected to make sure that enough reverse current can be generated through SCR₁. Because $I = CV/t$, the value of C is given by

$$C = \frac{1.5 t_{\text{off}} I_{\text{load}}}{V}$$

where V is the dc line voltage, I the load current, and t the turnoff time of SCR₁. Also, the requirement for R in Figure 13.12 (load resistance of SCR₂) is that R be one-tenth the value of the internal leakage resistance of SCR₂. This requirement ensures the commutating capacitor's being placed across SCR₁.

Another interesting dc application of the SCR and a PUT, a sequential flasher, is shown in Figure 13.13. Lamp L_1 lights after the switch S is closed. Lamp L_2 is lit after a set time determined by the PUT timer, which triggers SCR₂. The trigger pulse necessary for the gate of SCR₂ is generated across resistor R_2. This pulse is caused when PUT₂ discharges C_2. The third lamp is turned on by SCR₃, which is triggered by the timer PUT₃. Note that the power for PUT₃ and SCR₃ is applied via lamp L_2. This sequence permits the PUT₃ timer to begin only after lamp L_2 is lit.

Once all the lamps are lit, power has to be removed via switch S and the timing sequence starts over again. Switch S is a thermal breaker that opens after a fixed amount of current (the three lamps) flows through the device.

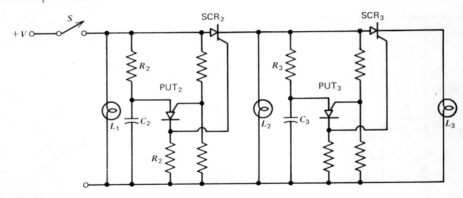

Figure 13.13
Sequencial flasher using PUTs and SCRs.

ac APPLICATIONS

The SCR is very commonly used for power control in ac applications. With the SCRs, power can be either dc or ac current that passes through the load. Figure 13.14 shows a circuit that uses a PUT timer for the triggering of two SCRs. The power delivered to the load is direct current. Figure 13.13a shows the circuit configuration and Figure 13.14b shows the waveforms for the circuit.

Figure 13.14
SCR power control: (a) circuit and (b) waveforms.

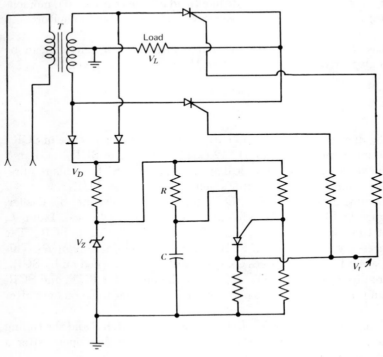

(a)

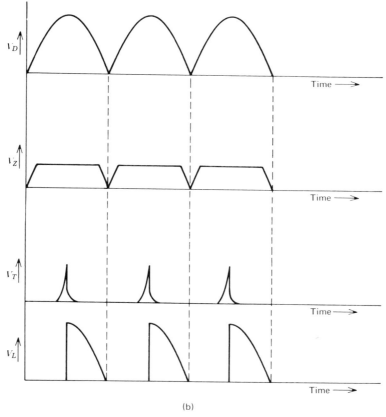

(b)

Figure 13.14 *(continued)*

The voltage applied to the PUT timer is shown as V_Z. Each alternation that is present (V_Z) causes the timer to generate a pulse that is fed to the gates of the SCRs. The length of time, which starts at the beginning of each alternation, that a pulse is generated is determined by the PUT timer. As can be seen from the figure, the triggering pulse from the PUT timer triggers the SCR. Once the SCR is triggered into conduction, power is delivered into the load. By varying the time of the triggering pulses, the power delivered to the load can be varied. This variation in power is shown in Figure 13.14 as V_L.

Because the PUT timer generates a pulse for each alternation, the SCR that is triggered is the one that has the correct polarity across it for firing. Each SCR is turned off, once it is fired, by the alternating voltage across it.

Figure 13.15 shows a SCR being triggered by an ac gating voltage. When a SCR is triggered by an ac voltage, a diode is placed in series with the gate of the SCR. With the positive gate voltage at the gate terminal (gate-to-cathode), the diode is forward-biased. When the gate voltage is negative, the diode becomes reverse-biased. The high resistance of the reverse-biased diode is in series with the gate junction of the SCR. The combination of these two resistances forms a voltage divider. The voltage between gate and cathode when the diode is reverse-

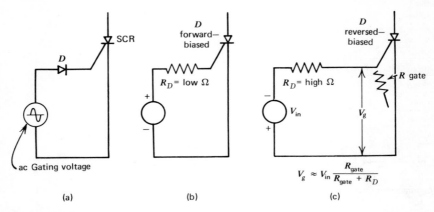

Figure 13.15
Gate protection: (a) circuit, (b) diode forward-biased, and (c) diode reverse-biased.

biased is given in Figure 13.15. If an excessive reverse-bias voltage were to appear across the gate junction of the SCR, the junction would breakdown (zener effect). The excessive current in the gate terminal could destroy the SCR. A diode placed across the gate-cathode of the SCR would also serve to protect the SCR (see Review Questions).

Figure 13.16 shows how a triac and a diac can be used to vary the ac power to a load. The triac can be made to conduct with both positive and negative polarities across it. This ability of the triac will permit ac current in the load. The simple rc phasing network controls the phase relationship between the voltage across the capacitor C and the ac input voltage. The voltage across C is used to trigger the triac. The phase between v_C and the input ac voltage determines when the triac fires. Thus the power delivered to the load can be varied. The diac shown in the circuit insures that the triac is triggered consistently once the value of R is set. Triacs can be purchased with the diac built into the triac.

There are many circuit applications using thyristors. It is suggested that the reader look through one of the many thyristor handbooks published by the manufacturers that produce SCRs, triacs, diacs, SCSs, and so forth. The purpose

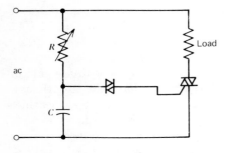

Figure 13.16
ac power control using the TRIAC and DIAC.

of this chapter is to give the reader a general idea of the workings of some of the thyristors manufactured and a few applications.

REVIEW QUESTIONS AND PROBLEMS

13-1   (a) Why is "silicon-controlled rectifier" (SCR) an appropriate name for the device?

(b) Why does the gate of the SCR lose its control over the device when the SCR is in the conduction state?

13-2   Show that the gate current is given by $I_G = \dfrac{V - V_{BE}}{R}$ in the following diagram.

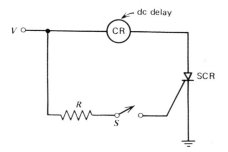

13-3   Show that the gate current shown in Figure 13.11 is given by

$$I_G = \frac{V_{BB} - V_{BE}}{R_B}$$

where

$$R_B = R_1 \| R_2 \quad \text{and} \quad V_{BB} = V \frac{R_2}{R_1 + R_2}$$

when the loop is broken.

13-4   With the SCS, explain the polarity necessary for $V_{G(1)}$ and $V_{G(2)}$ to cause the device to be turned on.

13-5   What is the waveform across the PNPN switch shown in the following figure?

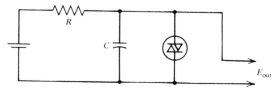

13-6   In the following figures, with the voltages shown, are the SCSs "on" or "off"?

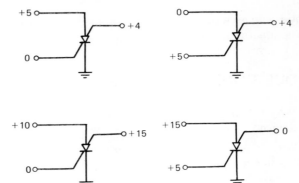

13-7   For the circuit shown, what is the time t for the waveform?

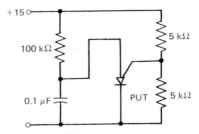

13-8   Sketch the waveform across the diac shown in the following figure.

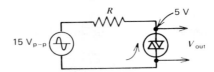

13-9   In the circuit shown in Figure 13.14, what is the range necessary for the PUT timer to permit approximately full and zero power?

13-10   List the various loads that would require (a) ac power control and (b) dc power control.

13-11   Explain why a triac can be equated to two SCRs connected in reverse parallel.

13-12   Design a sequential flasher (Figure 13.13) to operate from a 15-V supply. Assume that the PUT timers have a period of 1 sec. Select the lamps, SCRs, and PUTs and calculate the values for all the resistors and capacitors.

13-13   Explain why the circuits shown in the following figures protect the SCR's gate from breaking down.

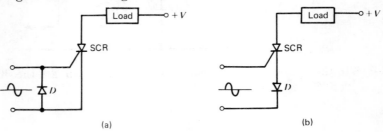

OBJECTIVES FOR CHAPTER 14, MULTISTAGE AND POWER AMPLIFIERS

At the completion of this chapter you can demonstrate your understanding of the material by answering these questions:

- What is meant by coupling?
- What are the various coupling techniques?
- What ia a Darlington Pair?
- What are class A, B, and C amplifiers?
- How is the efficiency of an amplifier related to its class of operation?
- How and why are heat sinks incorporated in amplifiers?
- What is power dissipation?

14 MULTISTAGE AND POWER AMPLIFIERS

Multistage amplifiers are used to obtain large gains. You should study multistage and power amplifiers in order to comprehend the various coupling techniques, classes of amplification, and heat sinks found in such circuit applications as audio-frequency (AF) and radio-frequency (RF) amplifiers.

With many circuit applications, an amplifier must have a high gain. The gain can be power, voltage, current, transconductance, and/or transresistance. A single-stage amplifier is limited in its gain. To increase the gain, two or more single stages can be connected or cascaded to from a multistage amplifier.

THE AMPLIFIER

Figure 14.1a illustrates several single amplifier stages cascaded. The input signal is applied to the first stage A_1 and the output is taken from the third stage A_3. The output of the first stage is A_1 times the input signal. The output of the second stage is equal to the product A_1A_2 times the input signal. The output of the third stage is the product $A_1A_2A_3$ times the input signal. Thus the total gain of the multistage amplifier is A_t and equals $A_1A_2A_3$. This relationship is shown in Figure 14.1b. Frequently, with multistage amplifiers, feedback techniques are incorporated to stabilize the overall gain.

The phase relationship between the input and output signal can be zero or 180° out of phase with each other, depending on the number of stages and the type of stage used (CE, CB, CC, CS, CD, or CG). The stages may be constructed of BJTs, FETs, or BJTs and FETs.

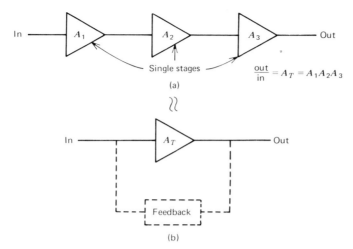

Figure 14.1
Amplifiers: (a) single stages cascaded and (b) total amplifier.

COUPLING

With the cascading of single-amplifier stages, the output of one stage is coupled or fed to the input of another stage. The manner in which the signals are fed from one stage to the other is determined by the type of coupling. The three major types of coupling are (a) direct, (b) *RC*, and (c) transformer.

Direct coupling. Direct coupling simply means that signals from one amplifier are fed to another amplifier directly. Figure 14.2 illustrates several direct-coupled amplifiers using BJTs and FETs. Figure 14.2*a* shows two NPN BJTs

Figure 14.2
Direct-coupled amplifiers: (a) two NPN BJTs, (b) a PNP and a NPN BJT, (c) a FET, NPN and PNP, and (d) difference amplifier.

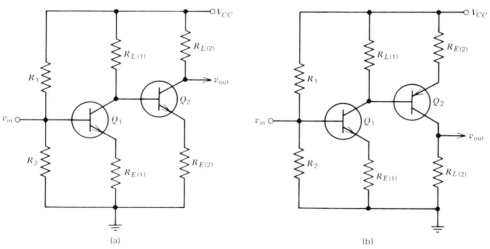

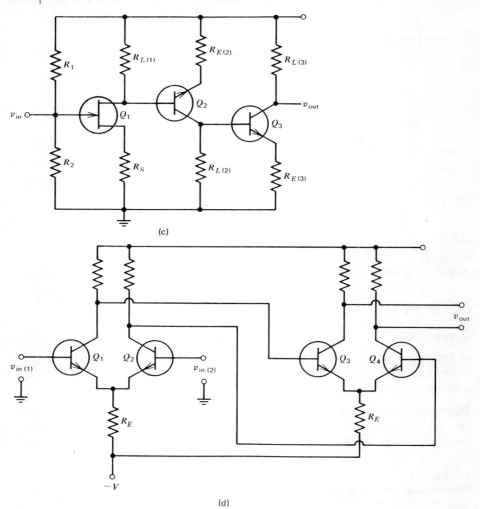

Figure 14.2 *(continued)*

connected. The output of Q_1 is fed to the input base of Q_2. The dc bias voltage requirements of Q_2 are fulfilled by the dc voltage at the collector of Q_1 to common. Notice that both BJT stages are common emitters. The ac load seen by the collector of Q_1 is the aprallel combination of $RL_{(1)}$ and the input resistance of the Q_2 stage. The dc load seen by Q_1 is simply $RL_{(1)}$. Because both amplifiers are common emitters, the input and output voltage is in phase.

Figure 14.2b illustrates two BJT amplifiers connected, one a NPN and the other a PNP. As can be reasoned from the figure, the input and output voltages are in phase.

Figure 14.2c illustrates a direct-coupled amplifier using two BJTs and a JFET. The JFET is used as the input stage of the amplifier to take advantage of its high input resistance. The input and output voltages are 180° out of phase with each other.

One of the major drawbacks of the three types of amplifiers discussed in Figure 14.2a, b, and c is the effect of temperature on the dc operating points of the amplifiers. Because the amplifier can amplify dc signals, any change in the quiescent point due to temperature is amplified. Figure 14.2d illustrates a circuit configuration that eliminates the effects of temperature on the quiescent point and is still direct-coupled. This circuit configuration is called a difference amplifier. Amplifiers Q_1 and Q_2 form a difference amplifier that is directly coupled to another difference amplifier (Q_3 and Q_4). The secret of the difference amplifier is that a change in temperature affects all the transistors in the amplifier.

As the ambient temperature increases, the leakage current of both transistors in the difference amplifier increases. Because the output is taken from both collectors, the difference is zero. For a detailed discussion of difference amplifiers see Chapter 15.

Darlington pair. Figure 14.3a shows the connection of two NPN transistors forming what is commonly referred to as a Darlington pair. The two BJTs function as a single transistor having a total beta equal to the product of the individual betas of both transistors (see the Review Questions). If Q_1 has an h_{FE} equal to 60 and Q_2 has a beta equal to 50, the total beta for the equivalent transistor is 3000. Unfortunately, the leakage current is also multiplied by the high beta. One way to overcome this dilemma is to make sure that both NPN BJTs are silicon units.

Figure 14.3b illustrates a circuit configuration using complementary transistors NPN/PNP to form an equivalent transistor that has a total beta equal to the product of the individual betas (see the Review Questions). With this circuit configuration, the offset voltage for the equivalent transistor is the V_{BE} of Q_2, whereas in the circuit shown in Figure 14.3a, the offset voltage is ($V_{BE(2)}$ + $V_{BE(1)}$). As an example, with the complementary pair, the offset voltage (assuming silicon units are used) would be 0.7 V. With the Darlington pair, the offset voltage would be 1.4 V. Thus the advantage of the complementary pair would would be a lower offset voltage compared to the Darlington pair. However, the disadvantage of the complementary pair is that it requires a NPN and a PNP BJT.

The frequency response of the direct-coupled amplifier extends from dc to some upper frequency limit determined by the internal capacitance of the BJTs or FETs used in the construction of the amplifier. (see Chapter 8). Of course, if feedback techniques are incorporated, the upper-frequency limit of the multistage amplifier can be extended (Chapter 9).

RC coupling. Single-stage amplifiers can be coupled with capacitors. This type of coupling is generally called RC coupling and is illustrated in Figure 14.4a. The output of Q_1 is fed to the input of Q_2 via C. The advantage of RC coupling is that it permits the transfer of only ac signals. Changes in the dc operating points of the various stages are not amplified.

The disadvantage of RC coupling is that it does not permit the amplification of low frequencies or direct current. The upper frequency limit of an RC-coupled amplifier is determined by the internal capacitance of the BJTs and/or FETs. However, the lower-frequency limit of the amplifier is determined by the value

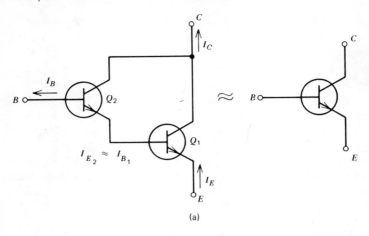

(a)

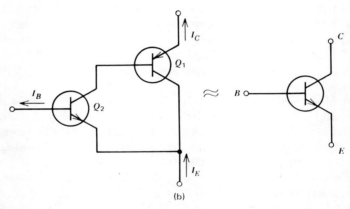

(b)

Figure 14.3
Super beta transistors: (a) Darlington pair and (b) complementary pair.

of the coupling capacitor and the input resistance. Figure 14.4b illustrates the value of the RC coupling network.

The total resistance R is a parallel combination of the dc biasing resistor R_B and the input resistance of Q_2 (h_{ie}). The frequency at which the voltage across R equals the voltage across C is called the half-power point and is equal to -3 dB. The -3 dB point occurs at a frequency that causes the resistance of R to equal the reactance of C, or

$$R = X_C = (\omega C)^{-1}$$

$$f_L = (2\pi RC)^{-1} \tag{1}$$

In order to extend the lower frequency limit for a multistage amplifier, the coupling capacitors are very large in value (μF) because of the low resistance of h_{ie}.

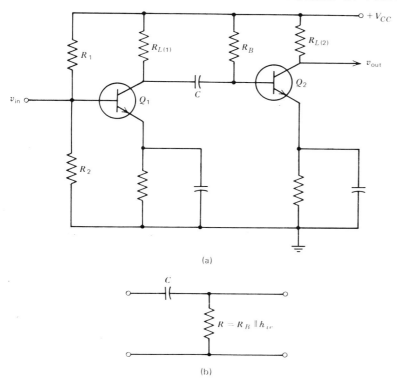

Figure 14.4
RC coupling: (a) circuit configuration and (b) equivalent circuit.

Transformer coupling. Transformers (tuned and untuned) can be used to couple the output of one stage to the input of another stage. With RF amplifiers, tuned or resonated tank circuits are used for coupling. With audio-frequency (AF) amplifiers, untuned transformers are used for coupling. Figure 14.5 illustrates a transformer-coupled multistage amplifier.

With RF amplifiers, a certain band of frequencies is amplified. This band of frequencies is called the bandwidth of the tank circuit (LC). For AF amplifiers, the transformers are untuned. The lower-frequency limit depends on the reactance of the transformer at low frequencies compared to the input resistance of the amplifier. The upper-frequency limit of the untuned AF amplifier is determined by the internal capacitance of the windings of the transformer.

CLASSES OF AMPLIFIERS

The class of an amplifier is defined by the duration of the collector or drain current flow with respect to the input signal. The input signal is assumed to be a full 360°.

Class A. A class A amplifier is defined as having an output current flow of 360°. Figure 14.6a illustrates a class A amplifier using a FET. Note that the

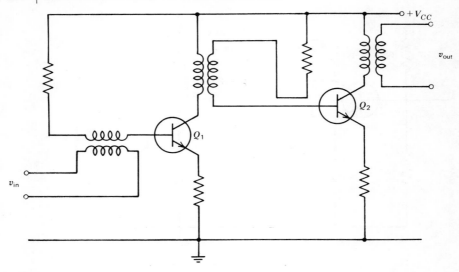

Figure 14.5
Transformer coupling.

drain current flows for a full 360° with respect to the input signal. Class A amplifiers are considered linear amplifiers; that is, the output signal is an exact replica of the input signal. The quiescent point is located in the linear region of the active device.

Class B. A class B amplifier is defined as having an output current flow of 180° with respect to the input signal. Figure 14.6*b* illustrates a class B amplifier using a FET. The drain current flows for approximately 180° with respect to the input signal. The quiescent point is located near the cutoff point. If two class B amplifiers are connected in a specific way, the combination can be linear (see power amplifiers in this chapter).

There are some amplifiers that have operating characteristics somewhere between a class A and a class B amplifier. These amplifiers are called class AB.

Class C. A class C amplifier is defined as having an output current flow of less than 180°. Figure 14.6*c* illustrates a class C amplifier using a FET. The drain current flows for less than 180° with respect to the input signal. The quiescent point is located well below the cutoff point.

EFFICIENCY

The efficiency of an amplifier (ac power output/dc power input) is related to its class of operation. The class A amplifier has the lowest efficiency, whereas the class C amplifier has the highest efficiency. The discussion of class A and B amplifiers for the remaining parts of this chapter assumes that the amplifiers are linear with a frequency response in the audio-frequency (AF) range. However, class A and class B amplifiers can be used for linear radio-frequency (RF)

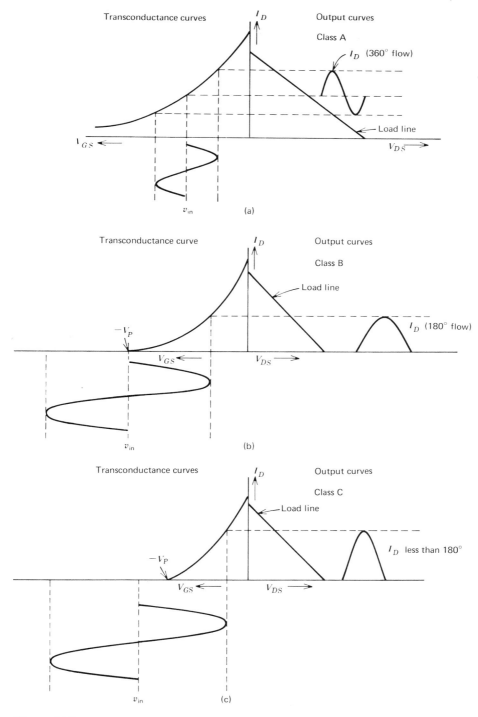

Figure 14.6
Amplifiers: (a) class A, (b) class B, and (c) class C.

amplifiers. The class C amplifier is generally used at only radio frequencies. The load for a class C amplifier is a resonant tank circuit (LC). The current spikes (output current less than 180°) supply energy to the tank circuit. The tank circuit, because of its fly-wheel effect, can supply a sinusoidal waveform.

The efficiency of an amplifier may be defined as its ac output power divided by its dc input power, or

$$\eta = \frac{P_{ac}}{P_{dc}} \tag{2}$$

where eta (η) changes from 0 to 1. The power dissipated (P_D) by the active device in an amplifier can be found by

$$P_D = P_{dc} - P_{ac} \tag{3}$$

or

$$P_D = P_{dc}(1 - \eta) \tag{4}$$

Ideally, an amplifier should convert all of the dc input power into ac output power. Equation 3 tells us that the dc input power not converted into ac output power is wasted. The transistor has to dissipate the difference between the dc and ac power. Thus the higher the efficiency of the amplifier, the greater the amount of dc power that is converted into ac power.

EXAMPLE

Assume that an amplifier has an efficiency of 50 percent. The ac output power is 25 W. Calculate the power dissipated by the active device in the amplifier and the dc input power delivered by the power supply.

The dc input power is found with Equation 2.

$$P_{dc} = \frac{P_{ac}}{\eta} = \frac{25 \text{ W}}{0.5}$$

$$P_{dc} = 50 \text{ W}$$

The power dissipated by the active device (wasted power) is found with Equation 4.

$$P_D = P_{dc}(1 - \eta)$$
$$= 50 \text{ W}(1 - 0.5)$$
$$P_D = 25 \text{ W}$$

The following sections demonstrate that the clsss of an amplifier determines its efficiency. A class A amplifier is the least efficient, whereas the class C amplifier is the most efficient.

CLASS A POWER AMPLIFIER WITH A LOAD RESISTOR

Figure 14.7 illustrates a circuit configuration for a class A amplifier along with its output characteristics curve. With an ideal transistor, the maximum rms of ac output power is equal to

$$P_{ac} = \frac{V_{CC}}{2\sqrt{2}} \times \frac{I_{C(max)}}{2\sqrt{2}}$$

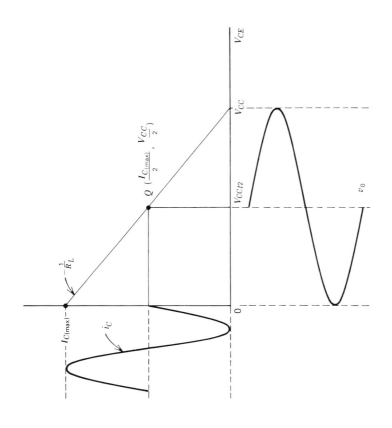

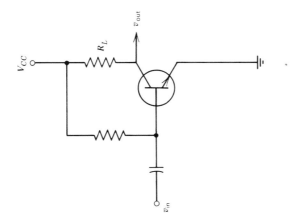

Figure 14.7
Operation of a class A amplifier.

$$P_{ac} = \frac{V_{CC}I_{C(max)}}{8} \tag{5}$$

The dc or average input power for this amplifier is found to be

$$P_{dc} = \frac{V_{CC}I_{C(max)}}{2} \tag{6}$$

With the use of Equation 2, the maximum efficiency of this amplifier, then, is calculated as

$$\eta = \frac{P_{ac}}{P_{dc}} = \frac{V_{CC}I_{C(max)}/8}{V_{CC}I_{C(max)}/2}$$

$$\eta = 0.25 \quad \text{or} \quad 25 \text{ percent}$$

Thus this type of amplifier converts (ideally) only 25 percent of the dc input power into ac output power. One way of increasing the efficiency of this class A amplifier is to elimiate the dc load resistor. This elimination can be accomplished by the output of the transistor being connected to a transformer.

CLASS A POWER AMPLIFIER WITH A TRANSFORMER

Figure 14.8 shows a BJT with a transformer for a load. Note that without an input signal, the dc load line is vertical. With the input signal at the base of the BJT, the ac load line is the result of transformer action. Because of the transformer, the maximum peak-to-peak output voltage is equal to twice the supply

Figure 14.8
Class A transformer amplifier.

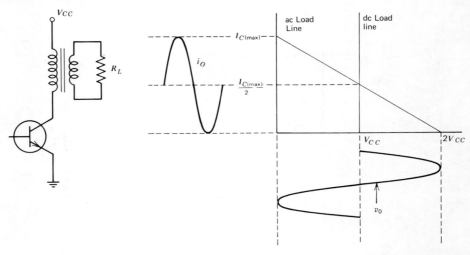

voltage, or $2V_{CC}$. The quiescent current for the BJT is equal to one-half the peak collector current, or $I_{C(max)}/2$. The ac output power is then equal to

$$P_{ac} = \frac{2V_{CC}}{2\sqrt{2}} \times \frac{I_{C(max)}}{2\sqrt{2}}$$

$$P_{ac} = \frac{V_{CC}I_{C(max)}}{4} \tag{7}$$

The dc or average input power for this amplifier is found to be

$$P_{dc} = \frac{V_{CC}I_{C(max)}}{2} \tag{8}$$

With the use of Equation 2, the maximum efficiency (ideally) of this amplifier can then be calculated to be

$$\eta = \frac{P_{ac}}{P_{dc}} = \frac{V_{CC}I_{C(max)}/4}{V_{CC}I_{C(max)}/2}$$

$$\eta = 0.5 \quad \text{or} \quad 50 \text{ percent}$$

The ac resistance seen by the transistor is the reflected resistance of R_L through the turns ratio of the transformer.

EXAMPLE

Assume that a transformer class A amplifier such as the one shown in Figure 14.8 has the following parameters: The transformer T has a turns ratio of $N_P/N_S = 20/1$, $V_{CC} = 15$ V, and $R_L = 8\ \Omega$. Calculate R_{ac}, P_{dc}, and P_D. Assume that the transistor and transformer are ideal devices.

Because of the transformer, the 8-Ω load resistor is reflected to the primary as

$$Z_P = Z_S(N_P/N_S)^2 = 8\ \Omega \left(\frac{20}{1}\right)^2$$

$$Z_P = 3200\ \Omega$$

The peak ac collector current is

$$I_{C(max)} = \frac{2V_{CC}}{R_L'} = \frac{2 \times 15\text{ V}}{3200\ \Omega}$$

$$I_{C(max)} = 9.375\text{ mA}$$

The maximum output voltage is $2V_{CC}$ or

$$2 \times 15\text{ V} = 30\text{ V}$$

The ac output power is calculated with Equation 7.

$$P_{ac} = \frac{V_{CC}I_{C(max)}}{4} = \frac{15\text{ V} \times 9.375\text{ mA}}{4}$$

$$P_{ac} = 35.2\text{ mW}$$

The dc input power is calculated with Equation 8.

$$P_{dc} = \frac{V_{CC}I_{C(max)}}{2} = \frac{15 \text{ V} \times 9.375 \text{ mA}}{2}$$

$$P_{dc} = 70.3 \text{ mW}$$

The power dissipated by the transistor is

$$P_D = P_{dc} - P_{ac}$$
$$= 70.3 \text{ mW} - 35.2 \text{ mW}$$
$$P_D = 35.1 \text{ mW}$$

CLASS B PUSH-PULL AMPLIFIER WITH A TRANSFORMER

Figure 14.9 illustrates a class B push-pull power amplifier. Transistors Q_1 and Q_2 are biased at cutoff. The input signals at the bases of Q_1 and Q_2 are 180° out of phase with respect to each other because of the center tap of the input transformer T_1. If the input signal is positive at the base of Q_1, the base of Q_2 is negative. If the input signal is negative at the base of Q_1, the base of Q_2 is positive. Thus the current through the top half of the output transformer T_2 is caused by the class B amplifier Q_1, and the current through the bottom half of the output transformer by the class B amplifier Q_2. The net result of these two transistors causes the output current through the transformer T_2 to be a complete sinusoidal waveform (if ideal transistors are used).

The total output power is then

$$P_{ac} = \frac{2V_{CC}}{2\sqrt{2}} \times \frac{2I_{C(max)}}{2\sqrt{2}}$$

or

$$P_{ac} = \frac{V_{CC}I_{C(max)}}{2} \qquad (9)$$

The dc or average input power supplied by the power supply is supply voltage V_{CC} times the average current of the two stages. As can be reasoned from Figure 14.9, the current supplied by the power supply is similar to the current across the load resistance for a full-wave rectifier, or $(2/\pi)I_{C(max)}$. Thus the dc input power is

$$P_{dc} = V_{CC} \frac{2}{\pi} I_{C(max))} \qquad (10)$$

Because of the constant voltage V_{CC} and the varying load current, Class B power amplifiers need a well regulated power supply for good linearity.

The maximum efficiency for this push-pull amplifier is then

$$\eta = \frac{P_{ac}}{P_{dc}} = \frac{[V_{CC}I_{C(max)}/2]}{[2V_{CC}I_{C(max)}/\pi]}$$

$$\eta = \frac{\pi}{4} = 0.785 \qquad \text{or} \qquad 78.5 \text{ percent}$$

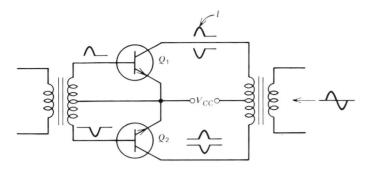

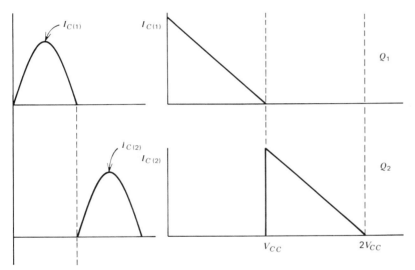

Figure 14.9
Transformer class B push-pull amplifier.

Figure 14.10 illustrates two circuit configurations that can be used to drive the input of a push-pull amplifier. Figure 14.10a illustrates a transformer having a tapped secondary. The two output signals are 180° out of phase with respect to each other; thus the drive requirement for the push-pull amplifier is met.

Figure 14.10b illustrates a transformerless phase splitter. The values of R_E and R_L are selected so that the voltage gain is approximately the same for both output signals. The voltage $v_{0(2)}$ is taken across the emitter resistor and is in phase with the input signal. The voltage $v_{0(1)}$ is taken across the load resistor and is 180° out of phase with the input signal.

CLASS B TRANSFORMERLESS PUSH-PULL AMPLIFIERS

Figure 14.11 illustrates some of the possible circuitry for push-pull amplifiers without output transformers. An output transformer is very costly and physically large in comparison with the devices making up solid-state amplifiers.

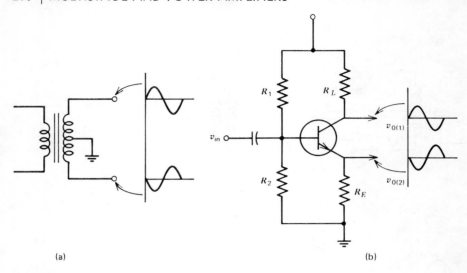

(a)   (b)

Figure 14.10
Phase splitters: (a) transformer and (b) BJT.

Figure 14.11*a* shows two NPN transistors connected in series, or totem pole fashion. The input signal is split and fed to the two bases, 180° out of phase. The load (i.e., an 8-Ω speaker) is connected between common and the output of the amplifier. If the base of Q_1 swings positive, the base of Q_2 swings negative, and vice versa. Thus each transistor amplifies one of the alternations of the input signal.

The circuit configuration shown in Figure 14.11*a* needs a phase splitter. The phase splitter can be eliminated if complementary transistors are used. Figure 14.11*b* illustrates a circuit configuration using complementary output transistors. Note that the input signal is fed to both bases in phase. As can be reasoned from the figure, each transistor amplifies one of the alternations of the input signal. A very simple means of increasing the output drive capability of an IC OP AMP is with the circuit configuration shown in Figure 14.11*b*.

The circuits illustrated in Figure 14.11*a* and *b* require a split supply (V_{CC} and $-V_{CC}$). If the load resistance is ac coupled, the split supply requirement can be eliminated. Figure 14.11*c* illustrates one possible circuit arrangement. The amplifier uses a single supply voltage V_{CC}. Capacitor C couples the output of the amplifier to the load resistance. The phase requirements for Q_3 and Q_4 are met by the complementary transistors Q_1 and Q_2. The coupling capacitor C is generally large in value so that the amplifier has a low frequency limit.

CLASS C AMPLIFIER

Figure 14.12 illustrates a class C RF amplifier. The load for the BJT is the tank circuit consisting of LC. The amplifier stage is biased well below cutoff. The bias necessary for this particular amplifier is obtained by the "diode clampling" action of the base-emitter junction of Q. If the input signal is removed, the bias

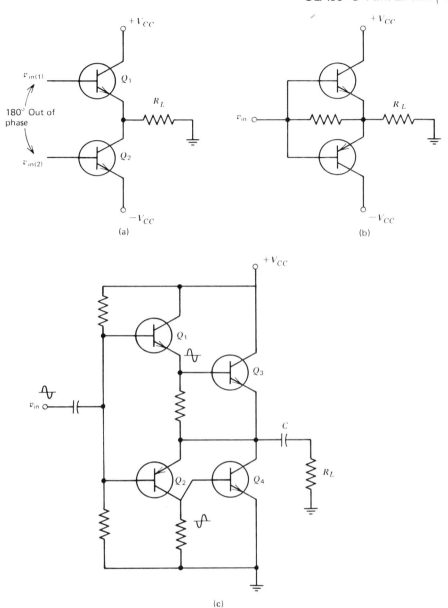

Figure 14.11
Transformerless power amplifiers: (a) NPN transistors, (b) complementary transistors, and (c) circuit configuration requiring no phase splitter.

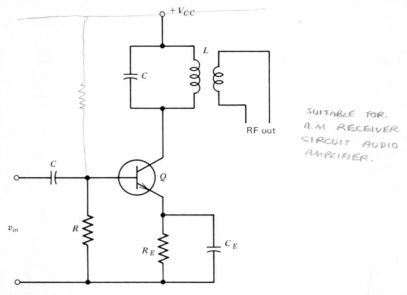

SUITABLE FOR.
A.M RECEIVER
CIRCUIT AUDIO
AMPLIFIER.

Figure 14.12
Circuit for a class C amplifier using diode bias clamping.

is gone. Therefore, the collector current is limited to a safe value by the emitter-resistor R_E. The capacitor C_E simply bypasses the emitter resistor at the RF frequency. The maximum efficiency for the stage is between 0.90 and 0.95 if an ideal transistor and tank circuit are used.

The waveform from the tank circuit is sinusoidal because the pulses of collector current from Q supply the tank circuit with energy. Between these current pulses, the "fly wheel" effect of the tank keeps the output waveform sinusoidal.

HEAT SINKS

When a transistor is used as an active device in an amplifier, oscillator, series voltage regulator, and so forth, the device dissipates heat. With the dissipation of heat, the junction temperature rises. The manufacturer gives the maximum junction temperature for the device (BJT, FET, diodes, SCR, etc.). For a silicon device, the maximum junction temperature is generally 150 °C, and for a germanium device, 75 °C. The physical construction and size of the device determine its maximum power dissipation.

Figure 14.13a illustrates Ohm's law for electrical circuits. Figure 14.13b illustrates the thermal equivalent for the electrical circuit. The difference in temperature $(T_1 - T_2)$ is equivalent to the potential V. The power dissipated P_D is equivalent to the current I. The thermal resistance Θ_{12} is equivalent to the electrical resistance R. Using this analogy $R = V/I$, we can see that the thermal resistance is

$$\Theta_{12} = \frac{T_1 - T_2}{P_D} \tag{11}$$

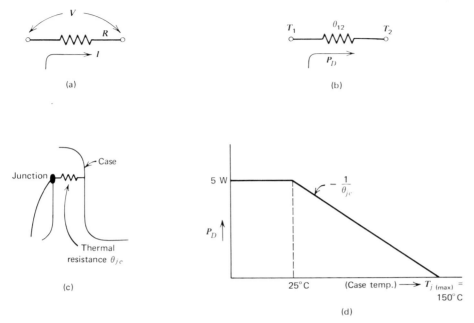

Figure 14.13
Thermal characteristics: (a) electrical circuit, (b) thermal circuit, (c) transistor, and (d) derating curve

Equation 11 can be applied to a transistor. The junction temperature T_j, the case temperature T_c, and the thermal resistance Θ_{jc} are labeled in Figure 14.13c. The magnitude of power dissipated by the transistor, diode, SCR, and so on, has to be dissipated so that the junction temperature does not exceed its maximum allowable limit. In order not to destroy the device, the manufacturer lists a maximum power dissipation under certain conditions. These conditions are illustrated in Figure 14.13d.

The transistor shown in Figure 14.13d can dissipate 5 W if the case temperature is not greater than 25 °C. If the case temperature is greater than 25 °C, the power dissipation for the transistor has to be derated linearly. The thermal resistance for the transistor is equal to

$$\Theta_{JC} = \frac{T_j - T_C}{P_D} \tag{12}$$

EXAMPLE

For the transistor shown in Figure 14.13d, calculate the maximum power the device can dissipate if the case temperature is maintained at 80 °C. From the figure, it can be seen that the thermal resistance Θ_{jc} is equal to

$$\Theta_{JC} = \frac{150 \text{ °C} - 25 \text{ °C}}{5 \text{ W}}$$

$$\Theta_{JC} = \frac{25 \text{ °C}}{\text{W}}$$

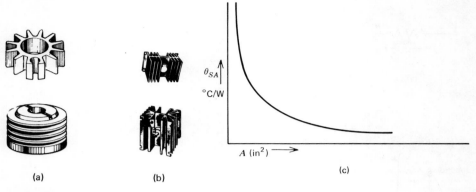

Figure 14.14
Heat sinks: (a) TO-5 case, (b) TO-3 case, and (c) thermal resistance.

Now, solving Equation 12 for the power dissipation P_D, we have

$$P_D = \frac{T_J - T_C}{\Theta_{JC}}$$

$$= \frac{150 \text{ °C} - 80 \text{ °C}}{25 \text{ °C/W}}$$

$$P_D = 2.8 \text{ W}$$

Note that the derating factor for a device is the reciprocal of its thermal resistance $(1/\Theta_{jc})$. As in the example problem worked above, Θ_{jc} equals 25 °C/W. The derating factor is the reciprocal of Θ_{jc} or $(1/25 \text{ °C/W}) = 0.04 \text{ W/°C}$. Thus for every degree above 25 °C, the maximum power dissipation decreases by 0.04 W (see Figure 14.13d).

A transistor with the thermal characteristics shown in Figure 14.13d can dissipate only 2.8 W if the case temperature for the device is maintained at 80 °C. If the case temperature is maintained at 25 °C, the transitor would be able to dissipate 5 W. One method of maintaining the case temperature at 25 °C, if the ambient or surrounding air is also 25 °C, would be to mount the transistor on an infinite heat sink.

A heat sink is a means of keeping the case temperature of the device at the same temperature as the surrounding air. If an infinite heat sink were possible, it would have a thermal resistance of 0 °C/W. A heat sink is constructed of material that absorbs and emits heat rapidly.

As a point of interest, good electrical conductors are also good thermal conductors. Because of economy, aluminum or an aluminum alloy is frequently used in the construction of heat sinks. Figures 14.14a and b illustrate some practical heat sinks.

The thermal resistance of a heat sink is a function of the surface area and the type of material that are used in its construction. Figure 14.14c graphically shows the relationship between the thermal resistance Θ_{SA} and the surface area A

of a thin piece of aluminum. An approximation for the relationship between Θ_{SA} and A is

$$\Theta_{SA} = \frac{K}{A_{(in)^2}} \tag{13}$$

where Θ_{SA} has units of °C/W, A is in inches squared, and K is a constant depending on the thickness and type of material used in the construction of the heat sink. As can be reasoned, the larger the surface area, the lower the thermal resistance. It is important to realize that the surface area exposed to the surrounding air need not be a flat piece of material. The total surface area of the heat sink may be folded to appear as fins.

Figure 14.15 illustrates the thermal circuit for a transistor mounted on a heat sink. Note that a thermal resistance exists between the case of the transistor and the heat sink intself (Θ_{CS}). In some circuit applications, because the collector of the transistor is electrically connected to the case of the transistor, the transistor has to be electrically isolated and yet thermally connected to the heat sink. (The sink might be the metal housing for the components.)

A Teflon or mica washer may be used to electrically isolate and thermally connect the case of the transistor to the heat sink. The washer should be smeared with silicon grease, which fills up the small air spaces between the case of the transistor and the washer and between the washer and the heat sink. The use of silicon grease minimizes the thermal resistance between the case and sink because air has a high thermal resistance.

EXAMPLE

Assume that a silicon power transistor has the following parameters: $\Theta_{JC} = 5$ °C/W and P_D (at 25 °C case temperature) = 25 W. The transistor is mounted

Figure 14.15
Equivalent circuit for a transistor mounted on a heat sink.

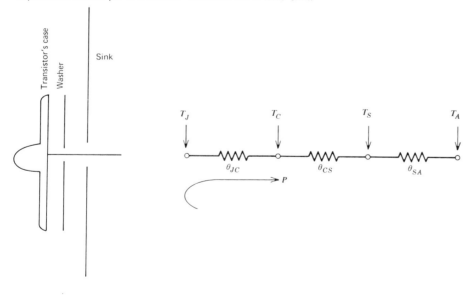

with a washer with silicon grease for a combined Θ_{CS} equal to 0.5 °C/W. The heat sink is to be constructed from a thin sheet of aluminum having a constant K equal to 100 (Equation 13). The power dissipated by the transistor is 10 W with an ambient temperature that will not be greater than 50 °C. Calculate the size of the heat sink required for this transistor.

Using Equation 12 to calculate the total thermal resistance Θ_{JA}, we have

$$\Theta_{JA} = \frac{T_j - T_A}{P_D} = \frac{150 \text{ °C} - 50 \text{ °C}}{10 \text{ W}}$$

$$\Theta_{JA} = 10 \text{ °C/W}$$

But Θ_{JA} equals

$$\Theta_{jA} = \Theta_{jC} + \Theta_{CS} + \Theta_{SA}$$

$$\frac{10 \text{ °C}}{\text{W}} = \frac{5 \text{ °C}}{\text{W}} + \frac{0.5 \text{ °C}}{\text{W}} + \Theta_{SA}$$

or

$$\Theta_{SA} = \frac{4.5 \text{ °C}}{\text{W}}$$

Substituting Θ_{SA} into Equation 13, we have

$$A_{(\text{in}^2)} = \frac{K}{\Theta_{SA}} = \frac{100 \text{ in}^2/\text{°C/W}}{4.5 \text{ °C/W}}$$

$$A_{(\text{in}^2)} = 22 \text{ in}^2$$

When the manufacturer lists the power dissipation for a device, it is important to realize whether a heat sink is necessary for the device. A power rating for a device is given with respect to a temperature of 25 °C, but the important point to realize is whether the power dissipation is with respect to ambient or case temperature.

For a 2N4400 the following parameters are given: $\Theta_{JC} = 0.137$ °C/mW, $\Theta_{JA} = 0.357$ °C/mW, P_D at 25 °C = 310 mW, and $T_{J(\text{max})} = 135$ °C (see the Appendix). If we assume that the 310 mW is the maximum power dissipation for the transistor in free air (without a heat sink), Θ_{JA} is found to be

$$\Theta_{jA} = \frac{T_{j(\text{max})} - T_A}{P_D}$$

$$= \frac{135 \text{ °C} - 25 \text{ °C}}{310 \text{ mW}}$$

$$\Theta_{jA} = \frac{0.355 \text{ °C}}{\text{mW}}$$

Our assumption was correct. The 310 mW was for the device without a heat sink because the manufacturer lists Θ_{JA} as 0.357 °C/mW. If the 2N4400 were to be mounted on a heat sink, the device would dissipate more than 310 mW.

EXAMPLE

Calculate the maximum power dissipated at 25 °C for the 2N4400 if it were mounted on an infinite heat sink ($\Theta_{SA} = 0$ °C/mW). Knowing Θ_{JC} is equal to 0.137 °C/mW and the maximum junction temperature is 135 °C, we calculate P_D as

$$P_D = \frac{T_j - T_C}{\Theta_{JC}}$$

$$= \frac{135 \text{ °C} - 25 \text{ °C}}{0.137 \text{ °C/mW}}$$

$$P_D = 803 \text{ mW}$$

POWER DISSIPATION OF DEVICES

As a rule of thumb, if a device is called a power device (i.e., power diode, power transistor, etc.), it is meant to be used with a heat sink. The power dissipated by a device depends on the circuit amplification. Some of the solid-state devices that have been discussed so far are transistors, diodes, FETs, UJTs, and PUTS.

Transistors. The total power dissipated by a transistor is the sum of the power dissipated in the collector-base and base-emitter junctions. However, because the base-collector junction dissipates much more power than the base-emitter junction, the power dissipated by the base-emitter junction can be disregarded.

In Chapter 12, the transistor served as a series voltage regulator. The power dissipated by the transistor was approximated as $(V_{\text{in}} - V_{\text{out}}) I_{\text{load}}$. The transistor used in the amplifiers discussed in this chapter has a power dissipation that is calculated by Equation 3, $P_D = P_{\text{dc}} - P_{\text{ac}}$.

Diodes. A power diode dissipates power when it is forward-biased. The manufacturer lists the power rating and thermal resistance for the diode. The power dissipated by the diode can be calculated by $P_D = I_D{}^2R_f + V_fI_D$, where I_D is the dc current through the device, R_f the forward resistance of the device, and V_f the barrier potential of the device (Si $\approx$ 0.7 V, Ge $\approx$ 0.3 V).

The maximum power dissipation of the zener diode is calculated by $P_D = I_Z{}^2_{(\text{max})}R_Z + V_ZI_{Z(\text{max})}$. With both the power and zener diodes, heat sinks may be required. The size of the heat sink for the diode is calculated in the same manner as that for the transistor.

Other devices. The FET, UJT, and PUT generally are not mounted on heat sinks. The devices are generally considered small signal devices, with power rating in the mW range. The SCRs, triacs, and so forth, may or may not be intended to be mounted on heat sinks. The power rating of these devices determines if a heat sink is required.

REVIEW QUESTIONS AND PROBLEMS

14-1   What is a multistage amplifier? Why are they necessary?

14-2   Discuss the coupling methods explored in this chapter.

14-3   With RC coupling, is there any relationship between the phase shift of the amplifier and the ratio of R to C? Phase shift is the number of degrees difference between the input and output signals of the amplifier.

14-4   How and why do temperature changes affect the direct-coupled amplifier? How does the difference amplifier overcome this difficulty?

14-5   Discuss the classes of amplifiers in terms of their operating points.

14-6   Why does a class C amplifier have a higher efficiency than a class A amplifier? Are both of these amplifiers linear?

14-7   What is the difference between a linear and a nonlinear amplifier?

14-8   Is it possible for a class B amplifier to be linear?

14-9   What is the relationship between power dissipation and efficiency?

14-10   What is a heat sink? Why is it necessary?

14-11   Does a heat sink have to be a good electrical conductor?

14-12   Discuss the difference between the thermal resistance for a power transistor labeled Θ_{JC} and Θ_{JA}.

14-13   Calculate the required area of a heat sink needed for a HEP 703 if it were to dissipate 15 W at a maximum ambient temperature of 40 °C.

OBJECTIVES FOR CHAPTER 15, OPERATIONAL AMPLIFIER

At the completion of this chapter you can demonstrate your understanding of the material by answering these questions:

- What is an OP AMP?
- How is an OP AMP used as a voltage amplifier, a summing amplifier, an integrator, a differentiator, and a wave generator?
- What are the functions of the inverting and the noninverting inputs for the OP AMP?

15 OPERATIONAL AMPLIFIERS

Because of advances in integrated circuit technology, the operational amplifier is being used in many more circuit applications than ever before. You should study the OP AMP in order to comprehend the many uses of this device in such applications as regulators, analog computers, wave generators, and control circuitry.

The operational amplifier (OP AMP) derives its name from its application in the analog computer. The OP AMP was (and still is) used as an adder, subtracter, mutiplier, divider, integrater, and differentiater. These mathematical functions gave the device its name: operational amplifier. Many applications of the OP AMP require feedback techniques; other applications use the OP AMP without feedback. Without feedback, the device is used as a comparator. With the advent of integrated circuit manufacturing techniques, the OP AMP has many, many circuit applications. Some of these circuit applications are discussed in this chapter and others are left for Chapter 20.

DESCRIPTION OF CIRCUITRY

The OP AMP is a high gain, dc amplifier with both positive and negative inputs. The voltage gain of the OP AMP is on the order of tens of thousands. This high voltage gain is referred to as the *open loop gain* of the device. The OP AMP is a multistage amplifier using direct coupling techniques. Multistage amplifiers are direct-coupled, difference-amplifier circuit configurations. This technique permits dc and ac amplification and insures that the OP AMP's Q point will not drift with temperature changes.

The two inputs of the OP AMP are labeled positive $(+)$ and negative $(-)$ with respect to the output terminal of the device. The positive $(+)$ input is a

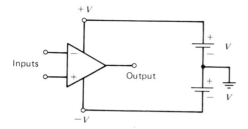

Figure 15.1
Symbol for OP AMP.

noninverting input. Thus, if a voltage appears at this input, the output of the device will be in phase with the input signal. The negative $(-)$ input is an inverting input. Thus, if a voltage appears at this input, the output of the device will be 180° out of phase with the input signal.

The schematic symbol for the OP AMP is shown in Figure 15.1. The $+V$ and $-V$ voltage sources are shown with respect to the ground potential. The input-terminal and output terminal voltages are measured with respect to ground. For simplicity, frequently the voltages $+V$ and $-V$ are not shown in schematic diagrams; nevertheless, they are a requirement for the device to function. In the discussion to follow in this chapter, the supply voltages $+V$ and $-V$ will be omitted. Generally, these voltages are obtained from zener diodes.

Figure 15.2 shows a diagram of a difference amplifier. From this diagram the two $(+)$ and $(-)$ inputs of the OP AMP will be explained. In the figure, Q_1 and Q_2 are fed current from the constant current pump I; $R_{L(1)}$ and $R_{L(2)}$ are the load resistances for the two transistors Q_1 and Q_2. The transistor Q_1 is made equal to transistor Q_2, and $R_{L(1)}$ is made equal to $R_{L(2)}$. With these components equal in both value and temperature characteristics, the output voltage V_0 will be zero

Figure 15.2
Discrete difference amplifier.

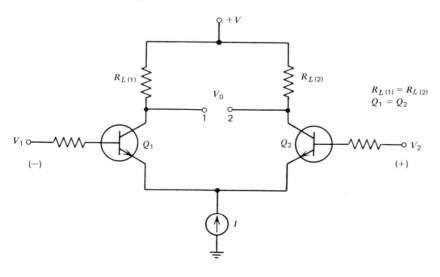

volts, with V_1 and V_2 equal to zero volts. If the temperature increases, because both transistors have identical characteristics, V_0 will remain equal to zero volts. The Q_1 and Q_2 collector current without base current is given

$$I_C = (h_{FE} + 1)\, I_{CO}.$$

Because I_{CO} is a function of temperature in both transistors, the collector currents through each transistor increase equally, causing V_0 to remain at zero volts.

Let us assume that the output voltage V_0 is measured with respect to terminal 2 or $V_0 = V_{12}$. Now, if a positive voltage is applied at V_2, the collector current of Q_2 is increased. Because the current pump is pumping a constant current to both transistors, if the collector current of Q_2 is increased, the collector current of Q_1 is decreased. These changes in collector current cause the output voltage $V_0(V_{12})$ to be positive or noninverting.

Now if V_1 is made positive ($V_2 = 0$ V), the collector current of Q_1 will increase. Because of the current pump, if the collector current of Q_1 increases, the collector current of Q_2 will decrease. Because $V_0(V_{12})$ is taken with respect to terminal 2, the output voltage will be negative with respect to V_1. Then V_1 can be labeled inverting or ($-$).

From Figure 15.2 it can be seen that the output voltge is the difference between the input voltages V_2 and V_1 times a constant k. The k is the amplification factor for the stage.

The circuit configuration shown in Figure 15.2 has to be modified to be used in an operational amplifier circuit because the OP AMP has to be able to amplify both positive and negative voltages with respect to ground. Figure 15.3a shows a diagram for an operational amplifier. Note that the stage amplifiers are difference amplifiers. Because the OP AMP has to amplify dc signals at the input terminals, direct coupling techniques are incorporated through the use of difference amplifiers. As stated above, the difference amplifier permits the output voltage to remain constant against temperature changes. Because of the use of integrated circuit techniques, the temperature and value characteristics are identical for the components used in the difference amplifiers making up the OP AMP.

Figure 15.3b shows the output versus input relationship for an OP AMP. Note that a very small difference between $V_{in}(-)$ and $V_{in}(+)$ (≈ 10 mV) will cause the OP AMP to have an output equal to the supply voltage V. Ideally, if $V_{in}(+) = V_{in}(-) = 0$ V, the output voltage will be zero. In a "real" OP AMP, this ideal condition is met by an offset voltage (see Offset Voltage in this chapter).

VOLTAGE AMPLIFIER

Figure 15.4 shows an OP AMP being used with shunt current feedback, (see Chapter 9). With this type of feedback, the input and output impedances are very low values of resistance. The input impedance is from the inverting terminal of the amplifier with respect to ground. The noninverting input terminal is at ground position.

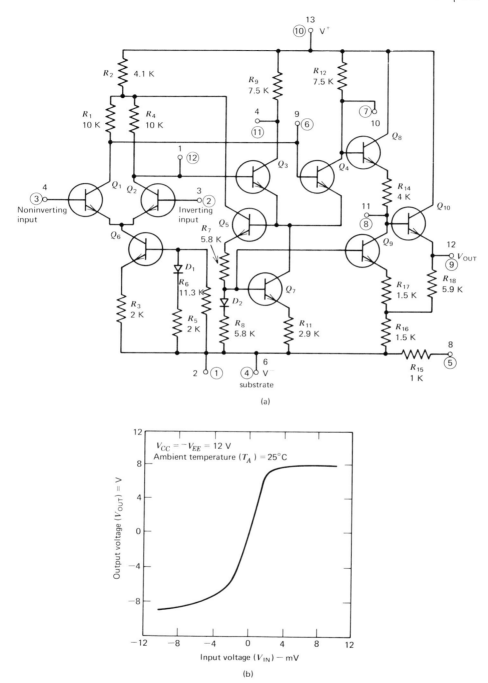

Figure 15.3
OP AMP: (a) circuit and (b) input curve (RCA CA 3016).

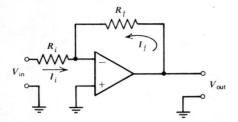

Figure 15.4
Inverting OP AMP.

The impedance seen by the source V_{in} in Figure 15.4 is given by approximately

$$Z_{inS} = R_i \tag{1}$$

Because the input impedance of the OP AMP is a very low value in this circuit configuration, we can approximate that $I_i = -I_f$, where

$$I_i = \frac{V_{in}}{R_i}$$

and

$$I_f = \frac{V_{out}}{R_f}$$

Combining the above two equations, we have

$$I_i = -I_f$$

or

$$\frac{V_{in}}{R_i} = -\frac{V_{out}}{R_f}$$

$$V_{out} = -\frac{R_f}{R_i} V_{in} \tag{2}$$

and

$$-A_{vf} = R_f/R_i \tag{3}$$

Equation 3 relates the values for R_i and R_f for a specified voltage gain called the closed loop voltage gain. As can be seen in Figure 15.4, the values of R_i determine the input impedance seen by the source, or

$$Z_{ins} = R_i \tag{4}$$

Solving Equation 3 for R_f, we have

$$R_f = |A_{vf}| R_i \tag{5}$$

Substituting Equation 5 into Equation 4, we have

$$R_f = |A_{vf}| Z_{ins} \tag{6}$$

Equation 6 gives the value of R_f as a function of Z_{ins} and A_{vf}.

EXAMPLE

Determine the necessary values of R_i and R_f in Figure 15.4 if the closed loop gain A_f is 5 and the input impedance seen by the source Z_{ins} is 100 kΩ. Using Equation 6, we have

$$R_i = Z_{\text{ins}}$$

$$R_i = 100 \text{ k}\Omega$$

Using Equation 5 to solve for R_f, we have

$$R_f = |A_{vf}| R_i = 5 \times 100 \text{ k}\Omega$$

$$R_f = 500 \text{ k}\Omega$$

With the derivation above, the output and input voltages were denoted as dc values V_{in} and V_{out}. Because the OP AMP can amplify both dc and ac voltages, V_{in} and V_{out} can be replaced with ac values denoted as v_{in} and v_{out}.

SUMMING AMPLIFIER

Figure 15.4 can be modified as shown in Figure 15.5, which shows three input voltage sources. This circuit configuration uses shunt current feedback and the sum of the input currents I_1, I_2, and I_3 are approximately equal to the output current $-I_f$, or

$$-I_f = I_1 + I_2 + I_3 \tag{7}$$

But, from the figure, we have $I_1 = V_{\text{in}(1)}/R_{i(1)}$, $I_2 = V_{\text{in}(2)}/R_{i(2)}$, $I_3 = V_{\text{in}(3)}/R_{i(3)}$, and $I_f = V_{\text{out}}/R_f$. Using these values of current, then, Equation 7 becomes

$$-\frac{V_{\text{out}}}{R_f} = \frac{V_{\text{in}(1)}}{R_{i(1)}} + \frac{V_{\text{in}(2)}}{R_{i(2)}} + \frac{V_{\text{in}(3)}}{R_{i(3)}} \tag{8}$$

Figure 15.5
Summing OP AMP.

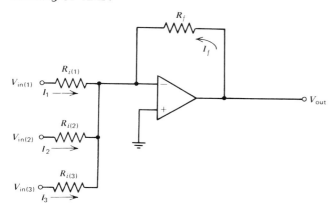

If $R_{i(1)} = R_{i(2)} = R_{i(3)} = R_i$, Equation 8 becomes

$$V_{out} = -\frac{R_f}{R_i}[V_{in(1)} + V_{in(2)} + V_{in(3)}] \qquad (9)$$

As can be seen from Equation 9, the output voltage of the amplifier is the sum of the individual input voltages. For this reason, the circuit configuration is sometimes called a *summing amplifier*. The input impedance seen by the sources is approximately

$$Z_{ins} = R_i \qquad (10)$$

In Figure 15.5, if $R_{i(1)} = 2R_{i(2)} = 4R_{i(3)}$, Equation 8 becomes

$$-\frac{V_{out}}{R_f} = \frac{V_{in(1)}}{R_i} + \frac{V_{in(2)}}{1/2R_i} + \frac{V_{in(3)}}{1/4R_i}$$

or

$$V_{out} = -\frac{R_f}{R_i}[V_{in(1)} + 2V_{in(2)} + 4V_{in(3)}] \qquad (11)$$

As can be seen from Equation 11, the input voltages ($V_{in(1)}$, $V_{in(2)}$, $V_{in(3)}$) are "weighted." Thus if $V_{in(1)} = V_{in(2)} = V_{in(3)} = 1$ V, the output voltage would be 7 V. This circuit configuration has many interesting applications in digital electronics as a digital-to-analog (D/A) converter (see Chapters 16 and 20). Note again that the dc values of the input and output voltages can be replaced with ac values in Equations 7 through 11.

CURRENT-TO-VOLTAGE CONVERTER

Figure 15.6 shows a circuit arrangement for a current-to-voltage converter. The input current, I_{in}, is converted to an output voltage, V_{out}. Also, $I_{in} = -I_f$, or

$$I_{in} = -\frac{V_{out}}{R_f}$$

or

$$V_{out} = -I_{in}R_f \qquad (12)$$

Equation 12 states that the output voltage is equal to the input current times the feedback resistance of R_f.

EXAMPLE

Calculate the necessary value of R_f in Figure 15.6 so that 1 V at the output terminal will indicate 1 ma of input current. Using Equation 12, we have

$$1 \text{ V} = -1.0 \times 10^{-3} \text{ A} \times R_f$$

or

$$R_f = 10^3 \ \Omega = 1 \text{ k}\Omega$$

If the input current were 2 mA, the output voltage would be 2 V, and so forth.

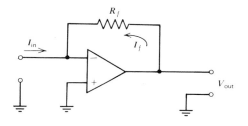

Figure 15.6
Current-to-voltage converter.

NONINVERTED OUTPUT

With the circuit arrangements discussed so far, the (+) positive or noninverting input was at ground potential and the (−) negative or inverting input was used. With this arrangement, the output voltage is 180° out of phase with the input voltage. A circuit arrangement that has no phase reversal between output and input voltages is shown in Figure 15.7.

In Figure 15.7 the feedback resistors R_f and R_i are returned to the inverting input of rhe OP AMP. The input voltage V_{in} is at the noninverting input of the OP AMP. With this circuit arrangement, the feedback voltage that is across R_i is in series and is subtracting from the input voltage V_{in}. As demonstrated in Chapter 9, because the input voltage is in series and is subtracting from the feedback voltage, the impedance seen by the source V_{in} is a very high value. From Figure 15.7, the closed loop voltage gain is given by

$$A_{vf} = \frac{A_v}{1 + \beta A_v} \approx \frac{1}{\beta} \tag{13}$$

where β is the voltage feedback factor,

$$\beta = \frac{R_i}{R_i + R_f} \tag{14}$$

Substituting Equation 14 into Equation 13 and solving for A_{vf}, we have

$$A_{vf} = \frac{1}{R_i/(R_i + R_f)} = \frac{R_i + R_f}{R_i}$$

$$A_{vf} = (1 + R_f/R_i) \tag{15}$$

or

$$V_{out} = V_{in}(1 + R_f/R_i) \tag{16}$$

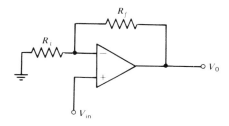

Figure 15.7
Non-inverting OP AMP.

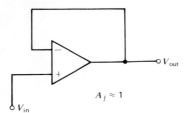

$A_f \approx 1$

Figure 15.8
Voltage follower.

VOLTAGE FOLLOWER

The circuit arrangement shown in Figure 15.7 can be modified to a circuit configuration that has a voltage gain of approximately one, high input impedance, low output impedance, and no phase reversal between input and output voltages. This circuit configuration is shown in Figure 15.8 and is called a voltage follower.

From the figure, 100 percent of the output voltage is feedback to the inverting input, causing the feedback factor β to be equal to one. The closed loop gain of the voltage follower is then

$$A_{vf} = \frac{A_v}{1 + \beta A_v} = \frac{1}{1/A_v + \beta}$$

But $\beta = 1$; then

$$A_{vf} \approx 1 \tag{17}$$

The voltage follower is frequently used in circuits as an isolator. The isolator prevents one circuit component from being "loaded" by another. As an example of the need of one circuit component to be isolated from another, consider the circuit shown in Figure 15.9a. The figure shows a constant current being pumped to a capacitor from a current pump. With the circuit arrangement, the voltage across the capacitor is linear with respect to time and is given by

$$V_C = \frac{I}{C} t \tag{18}$$

where I is the current pumped through capacitor C for a period of time t. The circuit shown in Figure 15.9a is sometimes used in timing configurations (see Chapter 17).

If V_C is to be used to drive another part of a timer circuit, the voltage across the capacitor has to be isolated from the rest of the circuit. If this isolation is not accomplished, the voltage across C is no longer linear. A voltage follower can be used to isolate the V_C because of its high input impedance and low output impedance.

Figure 15.9b shows another application of the voltage follower. In this circuit arrangement, the voltage follower is used to monitor the voltage across the capacitor C. The circuit arrangement is called a sample-hold circuit. Switch S, when closed, causes C to charge up to V_i or to sample V_i. With S open, the output of the voltage follower is equal to V_i. Because of the high input impedance of the voltage follower, the capacitor C retains its charge for a period of time. Hence the voltage follower is said to hold the voltage V_i.

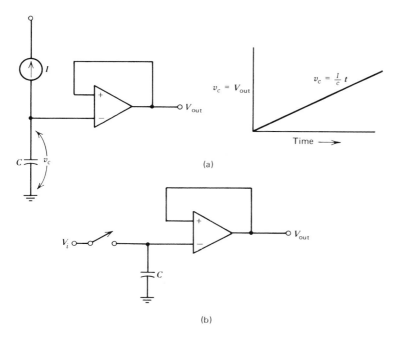

Figure 15.9
Applications for the voltage follower: (a) nonloading and (b) sample and hold circuit.

OFFSET VOLTAGE

As was mentioned early in this chapter, for the OP AMP to have a zero output voltage when both input voltages are zero, an offset voltage is necessary. This offset voltage arrangement is shown in Figure 15.10 for an inverting and a non-inverting amplifier.

In Figure 15.10a, R_3 is calculated to be equal to $R_i \| R_f$; both inputs have the same resistance value with respect to ground. The R_1 and R_2 are chosen to be much larger in value than R_i and R_f so that they do not interfere with the gain of the amplifier. Also, R_2 is adjusted, with V_{in} equal to zero volts so that the output voltage V_{out} is equal to zero volts. In Figure 15.10b, R_1 and R_2 are chosen to be much greater than R_i and R_f so that the gain of the amplifier does not change. Then R_2 is adjusted with V_{in} equal to zero volts so that the output voltage V_{out} is equal to zero volts. With the circuits discussed thus far, the offset voltage arrangement should be included.

COMPARATOR

The OP AMP when used without feedback is called a comparator. As its name implies, the comparator is used to compare two input voltages that are present

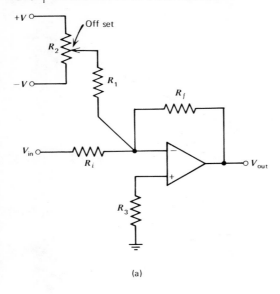

(a)

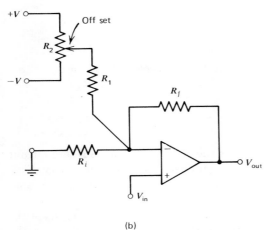

(b)

Figure 15.10
Voltage offset: (a) inverting OP AMP and (b) noninverting OP AMP.

at the inverting and noninverting input terminals of the OP AMP. Figure 15.11 shows a comparator. The output voltage is given by

$$V_{out} = k(V_2 - V_1) \qquad (19)$$

where k is the open loop gain of the OP AMP. Because this open loop gain is very large (tens of thousands), a small difference ($\approx$ 10 mV) between the voltages at the two inputs of the device will cause the output of the OP AMP to be saturated. The polarity of the differences between V_2 and V_1 determines if the output voltage of the OP AMP is either positive or negative. Generally, the output, when saturated, can be considered either the positive or negative supply voltage V.

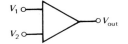

V_2 (Volts)	V_1 (Volts)	V_{out} (Volts)
0	+5	−V
0	−5	+V
+5	0	+V
−5	0	−V
+5	+4	+V
+4	+5	−V
−5	−4	−V
−4	−5	+V
−5	+5	−V
+5	−5	+V
0	0	0
+5	+5	0
−5	−5	0

Figure 15.11
Comparator.

The table shown in Figure 15.11 lists various combinations of input voltages and the resulting output voltage of the comparator. The reader, using Equation 19, is asked to verify the table with the inputs shown.

INTEGRATOR AND DIFFERENTIATOR

Figure 15.12 shows the circuit configuration of an integrator. The current of the capacitor C is fed to the inverting input of the OP AMP and the noninverting input is at ground potential. Because the voltage across the capacitor C is the output voltage V_{out}, V_{out} is given by

$$V_{out} = \frac{1}{C} \int I_f \, dt \tag{20}$$

where I_f is the feedback current of the capacitor C. But $I_f = -I_{in}$ and

$$I_{in} = \frac{V_{in}}{R} \tag{21}$$

Substituting I_{in} for I_f in Equation 20, we have

$$- V_{out} = \frac{1}{RC} \int V_{in} \, dt \tag{22}$$

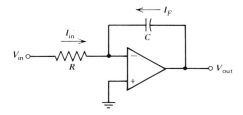

Figure 15.12
Integrator.

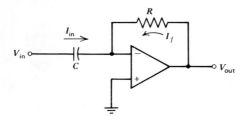

Figure 15.13
Differentiator.

Equation 22 shows the relationship of V_{out} as a function of the integral of the input voltage and time.

Figure 15.13 shows a circuit arrangement of a differentiator. The input current I_{in} is given by

$$I_{in} = C \frac{d}{dt} (V_{in})$$

and the feedback current I_f is given by

$$I_f = \frac{V_{out}}{R}$$

Setting these currents equal to each other, we have

$$I_{in} = -I_f$$

$$C \frac{d}{dt} (V_{in}) = -\frac{V_{out}}{R}$$

or

$$V_{out} = -RC \frac{d}{dt} (V_{in}) \tag{23}$$

Equation 23 tells us that the output voltage of the differentiator is equal to the product of RC times the time derivative of the input voltage V_{in}.

GYRATOR

Figure 15.14 illustrates one of the possible circuit configurations for a gyrator. A gyrator is a circuit configuration that exhibits inductance (L). The inductance L, because of the feedback principle of the OP AMPs, is generated with resistance (R) and capacitance (C). In Figure 15.14, the value of inductance seen looking into the device is given by the product CR^2 ($L = CR^2$).

The gyrator can be used in active filters. The advantage of the gyrator is that inductance is generated with only resistors and capacitors. Thus IC chips can be constructed having the property of inductance using OP AMPs, resistors, and capacitors.

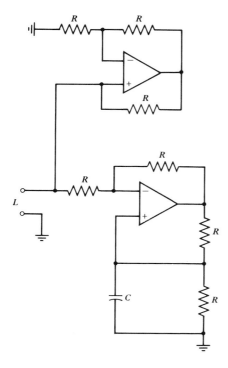

Figure 15.14
Gyrator.

OP AMP MULTIVIBRATOR

OP AMPs can be used to generate squarewaves. One type of square wave generator using an OP AMP, called a astable multivibrator, is shown in Figure 15.15. In the figure, the non-inverting input has a portion of the output voltage fed to it and is equal to $\pm\beta V_0$. The inverting input is fed the voltage that is across the capacitor C. The capacitor C has current fed to it by the feedback resistor R. The output voltage changes between $+V_0$ and $-V_0$ or $+V$ and $-V$, where V is the supply voltage for the OP AMP. The output voltage is equal to

$$V_0 = K\,[V(+) - V(-)]$$

If $V(+)$ is greater than $V(-)$, V_0 will be the positive V, and if $V(+)$ is less than $V(-)$, V_0 will be the negative V.

In Figure 15.15, the output voltage of an astable OP AMP MV swings between $+V$ and $-V$; therefore, the maximum time-varying voltage across the capacitor C is $\pm\beta V_0$, where $\beta = R_2/(R_1 + R_2)$. Assume that the capacitor has zero volts across it and that V_0 is $+V$. The feedback voltage V_0 is positive and the output voltage is positive. Now C is charging towards V_0 through the feedback resistor R. At a time t, the voltage across C will reach a value equal to the feedback voltage βV_0. When C reaches the voltage level of βV_0, the output of the OP AMP will swing to $-V_0$. With the output at $-V_0$, the feedback voltage at the noninverting input is V_0. The capacitor now charges from a $+V_0$ to a $-V_0$.

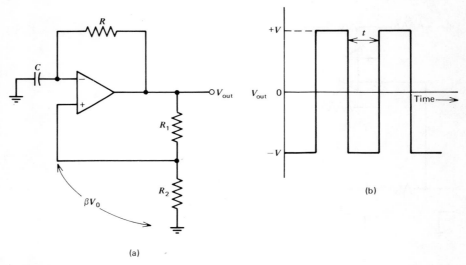

Figure 15.15
Astable OP AMP MV.

Once the voltage of the capacitor reaches $-V_0$, the output of the OP AMP will swing positive $(+V_0)$ and the oscillating process will continue.

To calculate the time necessary for the OP AMP to swing from a positive V to a negative V, the equation for the voltage across C will be used. The voltage across C is given by

$$v_c = V_F + (V_I - V_F)\, e^{-t/RC} \tag{24}$$

When v_c is the voltage across C in time t, V_F is the final voltage C could charge up to and V_I is the initial voltage C starts from. Letting $v_c = \beta V_0$, $V_I = -\beta V_0$, and $V_F = V_0$, we have

$$\beta V_0 = V_0 + (-\beta V_0 - V_0)\, e^{-t/RC} \tag{25}$$

or

$$\frac{1 - \beta}{1 + \beta} = e^{-t/RC} \tag{26}$$

Solving Equation 26 for t, we have

$$t = RC \ln\left(\frac{1 + \beta}{1 - \beta}\right) \tag{27}$$

where $\beta = R_2/(R_1 + R_2)$ and ln is the natural log.

Equation 27 gives the time t for one-half of the period of the square wave. The total period of the square wave is twice t or T.

$$T = 2RC \ln\left(\frac{1 + \beta}{1 - \beta}\right) \tag{28}$$

The frequence of the square wave is given by

$$f = \frac{1}{T} \tag{29}$$

EXAMPLE

Design an astable multivibrator having a square wave frequency of 5 kHz using an internally compensated 741C OP AMP. (see Appendix). The supply voltage is to be ±15 V and $R = R_1 = R_2 = 10$ kΩ. Using Equation 29 to calculate the time period T, we have

$$T = \frac{1}{f} = \frac{1}{5 \times 10^3} = 200 \ \mu s$$

$T = 200 \ \mu s$, and

$$\beta = \frac{R_2}{R_1 + R_2} = \frac{10 \ k\Omega}{20 \ k\Omega} = \frac{1}{2}$$

Using Equation 28 to calculate the value of C, we have

$$T = 2RC \ln \left(\frac{1 + \beta}{1 - \beta}\right)$$

$$200 \times 10^{-6} s = 2 \times 10^5 \ \Omega \times C \times \ln \left(\frac{1 + 1/2}{1 - 1/2}\right)$$

or

$$C \approx 0.001 \ \mu F$$

See Figures 15.15a and b. Note that if the value of β were to vary, the frequency of the square wave would vary (see Review Questions). The reader is referred to Chapter 19 for other types of oscillators incorporating the operational amplifier.

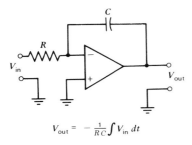

$$V_{out} = -\frac{1}{RC} \int V_{in} \, dt$$

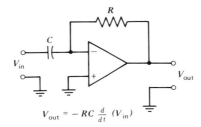

$$V_{out} = -RC \frac{d}{dt} (V_{in})$$

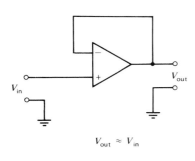

$$V_{out} \approx V_{in}$$

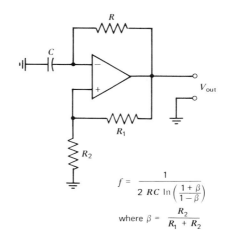

$$f = \frac{1}{2 \ RC \ \ln \left(\frac{1 + \beta}{1 - \beta}\right)}$$

$$\text{where } \beta = \frac{R_2}{R_1 + R_2}$$

Figure 15.16
OP AMP circuit configurations.

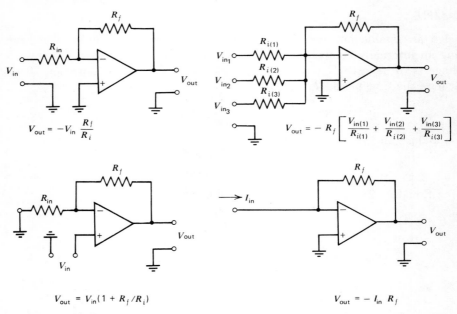

$$V_{out} = -V_{in} \frac{R_f}{R_i}$$

$$V_{out} = -R_f \left[\frac{V_{in(1)}}{R_{i(1)}} + \frac{V_{in(2)}}{R_{i(2)}} + \frac{V_{in(3)}}{R_{i(3)}} \right]$$

$$V_{out} = V_{in}(1 + R_f/R_i)$$

$$V_{out} = -I_{in} R_f$$

Figure 15.16 (*continued*)

COMPENSATION

Operational amplifiers, because of the very high voltage gain they are capable of, require compensation so that they do not oscillate. OP AMPs that require external compensation have terminals labeled "led" and "lag." Combinations of resistors and capacitors are connected to these "lead" and "lag" terminals to keep the OP AMP from oscillating.

Some operational amplifiers have internal compensation. These internally compensated OP AMPs require no external components to keep the OP AMP from oscillating. The OP AMPs used in the circuitry discussed in this chapter show no compensation. The circuits used in the example problems were built in the laboratory with the internally compensated 741C operational amplifiers.

The offset voltage techniques discussed may have to be used with OP AMP circuitry.

There are many excellent texts on the market devoted to OP AMPs and their applications. The reader may want to review the literature and find some of the interesting applications for OP AMPs. See Figure 15.16.

REVIEW QUESTIONS AND PROBLEMS

15-1 What is an operational amplifier?

15-2 List some applications for this type of amplifier?

15-3 What type of feedback is used in the inverting OP AMP?

15-4 What type of feedback is used in the noninverting OP AMP?

15-5   With the voltage follower, what percentage of feedback is used?

15-6   An inverting OP AMP is to have a closed loop gain of -10 with an input resistance of 47 kΩ. Calculate the necessary values of R_i and R_f.

15-7   A noninverting OP AMP is to have a closed loop gain of 10. Calculate the necessary values of R_i and R_f.

15-8   Design a current-to-voltage converter having a transfer function of 1 V per 4μA.

15-9   If the following input waveform is applied to an integrator, what will the output wave form look like?

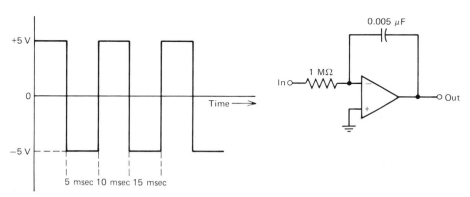

15-10 If the input waveform shown is applied to a differentiator, what will the output waveform look like?

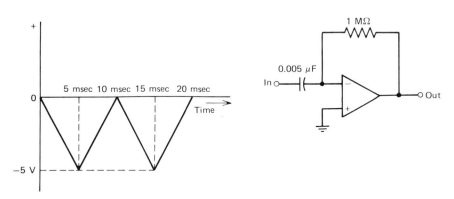

15-11 The output of a summing OP AMP is to have the equation

$$-V_{out} = V_{in_1} + 10\,V_{in_2} + 100\,V_{in_3}$$

Calculate the necessary values of input and feedback resistors.

15-12 What is a sample and hold circuit?

15-13 If the input of an integrator is a constant voltage, the output voltage across the feedback capacitor C is a linear rising voltage. Is there a constant current being pumped into C? Explain.

15-14 Design an astable OP AMP MV having a frequency of 100 Hz. Is there a limit to the frequency of the OP AMP MV?

15-15 If a diode were used in the feedback path for an inverting OP AMP, what would the output waveform look like?

15-16 For the circuit shown, plot the input voltage (positive) versus the output voltage.

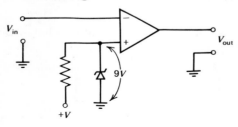

OBJECTIVES FOR CHAPTER 16, DIGITAL ELECTRONICS

At the completion of this chapter you can demonstrate your understanding of the material by answering these questions:

- What are astable, bistable, and monostable multivibrators?
- What are logic families?
- How is Boolean algebra related to logic gates?
- What are the five basic logic gates?
- What are a ROM and a RAM?
- How does a dual slope converter function?

16 DIGITAL ELECTRONICS

Because of advances in integrated circuit technology, digital circuitry is being used in more applications than ever before. You should study digital principles in order to comprehend the workings and theory in such circuit applications as data acquisition systems.

Digital electronics concerns itself with circuitry that has only two possible states or logic levels. These logic levels are represented by "1's" and "0's." As will be discussed, the logic level 1 might represent a voltage level of $+5$ V and the logic level 0 might represent a voltage level of 0 V. This concept of only two possible states may also represent a device being "on" or "off," hot or cold, saturated or cutoff, and so forth. For digital circuitry to function correctly, only the 1 and 0 logic levels are permitted. In contrast to digital systems, an analog system has an infinite variety of input and output levels.

With advancements in semiconductor technology, the field of digital electronics has grown rapidly. Because of reliability, cost, physical size, and ease of maintenance, more and more digital circuitry is being used in applications that were dominated by analog systems. Such applications in communications, instrumentation, calculators, computers, automobiles, and home appliances have all been influenced by digital electronics.

This chapter is concerned with the discussion of some of the common terms and circuitry associated with the rapidly growing area of digital electronics. Because of the introductory nature of this chapter, concepts are discussed whenever possible using "black boxes."

ASTABLE MULTIVIBRATORS

An astable multivibrator (MV) is basically a square wave generator or clock. It requires no input trigger to change its states. It is sometimes desirable to

320

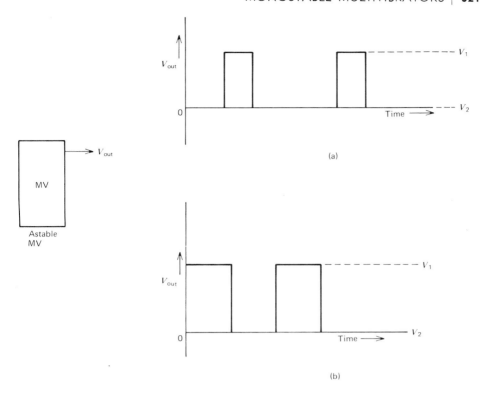

Figure 16.1
Output waveforms for an astable MV: (a) *asymmetrical output and* (b) *symmetrical output.*

synchronize the frequency of the astable with an external source frequency, the astable MV is then called a synchronized astable MV.

Astable MVs can produce symmetrical or asymmetrical waves (see Figures 16.1a and b). The output pulses alternate between two voltage levels, V_1 and V_2. These two voltage levels can be used to turn other electronic devices on and off.

Astable MVs can consist of two transistors. Each transistor switches between cutoff and saturation at a certain rate of frequency. If one transistor is cutoff, the other is saturated, and vice versa. As each transistor switch in the astable MV alternates between cutoff and saturation at a certain rate (circuit components determine the amount of time each state exists), the astable MV is said to have two unstable states. The output is usually taken from the collector of one of the transistors with respect to ground potential (see Figure 16.1a).

With astable MVs, time is an important parameter in discussing the output waveform. In many applications, the time that V_1 and V_2 exist becomes an important parameter in turning devices on and off for specific periods of time.

MONOSTABLE MULTIVIBRATORS

The monostable MV (or one-shot) has one stable state and one unstable state. It requires a rectangular tigger pulse to cycle the device. This cycle, once the

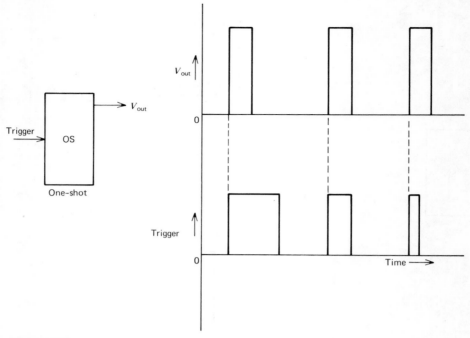

Figure 16.2
Waveforms of a one-shot being triggered with the leading edge of the triggering pulses.

trigger arrives, consists of the device going from the stable state to the unstable state and then back to the stable state. The length of time of the unstable state is determined *only* by circuit components and *not* by the duration of the triggering pulse. Because the duration of the unstable state is independent of the width of the trigger pulse, the one shot can be used to reshape pulses of different widths.

The one-shot basically consists of two transistor switches. Without a trigger present, one transistor switch is cut off and the other is saturated; this condition represents the stable state. When the triggering pulse is present, the transistor that was cut off goes into saturation and the transistor that was saturated goes into cutoff; this condition represents the unstable state.

Figure 16.2 shows a one-shot being triggered by a train of pulses of unequal width. As can be seen, the output waveform is of uniform width.

It should be pointed out that one-shots can be triggered by negative- or positive-going pulses. The negative-going pulse is the falling or lagging edge of a square wave, and the positive-going pulse is the rising or leading edge of a square wave. In Figure 16.2 the one-shot is being triggered by positive-going pulses. In Figure 16.3 the one-shot is being triggered by negative-going pulses.

BISTABLE MULTIVIBRATORS

The bistable MV consists of two transitor switches, and the circuit has two stable states. Without a triggering pulse, one transistor is cut off and the other

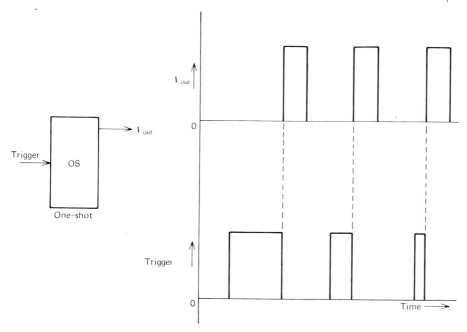

Figure 16.3

Waveforms of a one-shot being triggered with the lagging edges of the triggering pulses.

is saturated; this condition represents one stable state. With one trigger pulse, the transistor that was cut off goes into saturation and the transistor that was saturated goes into cutoff; this condition represents the other stable state. The bistable requires a square wave trigger to switch from one stable state to the other stable state. Once the trigger is removed, the device remains in the state to which it was switched; thus it has two stable states.

Figure 16.4 shows a bistable being triggered by a train of pulses. From the output waveform, it can be seen that the output frequency of the bistable is one-half of the frequency of the trigger. This relationship remains true whether the trigger waveform is symmetrical or asymmetrical.

As mentioned for monostables in the previous section, the bistable can also be triggered by leading or lagging edges of pulses. Circuit arrangement determines this positive of negative triggering. Figure 16.5 shows the output waveform with respect to both negative and positive triggers.

Bistable MVs are commonly known as flip-flops. Such terms as clocked flip-flops, reset-set flip-flops, clocked-reset-set flip-flops, JK flip-flops, ac and dc flip-flops, and master-slave flip-flops fall in the general category of flip-flops.

SCHMITT TRIGGER

A Schmitt trigger is an electrical switching device that has two output levels. Each of these levels is controlled by a specific input amplitude or level. The

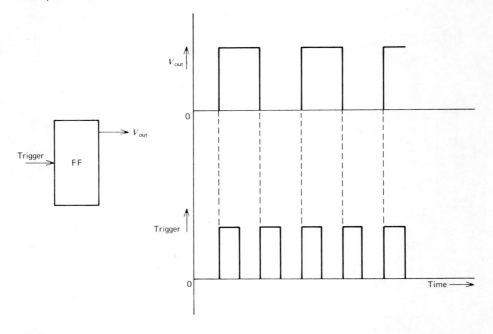

Figure 16.4
Waveforms of a bistable MV or flip-flop.

Schmitt trigger can be triggered by sine waves, sawtooth waves, and so forth (see Figure 16.6). Its two output levels are controlled by two specific input levels. These two input levels are designated the upper triggering level (UTL) and the lower triggering level (LTL).

The UTL is equal to or greater than the LTL. If the UTL is greater than the LTL, the output waveform has a hysteresis loop associated with it. In Figure 16.7, this hysteresis loop can be seen by plotting the output voltage versus the input voltage.

If the input to the Schmitt trigger is a sine wave or a sawtooth wave, the output waveform of the Schmitt trigger will be a square wave of the same frequency as the input waveform. This output then can be used to trigger other devices.

BASIC LOGIC GATES

The three basic logic gates are AND, OR, and NOT. Two additional gates can be constructed by the combination of the AND and NOT gates and the OR and NOT gates. The NAND gate is an AND gate driving a NOT gate. The NOR gate is an OR gate driving a NOT gate.

The NOR, OR, NAND, and AND gates have two or more inputs (variables) and one output. The NOT gate has one input (variable) and one output.

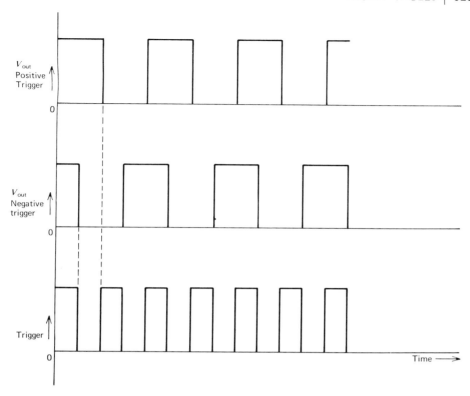

Figure 16.5
Waveforms showing both positive and negative triggering of a flip-flop.

LOGIC VARIABLES

Switching variables have *only* two possible states, or levels. These two possible levels are designated 0 and 1. The 0 and 1 could represent a switch being opened or closed, off or on.

When solid-state devices make up logic gates, voltage levels are designated as 0 and 1. Associated with these voltage levels are *positive* and *negative* logic. Positive logic is logic that the 1 is the most positive level. Negative logic is logic that the 1 is the most negative level.

As an example, if two voltage levels are available from a switching device, Figure 16.8 shows the assignment of positive and negative logic. As can be seen in the figure, the type of logic we are using becomes very important.

TRUTH TABLES

Truth tables are one of several possible methods of displaying Boolean algebraic expressions or equations. In conventional algebra, an expression or equation can be displayed graphically by a three-dimensional Cartesian coordinate system. An example of this system would be the plotting of the expression $x + y = z$.

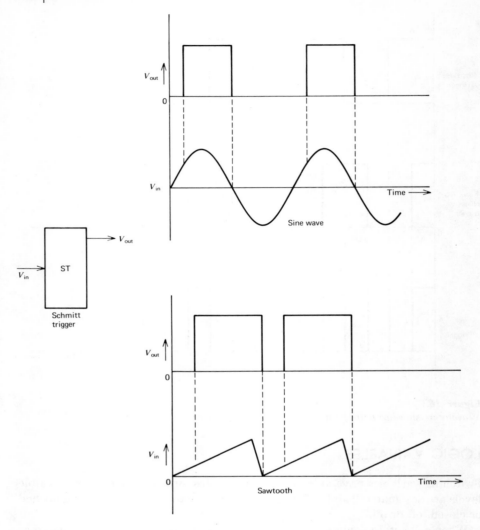

Figure 16.6
Waveforms of a Schmitt trigger being trigered by a sine wave and a sawtooth wave.

The dependent variable is z because it depends on the value of the x and y (called the independent variables). In the example just presented, the variables, dependent and independent, have a wide range or level of values.

In Boolean algebra, the variables have only two possible levels or values (states), designated 1 and 0. A truth table displays what combinations of independent variables will produce or generate an output, dependent variable.

LOGIC EXPRESSIONS

Symbols used in Boolean algebra denote the AND, OR, and NOT function generators. The plus sign (+) in conventional algebra denotes the OR gen-

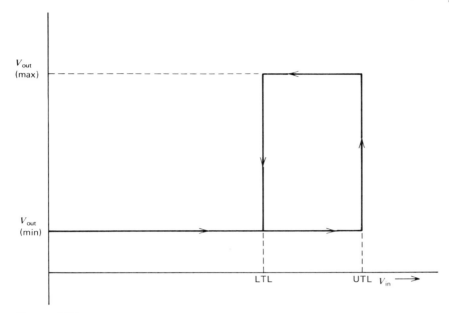

Figure 16.7
Hysteresis loop for a Schmitt trigger.

erator or function. The prime or bar (′) or (⁻) denotes the NOT generator or function. The blank ( ) denotes the AND generator or function.

In Boolean algebra, a variable *cannot* be raised to a power, added, divided, multiplied, subtracted, squared, and so forth. The AND, OR, and NOT functions (also the NOR and NAND) are the only operational generators in Boolean algebra.

AND GATE

Figure 16.9 shows SPST switches arranged to generate the AND function. Switch A (or B) being open, not making contact, represents the 0 state or level.

Figure 16.8
Positive and negative logic comparison.

	Level	Positive Logic	Negative Logic
Group A	+5 V	1	0
	0 V	0	1
Group B	+5 V	1	0
	−5 V	0	1
Group C	−10 V	0	1
	−5 V	1	0
Group D	+10 V	1	0
	+5 V	0	1

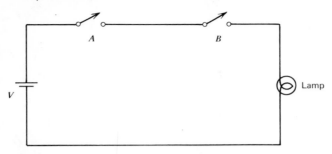

Figure 16.9
Simple AND gate.

The switch being closed, shorted, represents the 1 state or level. The lamp's being lit represents the 1 state. The lamp's not being lit represents the 0 state. This set of conditions represents positive logic. The independent variables are the switches A and B. The output variable (dependent) is represented by the lamp (L). Figure 16.10 shows the Boolean expression for the AND generator and its truth table.

From the figure, the diagram, and the truth table, it follows that there is an output (lamp is lit) only when both switches A and B are 1's. The number of combinations for the variables (independent) is 2^N, where N represents the number of variables and the 2 represents the two possible states for the variable (0 and 1).

If there were three switches (A, B, and C) connected in series, the number of combinations of variables would be 2^3, or 8. The Boolean expression would be $ABC = L$, and there would be an output (lamp lit) only when A AND B AND C were present as 1's.

Figure 16.11 shows the electronic block for the AND gate with two variables. The output level (R) would depend on the input independent variables A and B. The variables (A, B, and R) would have two possible voltage levels.

OR GATE

Figure 16.12 shows SPST switches connected to generate the OR function. The lamp will light when A OR B is present (1's). Figure 16.13 shows the logic gate

$AB = L$
expression

Variables

	A	B	L
	0	0	0
2^N	1	0	0
	0	1	0
	1	1	1

Truth Table

Figure 16.10
Expression and truth table for the AND gate.

A	B	R
0	0	0
1	0	0
0	1	0
1	1	1

Truth Table

Figure 16.11
Logic symbol and truth table for the AND gate.

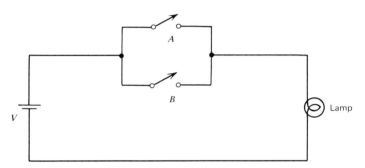

Figure 16.12
Simple OR gate.

symbol along with its truth table for the OR gate. The output variable R depends on the input independent variables A and B. The variable A, B, and R represent voltage levels.

NOT GATE

The electronic NOT gate is an inverter. If the single input variable is a 1, the output variable would be a 0. If the input variable is a 0, the output variable would be a 1. Figure 16.14 shows the electronic block and truth table for the NOT gate, or generator.

Figure 16.13
Logic symbol and truth table for the OR gate.

A	B	R
0	0	0
1	0	1
0	1	1
1	1	1

Truth Table

A ———▷○——— $R = \bar{A}$

Logic gate

A	R
0	1
1	0

Truth Table

Figure 16.14
Logic symbol and truth table for the
NOT gate.

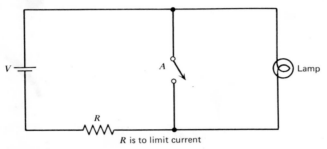

R is to limit current

Figure 16.15
Simple NOT gate.

A NOT gate can also be constructed with a switch and a lamp, as shown in Figure 16.15. If A is 1, or closed, the lamp is not lit ($L = 0$). If A is a 0, or open, the lamp is lit ($L = 1$). Figure 16.16 shows the truth table and Boolean expression for the circuit shown in Figure 16.15.

With the NOT gate, the input and output variables are complemented, or opposite if $A = 1$ and $L = 0$ and if $A = 0$ and $L = 1$.

Figure 16.17 shows a SPDT switch. From the switch itself the A and $\bar{A}$ (NOT A) functions are generated.

$A = L$
expression

A	L
0	1
1	0

Truth Table

Figure 16.16
Expression and truth table for the NOT
gate.

——————○———— $\bar{A}$

○———— A

SPST switch

A	A
0	1
1	0

Truth Table

Figure 16.17
SPDT switch generating complements.

NOT AND NAND GATES

The NOR gate (NOT-OR) is an OR gate driving an inverter (NOT gate). The truth table for the NOR gate is shown in Figure 16.18 along with its logic symbol.

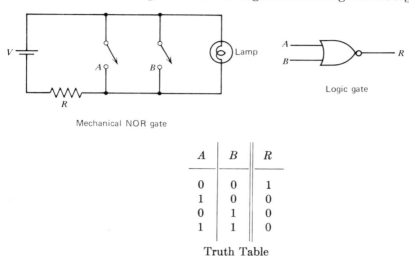

A	B	R
0	0	1
1	0	0
0	1	0
1	1	0

Truth Table

Figure 16.18
Logic NOR gate with truth table.

The NAND gate (NOT-AND) is an AND gate driving an inverter (NOT gate). The truth table for the NAND gate is shown in Figure 16.19 along with its logic symbol.

EXCLUSIVE OR GATE

Figure 16.20 illustrates an exclusive OR gate. Notice from the truth table that the output of this gate is a logic level 1 only if the inputs are complementary. Thus another name for the exclusive OR gate is an inequality detector or gate. If the output of the exclusive OR gate is inverted, the resulting gate becomes an equality gate or detector.

Figure 16.19
Logic NAND gate with truth table.

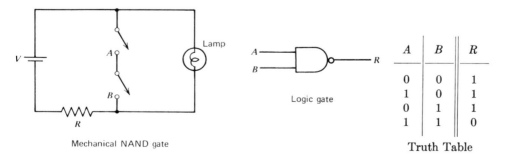

A	B	R
0	0	1
1	0	1
0	1	1
1	1	0

Truth Table

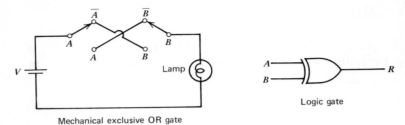

Mechanical exclusive OR gate

Logic gate

A	B	R
0	0	0
1	0	1
0	1	1
1	1	0

Truth Table

Figure 16.20
Exclusive OR gate with truth table.

LOGIC FAMILIES

The devices used in the construction of logic gates such as FFs determine the logic family. If AND and OR gates are built using diodes and resistors, the gates belong to the diode logic (DL) family. If AND, OR, NOT, NAND, and NOR gates are built using resistors and transistors, these gates belong to the resistor-transistor logic (RTL) family. If the various gates are constructed with special multiemitter transistors, the gates belong to the transistor-transistor logic (TTL) family. If the gates are constructed with diodes and transistors, the gates belong to the diode-transistor logic (DTL) family. If the various gates are constructed with only complementary (P- and N-type) metal oxide semiconductor field effect transistors, the gates belong to the CMOS[1] logic family. Figure 16.21 illustrates the various families discussed above. Each family has its advantages and disadvantages.

As a point of interest, the first logic families commonly used were DL, DTL, and RTL, in discrete and integrated form. Because of the large driving capability (fan-out) of TTL, it replaced DTL and RTL circuitry. Today, CMOS logic is starting to replace the popular TTL family. The advantages of CMOS compared to TTL are its small physical size and its low power requirement in ICs.

The number of gates that are in one monolithic IC determines if the chip fits in one of three possible categories. The first category is called small-scale integration (SSI) and is defined as a chip having 10 or fewer gates. The second category is medium-scale integration (MSI) and is defined as a chip having between 10 and 40 gates. The third category is large-scale integration (LSI) and is defined as a chip having more than 40 gates. The SSI circuitry can be

[1] Some manufacturers use the notation COS/MOS, which stands for complementary symmetry metal oxide semiconductor.

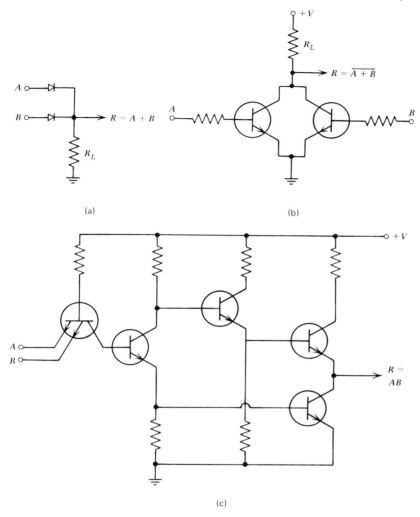

Figure 16.21
Logic gates: (a) DL OR gate, (b) RTL NOR gate, (c) TTL NAND gate, (d) DTL NAND gate, and (s) CMOS NAND gate.

constructed from any of the families of logic discussed; MSI circuitry is generally constructed from TTL and CMOS logic families; and LSI circuitry is exclusively constructed from the CMOS logic family. The MSI and LSI circuitry have made possible such systems as hand-held calculators, mini-computers, digital clocks, and hand-held digital multimeters.

DECADE COUNTING MODULE (DCM)

Because digital circuitry has only two possible states, counting must be accomplished with the binary, or base-two, system of the decade counting module

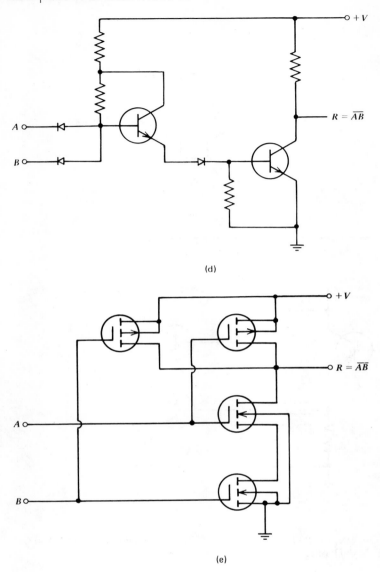

(d)

(e)

Figure 16.21 (*continued*)

(DCM). Counting in the binary system uses 1's and 0's in a weighted, position code. The weight and position for variables in the binary system are illustrated in Figure 16.22.

In Figure 16.22, the binary code with its equivalent decimal value is shown. The variables are labeled A, B, C, and D. The most significant bit is positioned at the left and labeled D. The least significant bit is positioned at the right and labeled A. Notice that if all the possible combinations for the four variables were used, a total of 16 combinations would be generated. These combinations are equivalent to the decimal 0 through 15. If only the first 10 possible combina-

		(MSB)			(LSB)

2^3 D	2^2 C	2^1 B	2^0 A	Decimal Equivalent
0	0	0	0	0
0	0	0	1	1
0	0	1	0	2
0	0	1	1	3
0	1	0	0	4
0	1	0	1	5
0	1	1	0	6
0	1	1	1	7
1	0	0	0	8
1	0	0	1	9

BCD Code

Figure 16.22
BCD code.

tions that are equivalent to the decimal 0 through 9 are used, the code is called the binary coded decimal or BCD code.

A series of FFs and gates can be connected to produce the BCD code. One type of IC that generates the BCD code is the 7490 and is called a decade counter. Figure 16.23a illustrates the package; Figure 16.23b shows the truth table for the 7490. For every input pulse or trigger, the output of the 7490 is changed by one in the BCD code. Thus the 7490 can count up to 10 triggers or events with its output in the BCD form.

BCD Count Sequence

Count	Output D	C	B	A
0	0	0	0	0
1	0	0	0	1
2	0	0	1	0
3	0	0	1	1
4	0	1	0	0
5	0	1	0	1
6	0	1	1	0
7	0	1	1	1
8	1	0	0	0
9	1	0	0	1

Reset/Count

Reset Inputs $R_{0(1)}$	$R_{0(2)}$	$R_{9(1)}$	$R_{9(2)}$	Output D	C	B	A
1	1	0	x	0	0	0	0
1	1	x	0	0	0	0	0
x	x	1	1	1	0	0	1
x	0	x	0	Count			
0	x	0	x	Count			
0	x	x	0	Count			
x	0	0	x	Count			

(a)

Figure 16.23
Signetics' 7490 decade counter: (a) truth table and (b) package.

W Package

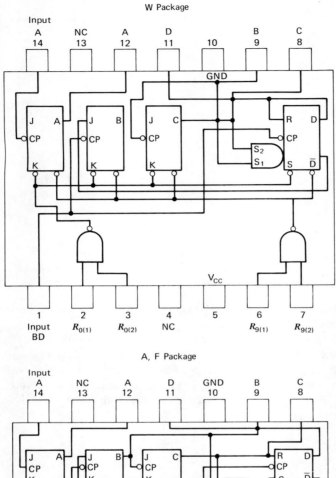

A, F Package

(b)

Figure 16.23 (continued)

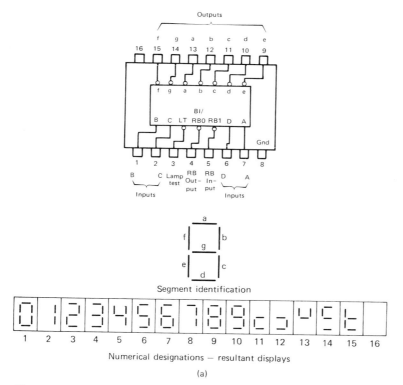

Figure 16.24
Signetics' 7447 decoder: (a) chip and (b) truth table.

Figure 16.24 shows an IC that is capable of taking the BCD and transforming it into its decimal equivalent when displayed with a seven-segment readout device. The chip is called a decoder/driver (7447).

Now if the decade counter, decode/driver, and readout display are connected, the whole device is called a decade counting module (DCM). A single DCM is capable of counting and displaying input pulses from 0 through 9. If the *D* output of one DCM is connected to the input of another DCM, the system can count and display events from 0 through 99.

With the appropriate gating circuitry, the DCMs can be used to build digital frequency meters, digital meters, digital clocks, and so forth. Chapter 20 discusses a simple temperature device that uses DCMs.

READ-ONLY MEMORY

The read-only memory (ROM) is an element that can store information in binary form (1 and 0). An example of a bipolar ROM is the 8223 ROM manufactured by the Signetics Corporation. The 8223 is a TTL 256 bit ROM organized in 32 words with 8 bits per word (see Figure 16.25a). The words are

Figure 16.24 (continued)

Decimal or Function	Inputs						Outputs							Note
	LT RBI	D	C	B	A	BI/RBO	a	b	c	d	e	f	g	
0	1 1	0	0	0	0	1	0	0	0	0	0	0	1	1
1	1 x	0	0	0	1	1	1	0	0	1	1	1	1	1
2	1 x	0	0	1	0	1	0	0	1	0	0	1	0	
3	1 x	0	0	1	1	1	0	0	0	0	1	1	0	
4	1 x	0	1	0	0	1	1	0	0	1	1	0	0	
5	1 x	0	1	0	1	1	0	1	0	0	1	0	0	
6	1 x	0	1	1	0	1	1	1	0	0	0	0	0	
7	1 x	0	1	1	1	1	0	0	0	1	1	1	1	
8	1 x	1	0	0	0	1	0	0	0	0	0	0	0	
9	1 x	1	0	0	1	1	0	0	0	1	1	0	0	
10	1 x	1	0	1	0	1	1	1	1	1	0	0	0	
11	1 x	1	0	1	1	1	1	1	0	0	1	1	0	
12	1 x	1	1	0	0	1	1	0	1	1	1	0	0	
13	1 x	1	1	0	1	1	0	1	1	0	1	0	0	
14	1 x	1	1	1	0	1	1	1	1	0	0	0	0	
15	1 x	1	1	1	1	1	1	1	1	1	1	1	1	
BI	x x	x	x	x	x	0	1	1	1	1	1	1	1	2
RBI	1 0	0	0	0	0	0	1	1	1	1	1	1	1	3
LT	0 x	x	x	x	x	1	0	0	0	0	0	0	0	4

(b)

Notes:

1. BI/BRO is wire-OR logic serving as blanking input (BI) and/or ripple-blanking output (RBO). The blanking input must be open or held at a logical 1 when output functions 0 through 15 are desired and ripple-blanking input (RBI) must be open or at a logical 1 during the decimal 0 input. X = input may be high or low.

2. When a logical 0 is applied to the blanking input (forced condition) all segment outputs go to a logical 1 regardless of the state of any other input condition.

3. When ripple-blanking input (RBI) is at a logical 0 and $A = B = C = D = $ logical 0, all segment outputs go to a logical 1 and the ripple-blanking output goes to a logical 0 (response condition).

4. When blanking input/ripple-blanking output is open or held at a logical 1, and a logical 0 is applied to lamp-test input, all segment outputs go to a logical 0.

selected by five binary address lines labeled $A_0, \ldots A_4$. The 8 output bits appear at the pins labeled $B_0, \ldots B_7$.

With this IC memory, the user can select what information is to be stored (called programming) at each word address. Once the memory is programmed, the IC permanently retains this information. As an example, the address 10010 might contain the information 10011101.

Figure 16.25*b* illustrates a 2462 ROM. The 2462 ROM is constructed from P-channel enhancement mode devices and contains 2048 bits of information in a 16-pin DIP. The 2048 bits are organized into 512 words with 4 bits per word.

ROMs can be used to store instructions and routines for calculators, "look-up" tables, code converters (converging one code to another), and so forth. I have

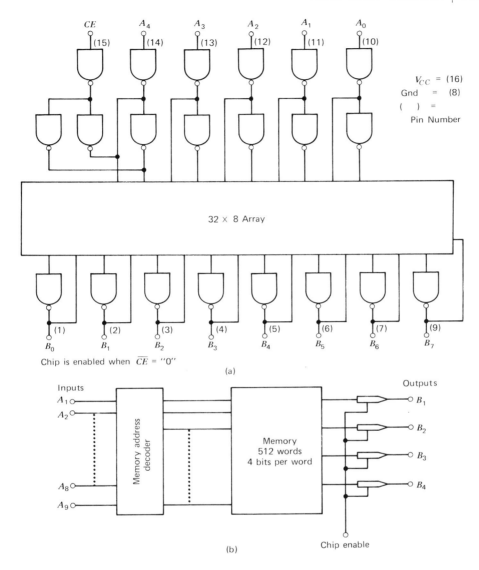

Figure 16.25
Signetics' ROMs: (a) 8223—256 bit ROM and (b) 2462—2048 bit ROM.

even used a ROM to store track patterns for my HO train layout, which has eight track switches. With these eight track switches, I have 16 possible track configurations that the train can follow. I programmed an 8223 to contain 16 track configurations (16 out of the 32 possible words) with the output of the ROM feeding the eight track switches (8 bits per word). To change track patterns, I simply address the ROM for the track pattern wanted and thus the track switches are changed.

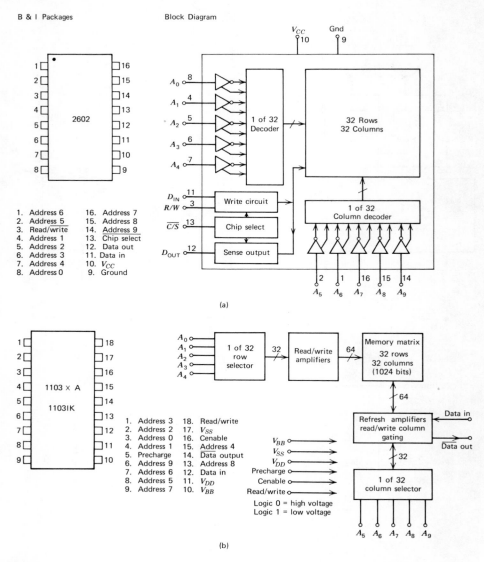

Figure 16.26
Signetics' RAMs: (a) 2602 static RAM and (b) 1103 dynamic RAM.

RANDOM ACCESS MEMORY

Another type of solid-state memory is the random access memory (RAM).
Unlike the ROM, the information stored in a RAM is not permanent. The information stored can be changed at any time by a write cycle. There are two general categories of RAMs: the static RAM and the dynamic RAM.

An example of a static RAM is the Signetics 2602 (see Figure 16.26a). This device has 1024 words with 1 bit per word. The RAM uses low threshold N-channel silicon gate technology in its construction and is completely compatible with TTL. The static RAM will store information indefinitely as long as power is maintained at the chip. The stored information is changed with the write cycle.

An example of a dynamic RAM is the Signetics 1103 (see Figure 16.26b). This device has 1024 words with 1 bit per word. The RAM is constructed from enhancement mode P-channel MOS devices integrated on a monolithic array. The information stored by a dynamic RAM must be "refreshed." The 1103 requires 32 read cycles to refresh all 1024 bits every 2 msec. Because the memory needs to be refreshed every 2 msec, additional clocking and gating circuity is necessary. The static RAM does not need to be refreshed; thus additional clocking and gating circuitry is not necessary. However, the dynamic RAM has less access and cycle time than does the static RAM. The dynamic RAM is a faster memory than the static RAM.

Because of the advancement in semiconductor technology, the dynamic RAM with large bit capacity has made feasible minicomputers and microprocessors. Computer memories have been drastically reduced in physical size, cost, and power consumption compared to the standard core memories. As an example, a 32,000-bit core memory can be replaced with eight chips, each a 4096 word, 1 bit per word, dynamic RAM.

CHARACTER GENERATOR

The character generator is a chip that, when properly addressed, will generate alpha-numeric characters. Figure 16.27 illustrates the Signetics 2513 character generator. This device can be thought of as a ROM that has already been programmed to generate 64 alpha-numeric characters when the output of the device is connected to an appropriate display device. The characters are addressed by the ASCII code and arranged in a 7 x 5 dot matrix. One of the many applications of this chip is the display of alpha-numeric characters on the face of a cathode ray tube (picture tube).

DUAL SLOPE ANALOG-TO-DIGITAL CONVERTER

One popular method used to convert an analog voltage to its digital equivalent is the dual slope analog-to-digital (A/D) converter. Figure 16.28a illustrates a diagram for the dual slope A/D converter. The switch shown in the figure is usually a solid-state switch, but a mechanical switch is demonstrated in the explanation of the circuit.

The DCMs are set at zero. The analog switch is positioned to read the input voltage V_{in}. With a constant input voltage, the integrator has an output voltage that is equal to

$$-V_{out} = \frac{V_{in}}{RC} T_1 \qquad (1)$$

As soon as the output voltage of the integrator goes negative, the output of the comparator goes positive, permitting the clock pulses to be counted by the

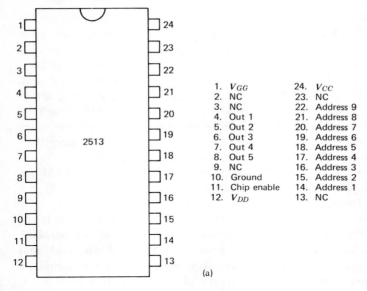

1. V_{GG}
2. NC
3. NC
4. Out 1
5. Out 2
6. Out 3
7. Out 4
8. Out 5
9. NC
10. Ground
11. Chip enable
12. V_{DD}

24. V_{CC}
23. NC
22. Address 9
21. Address 8
20. Address 7
19. Address 6
18. Address 5
17. Address 4
16. Address 3
15. Address 2
14. Address 1
13. NC

(a)

Figure 16.27
Signetics' 2513 character generator: (a) pin configuration and (b) output format.

DCMs. The maximum count of the DCMs shown in Figure 16.28b is 10,000. At the count of 10,000 in time T_1, the analog switch is positioned to the reference voltage V_{ref}. Because the polarity of the reference voltage is negative, the output of the integrator goes from a negative voltage towards zero in the time T_2.

The output of the integrator is now equal to

$$-V_{out} = \frac{V_{reff}}{RC} T_2 \tag{2}$$

Setting Equations 1 and 2 equal to each other, we have

$$V_{in} = V_{reff} \frac{T_2}{T_1} \tag{3}$$

The time T_1 is fixed and is equal to $N_1 \Delta t$ where Δt is the time pulses of the clock. The time T_2 depends on the input voltage V_{in} and is equal to $N_2 \Delta t$. The AND gate, once the comparator goes low, does not permit any more clock pulses to be counted. Thus the count N_2 on the DCMs is the voltage V_{in}. In the circuit shown in Figure 16.28b, N_1 is equal to 10,000. As can be reasoned from Equation 3, the reference voltage is very important in determining the error associated with this dual slope A/D converter.

OTHER CHIPS

There are many other circuits commonly available in SSI, MSI, and LSI form and specifically used in digital electronics. Such devices as shift registers (groups

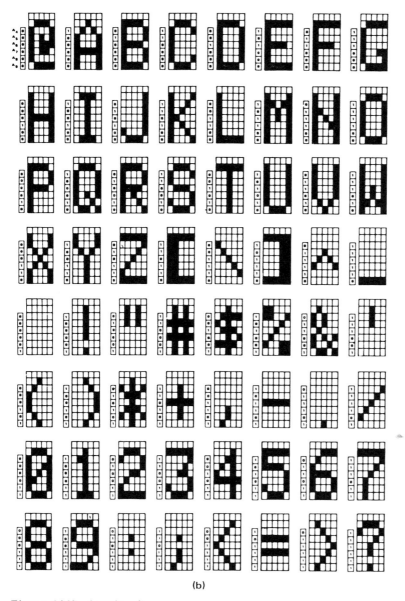

(b)

Figure 16.27 (*continued*)

of FFs), arithmetic units (adders, subtracters, multipliers, and dividers), and complete data acquisition systems (analog inputs are converted to digital outputs that are processed) are found in chips. The reader should look through the many catalogs that are available to get acquainted with some of the more commonly manufactured chips.

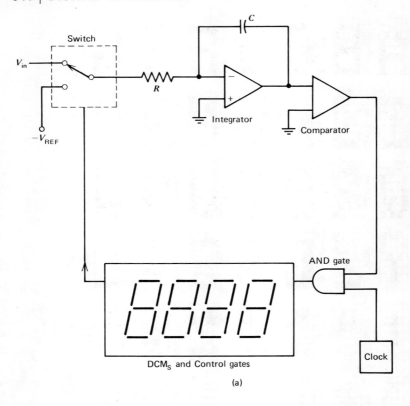

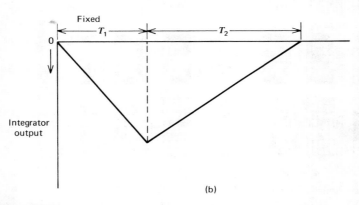

Figure 16.28
Dual slope A/D converter: (a) circuit and (b) output waveform of the integrator.

REVIEW QUESTIONS AND PROBLEMS

16-1   What is the difference between a symmetrical and asymmetrical waveform?

16-2   Why is the term *one-shot* an appropriate name for the monostable MV?

16-3   Describe the astable, bistable, and monostable MV in terms of their stable and unstable states.

16-4   Discuss the AND gate.

16-5   Discuss the OR gate.

16-6   Discuss the NOR and NAND gates.

16-7   What is a ROM? List some applications of the device.

16-8   What is a RAM? List some applications of the device.

16-9   Assume that a dual slope A/D converter is to be used with three DCMs. Select the reference voltage so that the maximum voltages displayed by the DCMs are (a) 9.99 V and (b) 0.999 V.

16-10   Conversion time for the dual slope A/D converter is the time it takes to convert the analog input voltage to its decimal equivalent. If the clock used in Problem 16-9 generated one pulse every 1 μsec, calculate its conversion time.

16-11   Using exclusive OR gates, design a circuit arrangement that would enable a person to turn a light on or off from three locations.

16-12   List some reasons why you think that a digital system is more reliable than an analog system. Would a digital system be easier to maintain and repair than an analog system?

OBJECTIVES FOR CHAPTER 17, TIMERS

At the completion of this chapter you can demonstrate your understanding of the material by answering these questions:

- How are a UJT and a PUT used as timers?
- How does a BJT timer function?
- What is an IC timer?
- Where are timers used?

17 TIMERS

Solid-state timers are rapidly replacing mechanical and electromechanical timers. You should study the various types of solid-state timers because of their frequent use in control circuitry applications.

Timers are used frequently in industrial control circuits. Electronic timing circuits have replaced the majority of mechanical timers. The various timers discussed in this chapter use the time relationship of voltage across a capacitor. Also discussed are unijunction transistors (UJT), programmable unijunction transistors (PUT), bipolar transistors, operational amplifiers, and integrated circuits used in timing devices.

UNIJUNCTION TRANSISTOR

Figure 17.1a shows an N-channel unijunction transistor (UJT). Figure 17.1b shows the equivalent circuit for the UJT; $R_{B(1)}$ is shown variable and $R_{B(2)}$ is shown fixed. Figure 17.2 shows an external supply voltage connected to a UJT and a variable input voltage V_E. If V_E is zero volts, a fixed amount of current flows through R_{B_1} and R_{B_2}. This fixed current is equal to

$$I = \frac{V_{CC}}{R_{B_1} + R_{B_2}} = \frac{V_{CC}}{R_{BB}}$$

The value of R_{BB} ranges between 4 and 10 kΩ in an UJT. The intrinsic standoff ratio is given by the manufacturer ($I_E = 0$ A) and is defined as

$$\eta = \frac{R_{B1}}{R_{BB}} = \frac{R_{B1}}{R_{B1} + R_{B2}} \tag{1}$$

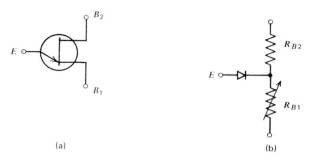

Figure 17.1
UJT: (a) symbol and (b) equivalent circuit.

In Figure 17.2 the diode is reverse-biased. The A side of the diode has a potential equal to the voltage across R_{B_1} or

$$V_{RB(1)} = V_{CC} \frac{R_{B1}}{R_{BB}} = \eta V_{CC} \qquad (2)$$

If V_E is increased, the diode becomes forward-biased (I_E increases). The holes from the emitter are injected into the N-channel from emitter to ground. The holes attract electrons from the common, causing the conductivity of $R_{B(1)}$ to increase (the resistance value of $R_{B(1)}$ decreases). This increase of conductivity results in the UJT's having a negative resistance.

The controlling parameters for the UJT are the emitter current (I_E) and voltage (V_E) and are shown in Figure 17.3. The device is useful because of the negative resistance between I_P and I_V. The I_P is the minimum value of emitter current necessary to "fire" the UJT; V_P is the minimum value of V_E to forward-bias the emitter junction of the UJT; and V_V and I_V are the operating values of the emitter voltage and current (once the UJT is fired) necessary to maintain the negative resistance of the UJT.

Figure 17.2
Triggering the UJT.

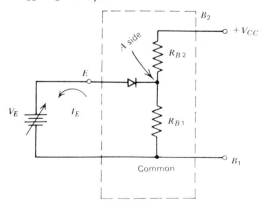

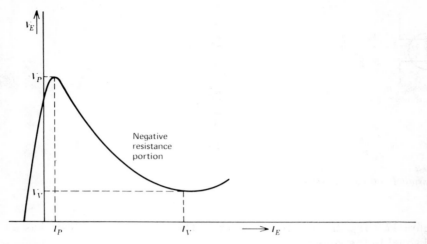

Figure 17.3
Characteristic curve for the UJT.

The I_P is given by the manufacturer and ranges between 1 and 100 μA; I_V is given by the manufacturer and ranges between 1 and 10 mA; and V_V is not given by the manufacturer but ranges between 1 and 5 V. The V_P is a function of R_{BB}, and V_{CC} and is found by

$$V_P = V_{\text{diode}} + V_{RB_1} = V_{\text{diode}} + \eta V_{CC}$$

If V_{diode} is much smaller than V_{CC}, then

$$V_P \approx \eta V_{CC} \tag{3}$$

Figure 17.4 shows a typical relaxation oscillator using the UJT. The voltage across C (v_C) increases via the current supplied through R and V_{CC}. When v_C reaches the peak value of the emitter voltage to forward-bias the UJT (V_P), $R_{B(1)}$ decreases quickly in value and discharges C until v_C reaches V_V. The sawtooth wave generated by the UJT is shown in Figure 17.5.

If V_V is much smaller than V_P and the discharge time is small, the approximate charge time (t) necessary for v_C to go from V_V to V_P can be determined. Using the charging equation for a capacitor, we have

$$v_C = V_P = V_{CC}(1 - e^{-t/RC})$$

But $V_P = \eta V_{CC}$, and then

$$\eta V_{CC} = V_{CC}(1 - e^{-t/RC})$$

Solving for time t, we have

$$t = RC \ln\left(\frac{1}{1 - \eta}\right) \tag{4}$$

where t is the approximate time taken for v_C to reach V_P. The only requirement in Equation 4 is that R be a small enough value to insure that I_P can be reached, or

$$R_{\max} = \frac{V_{CC} - V_P}{I_P} \tag{5}$$

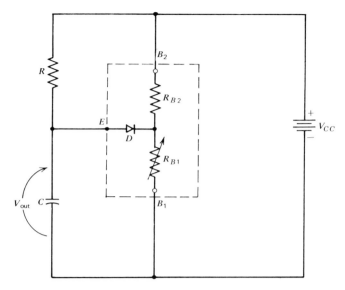

Figure 17.4
UJT oscillator.

Also, R has to keep the emitter current in the negative resistance portion of the UJT, or

$$R_{\min} = \frac{V_{CC} - V_V}{I_V} \tag{6}$$

If follows that R has to be between $R_{\max}$ and $R_{\min}$ so that the UJT oscillator will function.

Figure 17.5
Waveform for the UJT oscillator.

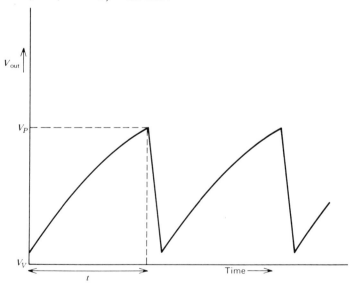

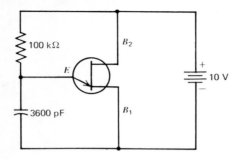

Figure 17.6
UJT oscillator used in example problem.

EXAMPLE

Using a UJT 2N2646 (V_{CC} = 10 V, η = 0.75, R_{BB} = 7 kΩ, I_P = 5 μA, and I_E = 5 mA), design a relaxation oscillator with a time period t of 500 μsec. See Figure 17.6.

Using Equation 3, we have

$$V_P \approx \eta V_{CC} = 0.75 \times 10 \text{ V} = 7.5 \text{ V}$$

Using Equation 5, we have

$$R_{\max} = \frac{V_{CC} - V_P}{I_P} = \frac{10 - 7.5}{5 \times 10^{-6}} = 0.5 \text{ MΩ}$$

Using Equation 6 and assuming that V_V = $2v$ (typical value)

$$R_{\min} = \frac{V_{CC} - V_V}{I_V} = \frac{10 - 2}{5 \times 10^{-3}} = 1.6 \text{ kΩ}$$

Select R to be 100 kΩ. Then Equation 4 becomes

$$500 \times 10^{-6} = 1 \times 10^5 \times C \times \ln\left(\frac{1}{1 - 0.75}\right)$$

or

$$C = 0.0036 \text{ μF}$$

The circuit shown in Figure 17.4 can be modified to produce spiked pulses. This modification is shown in Figure 17.7. Figure 17.8 shows the relationship between V_E and V_{out}. The width of the spike (V_{out}) is equal to the discharge time of the capacitor C. The peak voltage (V_P) is now found by

$$V_P = V_{CC}\left[\frac{(R_1 + R_{B_1})}{(R_1 + R_{BB})}\right] = V_{CC}\left[\frac{(R_1 + \eta R_{BB})}{(R_1 + R_{BB})}\right] \tag{7}$$

When v_C reaches the V_P, the UJT fires and

$$V_P = v_C = V_{CC}\left[\frac{(R_1 + \eta R_{BB})}{(R_1 + R_{BB})}\right] = V_{CC}(1 - e^{-t/RC}) \tag{8}$$

Solving Equation 8 for t, we have

$$t = RC \ln\left[\frac{(R_1 + R_{BB})}{R_{BB}(1 - \eta)}\right] \tag{9}$$

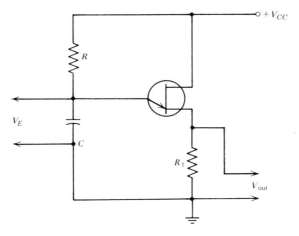

Figure 17.7
UJT oscillator having a base resistor.

where t is the approximate time between voltage spikes. If R_1 is smaller than R_{BB}, Equation 9 reduces to Equation 4. The circuit that the UJT is driving determines the value selection of R_1.

The voltage spikes that are generated across R_1 are caused by the UJT discharging the timing capacitor C. The voltage spikes across R_1 can be used to trigger a SCR in a timing configuration. This timer is shown in Figure 17.9. In the figure the UJT is triggering the gate of the SCR. The relay is the load for the SCR. Once the SCR is triggered by the voltage spike across R_1, the relay is pulled in. The relay is "reset" by opening the switch and thus removing voltage from the SCR. The diode is across the relay to insure that the emf generated by the relay does not destroy the SCR.

Figure 17.8
Waveforms for the UJT oscillation shown in figure 17.7.

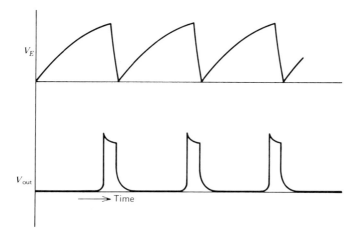

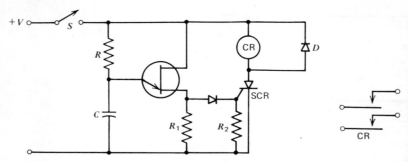

Figure 17.9
UJT/SCR timer.

PROGRAMMABLE UNIJUNCTION TRANSISTOR

The programmable unijunction transistor (PUT) is a PNPN switch with a gate. The reader may want to review the theory of the PUT discussed in Chapter 13. The PUT has a timing equation that is the same as Equation 4; however, in the PUT device the η is defined as

$$\eta = \frac{R_2}{R_1 + R_2} \tag{10}$$

In the UJT device the η is fixed and stated by the manufacturer. The advantage of the PUT over the UJT is that η can be varied in circuit configuration by the ratio of R_1 and R_2. Figure 17.10 shows a PUT timer triggering a SCR. The SCR is turned on by the timing pulse across R_3. The magnitude of the voltage spike across R_3 is approximately ηV_{cc}.

EXAMPLE

Assume that η in Figure 17.10 is to be 0.68 and that the value of $R_1 + R_2$ is to be 10 kΩ. Calculate the value of R_1 and R_2.

Figure 17.10
PUT/SCR timer.

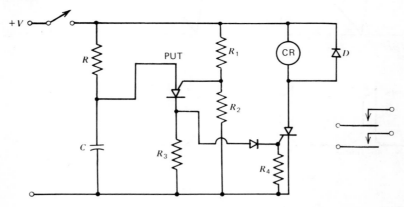

Using Equation 10, we have

$$R_2 = \eta(R_1 + R_2) = 0.68 \times 10 \text{ k}\Omega$$

$$R_2 = 6.8 \text{ k}\eta\Omega$$

Then

$$R_1 = (R_1 + R_2) - R_2 = 10 \text{ k}\Omega - 6.8 \text{ k}\Omega$$

$$R_1 = 3.2 \text{ k}\Omega$$

BIPOLAR TIMERS

Figure 17.11 shows a bipolar transistor used in a timer circuit. Transistor T in the figure has a relay for its load. With switch S in position A, the capacitor C charges to V_{CC}. The value of R is selected to limit the charging current of C with S in position A. If S is moved to position B, the capacitor C will dishcarge through R_B and h_{ie} of the transistor T. The moment S is thrown to position B, the base current of T is sufficient to pull the relay in. After a fixed amount of time, the base current of T is reduced to a minimal value that causes T to go into cutoff. The time that the relay remains "in" is determined by the values of C, R_B, and h_{ie}.

The disadvantage of the circuit shown is that if T were replaced, the timer would have to be recalibrated because of changes in the h_{FE}.

Figure 17.12 shows a circuit arrangement that causes C to charge up to a specific value of voltage to trigger the transistor T. In the figure, the switch S in position B will cause the transistor T to be in the cutoff state and capacitor C to be discharged. When switch S is moved to position A, the voltage across C will increase. After a fixed amount of time, v_C is large enough in magnitude to cause transistor T to be turned on (saturated); T being saturated will cause

Figure 17.11
BJT timer.

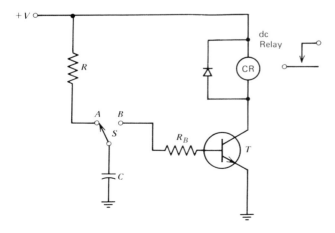

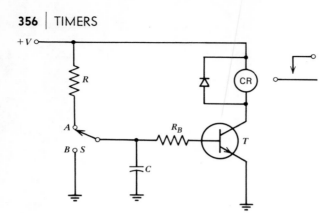

Figure 17.12
BJT timer.

the relay to be pulled in, assuming that R and R_B permit enough base current in T. With this timer, after S is moved to position A, the timing period it takes to pull the relay in starts.

OPERATIONAL AMPLIFIER TIMERS

Figure 17.13 shows a timer circuit arrangement that incorporates operational amplifiers (OP AMP) and a current pump. Transistor Q acts as a current pump. The values of the zener diode and R determine the amount of current pumped into capacitor C; R_Z limits the current to the zener diode. With the switch S in position A, the capacitor C is discharged. When S is moved to position B, the voltage across C builds up linearly according to $v_C = (I/C)t$, where I is the pump current, C the value of the timing capacitor, and t the time that the current I is pumped into C.

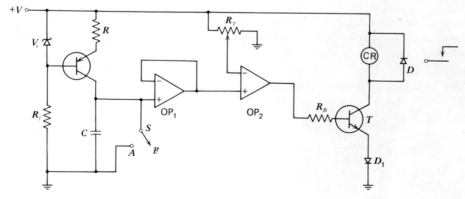

Figure 17.13
OP AMP timer.

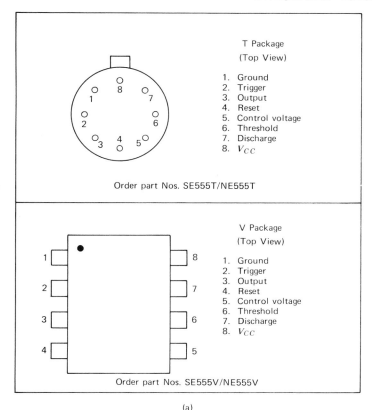

T Package
(Top View)

1. Ground
2. Trigger
3. Output
4. Reset
5. Control voltage
6. Threshold
7. Discharge
8. V_{CC}

Order part Nos. SE555T/NE555T

V Package
(Top View)

1. Ground
2. Trigger
3. Output
4. Reset
5. Control voltage
6. Threshold
7. Discharge
8. V_{CC}

Order part Nos. SE555V/NE555V

(a)

Figure 17.14
Signetics' 555 IC timer.

OP AMP$_1$ tracks the voltage across C and isolates C from the circuit. OP AMP$_2$ is used as a comparison amplifier. If the positive ($+$) input voltage is less than the voltage at the negative ($-$) input, the output of OP AMP$_2$ is negative causing T to be cutoff. When the $+$ input voltage of OP AMP$_2$ is slightly greater than the $-$ input voltage, the OP AMP$_2$ output goes positive, causing T to go into saturation, which causes the relay to be pulled in. With OP AMP$_2$ serving as a comparer amplifier between the voltage of C and the voltage at R_T, the time it takes to pull the relay in after S is thrown to position R can be varied by the setting of R_T.

The transistor T in the circuit could be replaced by a SCR. The circuit shown in Figure 17.13 was constructed using the 741 internally compensated operational amplifiers. The diode D_1 insures that the base-emitter junction of transistor T does not break down when T is reverse-biased.

INTEGRATED CIRCUIT TIMERS

Figure 17.14 shows the typical connections for a Signetics 555 Timer. Some of the features of this timer are (a) the timing ranges from microseconds to 1 hour;

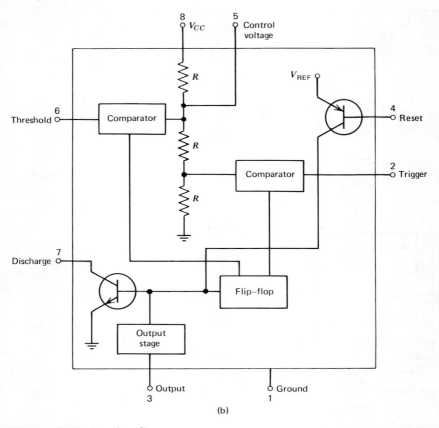

Figure 17.14 (*continued*)

(b) the temperature stability is 0.005 percent per °C; (c) it operates from a supply voltage of 4 to 15 V with a 1 percent change in timing; (d) the timing can be changed by a 10:1 ratio with an external control; and (e) the output of the timer can either sink or source 100 mA of load current. Figure 17.15 shows the electrical characteristics of the 555 Timer. Because the output of the timer can sink or source 100 mA of load current, the load for the timer can be a dc relay. Applications of this timer are shown and discussed in Chapter 20.

Figure 17.16 illustrates how a BJT can be connected to trigger the 555 timer. When switch S is closed, the BJTs collector voltage goes from a value of V_{CC} to zero volts. This negative-going pulse is used to trigger the 555 timer.

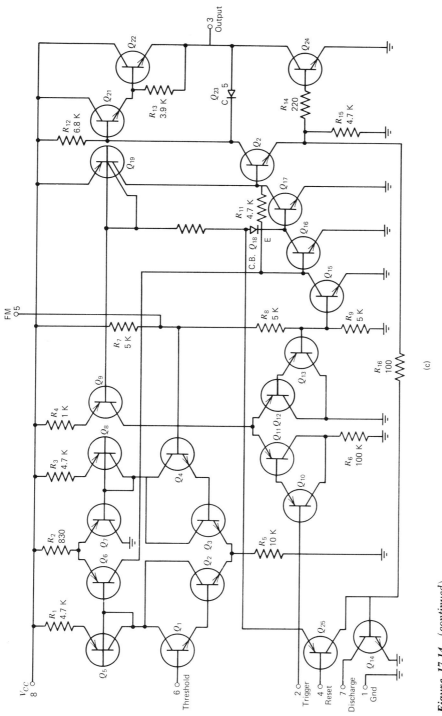

Figure 17.14 (continued)

Electrical Characteristics (25° unless otherwise specified)

Parameters	SE 555 T			NE 555 V			Units
	Min	Typ	Max	Min	Typ	Max	
Operating temperature range	−55		12.5	0		70	°C
Operating supply voltage	4.5		18	4.5		16	Volts
Operating supply current @ 5 V		3			3		mA
Operating supply current @ 15 V		8			8		mA
Timing error, 10 V[1]		0.5			1		%
Timing error, 10 V[2]		2			4		%
Time-temperature drift		30			50		ppm/°C
Time supply voltage drift		50			100		ppm/°C
Trigger voltage		1/3			1/3		V_{CC}
Trigger current		0.5			0.5		μA
Reset voltage		0.7			0.7		Volts
Reset current		0.1			0.1		mA
FM input impedance		3.3			3.3		kΩ
Bias level		2/3			2/3		V_{CC}
Deviation range		+30, −90			+30, −90		%
Output voltage drop (low)							
@ 10 mA		0.1			0.1		Volts
@ 100 mA		1.8			1.8		Volts
@ 200 mA		2.3			2.4		Volts
Output voltage drop (high)							
@ 100 mA		1.5			1.6		Volts
@ 200 mA		2.0			2.1		Volts
Rise time		100			100		nsec
Fall time		100			100		nsec

Notes:

1. R_A, R_B = 1 kOhm to 100 kOhm, t = 100 μsec.
2. R_A, R_B = 500 Ohm to 10 MOhm, t = 10 μsec.
3. External load to V_{CC}, 2.7 kOhm.

Figure 17.15
Electrical characteristics for the 555 timer.

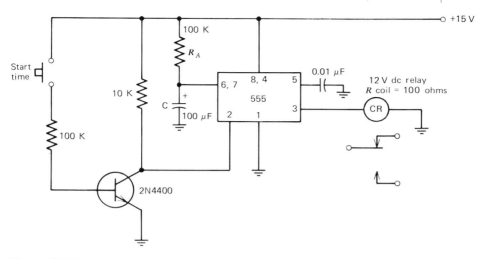

Figure 17.16
BJT triggering a 555 IC timer.

REVIEW QUESTIONS

17-1 What is a unijunction transistor?

17-2 What is a programmable unijunction transistor?

17-3 If the diode voltage of the UJT is considered, how would Equation 4 have to be modified?

17-4 Design a unijunction transistor timer driving a SCR having a t of 15 sec. Assume a voltage supply of 12 V and a relay that requires 100 mA of current to be pulled in. The UJT is to be a 2N2646 (see Appendix 2).

17-5 Repeat 17.4 using a PUT.

17-6 Is there any advantage in using a linear voltage rise rather than an exponential voltage rise across a capacitor in a timer circuit?

17-7 List some applications of timer circuits.

17-8 Why does C in Figure 17.13 need to be isolated?

17-9 Design a 5-min timer using the circuit shown in Figure 17.13. Assume a 30-V split supply ($+15$ and -15) for the OP AMPs, a 3.6-V zener diode, and a timing capacitor of 1000 μF.

OBJECTIVES FOR CHAPTER 18, PHOTOELECTRONIC DEVICES

At the completion of this chapter you can demonstrate your understanding of the material by answering these questions:

- What is light?
- What is a LED and how does it function?
- What are the various kinds of display devices commonly used?
- What is fiber optics?
- What is an optical coupler?

18 PHOTOELECTRONIC DEVICES

Photoelectronic devices are found in many electronic systems and instruments. You should study the various photoelectronic devices because of their use in such circuit applications as displays and light-activated controls.

Light and some of the common light-sensitive devices used in electronics are considered in this chapter. Several simple applications of light-sensitive devices are also provided.

LIGHT

Light behaves as both an electromagnetic wave and a particle. Experiments with interference patterns and the photoelectric effect demonstrate the duality of light. Young's double-slit experiment demonstrated that light behaves as waves because of the constructive and destructive interference patterns. To explain the photoelectric effect, Einstein developed the quantum theory of light, which states that light travels in bursts or bundles of energy. The bundles of light are called photons.

The energy of these photons was given by Planck as

$$E = h\nu \tag{1}$$

where h is Planck's constant and ν (nu) is the frequency. Because the frequency and wavelength are related by λ (lambda) $= C$ (speed of light)$/\nu$, Equation 1 becomes

$$E = \frac{hC}{\lambda} \tag{2}$$

Equation 2 relates the energy of these photons to their wavelength.

In 1913, Bohr, using Planck's and Einsteins's quantum theory, successfully explained the emission and absorption spectrum for hydrogen. Bohr's model of the atom is sometimes called a planetary model; the electrons revolve around the nucleus at particular energy levels.

If hydrogen gas is excited, it emits certain colors at specific wavelengths, called its emission spectrum. If white light passes through hydrogen, it absorbs certain colors at specific wavelengths, called its absorption spectrum. To explain this phenomenon, Bohr reasoned that the electrons traveling around the nucleus of the hydrogen atom can travel only at specific energy levels (see Figure 18.1), denoted by the orbital quantum number n ($n = 1, 2, 3, \ldots$). The energy levels are related to their quantum number by

$$E_n = - \frac{me^4}{8E_0{}^2n^2h^2} = - \frac{k}{n^2} \tag{3}$$

Figure 18.1
Bohr's model of the atom illustrating absorption and emission.

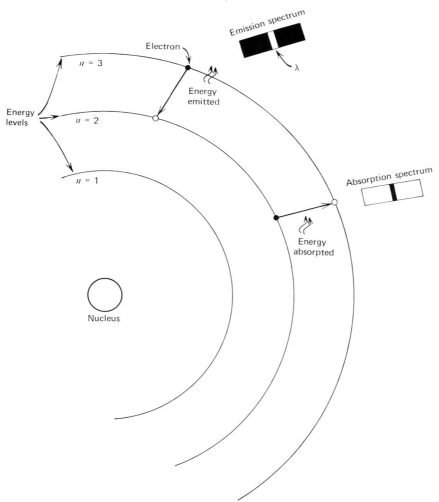

Bohr further theorized that because electrons can travel only in these energy levels around the nucleus, electrons that change from one energy level to another energy level (transition) will emit (or absorb) a quantum amount of energy. With the aid of Equations 2 and 3, we can determine the transitional energy and its related wavelength.

$$E_{n(\text{initial})} - E_{n(\text{final})} = h\nu \tag{4}$$

or

$$\frac{1}{\lambda} \approx 1.1 \times 10^7 \left(\frac{1}{n_f{}^2} - \frac{1}{n_i{}^2} \right) m^{-1} \tag{5}$$

where n_f is the final level and n_i is the initial level of the electron.

Although Bohr's theory of the atom does not take into account the spin of the electron, angular momentum, and so forth, it does demonstate the quantitized nature of the atom. Thus if energy is absorped by an atom, it is quantitized. If energy is emitted by an atom in the form of radiation, it is quantitized.

PHOTODIODE

The photodiode or light-sensitive diode (LSD) has a current that is a function of the amount and wavelength of light falling on its junction. The light-sensitive diode is reverse-biased. The energy given to the junction ($E = h\nu$) from the light causes additional hole-electron pairs to be generated. The additional electron-hole pairs increase the reverse-biased current through the diode. Thus the current through the diode is a function of the intensity and wavelength of light falling on the junction. The junction is exposed to light through some form of lens system. In inexpensive units, the lens system is simply constructed from plastic that surrounds the device.

LIGHT-EMITTING DIODE

When a diode is forward-biased, current carriers cross the junction. Electrons from the N-type material and holes from the P-type material recombine at the junction and give off energy. The energy given off is a function of the particles above ground state. This transition of energy states determines the wavelength and thus the frequency of the light emitted by the diode (see Figures 18.2a and b). The material used in the construction of the light-emitting diode (LED) is gallium arsenide phosphide (GsAsP).

Single LEDs commonly come in colors (wavelengths) of red, green, and yellow. LEDs have a life expectancy of over 250,000 h and are capable of operating at temperatures between -55 and 100 °C. These devices are also compatible with TTL logic voltages (5V). Because of the above listed advantages, the LED is replacing filament type indicators commonly referred to as pilot lamps.

Combinations of LEDs can be used for numeric display of characters from 0 to 9 (see Figure 18.2c). One of the common arrangements is in a seven-segment

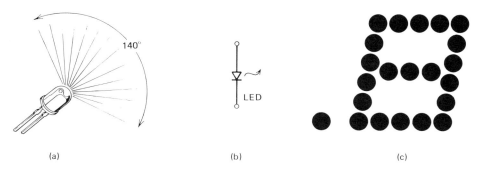

Figure 18.2
Light-emitting diode: (a) picture, (b) symbol, and (c) LED display.

display. The LEDs in this type of display may have a common cathode or common anode. Some of the applications of LED readouts are in calculators, digital watch displays, and digital instruments.

PHOTOSENSITIVE TRANSISTORS

If light is allowed to fall on the junction region of a transistor, its collector current will increase. The energy given to the junction by the light causes additional electron-hole pairs to form, increasing the collector current of the transistor. As with the photodiode, light is focused on the junction by a lens system of some type. The light-sensitive transistor (LST) is more sensitive to changes in light than is the photodiode. Figure 18.3a illustrates the schematic symbol for the light-sensitive transistor.

A Darlington pair connection can be used to increase the light sensitivity of a transistor (see Figure 18.3b). The Darlington pair is exposed to light by some simple lens system. This combination is more sensitive than a single transistor. Some of the many applications of this light detector are RPM meters, light-operated relays, light modulators, and optical couplers.

Figure 18.3
Light-sensitive transistors: (a) single BJT and (b) Darlington pair.

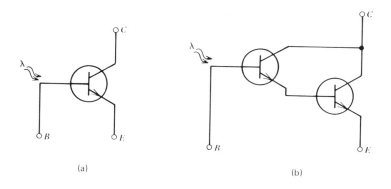

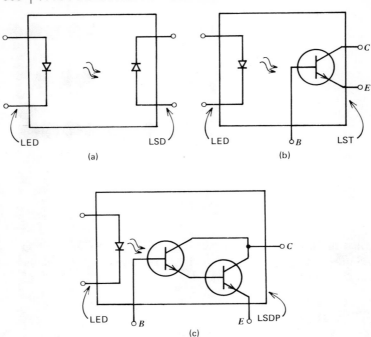

Figure 18.4
Optical couplers: (a) LED/LSD, (b) LED/LST, and (c) LED/LSDP.

OPTICAL COUPLERS

An optical coupler is constructed by optically connecting a light-emitting device to a light-sensitive detector in a light-tight enclosure. The light-emitting device is usually a LED. The light-sensitive detector can be a photodiode, a light-sensitive transistor, or a light-sensitive Darlington pair. The schematic symbols for these devices are shown in Figure 18.4.

Optical couplers are used in many applications as switches or relays. The major advantage of the optical coupler is its high isolation resistance between the light emitter and light detector. With this very high resistance, the two circuits that the device couple are isolated.

OTHER LIGHT-SENSITIVE DEVICES

Figure 18.5 illustrates a light-activated silicon controlled rectifier (LASCR), a light-activated silicon controlled switch (LASCS), and a light-sensitive field effect transistor (LSFET or photoFET). With the LASCR and LASCS, the devices are turned on by a specified amount of light. Once these devices are turned on, they can be made to stay on; thus they act as switches. The photoFET drain current is a function of the amount of light falling on the device. With all three devices, some form of a lens system is generally incorporated with the devices.

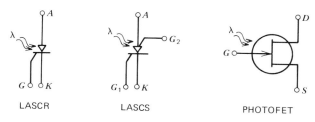

Figure 18.5
Symbols for the LASCR, LASCS, and photoFET.

INCANDESCENT DISPLAY

The RCA Numitron is an example of a seven-segment incandescent display device. This device has seven filaments that make up the segments in the display (see Figure 18.6). The eighth filament is the decimal point (DP) in the display. The Numitron has a common line to which one end of all the filaments are connected. The filament supply voltage is 5 V, which makes this display compatible with TTL gates. The display is very bright, and filters can be placed in front of the tube in order to display a particular color. All filament-type devices are sensitive to vibrations and overvoltage and can burn out.

FLUORESCENT DISPLAY

Another type of display device is the fluorescent display. The device has a filament that, because of thermionic emission, supplies electrons. These electrons are attracted to the seven anodes arranged in a seven-segment display

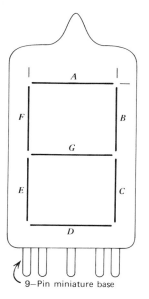

Figure 18.6
Construction of the RCA Numitron.

9—Pin miniature base

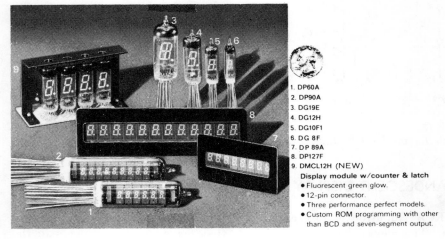

1. DP60A
2. DP90A
3. DG19E
4. DG12H
5. DG10F1
6. DG 8F
7. DP 89A
8. DP127F
9. DMCL12H (NEW)
   Display module w/counter & latch
   • Fluorescent green glow.
   • 12-pin connector.
   • Three performance perfect models.
   • Custom ROM programming with other
     than BCD and seven-segment output.

Figure 18.7
Fluorescent display devices (ISE Electronics).

format. With the application of a low voltage (15 to 35 V) at the anode with respect to common, electrons strike the anode and produce a fluorescent green glow (energy absorbed and radiated). Figure 18.7 illustrates some possible configurations for the fluorescent display device.

LIQUID CRYSTAL DISPLAY

Figure 18.8 shows one of the many configurations for the liquid crystal display (LCD). The LCD can be operated from low voltages, has extremely low power

Figure 18.8
Liquid crystal display LCD (Hamlin Inc.).

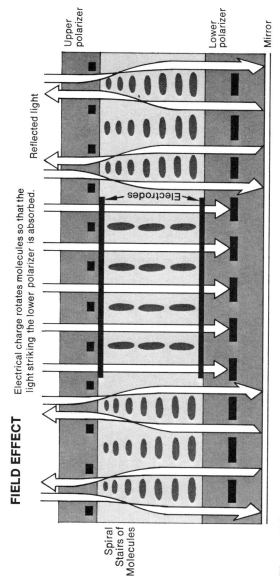

Figure 18.9
LCD field effect display.

consumption, and requires no filament; as ambient light becomes brighter, the display becomes brighter. However, the LCD can function over a limited temperature range and cannot be seen in the dark unless another light source is present.

The field effect LCD is the most popular and uses a nematic liquid crystal. When a potential is applied across the LCD, incident light is absorbed. With no potential, incident light is reflected from the device (see Figure 18.9). The electrical charge rotates the molecules in the nematic crystal so that the incident light is absorbed by the lower polarizer. The result of this absorption is a dark digit that contrasts with a light background. The reverse can be obtained, that is, a light digit that contrasts with a dark background.

The major advantages of the LCD are the very low power consumption and the large size of the display relative to other types of read-outs. The LCD is directly compatible with CMOS gates and has made possible digital wrist watches that have the time displayed continuously.

GAS DISCHARGE DISPLAYS

Gas discharge displays have been used in electronics for a long time. The Nixie tube was one of the first numeric display devices. This tube has 10 cathodes, each constructed in the form of a numeral (0 through 9) and each physically placed behind one another. With a high voltage (180 V) placed across the

Figure 18.10
Bar display (Burroughs Inc.): (a) display and (b) circuit for analog voltage measurements.

(a)

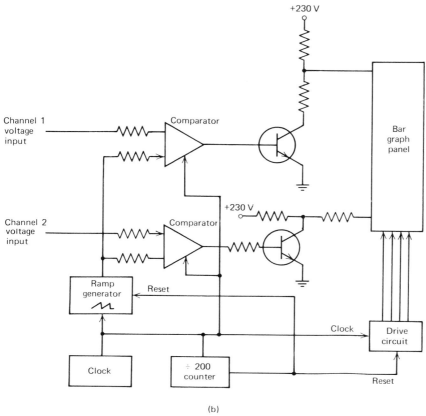

Figure 18.10   (*continued*)

cathode (numeral) and the common anode, an orange glow caused by the gas discharge is formed around the numeral.

Another interesting gas display device is the Self-Scan bar graph illustrated in Figure 18.10*a*. The display is a flat panel indicator consisting of two separate bar graphs. Each bar graph contains 200 elements (somewhat like a neon bulb). From a distance the elements appear to blend to appear as a continuous bar. The length of the bars and the rate at which they are scanned are controlled by a special three-phase clock (see Appendix 1). The bars are neon orange. Figure 18.10*b* illustrates a block diagram for the displaying of two analog voltages appearing at channels 1 and 2.

FIBER OPTICS

Fiber-optic light pipes are one means of transmitting light from one point to another. The inside coating of the tube, because of its low index of refraction, produces total internal reflection for the light inside the tube (see Figure 18.11*a*). Bundles of these tubes or fibers are placed together to construct light pipes.

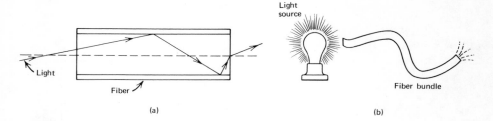

(a) (b)

Figure 18.11
Fiber optics: (a) light traveling inside fiber and (b) fibers bundled together to form a light pipe.

Thus, with the light reflecting inside the fibers, the light pipe can be bent (see Figure 18.11*b*).

A single light source is generally optically connected to one end of the light pipe; the other end of the light pipe emits the light. Some of the many applications of light pipes are light transmission lines, remote visual inspectors, and optical card readers for computers.

PHOTOCONDUCTIVE CELLS

The resistance of many materials changes when light falls on them. One such material is cadmium sulphide (CdS). Its resistance decreases as the amount of light falling on the material increases. Figure 18.12*a* shows the physical construction of a CdS cell, and Figure 18.12*b* illustrates the schematic symbol for the photocell. The "dark" resistance (no light) is determined by the material used and the amount of material making up the cell. It is important to realize that the peak response of various photocells (sensitivity to a specific band of frequencies) depends on the type of material used. As an example, some materials are more sensitive to infrared than to wavelengths in the visible spectrum.

APPLICATIONS

The light-operated relay, the capacitor discharger, the revolution meter, the card identifier, the infrared image detector, and the motionless light switch are some simple applications of light-sensitive devices.

Figure 18.12
Photocell: (a) construction and (b) symbol.

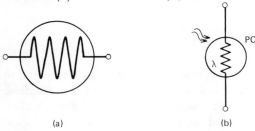

(a) (b)

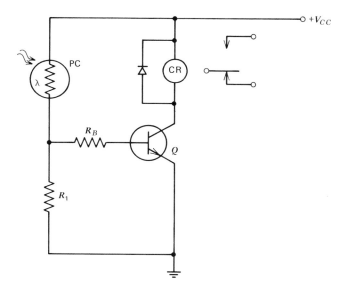

Figure 18.13
Light-activated relay.

Light-operated relay. Figure 18.13 illustrates a light-operated relay. Transistor Q is in series with a dc relay. The relay is pulled in when sufficient base current is supplied to the base of the transistor. The photocell is in series with resistor R_1 and the combination behaves as a voltage divider.

With no light falling on the PC, its resistance is high and prevents the necessary base current from pulling in the relay. With light falling on the PC, its resistance decreases sufficiently to permit base current to flow and thus to pull in the dc relay.

The diode across the relay suppresses cemf generated when the relay drops out. With this circuit arrangement, the relay is pulled in with a sufficient amount of light. If the PC and R_1 were reversed in position, the relay would be pulled in when there was no light. With this type of circuit arrangement, lights in a home could be turned on when it got dark outside.

Capacitor discharger. Figure 18.14 shows a circuit arrangement that discharges a timing capacitor when the capacitor is not in use (see Chapter 17). The timing circuit consists of capacitor C and resistor R. When switch S_w is closed, the voltage across C rises. This voltage is used as a triggering source for a timer. With S_w closed, transistor Q is saturated via the base current supplied by R_2 and V. With Q saturated, no current flows through the LED inside the optical coupler. With no light from the LED, the resistance of the light-sensitive transistor is very high and thus permits C to charge normally

When S_w is opened, Q goes into cutoff. With Q in cutoff, current flows through R_1 into the LED. The LED being on causes the light-sensitive transistor to have a very low resistance and thus discharges C. With this arrangement, the voltage across C always starts from zero volts for timing purposes.

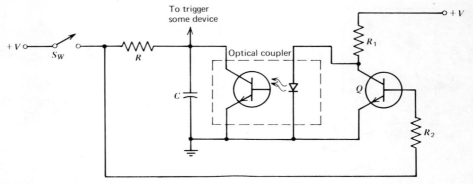

Figure 18.14
Capacitor discharge using an optical coupler.

Revolution meter. A light-sensitive transistor (LST) can be used to record rotational movement. Such a circuit arrangement is shown in Figure 18.15.

The rotating body can be a shaft, blades, and so forth. A reflective strip is marked on the shaft so that the light source is reflected at least once every revolution to the light-sensitive transistor. The pulses from the light-sensitive transistor are fed to a Schmitt trigger, which "squares" them up. The output of the Schmitt trigger is fed to a count-rate-meter (see Chapter 2). The value of the current indicated on the meter is equal to fC_1V (assuming $C_2 = C_1$) where f is the frequency of the pulses from the Schmitt trigger, C_1 the value of the input capacitor, and V the peak magnitude of the voltage pulses generated by the Schmitt trigger.

Card identifier. Figure 18.16 illustrates a card identifier. The card is inserted in a slot for identification purposes. The card has a specific number of holes, which are arranged in a certain pattern peculiar to each card. The arrangement of the location of the holes in the card and the number of holes are "read" by the device shown in Figure 18.16.

Figure 18.15
RPM device.

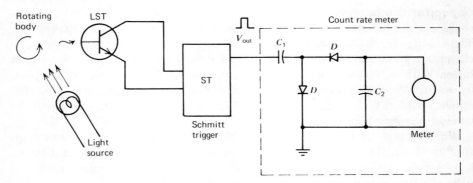

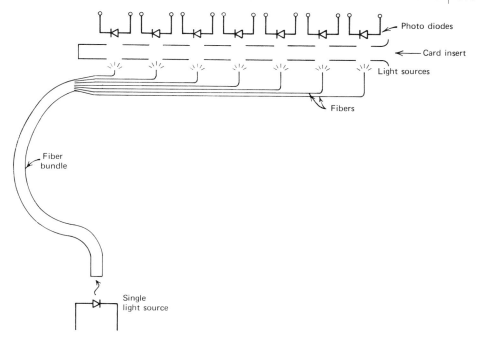

Figure 18.16
Optical card reader.

The output of the photodiodes is logically gated to read a specific card. A single light source is optically coupled to a light pipe. The output of the light pipe consists of individual fibers that are placed across from each of the photodiodes. When a card is inserted, a certain pattern is "read" by the card reader to identify the card. The system shown in the figure can be expanded to read IBM cards for a computer.

Motionless light switch. Figure 18.17 illustrates a simple circuit to "turn on the lights" automatically. The device consists of an impulse relay, a Darlington pair, a light-sensitive transistor, and a light source. The light source and light-sensitive transistor are located across from each other in a doorway. If the light source is not interrupted, the resistance of the LST is low because of the light falling on the LST. Under this condition, the voltage across capacitor C is a small value. With a small voltage across C, the Darlington pair is not permitted to conduct; thus no current can flow through the impulse relay. With one pulse, the contacts open; with another, the contacts close.

A person entering the room simply stands in the doorway for a specified number of seconds. As the person stands in the doorway, the light beam is broken, causing the resistance of the LST to be very high. With the high resistance of the LST, the capacitor charges up to the potential V in accordance with the time constant of RC. Once the voltage across C reaches a value sufficient to cause the Darlington pair to conduct, the impulse relay is pulsed.

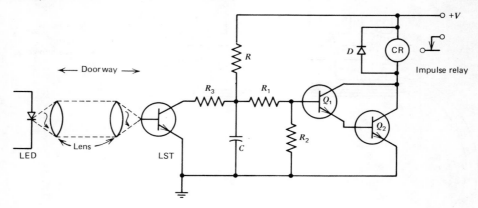

Figure 18.17
Motionless light switch.

Once the relay is pulsed, the contacts change position, turning the lights on (or off). The time constant is necessary so that the lights are not turned on (or off) every time a person breaks the light beam. The amount of time necessary for a person to stand in the doorway is determined by the value of RC and the voltage divider of R_1 and R_2. The lens system simply focuses the light source on the LST.

Infrared image detector. An infrared image detector contains a light-sensitive cathode. When energy in the infrared wavelength strikes this cathode, electrons are emitted. The electrons are attracted to and focused on a phosphorous-coated anode that acts as the viewing screen via a lens system for the image detector. As the electrons strike the phosphorous viewing screen, the screen absorbs their energy. The absorbed energy excites the phosphorous atoms which in turn emit energy in the visible spectrum which appears as yellowish-green light on the screen (see Figure 18.18). Thus objects that emit or reflect infrared wavelengths (invisible to the naked eye) can be seen with such an optical device as this image detector. Such a principle can be used to view people and objects with no visible light. For efficient operation, the detector requires very high anode potentials and a good lens system.

Figure 18.18
Image tube.

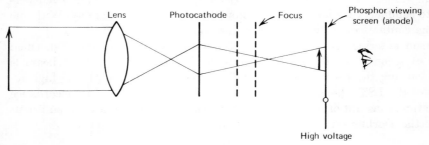

REVIEW QUESTIONS AND PROBLEMS

18-1   What is light?

18-2   What is the difference between a light-sensitive and a light-emitting device?

18-3   What is a photon?

18-4   Why does a LED have a longer life span than an incandescent bulb?

18-5   How and why does light affect the covalent bonds in semiconductor material?

18-6   What is an optical coupler? List some applications for this device.

18-7   List some of the many applications for display devices.

18-8   Compare the display devices discussed in this chapter.

18-9   What type of display does your calculator use?

18-10 What is a light pipe and how is it constructed? List some applications for the light pipe and/or fibers.

18-11 List and discuss various types of light-emitting devices.

18-12 List and discuss various types of light-sensitive devices.

18-13 Design a motionless light switch. The device designed will turn the lights on or off when a person stands in the doorway for a period of 4 sec.

18-14 If a light-sensitive device is exposed to ordinary, 60 Hz incandescent light, will the output of the device be 60 Hz? Explain.

18-15 Why is a lens system used with light-sensitive devices?

OBJECTIVES FOR CHAPTER 19, OSCILLATORS

At the completion of this chapter you can demonstrate your understanding of the material by answering these questions:

- What is the difference between sinusoidal and nonsinusoidal oscillators?
- What are the differences between Hartley and Coalpitts oscillators?
- What is a crystal oscillator?
- What are the differences between crystal and LC oscillators?
- What are the various devices that can be used in the construction of an oscillator?
- How do the PUT and UJT function as oscillators?
- What are phase shift oscillators?

19 OSCILLATORS

There are a variety of oscillator circuit configurations classified as sinusoidal and nonsinusoidal. You should study the various types of oscillator circuit configurations in order to become familiar with applications of the oscillator in communication, instrumentation, and industrial control circuitry.

An oscillator converts the dc input supply voltage to an output waveform having a frequency. The output waveform configuration of the oscillator determines whether the oscillator belongs to the sinusoidal or nonsinusoidal category. A sinusoidal oscillator generates a wave form that looks like a sine (or cosine) wave. All other oscillators that generate waveforms that do not look like a sine (or cosine) belong to the nonsinusoidal category. Some of the waveforms that belong to the nonsinusoidal category are the square, triangular, and sawtooth waveforms. This chapter is divided into sinusoidal and nonsinusoidal oscillator circuit configurations. The active devices used in the circuit configurations discussed are the OP AMP, BJT, UJT, PUT, and FET.

LC-SINUSOIDAL OSCILLATORS

As was discussed in Chapter 9, the general equation for an amplifier with feedback is given by

$$A_{vf} = \frac{A_v}{1 - \beta_v A_v} \tag{1}$$

where A_{vf} is the closed loop gain, A_v the open loop gain, and β (beta) the feedback factor. If in Equation 1 the denominator is zero, the amplifier behaves as an oscillator. For the amplifier to behave as an oscillator, the magnitude of the denominator in Equation 1 would be

$$|1 - \beta A_v| = 0 \tag{2}$$

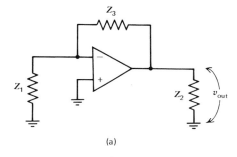

(a)

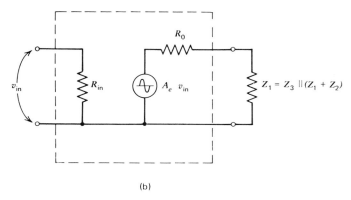

(b)

Figure 19.1
Oscillator: (a) OP AMP and (b) equivalent circuit.

Equation 2 is sometimes referred to as the Barkhausen criterion for oscillation. The equation implies that the term βA_v be equal to 1 and real (j terms equal 0).

To derive general equations for the requirements of a circuit configuration to oscillate, we shall use an OP AMP. These equations will be used for the majority of the oscillators discussed in this chapter.

Figure 19.1a illustrates an OP AMP feeding an impedance Z_2. A portion of the output voltage is fed back to the input of the OP AMP through the impedance network consisting of Z_1 and Z_3. Because of the selection of the various impedances (Z_1, Z_2, and Z_3), the feedback voltage is positive or in phase and is called regenerative feedback.

The impedance that the OP AMP is feeding is Z_2 paralleled with the series combination of Z_1 and Z_3, or $Z_T = Z_2 || (Z_1 + Z_3)$. Figure 19.1b illustrates the circuit model for the OP AMP. The volage gain for the OP AMP is given by

$$A_v = A \frac{Z_T}{R_0 + Z_T} \tag{3}$$

where R_0 is the output resistance of the OP AMP. The voltage fed back to the input of the OP AMP is given by

$$\beta_v = \frac{Z_1}{Z_1 + Z_3} \tag{4}$$

The Barkhausen criterion for oscillation requires that

$$\beta_v A_v = 1 \tag{5}$$

Substituting Equations 3 and 4 into Equations 5, we have

$$\left(\frac{Z_1}{Z_1 + Z_3}\right)\left(A_v \frac{Z_T}{Z_T + R_0}\right) = 1 \tag{6}$$

The total impedance seen by the OP AMP is $Z_2 \| (Z_1 + Z_3)$. Substituting this relationship into Equation 6, we have

$$A_v \frac{Z_1 Z_2}{Z_Z(Z_1 + Z_3) + R_0(Z_1 + Z_2 + Z_3)} = 1 \tag{7}$$

If we assume for the moment that the impedances are purely reactive, we have

$$Z_1 = jX_1 \qquad Z_2 = jX_2 \qquad Z_3 = jX_3$$

Substituting the above reactance equations into Equation 7 and recalling that $j^2 = -1$, we have

$$\frac{-A_v X_1 X_2}{-X_2(X_1 + X_3) + jR_0(X_1 + X_2 + X_3)} = 1 \tag{8}$$

The Barkhausen criterion required that βA_v be equal to 1 and real. For βA_v to be real, the j terms in Equation 8 have to be equal to 0. Because R_0 is certainly real, the sum of the reactances must be chosen to be 0, or

$$X_1 + X_2 + X_3 = 0 \tag{9}$$

Substituting Equation 9 into Equation 8, we have

$$\frac{A_{vf} X_1 X_2}{X_2(X_1 + X_3)} = 1$$

but because $X_1 + X_3 = -X_2$, we have

$$-A_{vf} = \frac{X_2}{X_1} \tag{10}$$

Equation 10 tells us that the gain of the amplifier must have a 180° phase shift and that the reactance of the elements X_1 and X_2 must have the same sign. Thus X_1 and X_2 must both be inductors or capacitors. It follows from Equation 9 that, if X_1 and X_2 must have the same sign, X_3 must have the opposite sign. If both X_1 and X_2 are inductors, X_3 is a capacitor; the circuit configuration is called a Hartley oscillator. If both X_1 and X_2 are capacitors, X_3 is an inductor; the circuit configuration is called a Colpitts oscillator.

OP AMP: HARTLEY AND COLPITTS OSCILLATORS

Figure 19.2 illustrates Hartley and Colpitts oscillator configurations. The gain of the OP AMP used in these configurations is given by

$$-A_{vf} = \frac{R_f}{R_i} \tag{11}$$

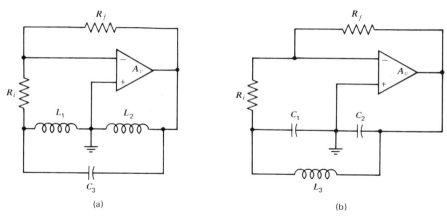

Figure 19.2
OP AMP oscillators: (a) Hartley and (b) Colpitts.

If in Figure 19.2b we assume no mutual coupling between the two inductors L_1 and L_2, the gain necessary for the circuit to oscillate is given by Equation 10, or

$$-A_{vf} = \frac{j\omega L_2}{j\omega L_1} = \frac{L_2}{L_1} \tag{12}$$

Now equating Equation 12 with Equation 11, we have

$$\frac{R_f}{R_i} = \frac{L_2}{L_1} \qquad M = 0 \tag{13}$$

If in Figure 19.2a we assume that a mutual coupling does exist between the inductors L_1 and L_2 (which is usually the case), Equation 12 has to be modified to Equation 14.

$$\frac{R_f}{R_i} = \frac{L_2 + M}{L_1 + M} \qquad M \neq 0 \tag{14}$$

The approximate oscillating frequency can be derived from Equation 9, or

$$f_0 = \frac{1}{2\pi\sqrt{L_T C_3}} \tag{15}$$

where $L_T = L_1 + L_2 + 2M$.

In Figure 19.2b, the gain necessary for the circuit to oscillate is given by Equation 10, or

$$-A_v = \frac{1/j\omega C_2}{1/j\omega C_1}$$
$$-A_v = \frac{C_1}{C_2} \tag{16}$$

Equating Equation 16 to Equation 11, we have

$$\frac{R_f}{R_i} = \frac{C_1}{C_2} \tag{17}$$

The frequency of oscillation for the Colpitts oscillator is approximately

$$f_0 = \frac{1}{2\pi\sqrt{L_3 C_T}} \tag{18}$$

where $C_T = C_1 C_2/(C_1 + C_2)$.

EXAMPLE

Calculate the resonance frequency for a Colpitts oscillator using an OP AMP and the necessary values of R_i and R_f if the inductor L is 100 μH and $C_1 = C_2$ equals 1000 pF. Using Equation 18 to calculate f_0, we have

$$f_0 = \frac{1}{2\pi\sqrt{L_3 C_T}}$$

$$= \frac{1}{2\pi\sqrt{100 \ \mu\text{H} \times 500 \ \text{pF}}}$$

$$f_0 = 712 \ \text{kHz}$$

Using Equation 17, we have

$$\frac{R_f}{R_i} = \frac{C_1}{C_2} = \frac{1000 \ \text{pF}}{1000 \ \text{pF}} = 1$$

Let $R_f = R_i = 10 \ \text{k}\Omega$.

L/C RATIO

The L/C ratio of the tank circuit of the oscillator determines the Q of the LC circuit. For a given resonance frequency, the product of LC can be made up of infinite combinations of L and C. If L is chosen to be large, the Q of the tank circuit is high.

BJT AND FET HARTLEY OSCILLATORS

The gain for a BJT or FET is related to the h_{fe} or g_m of the device. Because the h_{fe} or g_m of the device changes from unit to unit, an oscillator using a BJT or FET generally has a form of automatic gain adjustment in the form of bias clamping (see Chapter 2). If an oscillator incorporates bias clampng, its output will not become distorted because of variations in h_{fe} or g_m. The bias clamping used by a FET is also called gate leak bias.

Figure 19.3 illustrates some possible circuit configurations for the Hartley oscillator using BJTs and FETs. In Figures 19.3a and b, the bias clamping consists of the resistor R and the capacitor C together with the input junction of the device. The input junction acts like a diode to produce a biasing voltage across the resistor R (see Chapter 2). If the h_{fe} or g_m varies, the bias changes according to the signal across the tank circuit, thus affecting the gain of the stage. This automatic adjusting of the gain causes the output waveform not to be distorted; thus the stage operates in the class C region.

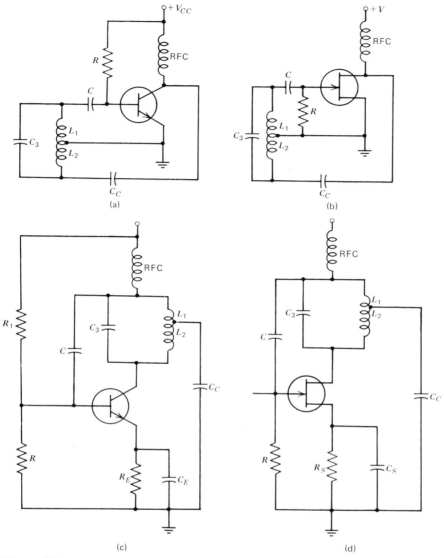

Figure 19.3
Hartley oscillators: (a) BJT shunt-fed, (b) FET shunt-fed, (c) BJT series-fed, and
(d) FET series-fed.

As a general rule of thumb, the time constant of RC is selected to be 10 times
the period of oscillation, or

$$RC \approx 10 \times \frac{1}{f_0} \tag{19}$$

where f_0 is the resonant frequency of the oscillator.

Figure 19.3a illustrates a BJT used in a Hartley circuit configuration. The
output of the BJT is fed to the bottom of the tank circuit via the coupling

capacitor C_c. The reactance of C_c should be at least one-tenth the resistance of the tank circuit at resonance. The RFC (radio frequency choke) provides a high ac reactance at the oscillator frequency to isolate the collector of the BJT from the power supply and a low dc resistance to supply voltage at the collector of the BJT. The output of the BJT (collector) feeds the reactance L_2 together with the feedback network consisting of L_1 and C_3.

The tap on the total inductor is generally located at a 10 percent point. Thus $0.1L_T$ equals L_1 and $0.9L_T$ equals L_T (approximately). With this arrangement, the transistor should have a minimum h_{fe} of 10. If the BJT has a higher h_{fe} than 10, the bias clamping will adjust the oscillator so that the output waveform is not distorted.

Figure 19.3b illustrates a FET used in a Hartley oscillator configuration. The minimum transconductance g_m that the FET requires to oscillate is given by

$$g_m \geq \frac{L_2}{r_{ds}L_1} \qquad (20)$$

(see the Review Questions).

Both circuit configurations shown in Figures 19.3a and b are called shunt-fed oscillators. Shunt-fed implies that no dc current for the BJT or FET flows through the inductor L (L_1 and L_2).

Figures 19.3c and d are other Hartley circuit configurations called series-fed oscillators. Series-fed implies that dc current for the BJT or FET does flow through the inductor L. The minimum requirements for the h_{fe} and g_m for the BJT or FET are the same in all four circuit configurations.

BJT AND FET COLPITTS OSCILLATORS

Figure 19.4 illustrates some possible circuit configurations for the Colpitts oscillator configuration using BJTs and FETs. Figures 19.4a and b are shunt-fed circuit configurations. Automatic biasing is accomplished by the resistor R and the capacitor C along with the input junctions of the devices. With the configuration shown in Figure 19.4a, the values of the capacitors C_1 and C_2 are generally selected to be equal. With equal values of capacitance, the minimum h_{fe} of the BJT is 1. If h_{fe} is greater than 1, the bias clamping network will cause the stage to work class C, having an undistorted output waveform.

Figure 19.4b illustrates a FET used in a Colpitts oscillator. The minimum requirement of the transconductance g_m is given by

$$g_m \geq \frac{C_1}{r_{ds}C_2}$$

(see the Review Questions). Figures 19.4c and d illustrate series-fed Colpitts oscillator circuit configurations using a BJT and a FET. Notice that the dc current for the BJT and the FET flows through the inductor L.

With all the circuit configurations illustrated in Figure 19.4, the BJT or FET feeds the reactance of C_2 and the feedback network consists of the reactances of L_3 and C_1. The frequency of oscillation is given by Equation 18.

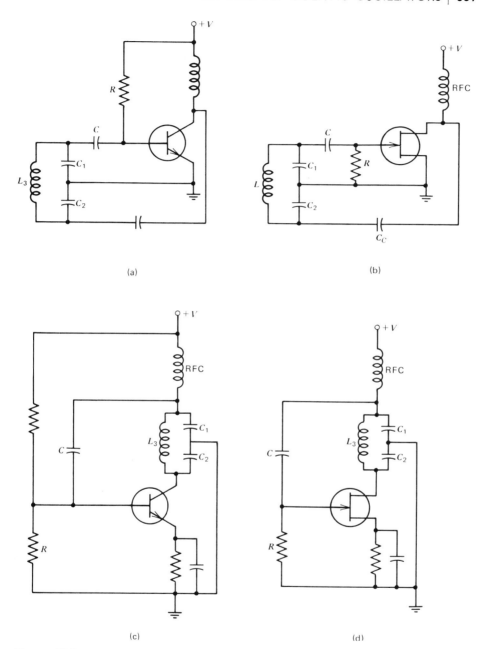

(a)

(b)

(c)

(d)

Figure 19.4
Colpitts oscillators: (a) BJT shunt-fed, (b) FET shunt fed, (c) BJT series fed, and
(d) FET series-fed.

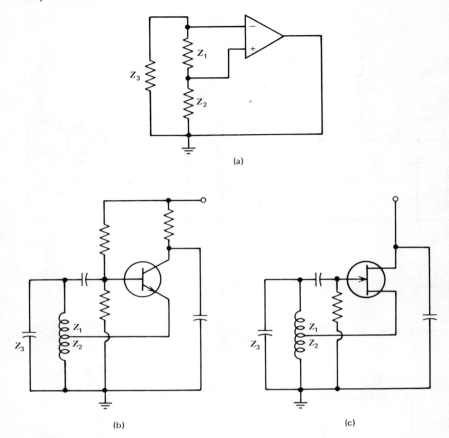

Figure 19.5
Other Hartley configurations: (a) equivalent circuit, (b) BJT, and (c) FET.

OTHER CIRCUIT CONFIGURATIONS FOR THE HARTLEY OSCILLATOR

Figure 19.5 illustrates two other circuit configurations for the Hartley oscillator. The circuit model for the oscillator is shown in Figure 19.5a; the oscillators using a BJT or a FET are illustrated in Figures 19.5b and c. Both of these circuit configurations are series-fed oscillators. Note the difference between these circuits and the ones shown in Figures 19.1 and 19.3a and b.

The 10 percent tap for the inductor is placed at the lower end of the tank circuit, compared to the upper end for the oscillators shown in Figures 19.3a and b. The equations discussed so far would have to be modified because of the circuit model used for these oscillators.

ARMSTRONG OSCILLATORS

Figure 19.6 illustrates a BJT and a FET used in circuit configurations commonly called Armstrong oscillators. The tank circuit consists of inductor L_1 and

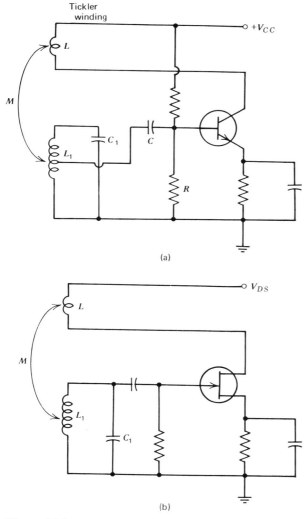

Figure 19.6
Armstrong oscillators: (a) BJT and (b) FET.

capacitor C_1. The tickler winding L couples energy from the output of the BJT or FET into the tank circuit of L_1 and C_1. The losses of the tank circuit are supplied by the BJT or FET amplifier and are fed back in phase with the tank circuit (regeneration). The physical coupling between the tickler winding and the tank L_1 determines the magnitude of energy fed back.

The bias clamping network of R and C insures that the output of the oscillator is sinusoidal and not distorted. As illustrated in Figure 19.6a, the inductor L_1 is tapped. The tap location serves as an impedance matching device for the input of the BJT. Note that because of the high input impedance of the FET, a tap is not necessary for the tank circuit. The frequency or oscillation is determined by $f_0 = 1/(2\pi\sqrt{LC})$.

CRYSTAL SINUSOIDAL OSCILLATORS

The frequencies of the LC sinusoidal oscillators discussed so far have been determined by the value of the inductor(s) and capacitor(s) in the tank circuit. As can be reasoned, a small change in the value of L and/or C will change the frequency of oscillation. Thus the frequency stability of the LC oscillator is dependent upon how small the changes in the value of L and/or C are kept. The value of L and/or C changes with temperature, vibration, and current through the tank circuit. In order to overcome the frequency instability of the LC oscillator, a quartz crystal can be used for the frequency determining network of an oscillator.

A quartz crystal causes an oscillator to have excellent frequency stability. Quartz crystals are transducers, converting mechanical energy into electrical energy, and vice versa. This conversion process is called the piezoelectric effect. If an ac voltage is placed across the face of a crystal, the crystal will physically vibrate. At a particular frequency, the vibration of the crystal is maximum and is called its natural resonating frequency. Because the crystal has a natural resonant frequency, the circuit model for a crystal is a tank circuit comprised of an inductor, capacitors, and a resistance. This circuit model is illustrated in Figure 19.7.

In the figure, the L and C represent the mass and elasticity of the crystal; R represents the losses in its vibrational mode; and C_1 represents the capacitance between the faces of the crystal due to the holder.

Figure 19.7c represents a plot of reactance versus frequency of the crystal. As can be seen, the crystal has two resonant frequencies. One frequency is called the series resonant frequency labeled f_s and is caused by the combination

Figure 19.7
Crystal: (a) symbol, (b) equivalent circuit, and (c) curves.

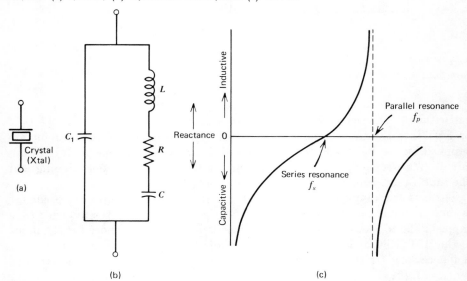

of L, C, and R in the circuit model. The other resonant frequency is called the parallel resonant frequency labeled f_p and is caused by the series combination of C_1, C, and LR. The parallel frequency is slightly higher than the series frequency.

$$f_s = \frac{1}{2\pi\sqrt{LC}} \tag{21}$$

$$f_p = \frac{1}{2\pi\sqrt{L[CC_1/(C + C_1)]}} \tag{22}$$

The bandwidth or difference between f_s and f_p is given by

$$f_p - f_s = f_s\left(\frac{C}{2C_1}\right) \tag{23}$$

(see the Review Questions).

The crystal in its parallel mode f_p has a very high resistance at resonance. The crystal in its series mode f_s has a very low resistance at resonance. The Q at f_s and f_p is very high (10^4 typically). The fundamental frequency of a crystal is limited to approximately 30 MHz. However, a crystal can have overtones or harmonics of the fundamental frequency. The manner in which the crystal is cut determines whether the vibrations are maximum with the fundamental or overtone frequency. With overtones, the crystal can be operated well into the 100 MHz to 200 MHz range.

EXAMPLE

Assume that $L = 1$ H, $C = 0.02$ pF, and $C_1 = 5$ pF. Calculate f_s and f_p. Using Equations 21, we have

$$f_s = \frac{1}{2\pi\sqrt{1 \times 0.02 \times 10^{-12}}}$$

$$f_s = 1.125 \text{ MHz}$$

$$f_p = \frac{1}{2\pi\sqrt{1 \times \dfrac{5 \times 10^{-12} \times 0.02 \times 10^{-12}}{5 \times 10^{-12} + 0.02 \times 10^{-12}}}}$$

$$f_p = 1.127 \text{ MHz}$$

CRYSTAL OSCILLATOR CONFIGURATION

As can be seen from the previous example problem, the difference between f_p and f_s is small. The placement of the crystal in the oscillator circuit configuration determines if f_s or f_p dominates.

Figure 19.8 illustrates several crystal controlled oscillators that function close to the series resonant frequency f_s. These circuit configurations are sometimes referred to as Pierce oscillators. Note that the circuits shown in the figure require no tank circuit.

The crystal is functioning near f_s and is slightly inductive (Z_3 in Figure 19.1). The internal capacitance of the BJT or FET serves as Z_1 and Z_2. Thus, Z_1 is the

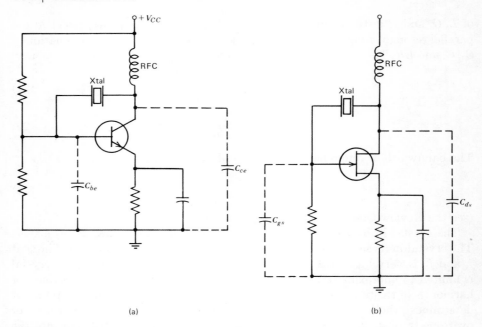

Figure 19.8
Crystal oscillators using f_s: (a) BJT and (b) FET.

base-to-emitter capacitance (C_{be}) or the gate-to-source capacitance (C_{gs}) and Z_2 is the collector-to-emitter capacitance (C_{ce}) or the drain-to-source capacitance (C_{ds}).

Figure 19.9 illustrates several crystal controlled oscillators that function near the parallel resonant frequency f_p. Figure 19.9a illustrates a Miller oscillator circuit configuration. The crystal is slightly inductive near f_p (Z_1 in Figure 19.1). The tank circuit of LC is tuned slightly off resonance so that it is inductive (Z_2). The internal capacitance of the FET (C_{dg}) serves as the feedback (Z_3). Figure 19.9b illustrates a modified Colpitts oscillator. The capacitance $(C_1$ and $C_2)$ serves as feedback elements, and the crystal is slightly inductive.

Very good frequency stability can be obtained when a crystal is used to determine the frequency of an oscillator. Thus changes in supply voltage, temperature, and other device parameters have very little effect on the frequency of oscillation. If the effects of temperature on the frequency of a crystal need to be minimized, the crystal can be placed inside a crystal oven. The crystal oven is operated at a higher temperature than the ambient temperature of the circuit configuration. The temperature of the crystal oven is maintained at a specified level by control circuitry.

SLIGHT CHANGES IN CRYSTAL FREQUENCY

The frequency f_p of a crystal can be changed slightly by placing a small value of capacitance across the crystal. In mobile transmitters that are crystal controlled,

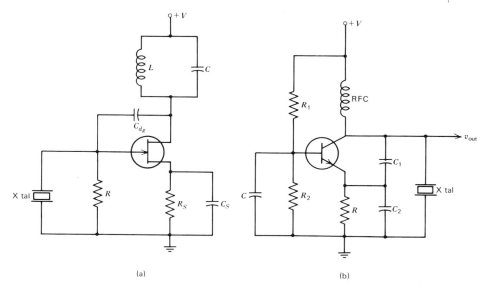

Figure 19.9
Crystal oscillators using f_p: (a) FET and (b) BJT.

very often a small trimmer capacitance is wired across the crystal holder. The frequency of the transmitter can be changed slightly or "trimmed" by varying the capacitance of the trimmer.

RC PHASE SHIFT OSCILLATORS

Sinusoidal waveforms can be generated with the use of resistors and capacitors arranged in a circuit configuration called a phase shift oscillator. Figure 19.10 illustrates the requirements of a phase shift oscillator. The amplifier (A_v) has a 180° phase difference between its input and output signals. The phase shift network has a 180° phase shift at a specified frequency. Thus a phase shift totaling 360° is achieved, causing regeneration. The gain for the amplifier is set equal to the loss in the phase shift network.

Figure 19.11 illustrates a few of the circuit arrangements of resistors and capacitors to achieve the required 180° phase shift at a particular frequency. The derivation of frequency and loss is shown in the following equations.

The mesh equations for the network shown in Figure 19.11a can be written as

$$v_{\text{in}} = i_1\left(R + \frac{1}{sc}\right) - i_2(R) - i_3(0) \tag{24}$$

$$0 = i_1(R) + i_2\left(2R + \frac{1}{sc}\right) - i_3(R) \tag{25}$$

$$0 = i_1(0) - i_2(R) + i_3\left(2R + \frac{1}{sc}\right) \tag{26}$$

where $s = j\omega$.

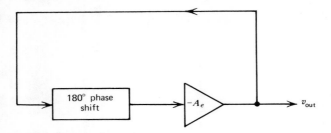

Figure 19.10
Block diagram for a phase shift oscillator.

Solving these equations for i_3, we have

$$i_3 = \frac{v_{in} R^2}{R^3 - \dfrac{5R}{(\omega C)^2} + j\left[\dfrac{1}{(\omega C)^3} - \dfrac{6R^2}{\omega C}\right]} \qquad (27)$$

Now the output voltage is given as $i_3 R$, or

$$v_{out} = \frac{v_{in} R^3}{R^3 - \dfrac{5R}{(\omega C)^2} + j\left[\dfrac{1}{(\omega C)^3} - \dfrac{6R^2}{\omega C}\right]} \qquad (28)$$

Figure 19.11
Phase shifting networks.

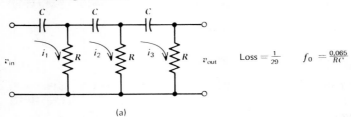

$$\text{Loss} = \frac{1}{29} \qquad f_0 = \frac{0.065}{RC}$$

(a)

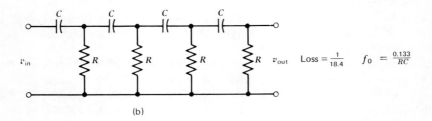

$$\text{Loss} = \frac{1}{18.4} \qquad f_0 = \frac{0.133}{RC}$$

(b)

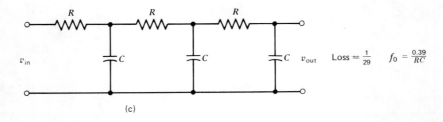

$$\text{Loss} = \frac{1}{29} \qquad f_0 = \frac{0.39}{RC}$$

(c)

For regenerative feedback the j term in Equation 28 must equal zero, or

$$\frac{1}{(\omega C)^3} = \frac{6R^2}{\omega C}$$

$$(\omega C)^2 = \frac{1}{6R^2}$$

$$f_0 = \frac{1}{2\pi\sqrt{6}RC} = \frac{0.065}{RC} \tag{29}$$

Thus the frequency of oscillation is given by Equation 29. In order to determine the loss of the network, we must use Equation 28. Substituting $(\omega C)^2$ from Equation 29 into Equation 28 and solving for the gain of the network, we have

$$v_{\text{out}} = \frac{v_{\text{in}}R^3}{R^3 - 5R(6R^2)}$$

$$\frac{v_{\text{out}}}{v_{\text{in}}} = -\frac{1}{29} \tag{30}$$

Equation 30 tells us that the loss of the network at f_0 is $\frac{1}{29}$ and the phase shift is 180°. The mesh equations can be applied to the other circuit configurations shown in Figure 19.11 to obtain the loss and frequency of the networks given in the figure (see the Review Questions).

RC PHASE SHIFT OSCILLATOR CONFIGURATIONS

Figure 19.12 illustrates RC phase shift oscillators using an OP AMP, a BJT, and a FET. In Figure 19.12a an OP AMP is used in the phase shift oscillator. The gain of the OP AMP is 29 and is determined by the ratio of R_f to R_i. The value of R_i should be selected so that it is much greater than R.

Figure 19.12b illustrates a FET used in a phase shift oscillator. The gain of the FET amplifier is 18.4 and can be approximated by $g_m R_L$. The FET amplifier is class A and is biased by resistor R_s. Note that the phase shifting network consists of four RC combinations. These combinations are selected because the FET amplifier needs a gain of only 18.4. A voltage gain of 29 for a single-stage FET amplifier is difficult to obtain.

Figure 19.12c illustrates a BJT used in a phase shift oscillator. The gain of the BJT amplifier is selected to be 29. The parallel combination of the biasing resistors R_1 and R_2, together with h_{ie} of the BJT, has to be equal to the value of R. Because of the variations of h_{ie} between units, the input resistance of the circuit can be changed by exposing a portion of the emitter-resistor R_E. Thus R_E reflected into the base of the BJT is varied by changing the position of the wiper arm on the potentiometer and allowing the total input resistance of the BJT amplifier to be equal to R.

The phase shift oscillator generally is used in a frequency range of less than 1 hertz to hundreds of kiloHertz. The phase shift oscillator has the advantage of size over the LC oscillator at lower frequencies. As an example, a 1-kHz phase

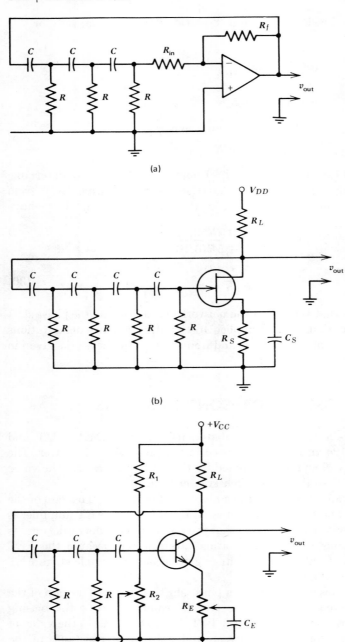

Figure 19.12
RC phase shift oscillators: (a) OP AMP, (b) FET, and (c) BJT.

shift oscillator would require only resistors and capacitors, whereas a 1 kHz LC oscillator would requird a relatively large value of inductance L for the frequency-determining network, which is costly, large in size, and quite heavy. Another advantage of the RC oscillator over the LC oscillator at low frequencies is the ease of changing frequencies. Simply changing the values of resistance will change the output frequency of the RC oscillator.

WEIN-BRIDGE OSCILLATORS

Figure 19.13a illustrates one of the many possible circuit configurations for the Wein-bridge oscillator. The frequency determining network is constructed of a series RC combination connected from the output of the OP AMP to the non-inverting input of the OP AMP and of a parallel RC combination from the non-inverting input of the OP AMP to common. The RC combinations, together with the feedback provided by R_f and R_i, cause a phase shift of 360° between the output and input of the OP AMP. The ratio of R_f to R_i determines the gain of the OP AMP, which compensates for the losses of the frequency-determining network.

The input voltage at the noninverting input of the OP AMP in Figure 19.13b can be calculated by the voltage division theorem, or

$$v_{in} = v_{out} \frac{Z_1}{Z_1 + Z_2} \tag{31}$$

$$v_{in} = v_{out} \left[\frac{1}{sRC + (1/sRC) + 3} \right] \tag{32}$$

where $Z_1 = R \| 1/sC$, $Z_2 = R + 1/sC$, and $s = j\omega$.

Substituting $j\omega$ for s, we have

$$v_{in} = v_{out} \left[\frac{1}{3 + j[\omega RC - (1/\omega RC)]} \right] \tag{33}$$

In order for the circuit to oscillate, the j term must be zero, or

$$j \left(\omega RC - \frac{1}{\omega RC} \right) = 0 \tag{34}$$

$$f_0 = \frac{1}{2\pi RC} \tag{35}$$

Equation 35 gives the frequency at which the circuit will oscillate.

Setting the j term in Equation 34 equal to zero, we see that the loss of the network is

$$v_{in} = v_{out} \left(\frac{1}{3} \right) \tag{36}$$

The input voltage at the inverting terminal of the OP AMP is given by

$$v_{in} = v_{out} \left(\frac{R_i}{R_i + R_f} \right) \tag{37}$$

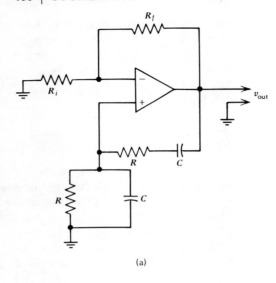

(a)

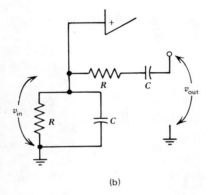

(b)

Figure 19.13
Wein-bridge oscillator: (a) circuit and (b) frequency selective network.

In order to calculate the values necessary for R_i and R_f, we must set Equations 36 and 37 equal to each other, or

$$v_{out}\left(\frac{1}{3}\right) = v_{out}\left(\frac{R_i}{R_i + R_f}\right)$$

or

$$R_f \geq 2R_i \qquad\qquad (38)$$

As can be reasoned from Equation 38, if the gain of the OP AMP is too large, the output waveform will be distorted. If the gain of the OP AMP is too small, the circuit will not oscillate. For stable operation of the oscillator, R_f is maintained slightly larger than $2R_i$. In order to compensate for the aging of component values, R_f can be replaced with a thermistor having a negative temperature coefficient, or R_i can be replaced with a temperature-sensitive resistor having a positive temperature coefficient. Thus as the output voltage increases or decreases, the value of the feedback resistor changes so that the gain of the amplifier is adjusted to maintain a distortionless output waveform. With some Wein-bridge configurations, R_i is a 7-W incandescent lamp.

NONSINUSOIDAL OSCILLATORS

A few of the waveforms that belong to the category of nonsinusoidal are square, sawtooth, and triangular waveforms. The rest of this chapter discusses the generation of these nonsinusoidal waveforms.

UJT and PUT sawtooth generators. Figure 19.14a illustrates a simple circuit configuration using a unijunction transistor (UJT) to generate a sawtooth

Figure 19.14
Oscillators: (a) UJT and (b) PUT.

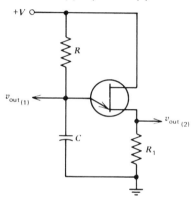
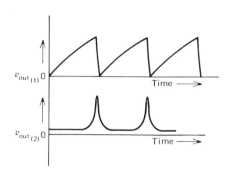

(a)

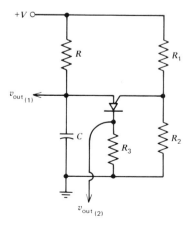
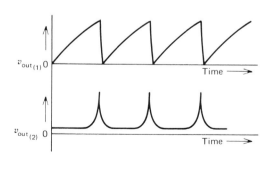

waveform. The theory of the UJT was discussed in Chapter 17. As can be seen from the figure, the waveform across the timing capacitor C is an approximate sawtooth. The period of the waveform is given by

$$t \approx RC \ln \left(\frac{1}{1 - \eta}\right) \tag{39}$$

(see Equation 4 in Chapter 17). One of the limitations of the UJT is that η (eta) is fixed by the manufacturer. If the fixed η is a handicap, a programmable unijunction transistor (PUT) can be incorporated in the circuit design.

Figure 19.14b illustrates a simple circuit configuration using the PUT. Because the PUT can be programmed for various values of η, the device can be used over a much greater frequency range than the UJT. The values of R_1 and R_2 determine η.

$$\eta = \frac{R_2}{R_1 + R_2} \tag{40}$$

For a free-running oscillator at a frequency determined by Equation 39, the total resistance of $R_1 + R_2$ is generally less than 100 Ω. If the sum of $R_1 + R_2$ is much greater than 100 Ω, the PUT will remain in its conduction state after being turned on by the capacitor voltage at its anode.

In the circuit configurations illustrated in Figure 19.14, the sawtooth waveforms are nonlinear. One method of generating a linear sawtooth waveform is illustrated in Figure 19.15a, where Q_1, D_z, and R_1 form a discrete current pump using a BJT. The voltage across the capacitor C is linear because of the constant current generated by the BJT current pump. The voltage across a capacitor with a constant current I is given by

$$V_C = \frac{I}{C} t \tag{41}$$

where t is the amount of time I is pumped into C.

The triggering point for a UJT is given by

$$V_P = V_{\text{diode}} + \eta V_{CC}$$

$$V_P \approx \eta V_{CC} \tag{42}$$

If Equations 41 and 42 are set equal to each other, we have

$$V_C = V_P$$

or

$$t = \frac{\eta C V_{CC}}{I} \tag{43}$$

where t is the period of the linear sawtooth waveform.

A simple application of a linear sawtooth is shown in Figure 19.15b. The frequency of the Hartley oscillator is changed by the depletion capacitance of the voltage-variable capacitor diode. The magnitude of change in the capacitance of the diode is determined by the voltage magnitude of the linear sawtooth across capacitor C. Thus the range of frequencies that the Hartley oscillator generates

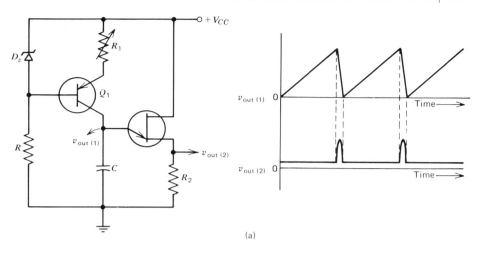

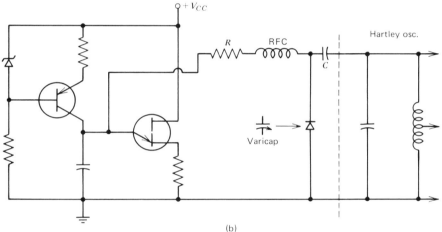

Figure 19.15
Sawtooth generator: (a) circuit and (b) application.

is determined by the magnitude of the sawtooth voltate; the frequency of the sawtooth determines the rate at which the Hartley "sweeps" this frequency range. The circuit arrangement is commonly called a FM "sweep" generator.

Square-wave generator. Figure 19.16 illustrates a circuit configuration for a square-wave generator. This particular circuit arrangement is called an astable multivibrator. The astable multivibrator is constructed with two transistor switches Q_1 and Q_2. The switches have two possible states: saturation and cutoff. The circuit functions because the timing capacitors C_1 and C_2 permit only one of the switches to be saturated (or cutoff) at a time. Thus if Q_1 is cutoff, Q_2 is saturated. The voltage across the timing capacitors determines which transistor is saturated (or cutoff) and the amount of time each transistor is saturated (or cutoff).

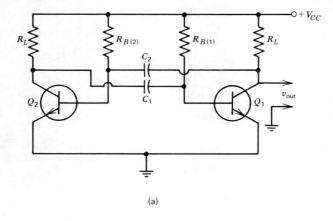

(a)

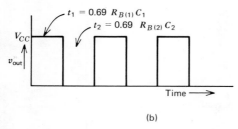

(b)

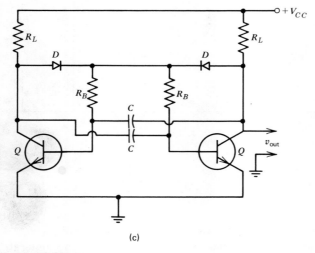

(c)

Figure 19.16
Astable MV: (a) circuit, (b) waveform, and (c) self-starting.

If we assume for the moment that Q_1 is saturated and Q_2 is cutoff, the voltage across C_1 would charge up through R_L to a value of $(V_{CC} - V_{BE(sat)})$. The voltage across C_2, which is holding Q_2 in its cutoff state, can then charge up to a voltage of $(V_{CC} - V_{CE(sat)})$. When the voltage across C_2 reaches the saturation voltage of the base-emitter $(V_{BE(sat)})$, Q_2 goes from cutoff to saturation. The transition of Q_2 causes Q_1 to go from saturation to cutoff. The amount of time t that Q_2 and Q_1 are in cutoff is given by

$$t_1 = R_{B(1)}C_1 \ln \left\{ \frac{2V_{CC} - [V_{BE(\text{sat})} + V_{CE(\text{sat})}]}{V_{CC} - [V_{BE(\text{sat})} + V_{CE(\text{sat})}]} \right\} \tag{44}$$

$$t_2 = R_{B(2)}C_2 \ln \left\{ \frac{2V_{CC} - [V_{BE(\text{sat})} + V_{CE(\text{sat})}]}{V_{CC} - [V_{BE(\text{sat})} + V_{CE(\text{sat})}]} \right\} \tag{45}$$

If $V_{BE(\text{sat})}$ and $V_{CE(\text{sat})}$ are small compared to V_{CC}, then Equations 44 and 45 become

$$t_1 = R_{B(1)}C_1 \ln 2 = 0.69 \ R_{B(1)}C_1 \tag{46}$$

$$t_2 = R_{B(2)}C_2 \ln 2 = 0.69 \ R_{B(2)}C_2 \tag{47}$$

The frequency of the waveform is the reciprocal of the period, or

$$f = \frac{1}{T} = \frac{1}{t_1 + t_2} \tag{48}$$

If $R_{B(1)} = R_{B(2)} = R_B$ and $C_1 = C_2 = C$, the freauency is given by

$$f = \frac{1}{1.39 \ R_B C} \tag{49}$$

Figure 19.16c illustrates a self-starting astable MV. The diodes D_1 and D_2 insure that both transistors are not saturated or cutoff at the same time. In the figure, if both transistors are at cutoff, both diodes are forward-biased, permitting base current to flow to both transistors. If both transistors are saturated, the diodes are reverse-biased, cutting off both transistors by not permitting base current to flow to the transistors. Thus the diodes prevent both transistors from being in saturation or cutoff at the same time. The only states that can exist are if Q_1 is saturated, Q_2 is cutoff, and if Q_1 is cutoff, Q_2 is saturated.

Triangular waveform generator. Figure 19.17 illustrates a circuit configuration using OP AMPs to generate a triangular waveform. OP AMP$_1$ is connected as an astable MV; OP AMP$_2$ is connected as an integrator (see Chapter 15). The zener diodes clamp the square wave from the MV to a level of ± 10 V. The integrator has an output voltage equal to $-V_{\text{out}} = (V_{\text{in}}/RC)\,t$ where t is the amount of time the input voltage is a $+V$ or a $-V$. When the input voltage to the integrator is negative, the output of the integrator has a positive slope. When the input voltage is positive, the output of the integrator has a negative slope.

EXAMPLE

Assume that the square-wave generator shown in Figure 19.17 has a period of 1 msec. Calculate the necessary values of R and C so that the triangular waveform has a peak-to-peak voltage of 10 V. With the square wave having a total period of 1 msec, the output is a $+10$ V for 500 μsec and -10 V for 500 μsec. The output voltage of the integrator increases to a maximum $+10$ V and -10 V.

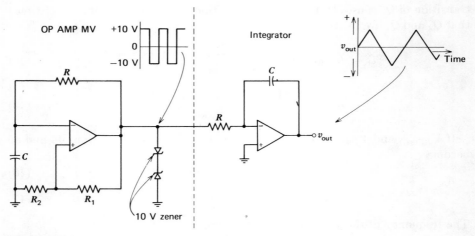

Figure 19.17
Triangular waveform generator.

Calculating the RC values, we have

$$-V_{out} = \frac{V_{in}}{RC} t$$

$$-10\ V = -\frac{10\ V}{RC}\ 500 \times 10^{-6}\ sec$$

or

$$RC = 5 \times 10^{-4}\ sec$$

If we let $C = 0.001\ \mu F$, R becomes

$$R = \frac{5 \times 10^{-4}\ sec}{10^{-9}} = 5 \times 10^{5}\ \Omega$$

$$R = 0.5\ M\Omega$$

REVIEW QUESTIONS AND PROBLEMS

19-1   What is an oscillator? List and discuss the types of oscillators found in this chapter.

19-2   Design a Hartley OP AMP oscillator for a frequency of 2MHz.

19-3   What is the difference between a Hartley and Colpitts oscillator?

19-4   Design a Colpitts OP AMP oscillator having a frequency of 2 MHz.

19-5   With an oscillator, why is some form of automatic biasing necessary?

19-6   What is the difference between a series- and shunt-fed oscillator?

19-7   A 1-MHz BJT oscillator has a biasing network consisting of R ahd C. If R is equal to 10 kΩ, what value should C be?

19-8 Derive Equation 20.

19-9 Derive the minimum g_m needed for a FET Colpitts oscillator.

19-10 What advantage does a crystal oscillator have over a LC oscillator?

19-11 Drive Equation 23.

19-12 Derive Equations 29 and 30.

19-13 Derive the gain and frequency equations illustrated in Figure 19.11.

19-14 Design a phase shift oscillator having a frequency of 600 Hz. Use the 2N4400 BJT and assume a emitter current of 1.5 mA.

19-15 Design an astable symmetrical MV having a frequency of 10 kHz. Assume a saturation current of 2 mA and a supply voltage of 10 V. Use the 2N4400 BJT for the transistors in the circuit configuration.

19-16 Show that the minimum beta for a BJT astable MV is 7.25. (*Hint*: Five time constants are required for a capacitor to be fully charged.)

OBJECTIVES FOR CHAPTER 20, ELECTRONIC SYSTEMS

The objective of this chapter is to let the reader "see" the many applications of BJTs, FETs, thyristors, and chips. Circuit applications include oscillators, amplifiers, current pumps, voltage regulators, and digital counters.

20 ELECTRONIC SYSTEMS

Chapter 20 is a presentation of simple electronic systems that incorporate the theory and discussions covered in the first 19 chapters of the text. This chapter is the "heart" of the text, enabling the reader to see how the material in the previous chapters can be used in practical circuits. The systems discussed have an introductory paragraph indicating how the need for the particular system came about.

The review questions give the reader additional information about the circuits discussed in this chapter. The data sheets for the chips discussed are in the appendix.

SECURITY SYSTEM

With the prospect of moving into a new home, I wanted to invent an inexpensive security system that would sound an alarm as well as indicate the point of entry into the house. The system described is unusual because it uses very few components to accomplish this aim.

The security system consists of a control box located in the master bedroom and magnetic reed switches protecting each door and window. On the control box is a layout of the three floors of the house with a lamp to indicate the position of each door and window. Inside the control box is a SCR, a transistor, a 12-V dc relay, a zener diode, three resistors, nine lamps, and a simple power supply. The circuit configuration of the security system is shown in Figure 20.1.

The surveillance loop is constructed of reed switches connected in series at the control box. Each magnetic reed switch is connected to the control box by a pair of wires. As can be seen in the figure, each reed switch has a lamp connected across it. Transistor Q is a PNP power transistor functioning as a current

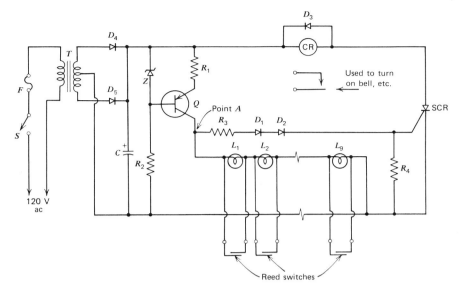

Figure 20.1
Intrusion alarm.

pump, which pumps approximately 50 mA of current. The amount of current pumped by Q is determined by the zener diode and resistor R_1.

The magnetic reed switches are connected in series and are normally closed. With all the reed switches normally closed, the collector of transistor Q is shorted to common. The 50 mA of pump current flows through the low resistance path of the surveillance loop and not through the lamps. If any of the reed switches is opened (a door or window is opened), 50 mA of current is pumped through the lamp that is connected across the reed switch that was opened. (The lamps are rated at 6 V, 50 mA.) This current causes the lamp to light. The other lamps will not light because the reed switches across them are closed.

Once any of the lamps is lit, approximately 6 V appears across the lamp. This 6 V also appears at point A in the circuit. This 6 V at point A will cause the diodes D_1 and D_2 to become forward-biased, triggering the SCR. Once the SCR is triggered, the relay CR will be pulled in and will remain in because of the SCR. The contacts of the relay can be used to turn on overhead lamps in the bedroom or to feed voltage to a large bell. The system is turned off when switch S is opened, thus removing voltage from the SCR.

In Figure 20.1, the resistor R_3 limits the gate current of the SCR once the 6 V appears at point A in the circuit. The diodes D_1 and D_2 are connected in series. This series connection will prevent the SCR from being triggered by the small voltage drop across the series-connected reed switches. The diode D_3 prevents the cemf of the relay from destroying the SCR when the switch S is opened to turn off the security system.

Figure 20.2 shows how the lamps are located on the control box. The lamps are positioned on the diagram of the floor plan of the house, indicating the place-

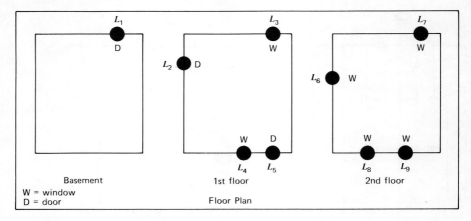

Figure 20.2
Placement of indicators.

ment of each reed switch that protects the various doors and windows. Once an entry is made, the alarm is turned on and the lamp is lit, indicating on the diagram the position of entry. Because the circuitry is simple, the entry lamps will be turned off if the door or window is closed again, but the alarm will remain on. This condition is not a major drawback because the door or window will probably remain open for a relatively long period of time. Once the alarm goes off in the master bedroom, a quick glance at the control box will indicate the point of entry.

A reed switch will protect only the opening of a door or window. Figure 20.3 shows how breakage of glass can be detected by the security system. The conducting strips are low resistance and are connected in series with the reed switch. A single pair of wires is connected to the combination, and the wire is run to the control box. This combination can be made at each window.

The security system is checked out by simply opening the doors and windows one at a time and checking that the appropriate lamp is lit and that the alarm is triggered each time. If more than one door or window is opened at one time, the alarm goes on and remains on, but the lamps indicating the points of entry will glow at reduced brilliance. The probability of more than one door or window being opened at one time is small.

Figure 20.3
Window monitor.

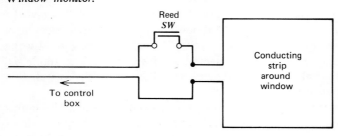

The reader is asked to design the various components shown in Figure 20.1. The original system used a 26-V center-taped transformer, a 12-V dc relay, a power PNP transistor for Q, and 6-V, 50-mA lamps.

CAR ALARM

A friend of mine had his car broken into recently; many of his tapes and his FM radio were stolen. His car was parked in front of his house at the time. After hearing of his experience, I suggested a simple security system that could be installed in his car. This security system is illustrated in Figure 20.4. The system consists of a unijunction transistor (UJT) connected as a timer, triggering a silicon controlled rectifier (SCR).

The dashboard lamp is the trigger for the system. This lamp is turned "on" (supplied voltage) with the opening of the front doors of the car. When the car door is opened, the UJT timer is supplied voltage from the lamp circuit. The UJT timer generates a voltage pulse across the R_1 resistor in the base lead of the UJT approximately 20 sec after battery voltage is supplied to the timer. This voltage pulse is then used to trigger the SCR. The SCR is connected to the horn relay of the car. Once the SCR is triggered "on," the horn relay will remain on until voltage is removed from the SCR by depressing the reset switch. The diodes in the gate circuit of the SCR insure that the SCR will not trigger until the voltage pulse from the UJT timer is generated. The diode across the SCR protects the SCR from the counter emf generated by the horn relay.

Figure 20.5 illustrates how the simple security system is connected within the car to the dashboard lamp and the horn relay. The horn and lamp leads from the

Figure 20.4
Car alarm.

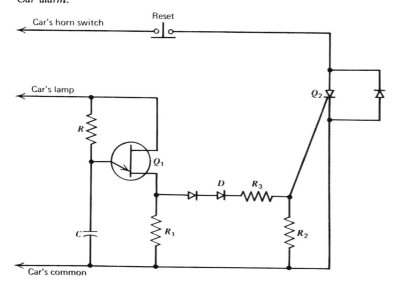

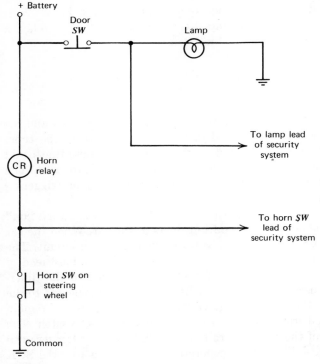

Figure 20.5
Connections for the car alarm.

security system must be connected to the proper side of the horn switch and lamp, as shown in Figure 20.5. The horn switch that is on the steering wheel will still function properly with the security system connected.

If the car door remains open for 20 sec or longer, the horn will be locked "on." If the car door is opened and then shut in less than 20 sec, the UJT timer will not trigger the SCR. The 20 sec was chosen because it is ample time for a person to open the car door, get inside, and then shut the car door. As can be seen from the diagram, if the dashboard lamp is burned out, the system will still work properly.

The circuit shown in Figure 20.4 was wired without a circuit board. Wiring was done point-to-point. The wired circuit was placed in a small plastic box with the reset button mounted on the plastic box. The three leads from the system were brought out from the plastic box and connected under the dashboard. When the system is first connected, the horn will probably lock "on," but depressing the reset button will turn it "off." The plastic box can be hidden under the dash or in the glove compartment. The system is checked by simply opening the door and listening for the horn to lock "on." Closing the door within the 20 sec of delay should not trigger the horn "on."

The reader is asked to design the various components shown in Figure 20.4. The original system used a UJT (2N4626).

INITIAL GENERATOR

In looking for an unusual Christmas gift for a "ham" friend of mine, I decided to design and to build him a Morse Code device that would generate his initials. I ruled out the use of a small tape recorder and used monostable multivibrators (one-shots).

Figure 20.6 shows the block diagram of this device. The Morse Code Generator (MCG) feeds one of the two inputs of a logic AND gate. The other input to the AND gate is a 600-Hz square-wave generator (astable MV). With the MCG at a logic "1" level (+V), the AND gate enables the 600-Hz signal to be fed to the audio amplifier. With the proper sequence of pulses of different duration generated from the MCG, the output of the audio amplifier will be a series of dots (.) and dashes (-).

Figure 20.7 shows the logic block diagram of the MCG. The MCG is unusual because it uses one-shots constructed from CMOS NOR gates as inverters. The NOR gate 1 starts the sequence. With a logic 0 (zero volts) at V_{in}, the output of NOR gate 1 is at the 1 level (+V), and C_2 charges to V_{CC}. With a logic 1 at V_{in}, C_2 (left plate) is placed at ground potential because the output of NOR gate 1 is at 0 level. This action causes C_2 to cutoff NOR gate 2 for a period of time approximately equal to $0.7R_2C_2$.

When the NOR gate 2 is cutoff ("1" level), a positive potential appears at the anode of diode D_2. This positive potential forward-biases D_2 and represents a logic "1" at the input of the AND gate (Figure 20.6). Thus a 600-Hz signal is fed to the audio amplifier for a period of time $(0.7R_2C_2)$.

While NOR gate 2 is at the logic "1" level, C_3 is charging up to approximately V_{CC}. When NOR gate 2 goes from the "1" level to the "0" level, C_3 causes NOR gate 3 to go from the "0" level to the "1" level. This process of generating pulses continues for each of the NOR gates connected as one-shots. Thus when NOR gate 2 produces a pulse of fixed duration, NOR gate 3 produces a pulse of fixed duration *after* NOR gate 2 produces its pulse. After NOR gate 3 produces a pulse of fixed duration, NOR gate 4 produces its pulse. This generation of pulses continues until all the one-shot generate a pulse.

These one-shots are connected to the AND gate through a diode OR gate. The timing of each one-shot is determined by R and C.

Figure 20.6
Initial generator.

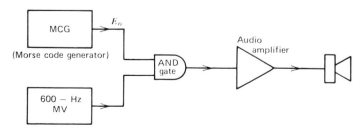

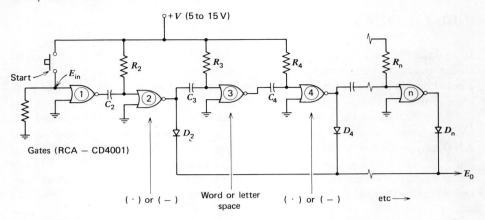

Figure 20.7
Schematic of the initial generator.

The individual one-shots are labeled dot, dash, letter space, and space between the dot and dash. The dot and dash one-shots are the only one-shots that are ORed together to turn on the AND gate and to allow the 600-Hz signal to be fed to the amplifier. The number of one-shots is determined by the letters to be generated.

I used RCA CD4001 quad two-input NOR gates, although a CD4009 could be used. The 600-Hz astable MV is self-starting so that its output is present at all times. This astable is shown in Figure 20.8. Figure 20.9 shows the output of the AND gate generating the letters "RWS." Any amplifier could be used to feed a loud speaker. In the original Morse Code device, I used a discrete Class B amplifier so that the current on Stand-By was minimum.

Figure 20.8
MCG and MV.

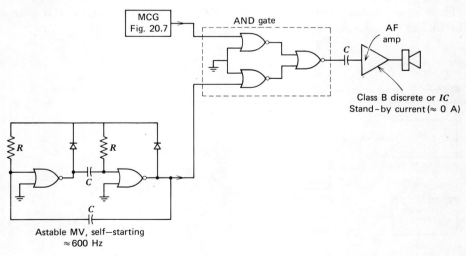

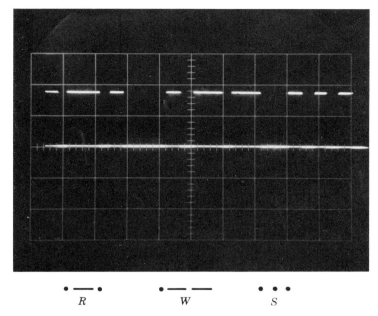

Figure 20.9
Photo of the device generating "RWS."

I breadboarded the device using a vector board and two 9-V batteries connected in parallel for the supply voltage. With the device idling, the current drain is 5.0 mA. A N.O. push-button switch is used to start the sequence of pulses. (See Figure 20.7.) When power is initially applied, a series of pulses will be heard. Once these pulses are generated, momentarily pressing the start-button will generate the letters the device is designed to produce.

The reader is asked to design a generator to generate his own initials.

ELECTRONIC DOORBELL

Being a technician interested in electronics, I wanted to replace my mechanical doorbell with a solid-state doorbell. The system discussed below is designed to replace the mechanical ("ding-dong") doorbell.

The device basically consists of two RC phase shift oscillators that are pulsed or triggered in sequence and whose amplitude decays exponentially. Figure 20.10 illustrates the device. The two oscillators have different natural running refquencies: one high and the other low. With the high and low frequencies and the decaying amplitude, the device simulates the mechanical "ding-dong."

Figure 20.11 illustrates how the decaying effect is achieved. Without a trigger pulse at point A, the transistor T is cutoff and the voltage across the capacitor C is the supply voltage V_{CC}. With the capacitor charged to V_{CC}, no voltage is present at the RC oscillator (V_{osc}). When a trigger pulse is present at point A momentarily, the transistor T becomes saturated, causing C to discharge to

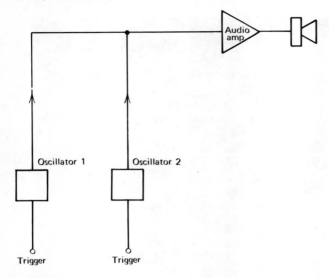

Figure 20.10
Block diagram of the electronic doorbell.

approximately zero volts. With zero volts across C, the supply voltage V_{CC} appears at the input supply of the oscillator or $V_{CC} = V_{osc}$. With $V_{CC} = V_{osc}$, the oscillator functions normally.

When the trigger pulse is removed, the voltage across the capacitor begins to charge back to the supply voltage V_{CC}. The rate at which the voltage across C charges is determined by the value of C and the resistance of the RC phase shift oscillator. Because the capacitor charges exponentially and $V_{osc} = V_{CC} - V_C$, the amplitude of the output of the oscillator decays exponentially. This effect simulates the "ding-dong" of a mechanical doorbell.

The output of both oscillators is fed into an audio amplifier, as shown in Figure 20.12. Note that with no trigger pulses at points A and B, the oscillators

Figure 20.11
Decay of the oscillator.

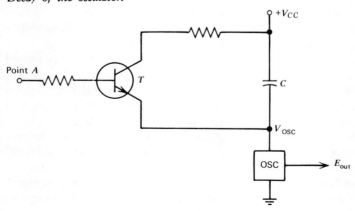

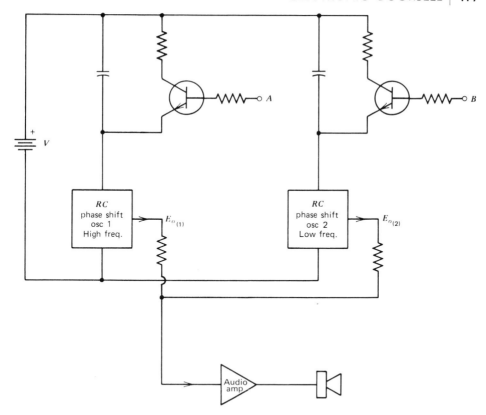

Figure 20.12
Electronic doorbell.

do not function. The trigger pulses that trigger the two oscillators have to be sequenced. The first trigger pulse pulses the high RC oscillator; then, after the high oscillator decays, the low-frequency oscillator is triggered. A simple method of triggering the two oscillators is illustrated in Figure 20.13.

Figure 20.13 shows two differentiating circuits. When the contacts are moved from the position shown in the figure, a positive pulse or spike appears across resistor R_A. This positive pulse triggers (momentarily) the high-frequency oscillator. When the contacts are returned to the position shown in Figure 20.13, a positive pulse appears across resistor R_B, which is used to trigger the low-frequency oscillator. The duration of these spiked pulses is determined by the values of R and C. The zener diodes insure well-shaped pulses for triggering purposes.

A second method of triggering the oscillators in sequence is shown in Figure 20.14. This method uses monostable multivibrators (one-shots). Once the switch is depressed, two pulses are generated from the one-shots that are used to trigger the two oscillators sequentially.

The reader is asked to design the components of the system. The original system used a high-frequency sine wave of 600 Hz and a low frequency of

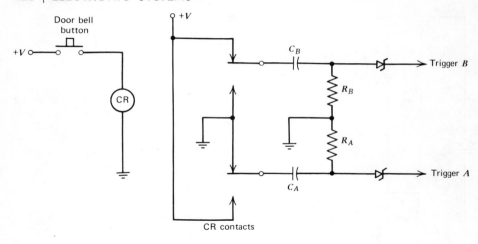

Figure 20.13
Triggering the doorbell.

200 Hz (see Chapter 19). The RC oscillators were discretely built, using 2N4400 transistors. The IC one-shots were constructed using CMOS NOR gates.

TELEPHONE ANSWERING DEVICE

I was asked to design a telephone answering device that enabled a cable TV company to switch channels on its system using the tones generated by a telephone. The answering portion of the total system is shown in Figure 20.15.

The first requirement of the total system is that it be able to answer the telephone and to "hold the lines" for a specified length of time. Once the 20-Hz ringing voltage is applied to the system, a dc load resistance of approximately 150 Ω is placed across the telephone lines to stop the ringing voltage. In Figure 20.15, the count-rate circuit, the triggering circuit, the IC timer, and the load with an audio transformer are shown.

As the ringing voltage appears at the input lines, the count-rate circuit pumps current into C_2. After a specified number of rings, the voltage across C_2 builds up to a potential that causes the diac to trigger. Once the diac triggers, a voltage pulse appears across the voltage divider resistor R_3. This voltage spike triggers the IC timer. The IC timer pulls in the relay CR for a specified length of time. With the CR pulled in, the auxiliary contacts of the CR place a load across the telephone lines. This load consists of the resistor R_6 and the primary of the audio transformer T. The total dc resistance of T and R_6 should be on the order of 150 Ω.

After the timer has "timed-out," the CR drops out and the answering device is ready to accept another call.

The reader can readily understand the design of the various components by analyzing each subsystem individually. The current pumped into capacitor C_2 is given by $I = C_1 \times V \times F_{\text{req}}$. (See the section on the count-rate meter.) The

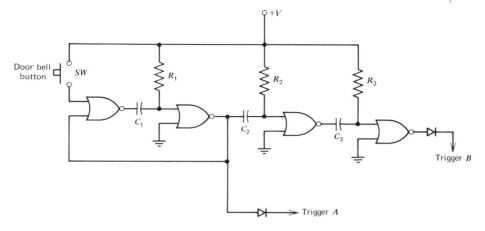

Figure 20.14
Triggering the doorbell.

voltage rise across C_2 is calculated by $v_c = (I/C)t$ where I is the current pumped into C_2, C is the value of C_2, and t is the total time that the ringing current is present. The voltage across C_2 builds up to the triggering point of the diac. Note that the diac could be replaced by a neon bulb.

The voltage divider (R_2 and R_3) insures that the input pulse to the IC timer is limited to a safe value of amplitude voltage. The time of the IC timer is determined by the value of the R and C as shown in the circuit.

The reader is asked to design the system to answer the phone after three rings and to "hold the line" for a period of 45 sec. The system that I built incorporated a 12-V relay for CR, a Signetics 555 as the IC timer, a 2N4400 transistor for Q, and a power supply that was appropriate for the various voltages needed in the total system.

HO TRAIN CONTROL

Last Christmas I received my first HO train set for a present. The set contained an engine with three cars, enough track for a 4-ft diameter circle, and a transformer. Last Christmas I decided to design and build a solid-state control unit (see Figure 20.16).

The controls for the train are a forward-reverse switch, a stop button, an increase speed button, and a decrease speed button. The unusual feature of the speed control buttons is that the length of time that the buttons are depressed determines the speed of the train. Once a button is released, the train remains at the speed at which the button was released.

The three control buttons have lamps inside them. The brilliancy of the lamps gives the operator a visual indication of the speed of the train.

Figure 20.17 illustrates the schematic diagram for the solid-state control unit. The unit consists of two OP AMPs, two current pumps, a rectifying circuit, and a transformer.

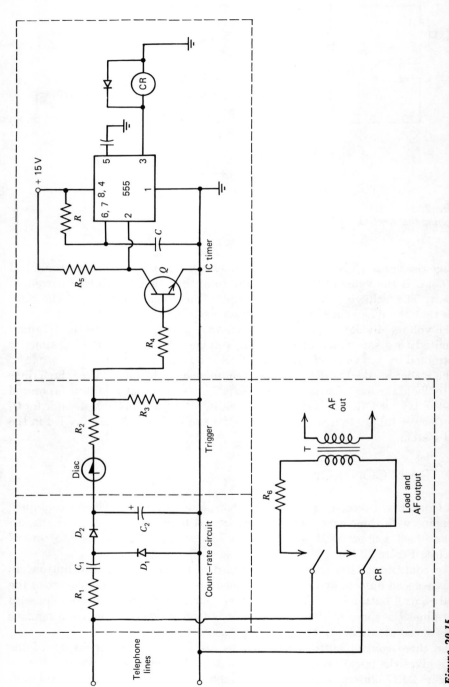

Figure 20.15
Telephone answerer.

Figure 20.16
Photo of the HO train control.

The OP AMP_1, along with the series voltage regulating transistors Q and Q_1, serves as a voltage regulator. The output voltage of this regulator is determined by the feedback voltage across R_1 and the voltage across capacitor C. The OP AMP_2 is connected as a voltage follower so that the input of OP AMP_1 does not load the voltage across capacitor C.

The voltage across capacitor C is the reference voltage for the regulator and is determined by the amount of time a current is pumped into the capacitor. Two current pumps are used to increase or decrease the voltage across C.

The two current pumps are constructed with $JFET_1$ and $JFET_2$. The $JFET_1$ pumps a positive current in C, and $JFET_2$ pumps a negative current into C when the increase or decrease pushbutton is depressed. A positive current pumped into C cause the voltage across C to increase. A negative current pumped into C causes the voltage across C to decrease. Because the reference voltage for the regulator is the voltage across C, its voltage determines the output voltage feed to the tracks of the HO train.

Depressing the increase speed switch SW_1 causes the output voltage to increase and depressing the decrease speed switch SW_2 causes the output voltage to decrease. Releasing the switch (SW_1 or SW_2) causes the output voltage to remain at the voltage level at which the button was released. The stop switch (SW_3), when depressed, causes the voltage across C to be reduced to zero. With zero volts across C, the output voltage is zero.

A visual indication of the train's speed is obtained from the brilliancy of the lamps inside the three control switches (SW_1, SW_2, SW_3). Each lamp is a 6-V unit. The three lamps are connected in series and placed across the output terminals of the control unit. Thus the greater the output voltage feed to the tracks, the brighter the lamps.

The diode current limiter (D_5, D_6, D_7, R_3) limits the current feed to the tracks to approximately 1 A. This limiting is necessary to prevent damage to the control unit in case shorts are placed across the tracks.

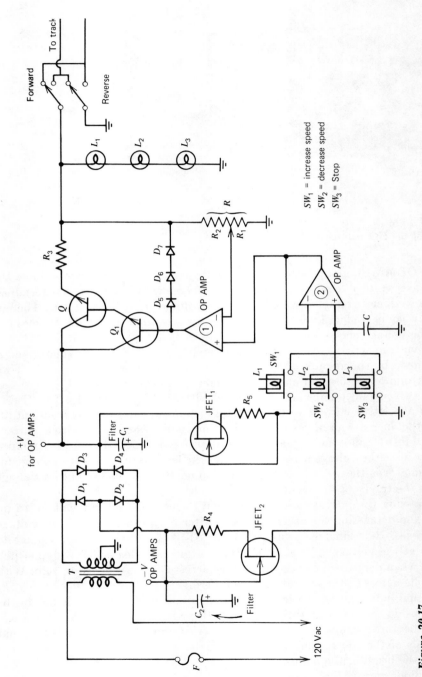

Figure 20.17
Schematic of the HO train control.

The layout of parts is not critical. I built the device in a bakelite housing. The majority of the components were placed on a PC board. The regulating transistor Q was fixed to a piece of aluminum used as a heat sink. The unit shown can deliver approximately 1000 mA (enough for two small HO trains.)

With the original unit, the output voltage changes approximately 2 V/sec with the button depressed. This rate is determined by the value of the resistors R_4 and R_5. Increasing these resistance values decreases the rate of voltage change per time.

TEMPER TIMER

Everyone knows the expression "count to 10" before you blow your top. I decided to dramatize this expression with the device shown in Figure 20.18. The device visually displays a count of 10 and audibly tells when the count of 10 is finished. The user of the Temper Timer simply throws a switch. The display readout device visually shows a count of 10 in approximately 10 sec. After the display readout counts to 9, the device audibly emits a pulsating tone that remains on until the power switch is turned off. The emitted tone tells you it's time to "blow your top."

The Temper Timer basically consists of a one pulse per second clock, a decade counting module (DCM consisting of a counter, decoder, and display readout device), and a triggered Sonalert device (see Figure 20.19).

When switch S_W is thrown on, the battery is supplied to the device. The resistor R and capacitor C form a differentiating circuit that is connected to the reset terminal of the DCM. This circuit arrangement makes sure that the DCM starts its count at ZERO every time the power switch S_W is thrown on.

Figure 20.18
Photo of the temper timer.

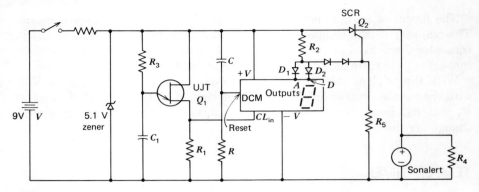

Figure 20.19
Schematic of the temper timer.

After power is applied, the unijunction timing oscillator (Q_1) generates one pulse per second, which is fed to the clock input of the DCM via R_1. The DCM will then display a count from 0 through 9. Once 9 is reached, the diode AND gate, consisting of D_1 and D_2, decodes 9 and its output goes "high." This "high" output is used to trigger the silicone controlled rectifier (Q_2). Once the SCR is triggered on, the Sonalert device is turned on, emitting a pulsating tone. The pulsating tone will remain audible until the power switch is turned off. The 1 kΩ resistor (R_4) across the Sonalert device ensures that the SCR will remain latched because of the holding current necessary for the SCR.

The placement of parts is not critical. I wired the timer and triggering circuits on the same PC board that is used by the DCM. The Numitron readout device was not soldered to the PC board. Instead, leads from the PC board to a nine-pin miniature socket were used. This technique allows the readout device to be mounted on the panel that sefved as a cover for the Temper Timer. I cut a rectangular hole in the cover panel so that the readout display could be seen. The Numitron display was attached to the cover panel by a homemade aluminum bracket.

The battery pack consists of four "C" cells connected in series. When the device is turned on, the average current drain is 225 mA. The size of the batteries and the mA-h rating would depend on how often the device is to be used and the size of the housing for the device.

With the device as shown, the housing is a plastic box ($6\frac{1}{4} \times 3\frac{3}{4} \times 2$ inches). The cover panel was cut from aluminum. Note that a different type of readout display could be used, such as an LED readout device.

DIGITAL TEMPERATURE INDICATOR

Figure 20.20 illustrates a digital temperature indicator. This device displays the temperature sensed by a thermister in degrees Fahrenheit. The device uses a 566 function generator whose frequency is a function of temperature. The output of the function generator is registered on DCMs. As will be demonstrated, the

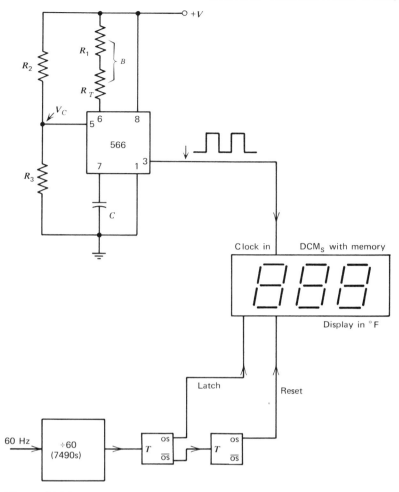

Figure 20.20
Digital temperature display device.

frequency of the function generator ranges from 1032 to 1212 Hz. With the three DCMs used, the display will readout numbers from 32 to 212 Hz, which are interpreted as degrees Fahrenheit.

The DCMs used have a memory or latch. The memory stores the last count in the counter before the latch pulse is generated. With this arrangement, the display is updated every second without seeing the display count. After the latch pulse (dumps the counter into memory, which is displayed), the reset pulse is generated to clear the counter for the next count cycle, which lasts 1 sec. The one-shots in the figure generate these latch and reset pulses. Note that the latch pulse is generated first and then the reset pulse.

The 1-sec pulses are obtained by dividing the 60-Hz line frequency by 60. The dividing is done by using two decade counters (7490) connected as a divide by 10 and a divide by 6 counter.

The frequency of the function generator is given by

$$f_0 = \frac{k}{RC} \tag{1}$$

where

$$k = \frac{2(V - V_C)}{V}$$

Note that the frequency increases as the value of R decreases, and vice versa. If a thermister is used as the temperature-sensing device whose resistance changes with temperature, the frequency of the function generator can be changed with temperature.

In order to calculate how the frequency of the function generator changes with changes in resistance, Equation 1 is differentiated with respect to R, or

$$\frac{df}{dR} = -\frac{f_o}{R}$$

or

$$R = -\frac{\Delta R}{\Delta f} f_o \tag{2}$$

The negative sign indicates that as the resistance increases, the frequency decreases. Let us assume that a linear thermister is available that has a resistance of 1000 Ω at 212 °F and has a resistance of 2800 Ω at 32 °F. The change in resistance ΔR is then equal to 1800 Ω.

The frequency of the function generator is chosen to be 1212 Hz at 212 °F and 1032 Hz at 32 °F. From this information, then, Δf is equal to 180 Hz. Using Equation 2 and taking into account the negative sign, we see that the resistance R needed at 212 °F is then

$$R_{212\,°F} = 10 f_{32\,°F}$$

$$= 10 \times 1032$$

$$R_{212\,°F} = 10{,}320 \ \Omega$$

The resistance R needed at 32 °F is then

$$R_{32\,°F} = 10 f_{212\,°F}$$

$$= 10 \times 1212$$

$$R_{32\,°F} = 12{,}120 \ \Omega$$

The value of R is the sum of R_1 and the resistance of the thermister R_T, or

$$R = R_1 + R_T$$

$$12{,}120 \ \Omega = R_1 + 2800 \ \Omega$$

or

$$R_1 = 9320 \ \Omega$$

As a check

$$10{,}320\ \Omega\ =\ R_1\ +\ 1000\ \Omega$$

or

$$R_1\ =\ 9320\ \Omega$$

The value of the capacitor can be obtained by using Equation 1, or

$$C\ =\ \frac{k}{fR}\ =\ \frac{k}{(1032)(12{,}120)}$$

or

$$C\ =\ \frac{k}{(1212)(10{,}320)}$$

where K is determined by the value of the supply voltage V and V_C.

With the temperature probe at a known temperature, the voltage V_C can be adjusted to display the correct count; V_C is changed by the ratio of R_2 and R_3. This adjustment is necessary because of the tolerance of the capacitor C. The unit was built with the one-shots being constructed from gate delay OS. The DCMs had Numitron readout displays and were constructed using 7490s, 7447s, and 7475s.

REVIEW QUESTIONS AND PROBLEMS

20-1 For the device illustrated in Figure 20.1, calculate the values and power reratings for all the components. Assume that the lamps (L_1 through L_9) are rated at 6 V at 50 mA. The zener diode is a 3.6 V unit and the relay is a 12-V, 100-Ω unit.

20-2 For the device illustrated in Figure 20.4, calculate the values and power ratings for all the components. Assume that the UJT is a 2N2646, the horn relay current is 150 mA, and the delay time is 20 sec.

20-3 Design an initial generator to generate your initials. Assume that the capacitors used in the one-shots are 0.05 μF.

20-4 Design an electronic door bell. The oscillators are constructed using 2N4400 BJTs. The power supply is a 15-V unit, and the decay time for the oscillators is 2 sec.

20-5 Calculate the components for the circuit shown in Figure 20.15. The ringing voltage is 70 V-rms at 20 Hz. The device is designed to answer after three rings (each ring is 1 sec) and hold the line for 45 sec. The BJTs are 2N4400s with the IC timer being the 555 unit.

20-6 Design a HO train control unit. Use the 741C OP AMPs, HEP 801 JFETs and a 26.8 VCT transformer rated at 1.0 A. Design the unit to have an output voltage rate of 1.0 V/sec.

20-7 Shown in Figure 20.20 is the digital temperature device. Design this unit to read temperatures from 32 to 120 °F. The thermister is the YSI 44201. The YSI 44201 has a resistance of 2764.7 Ω at 32 °F and 1921.6 Ω at 120 °F. Assume a supply voltage of 5 V.

MANUFACTURER'S SPECIFICATION SHEETS— CHIPS

signetics

HIGH PERFORMANCE OPERATIONAL AMPLIFIER | µA741

LINEAR INTEGRATED CIRCUITS

DESCRIPTION

The µA741 is a high performance operational amplifier with high open loop gain, internal compensation, high common mode range and exceptional temperature stability. The µA741 is short-circuit protected and allows for nulling of offset voltage.

FEATURES

- **INTERNAL FREQUENCY COMPENSATION**
- **SHORT CIRCUIT PROTECTION**
- **OFFSET VOLTAGE NULL CAPABILITY**
- **EXCELLENT TEMPERATURE STABILITY**
- **HIGH INPUT VOLTAGE RANGE**
- **NO LATCH-UP**

ABSOLUTE MAXIMUM RATINGS

	µA741C	µA741
Supply Voltage	±18V	±22V
Internal Power Dissipation (Note 1)	500mW	500mW
Differential Input Voltage	±30V	±30V
Input Voltage (Note 2)	±15V	±15V
Voltage between Offset Null and V⁻	±0.5V	±0.5V
Operating Temperature Range	0°C to +70°C	-55°C to +125°C
Storage Temperature Range	−65°C to +150°C	−65°C to +150°C
Lead Temperature (Solder, 60 sec)	300°C	300°C
Output Short Circuit Duration (Note 3)	Indefinite	Indefinite

Notes
1. Rating applies for case temperatures to 125°C; derate linearly at 6.5mW/°C for ambient temperatures above +75°C.
2. For supply voltages less than ±15V, the absolute maximum input voltage is equal to the supply voltage.
3. Short circuit may be to ground or either supply. Rating applies to +125°C case temperature or +75°C ambient temperature.

PIN CONFIGURATIONS

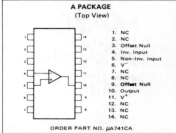

A PACKAGE (Top View)

1. NC
2. NC
3. Offset Null
4. Inv. Input
5. Non-Inv. Input
6. V⁻
7. NC
8. NC
9. **Offset Null**
10. Output
11. V⁺
12. NC
13. NC
14. NC

ORDER PART NO. µA741CA

T PACKAGE

1. Offset Null
2. Inverting Input
3. Non-Inverting Input
4. V⁻
5. Offset Null
6. Output
7. V⁺
8. NC

ORDER PART NOS. µA741T/µA741CT

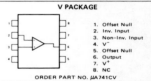

V PACKAGE

1. Offset Null
2. Inv. Input
3. Non-Inv. Input
4. V⁻
5. Offset Null
6. Output
7. V⁺
8. NC

ORDER PART NO. µA741CV

EQUIVALENT CIRCUIT

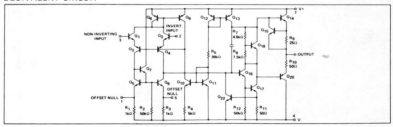

signetics

TIMER | **555**

LINEAR INTEGRATED CIRCUITS

DESCRIPTION

The NE/SE 555 monolithic timing circuit is a highly stable controller capable of producing accurate time delays, or oscillation. Additional terminals are provided for triggering or resetting if desired. In the time delay mode of operation, the time is precisely controlled by one external resistor and capacitor. For a stable operation as an oscillator, the free running frequency and the duty cycle are both accurately controlled with two external resistors and one capacitor. The circuit may be triggered and reset on falling waveforms, and the output structure can source or sink up to 200mA or drive TTL circuits.

FEATURES

- TIMING FROM MICROSECONDS THROUGH HOURS
- OPERATES IN BOTH ASTABLE AND MONOSTABLE MODES
- ADJUSTABLE DUTY CYCLE
- HIGH CURRENT OUTPUT CAN SOURCE OR SINK 200mA
- OUTPUT CAN DRIVE TTL
- TEMPERATURE STABILITY OF 0.005% PER $^{\circ}$C
- NORMALLY ON AND NORMALLY OFF OUTPUT

APPLICATIONS

PRECISION TIMING
PULSE GENERATION
SEQUENTIAL TIMING
TIME DELAY GENERATION
PULSE WIDTH MODULATION
PULSE POSITION MODULATION
MISSING PULSE DETECTOR

PIN CONFIGURATIONS

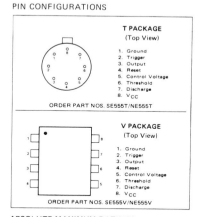

T PACKAGE
(Top View)

1. Ground
2. Trigger
3. Output
4. Reset
5. Control Voltage
6. Threshold
7. Discharge
8. V_{CC}

ORDER PART NOS. SE555T/NE555T

V PACKAGE
(Top View)

1. Ground
2. Trigger
3. Output
4. Reset
5. Control Voltage
6. Threshold
7. Discharge
8. V_{CC}

ORDER PART NOS. SE555V/NE555V

ABSOLUTE MAXIMUM RATINGS

Supply Voltage	+18V
Power Dissipation	600 mW
Operating Temperature Range	
NE555	0°C to +70°C
SE555	-55°C to +125°C
Storage Temperature Range	-65°C to +150°C
Lead Temperature (Soldering, 60 seconds)	+300°C

BLOCK DIAGRAM

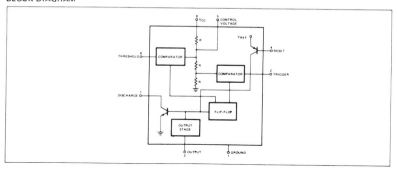

LINEAR INTEGRATED CIRCUITS ▪ 555

ELECTRICAL CHARACTERISTICS (T_A = 25°C, V_{CC} = +5V to +15 unless otherwise specified)

PARAMETER	TEST CONDITIONS	SE 555			NE 555			UNITS
		MIN	TYP	MAX	MIN	TYP	MAX	
Supply Voltage		4.5		18	4.5		16	V
Supply Current	V_{CC} = 5V R_L = ∞		3	5		3	6	mA
	V_{CC} = 15V R_L = ∞		10	12		10	15	mA
	Low State, Note 1							
Timing Error	R_A, R_B = 1KΩ to 100KΩ							
Initial Accuracy	C = 0.1 μF Note 2		0.5	2		1		%
Drift with Temperature			30	100		50		ppm/°C
Drift with Supply Voltage			0.005	0.02		0.01		%/Volt
Threshold Voltage			2/3			2/3		X V_{CC}
Trigger Voltage	V_{CC} = 15V	4.8	5	5.2		5		V
	V_{CC} = 5V	1.45	1.67	1.9		1.67		V
Trigger Current			0.5			0.5		μA
Reset Voltage		0.4	0.7	1.0	0.4	0.7	1.0	V
Reset Current			0.1			0.1		mA
Threshold Current	Note 3		0.1	.25		0.1	.25	μA
Control Voltage Level	V_{CC} = 15V	9.6	10	10.4	9.0	10	11	V
	V_{CC} = 5V	2.9	3.33	3.8	2.6	3.33	4	V
Output Voltage Drop (low)	V_{CC} = 15V							
	I_{SINK} = 10mA		0.1	0.15		0.1	.25	V
	I_{SINK} = 50mA		0.4	0.5		0.4	.75	V
	I_{SINK} = 100mA		2.0	2.2		2.0	2.5	V
	I_{SINK} = 200mA		2.5			2.5		
	V_{CC} = 5V							
	I_{SINK} = 8mA		0.1	0.25				V
	I_{SINK} = 5mA					.25	.35	
Output Voltage Drop (high)								
	I_{SOURCE} = 200mA		12.5			12.5		
	V_{CC} = 15V							
	I_{SOURCE} = 100mA							
	V_{CC} = 15V	13.0	13.3		12.75	13.3		V
	V_{CC} = 5V	3.0	3.3		2.75	3.3		V
Rise Time of Output			100			100		nsec
Fall Time of Output			100			100		nsec

NOTES:

1. Supply Current when output high typically 1mA less.

2. Tested at V_{CC} = 5V and V_{CC} = 15V

3. This will determine the maximum value of $R_A + R_B$. For 15V operation, the max total R = 20 megohm.

EQUIVALENT CIRCUIT

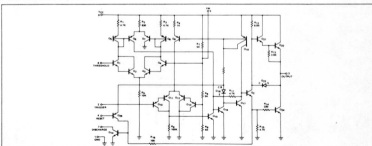

LINEAR INTEGRATED CIRCUITS ■ 555

MONOSTABLE OPERATION

In this mode of operation, the timer functions as a one-shot. Referring to Figure 1a the external capacitor is initially held discharged by a transistor inside the timer.

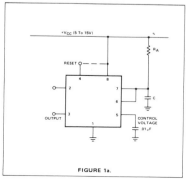

FIGURE 1a.

Upon application of a negative trigger pulse to pin 2, the flip–flop is set which releases the short circuit across the external capacitor and drives the output high. The voltage across the capacitor, now, increases exponentially with the time constant $\tau = R_A C$. When the voltage across the capacitor equals 2/3 V_{CC}, the comparator resets the flip–flop which in turn discharges the capacitor rapidly and drives the output to its low state. Figure 1b shows the actual waveforms generated in this mode of operation.

The circuit triggers on a negative going input signal when the level reaches 1/3 V_{CC}. Once triggered, the circuit will remain in this state until the set time is elapsed, even if it is triggered again during this interval. The time that the

output is in the high state is given by $t = 1.1\,R_A C$ and can easily be determined by Figure 1c. Notice that since the charge rate, and the threshold level of the comparator are both directly proportional to supply voltage, the timing interval is independent of supply. Applying a negative pulse simultaneously to the reset terminal (pin 4) and the trigger terminal (pin 2) during the timing cycle discharges the external capacitor and causes the cycle to start over again. The timing cycle will now commence on the positive edge of the reset pulse. During the time the reset pulse is applied, the output is driven to its low state.

When the reset function is not in use, it is recommended that it be connected to V_{CC} to avoid any possibility of false triggering.

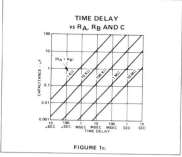

**TIME DELAY
vs R_A, R_B AND C**

FIGURE 1c.

ASTABLE OPERATION

If the circuit is connected as shown in Figure 2a (pins 2 and 6 connected) it will trigger itself and free run as a multivibrator. The external capacitor charges through R_A and R_B and discharges through R_B only. Thus the duty cycle may be precisely set by the ratio of these two resistors.

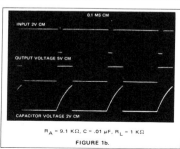

R_A = 9.1 KΩ, C = .01 μF, R_L = 1 KΩ

FIGURE 1b.

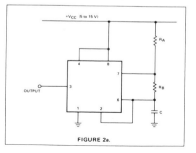

FIGURE 2a.

(Cont'd)

In this mode of operation, the capacitor charges and discharges between 1/3 V_{CC} and 2/3 V_{CC}. As in the triggered mode, the charge and discharge times, and therefore the frequency are independent of the supply voltage.

Figure 2b shows actual waveforms generated in this mode of operation.

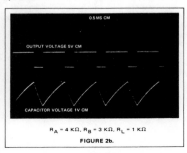

0.5 MS CM

OUTPUT VOLTAGE 5V CM

CAPACITOR VOLTAGE 1V CM

$R_A = 4\,K\Omega, R_B = 3\,K\Omega, R_L = 1\,K\Omega$

FIGURE 2b.

The charge time (output high) is given by:

$$t_1 = 0.693\,(R_A + R_B)\,C$$

and the discharge time (output low) by:

$$t_2 = 0.693\,(R_B)\,C$$

Thus the total period is given by:

$$T = t_1 + t_2 = 0.693\,(R_A + 2R_B)\,C$$

The frequency of oscillation is then:

$$f = \frac{1}{T} = \frac{1.44}{(R_A + 2R_B)\,C}$$

and may be easily found by Figure 2c.

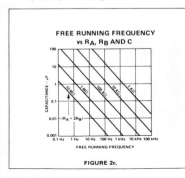

FREE RUNNING FREQUENCY
vs R_A, R_B AND C

CAPACITANCE — μF

$(R_A + 2R_B)$

FREE RUNNING FREQUENCY

FIGURE 2c.

The duty cycle is given by:

$$D = \frac{R_B}{R_A + 2R_B}$$

MISSING PULSE DETECTOR

Using the circuit of Figure 3a, the timing cycle is continuously reset by the input pulse train. A change in frequency, or a missing pulse, allows completion of the timing cycle which causes a change in the output level. For this application, the time delay should be set to be slightly longer than the normal time between pulses. Figure 3b shows the actual waveforms seen in this mode of operation.

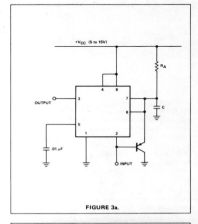

$+V_{CC}$ (5 to 15V)

OUTPUT

.01 μF

INPUT

FIGURE 3a.

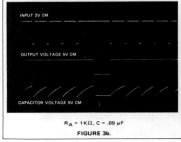

INPUT 2V CM

OUTPUT VOLTAGE 5V CM

CAPACITOR VOLTAGE 5V CM

$R_A = 1\,K\Omega, C = .09\,\mu F$

FIGURE 3b.

LINEAR INTEGRATED CIRCUITS ■ 555

(Cont'd)

FREQUENCY DIVIDER

If the input frequency is known, the timer can easily be used as a frequency divider by adjusting the length of the timing cycle. Figure 4 shows the waveforms of the timer in Figure 1a when used as a divide by three circuit. This application makes use of the fact that this circuit cannot be retriggered during the timing cycle.

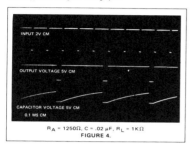

INPUT 2V CM

OUTPUT VOLTAGE 5V CM

CAPACITOR VOLTAGE 5V CM
0.1 MS CM

R_A = 1250Ω, C = .02 μF, R_L = 1KΩ
FIGURE 4.

PULSE WIDTH MODULATION (PWM)

In this application, the timer is connected in the monostable mode as shown in Figure 5a. The circuit is triggered with a continuous pulse train and the threshold voltage is modulated by the signal applied to the control voltage terminal (pin 5). This has the effect of modulating the pulse width as the control voltage varies. Figure 5b shows the actual waveforms generated with this circuit.

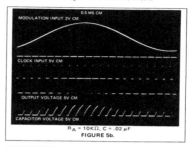

0.5 MS CM

MODULATION INPUT 2V CM

CLOCK INPUT 5V CM

OUTPUT VOLTAGE 5V CM

CAPACITOR VOLTAGE 5V CM

R_A = 10KΩ, C = .02 μF
FIGURE 5b.

PULSE POSITION MODULATION (PPM)

This application uses the timer connected for astable (free-running) operation, Figure 6a, with a modulating signal again applied to the control voltage terminal. Now the pulse position varies with the modulating signal, since the threshold voltage and hence the time delay is varied. Figure 6b shows the waveforms generated for triangle wave modulation signal.

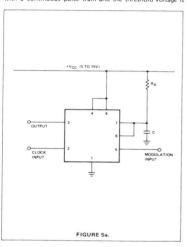

+V_{CC} (5 TO 15V)

R_A

OUTPUT

CLOCK INPUT

C

MODULATION INPUT

FIGURE 5a.

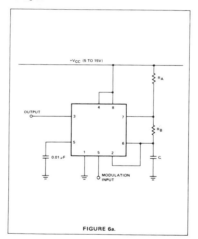

+V_{CC} (5 TO 15V)

R_A

OUTPUT

R_B

0.01 μF

C

MODULATION INPUT

FIGURE 6a.

Signetics

FUNCTION GENERATOR | 566

LINEAR INTEGRATED CIRCUITS

DESCRIPTION

The SE/NE 566 Function Generator is a voltage controlled oscillator of exceptional stability and linearity with buffered square wave and triangle wave outputs. The frequency of oscillation is determined by an external resistor and capacitor and the voltage applied to the control terminal. The oscillator can be programmed over a ten to one frequency range by proper selection of an external resistance and modulated over a ten to one range by the control voltage, with exceptional linearity.

FEATURES

- **WIDE RANGE OF OPERATING VOLTAGE**
  (10 to 24 volts)
- **VERY HIGH LINEARITY OF MODULATION**
- **EXTREME STABILITY OF FREQUENCY**
  (100 ppm/°C typical)
- **HIGHLY LINEAR TRIANGLE WAVE OUTPUT**
- **HIGH ACCURACY SQUARE WAVE OUTPUT**
- **FREQUENCY PROGRAMMING BY MEANS OF A**
  RESISTOR, CAPACITOR, VOLTAGE OR CURRENT
- **FREQUENCY ADJUSTABLE OVER 10 TO 1**
  RANGE WITH SAME CAPACITOR

APPLICATIONS

TONE GENERATORS
FREQUENCY SHIFT KEYING
FM MODULATORS
CLOCK GENERATORS
SIGNAL GENERATORS
FUNCTION GENERATORS

PIN CONFIGURATION (Top View)

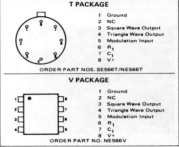

T PACKAGE

1	Ground
2	NC
3	Square Wave Output
4	Triangle Wave Output
5	Modulation Input
6	R_1
7	C_1
8	V+

ORDER PART NOS. SE566T/NE566T

V PACKAGE

1	Ground
2	NC
3	Square Wave Output
4	Triangle Wave Output
5	Modulation Input
6	R_1
7	C_1
8	V+

ORDER PART NO. NE566V

BLOCK DIAGRAM

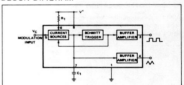

EQUIVALENT CIRCUIT

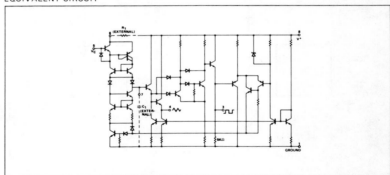

LINEAR INTEGRATED CIRCUITS ■ SE/NE566

ABSOLUTE MAXIMUM RATINGS (Limiting values above which serviceability may be impaired)

Maximum Operating Voltage	26V
Storage Temperature	$-65°C$ to $150°$ C
Power Dissipation	300mW

ELECTRICAL CHARACTERISTICS (25°C, 12 Volts, unless otherwise stated)

CHARACTERISTICS	SE566			NE566			UNITS
	MIN.	TYP.	MAX.	MIN.	TYP.	MAX.	
GENERAL							
Operating Temperature Range	-55		125	0		70	°C
Operating Supply Voltage			24			24	Volts
Operating Supply Current		7	12.5		7	12.5	mA
VCO (Note 1)							
Maximum Operating Frequency		1			1		MHz
Frequency Drift with Temperature		100			200		ppm/°C
Frequency Drift with Supply Voltage		1			2		%/volt
Control Terminal Input Impedance (Note 2)		1			1		MΩ
FM Distortion (± 10% Deviation)		0.2	0.75		0.2	1.5	%
Maximum Sweep Rate		1			1		MHz
Sweep Range		10:1			10:1		
OUTPUT							
Triangle Wave Output							
Impedance		50			50		Ω
Voltage	2	2.4		2	2.4		Volts pp
Linearity		0.2			0.5		%
Square Wave Output							
Impedance		50			50		Ω
Voltage	5	5.4		5	5.4		Volts pp
Duty Cycle	45	50	55	40	50	60	%
Rise Time		20			20		nsec
Fall Time		50			50		nsec

NOTES:

1. The external resistance for frequency adjustment (R_1) must have a value between 2KΩ and 20KΩ.

2. The bias voltage (Vc) applied to the control terminal (pin 5) should be in the range 3/4 $V^+ \leqslant V_C \leqslant V^+$.

TYPICAL PERFORMANCE CHARACTERISTICS

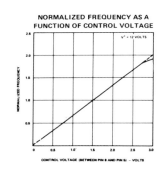

NORMALIZED FREQUENCY AS A
FUNCTION OF CONTROL VOLTAGE

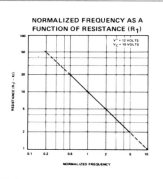

NORMALIZED FREQUENCY AS A
FUNCTION OF RESISTANCE (R_1)

TYPICAL PERFORMANCE CHARACTERISTICS (Cont'd.)

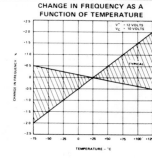

CHANGE IN FREQUENCY AS A FUNCTION OF TEMPERATURE

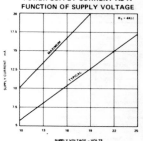

POWER SUPPLY CURRENT AS A FUNCTION OF SUPPLY VOLTAGE

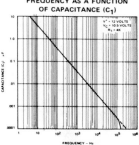

FREQUENCY AS A FUNCTION OF CAPACITANCE (C_1)

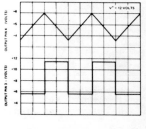

VCO OUTPUT WAVEFORMS

OPERATING INSTRUCTIONS

The SE/NE 566 Function Generator is a general purpose voltage controlled oscillator designed for highly linear frequency modulation. The circuit provides simultaneous square wave and triangle wave outputs at frequencies up to 1 MHz. A typical connection diagram is shown in Figure 1. The control terminal (pin 5) must be biased externally with a voltage (V_C) in the range

$$3/4\ V^+ \leqslant V_C \leqslant V^+$$

where V_{CC} is the total supply voltage. In Figure 1, the control voltage is set by the voltage divider formed with R_2 and R_3. The modulating signal is then ac coupled with the capacitor C_2. The modulating signal can be direct coupled as well, if the appropriate dc bias voltage is applied to the control terminal. The frequency is given approximately by

$$f_o \simeq \frac{2(V^+ - V_C)}{R_1 C_1 V^+}$$

and R_1 should be in the range $2K < R_1 < 20K\Omega$. A small capacitor (typically $0.001\mu f$) should be connected between pins 5 and 6 to eliminate possible oscillation in the control current source.

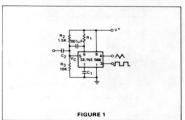

FIGURE 1

LINEAR INTEGRATED CIRCUITS ▪ SE/NE566

OPERATING INSTRUCTIONS (Cont'd.)

If the VCO is to be used to drive standard logic circuitry, it may be desirable to use a dual supply of ±5 volts as shown in Figure 2. In this case the square wave output has the proper dc levels for logic circuitry. RTL can be driven directly from pin 3. For DTL or T^2L gates, which require a current sink of more than 1 mA, it is usually necessary to connect a 5KΩ resistor between pin 3 and negative supply. This increases the current sinking capability to 2 mA. The third type of interface shown uses a saturated transistor between the 566 and the logic circuitry. This scheme is used primarily for T^2L circuitry which requires a fast fall time (< 50 nsec) and a large current sinking capability.

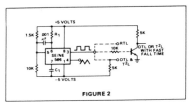

FIGURE 2

Signetics

TONE DECODER PHASE LOCKED LOOP | 567

LINEAR INTEGRATED CIRCUITS

DESCRIPTION

The SE/NE 567 tone and frequency decoder is a highly stable phase-locked loop with synchronous AM lock detection and power output circuitry. Its primary function is to drive a load whenever a sustained frequency within its detection band is present at the self-biased input. The bandwidth center frequency, and output delay are independently determined by means of four external components.

FEATURES

- WIDE FREQUENCY RANGE (.01Hz TO 500kHz)
- HIGH STABILITY OF CENTER FREQUENCY
- INDEPENDENTLY CONTROLLABLE BANDWIDTH (0 TO 14 PERCENT)
- HIGH OUT-BAND SIGNAL AND NOISE REJECTION
- LOGIC-COMPATIBLE OUTPUT WITH 100mA CURRENT SINKING CAPABILITY
- INHERENT IMMUNITY TO FALSE SIGNALS
- FREQUENCY ADJUSTMENT OVER A 20 TO 1 RANGE WITH AN EXTERNAL RESISTOR

APPLICATIONS

TOUCH TONE® DECODING
CARRIER CURRENT REMOTE CONTROLS
ULTRASONIC CONTROLS (REMOTE TV, ETC.)
COMMUNICATIONS PAGING
FREQUENCY MONITORING AND CONTROL
WIRELESS INTERCOM
PRECISION OSCILLATOR

BLOCK DIAGRAM

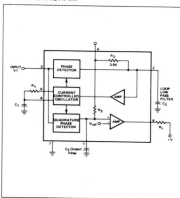

PIN CONFIGURATION

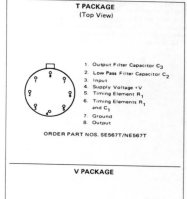

T PACKAGE (Top View)

1. Output Filter Capacitor C_3
2. Low Pass Filter Capacitor C_2
3. Input
4. Supply Voltage +V
5. Timing Element R_1
6. Timing Elements R_1 and C_1
7. Ground
8. Output

ORDER PART NOS. SE567T/NE567T

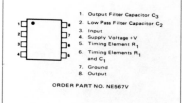

V PACKAGE

1. Output Filter Capacitor C_3
2. Low Pass Filter Capacitor C_2
3. Input
4. Supply Voltage +V
5. Timing Element R_1
6. Timing Elements R_1 and C_1
7. Ground
8. Output

ORDER PART NO. NE567V

ABSOLUTE MAXIMUM RATINGS:

Operating Temperature	0°C to 70°C NE567
	-55°C to 125°C SE567
Operating Voltage	10V
Positive Voltage at Input	0.5V above Supply Voltage (Pin 4)
Negative Voltage at Input	-10 VDC
Output Voltage (collector of output transistor)	15 VDC
Storage Temperature	-65°C to 150°C
Power Dissipation	300mW

TYPICAL CHARACTERISTIC CURVES

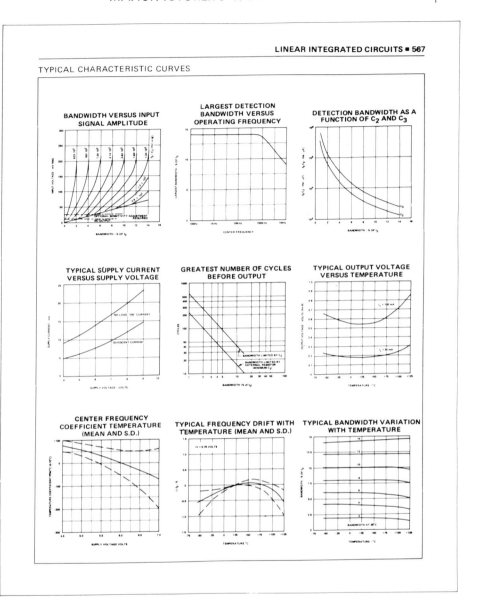

LINEAR INTEGRATED CIRCUITS ■ 567

TYPICAL CHARACTERISTIC CURVES (Cont'd.)

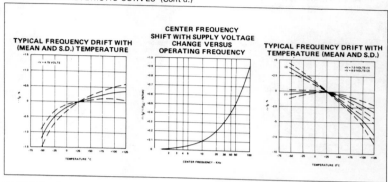

SCHEMATIC DIAGRAM

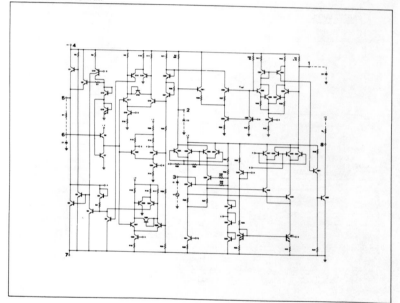

DESIGN FORMULAS

$$f_0 \simeq \frac{1.1}{R_1 C_1}$$

$$BW \simeq 1070 \sqrt{\frac{V_i}{f_0 C_2}} \quad \text{in \% of } f_0, \ V_i \lesssim 200mV$$

Where

V_i = Input Voltage (mV)
C_2 = Low-Pass Filter Capacitor (μF)

PHASE LOCKED LOOP TERMINOLOGY

CENTER FREQUENCY (f_0)

The free-running frequency of the current controlled oscillator (CCO) in the absence of an input signal.

DETECTION BANDWIDTH (BW)

The frequency range, centered about f_0, within which an input signal above the threshold voltage (typically 20mV rms) will cause a logical zero state on the output. The detection bandwidth corresponds to the loop capture range.

LARGEST DETECTION BANDWIDTH

The largest frequency range within which an input signal above the threshold voltage will cause a logical zero state on the output. The maximum detection bandwidth corresponds to the loop lock range.

DETECTION BAND SKEW

A measure of how well the largest detection band is centered about the center frequency, f_0. The skew is defined as $(f_{max} + f_{min} - 2f_0)/f_0$ where f_{max} and f_{min} are the frequencies corresponding to the edges of the detection band. The skew can be reduced to zero if necessary by means of an optional centering adjustment.

TYPICAL RESPONSE

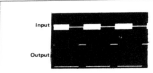

Input

Output

Response to 100mV RMS tone burst.
R_L = 100 ohms.

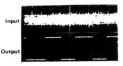

Input

Output

Response to same input tone burst with wideband noise.
$\frac{S}{N}$ = -6db R_L = 100 ohms

Noise Bandwidth = 140 Hz

OPERATING INSTRUCTIONS

Figure 1 shows a typical connection diagram for the 567. For most applications, the following three-step procedure will be sufficient for choosing the external components R_1, C_1, C_2 and C_3.

1. . Select R_1 and C_1 for the desired center frequency. For best temperature stability, R_1 should be between 2K and 20K ohm, and the $R_1 C_1$ product should have sufficient stability, over the projected temperature range to meet the necessary requirements.

2. Select the low-pass capacitor, C_2, by referring to the Bandwidth versus Input Signal Amplitude graph. If the input amplitude variation is known, the appropriate value of $f_0 C_2$ necessary to give the desired bandwidth may be found. Conversely, an area of operation may be selected on this graph and the input level and C_2 may be adjusted accordingly. For example, constant bandwidth operation requires that input amplitude be above 200mVrms. The bandwidth, as noted on the graph, is then controlled solely by the $f_0 C_2$ product (F_0 (Hz), C_2 (μfd),).

3. The value of C_3 is generally non-critical. C_3 sets the band edge of a low pass filter which attenuates frequencies outside the detection band to eliminate spurious outputs. If C_3 is too small, frequencies just outside the detection band will switch the output stage on and off at the beat frequency, or the output may pulse on and off during the turn-on transient. If C_3 is too large, turn-on and turn-off of the output stage will be delayed until the voltage on C_3 passes the threshold voltage. (Such a delay may be desirable to avoid spurious outputs due to transient frequencies.) A typical minimum value for C_3 is $2C_2$.

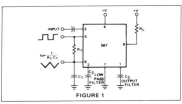

FIGURE 1

AVAILABLE OUTPUTS (Figure 2)

The primary output is the uncommitted output transistor collector, pin 8. When an in-band input signal is present, this transistor saturates; its collector voltage being less than 1.0 volt (typically 0.6V) at full output current (100mA).

The voltage at pin 2 is the phase detector output, a linear function of frequency, over the range of 0.95 to 1.05 f_0. with a slope of about 20mV/% frequency deviation. The average voltage at pin 1 is, during lock, a function of the in-band input amplitude in accordance with the transfer characteristic given. Pin 5 is the controlled oscillator square wave output of magnitude $(V^+ - 2Vbe) \approx (V^+ - 1.4V)$ having a dc average of $V^+/2$. A 1KΩ load may be driven from pin 5. Pin 6 is an exponential triangle of 1 volt peak-to-peak

LINEAR INTEGRATED CIRCUITS ■ 567

AVAILABLE OUTPUTS (Cont'd.)

with an average dc level of V^+ /2. Only high impedance loads may be connected to pin 6 without affecting the CCO duty cycle or temperature stability.

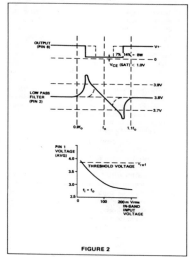

FIGURE 2

OPERATING PRECAUTIONS

A brief review of the following precautions will help the user attain the high level of performance of which the 567 is capable.

1. Operation in the high input level mode (above 200mV) will free the user from bandwidth variations due to changes in the in-band signal amplitude. The input stage is now limiting, however, so that out-band signals or high noise levels can cause an apparent bandwidth reduction as the in band signal is suppressed. Also, the limiting action will create in-band components from sub-harmonic signals, so the 567 becomes sensitive to signals at $f_0/3$, $f_0/5$, etc.

2. The 567 will lock onto signals near $(2n+1)$ f_0, and will give an output for signals near $(4n+1)$ f_0 where $n = 0, 1, 2$, etc. Thus, signals at 5 f_0 and 9 f_0 can cause an unwanted output. If such signals are anticipated, they should be attenuated before reaching the 567 input.

3. Maximum immunity from noise and out-band signals is afforded in the low input level (Below 200mVrms) and reduced bandwidth operating mode. However, decreased loop damping causes the worse-case lock-up time to increase, as shown by the Greatest Number of Cycles Before Output vs. Bandwidth graph.

4. Due to the high switching speeds (20ns) associated with 567 operation, care should be taken in lead routing. Lead lengths should be kept to a minimum. The power supply should be adequately bypassed close to the 567 with an 0.01μF or greater capacitor; grounding paths should be carefully chosen to avoid ground loops and unwanted voltage variations. Another factor which must be considered is the effect of load energization on the power supply. For example, an incandescent lamp typically draws 10 times rated current at turn-on. This can cause supply voltage fluctuations which could, for example, shift the detection band of narrow-band systems sufficiently to cause momentary loss of lock. The result is a low-frequency oscillation into and out of lock. Such effects can be prevented by supplying heavy load currents from a separate supply, or increasing the supply filter capacitor.

SPEED OF OPERATION

Minimum lock-up time is related to the natural frequency of the loop. The lower it is, the longer becomes the turn-on transient. Thus, maximum operating speed is obtained when C_2 is at a minimum. When the signal is first applied, the phase may be such as to initially drive the controlled oscillator away from the incoming frequency rather than toward it. Under this condition, which is of course unpredictable, the lock-up transient is at its worst and the theoretical minimum lock-up time is not achievable. We must simply wait for the transient to die out.

The following expressions give the values of C_2 and C_3 which allow highest operating speeds for various band center frequencies. The minimum rate at which digital information may be detected without information loss due to the turn-on transient or output chatter is about 10 cycles per bit, corresponding to an information transfer rate of $f_0/10$ baud.

$$C_2 = \frac{130}{f_0}\mu F$$

$$C_3 = \frac{260}{f_0}\mu F$$

in cases where turn-off time can be sacrificed to achieve fast turn-on, the optional sensitivity adjustment circuit can be used to move the quiescent C_3 voltage lower (closer to the threshold voltage). However, sensitivity to beat frequencies, noise and extraneous signals will be increased.

OPTIONAL CONTROLS

The 567 has been designed so that, for most applications, no external adjustments are required. Certain applications, however, will be greatly facilitated if full advantage is taken of the added control possibilities available through the use of additional external components. In the diagrams given, typical values are suggested where applicable. For best results resistors used, except where noted, should have the same temperature coefficient. Ideally, silicon diodes would be low-resistivity types, such as forward-biased low-voltage zeners or forward-biased transistor base-emitter junctions. However, ordinary low-voltage diodes should be adequate for most applications.

SENSITIVITY ADJUSTMENT

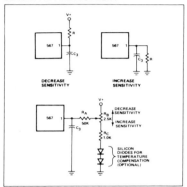

When operated as a very narrow band detector (less than 8 percent), both C_2 and C_3 are made quite large in order to improve noise and outband signal rejection. This will inevitably slow the response time. If, however, the output stage is biased closer to the threshold level, the turn-on time can be improved. This is accomplished by drawing additional current to terminal 1. Under this condition, the 567 will also give an output for lower-level signals (10m or lower).

By adding current to terminal 1, the output stage is biased further away from the threshold voltage. This is most useful when, to obtain maximum operating speed, C_2 and C_3 are made very small. Normally, frequencies just outside the detection band could cause false outputs under this condition. By desensitizing the output stage, the outband beat notes do not feed through to the output stage. Since the input level must be somewhat greater when the output stage is made less sensitive, rejection of third harmonics or in-band harmonics (of lower frequency signals) is also improved.

CHATTER PREVENTION

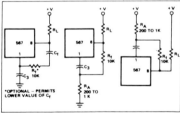

Chatter occurs in the output stage when C_3 is relatively small, so that the lock transient and the AC components at the quadrature phase detector (lock detector) output cause the output stage to move through its threshold more than once. Many loads, for example lamps and relays, will not respond to the chatter. However, logic may recognize the chatter as a series of outputs. By feeding the output stage output back to its input, (pin 1) the chatter can be eliminated. Three schemes for doing this are given above. All operate by feeding the first output step (either on or off) back to the input, pushing the input past the threshold until the transient conditions are over. It is only necessary to assure that the feedback time constant is not so large as to prevent operation at the highest anticipated speed. Although chatter can always be eliminated by making C_3 large, the feedback circuit will enable faster operation of the 567 by allowing C_3 to be kept small. Note that if the feedback time constant is made quite large, a short burst at the input frequency can be stretched into a long output pulse. This may be useful to drive, for example, stepping relays.

DETECTION BAND CENTERING (OR SKEW ADJUSTMENT

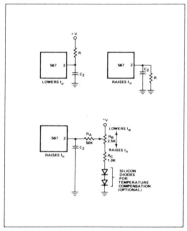

When it is desired to alter the location of the detection band (corresponding to the loop capture range) within the largest detection band (lock range), the circuits shown above can be used. By moving the detection band to one edge of the range, for example, input signal variations will expand the detection band in only one direction. This may prove useful when a strong but undesirable signal is expected on one side or the other of the center frequency. Since R_B also alters the duty cycle slightly, this method may be used to obtain a precise duty cycle when the 567 is used as an oscillator.

TYPICAL APPLICATIONS

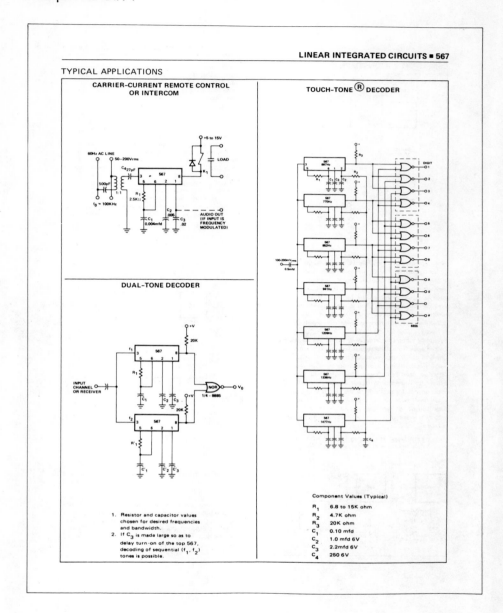

CARRIER-CURRENT REMOTE CONTROL
OR INTERCOM

TOUCH-TONE ® DECODER

DUAL-TONE DECODER

1. Resistor and capacitor values
chosen for desired frequencies
and bandwidth.
2. If C_3 is made large so as to
delay turn-on of the top 567,
decoding of sequential (f_1, f_2)
tones is possible.

Component Values (Typical)

R_1	6.8 to 15K ohm
R_2	4.7K ohm
R_3	20K ohm
C_1	0.10 mfd
C_2	1.0 mfd 6V
C_3	2.2mfd 6V
C_4	250 6V

LINEAR INTEGRATED CIRCUITS ■ 567

TYPICAL APPLICATIONS (Cont'd.)

OSCILLATOR WITH QUADRATURE OUTPUT	OSCILLATOR WITH DOUBLE FREQUENCY OUTPUT	PRECISION OSCILLATOR WITH 20nsec SWITCHING

OSCILLATOR WITH QUADRATURE OUTPUT

CONNECT PIN 3 TO 2.8V TO INVERT OUTPUT

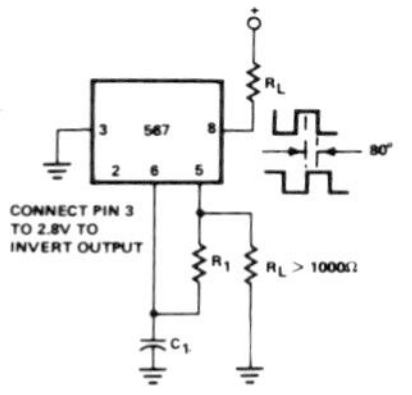

$R_L > 1000\Omega$

OSCILLATOR WITH DOUBLE FREQUENCY OUTPUT

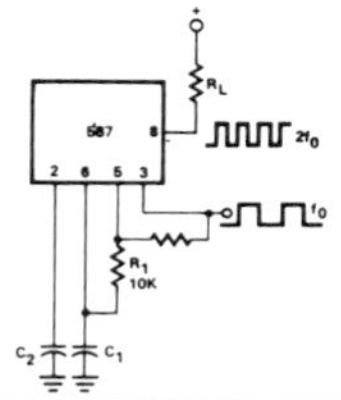

R_1 10K

PRECISION OSCILLATOR WITH 20nsec SWITCHING

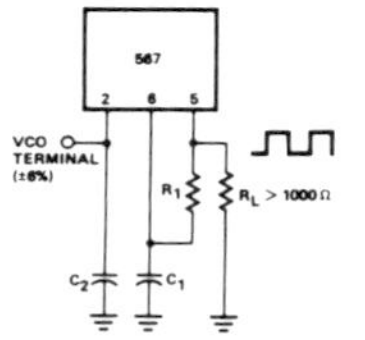

VCO TERMINAL (±6%)

$R_L > 1000\Omega$

PULSE GENERATOR WITH 25% DUTY CYCLE	PRECISION OSCILLATOR TO SWITCH 100ma LOADS	PULSE GENERATOR

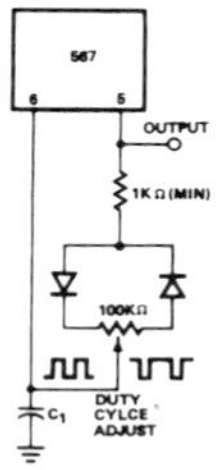

PULSE GENERATOR WITH 25% DUTY CYCLE

10KΩ

R_1

C_1

PRECISION OSCILLATOR TO SWITCH 100ma LOADS

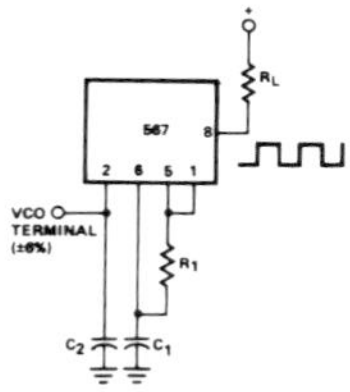

VCO TERMINAL (±6%)

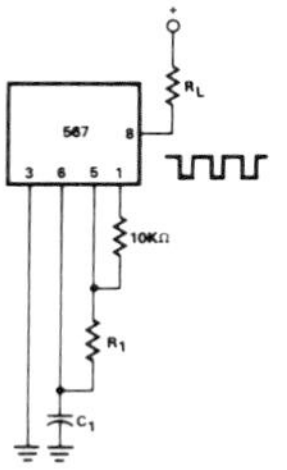

R_1

PULSE GENERATOR

OUTPUT

1KΩ (MIN)

100KΩ

DUTY CYCLE ADJUST

24% BANDWIDTH TONE DECODER

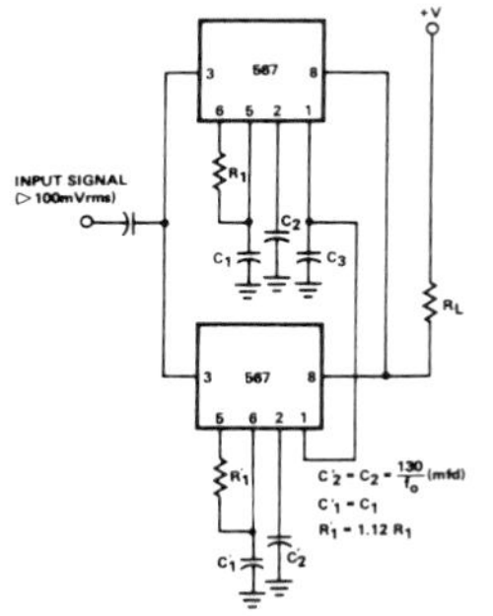

INPUT SIGNAL (>100mVrms)

R_L

$C_2' = C_2 = \frac{130}{f_0}$ (mfd)

$C_1' = C_1$

$R_1' = 1.12 R_1$

0° TO 180° PHASE SHIFTER

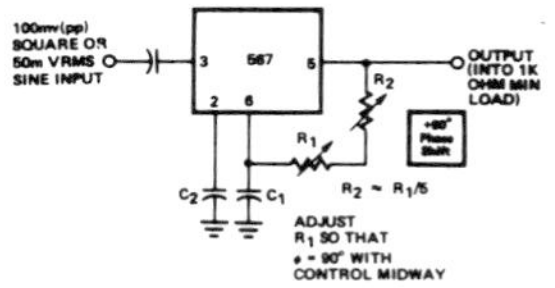

100mv(pp) SQUARE OR 50Ω VRMS SINE INPUT

OUTPUT (INTO 1K OHM MIN LOAD)

R_2

$R_2 = R_1/5$

ADJUST R_1 SO THAT $\theta = 90°$ WITH CONTROL MIDWAY

signetics

QUADRUPLE 2-INPUT POSITIVE NAND GATE | **S5400 N7400**

S5400–A,F,W • N7400–A,F

DIGITAL 54/74 TTL SERIES

SCHEMATIC (each gate)

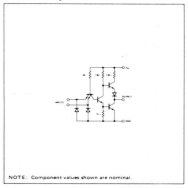

NOTE: Component values shown are nominal.

PIN CONFIGURATIONS

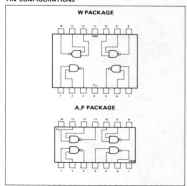

W PACKAGE

A,F PACKAGE

RECOMMENDED OPERATING CONDITIONS

		MIN	NOM	MAX	UNIT
Supply Voltage V_{CC}: S5400 Circuits		4.5	5	5.5	V
N7400 Circuits		4.75	5	5.25	V
Normalized Fan-Out from each Output, N				10	
Operating Free-Air Temperature Range, T_A: S5400 Circuits		-55	25	125	°C
N7400 Circuits		0	25	70	°C

ELECTRICAL CHARACTERISTICS (over recommended operating free-air temperature range unless otherwise noted)

	PARAMETER	TEST CONDITIONS*		MIN	TYP**	MAX	UNIT
$V_{in(1)}$	Logical 1 input voltage required at both input terminals to ensure logical 0 level at output	V_{CC} = MIN		2			V
$V_{in(0)}$	Logical 0 input voltage required at either input terminal to ensure logical 1 level at output	V_{CC} = MIN				0.8	V
$V_{out(1)}$	Logical 1 output voltage	V_{CC} = MIN, I_{load} = −400µA	V_{in} = 0.8V,	2.4	3.3		V
$V_{out(0)}$	Logical 0 output voltage	V_{CC} = MIN, I_{sink} = 16mA	V_{in} = 2V,		0.22	0.4	V
$I_{in(0)}$	Logical 0 level input current (each input)	V_{CC} = MAX,	V_{in} = 0.4V			−1.6	mA
$I_{in(1)}$	Logical 1 level input current (each input)	V_{CC} = MAX, V_{CC} = MAX,	V_{in} = 2.4V V_{in} = 5.5V			40 / 1	µA / mA
I_{OS}	Short circuit output current†.	V_{CC} = MAX	S5400 N7400	−20 −18		−55 −55	mA

signetics

BCD-TO-SEVEN SEGMENT DECODER/DRIVER

N7446
N7447

N7446-B • N7447-B

DIGITAL 54/74 TTL SERIES

DESCRIPTION

The 7446 and 7447 BCD-to-Seven Segment Decoder/Driver are TTL monolithic devices consisting of the necessary logic to decode a BCD code to seven segment readout plus selected signs.

Incorporated in this device is a blanking circuit allowing leading and trailing zero suppression. Also included is a lamp test control to turn on all segments.

The 7446 and 7447 provide bare collector output transistors for directly driving lamps. The output transistor breakdown of the 7446 is 30 volts and the 7447 is 15 volts.

PIN CONFIGURATION

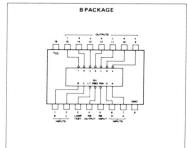

B PACKAGE

LOGIC DIAGRAM

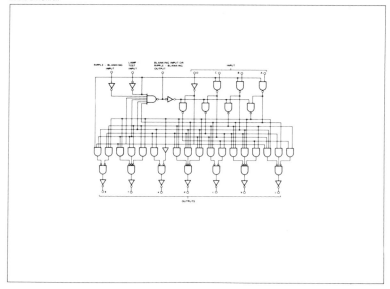

DIGITAL 54/74 TTL SERIES ■ N7446, N7447

TRUTH TABLE

DECIMAL OR FUNCTION	INPUTS							OUTPUTS							NOTE
	LT	RBI	D	C	B	A	BI/RBO	a	b	c	d	e	f	g	
0	1	1	0	0	0	0	1	0	0	0	0	0	0	1	1
1	1	x	0	0	0	1	1	1	0	0	1	1	1	1	1
2	1	x	0	0	1	0	1	0	0	1	0	0	1	0	
3	1	x	0	0	1	1	1	0	0	0	0	1	1	0	
4	1	x	0	1	0	0	1	1	0	0	1	1	0	0	
5	1	x	0	1	0	1	1	0	1	0	0	1	0	0	
6	1	x	0	1	1	0	1	1	1	0	0	0	0	0	
7	1	x	0	1	1	1	1	0	0	0	1	1	1	1	
8	1	x	1	0	0	0	1	0	0	0	0	0	0	0	
9	1	x	1	0	0	1	1	0	0	0	1	1	0	0	
10	1	x	1	0	1	0	1	1	1	0	0	1	1	0	
11	1	x	1	0	1	1	1	1	0	1	1	1	0	0	
12	1	x	1	1	0	0	1	0	1	1	0	1	0	0	
13	1	x	1	1	0	1	1	1	0	1	0	0	0	0	
14	1	x	1	1	1	0	1	1	1	1	1	1	1	1	
15	1	x	1	1	1	1	1	1	1	1	1	1	1	1	
BI	x	x	x	x	x	x	0	1	1	1	1	1	1	1	2
RBI	1	0	0	0	0	0	0	1	1	1	1	1	1	1	3
LT	0	x	x	x	x	x	1	0	0	0	0	0	0	0	4

NOTES:

1. BI/BRO is wire-OR logic serving as blanking input (BI) and/or ripple-blanking output (RBO). The blanking input must be open or held at a logical 1 when output functions 0 through 15 are desired and ripple-blanking input (RBI) must be open or at a logical 1 during the decimal 0 input. X = input may be high or low.

2. When a logical 0 is applied to the blanking input (forced condition) all segment outputs go to a logical 1 regardless of the state of any other input condition.

3. When ripple-blanking input (RBI) is at a logical 0 and A = B = C = D = logical 0, all segment outputs go to a logical 1 and the ripple-blanking output goes to a logical 0 (response condition).

4. When blanking input/ripple-blanking output is open or held at a logical 1, and a logical 0 is applied to lamp-test input, all segment outputs go to a logical 0.

SEGMENT IDENTIFICATION

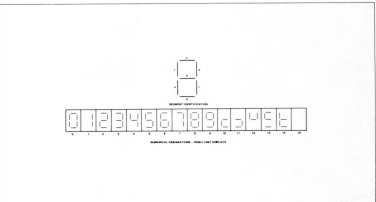

SEGMENT IDENTIFICATION

NUMERICAL DESIGNATIONS — RESULTANT DISPLAYS

signetics

DECADE COUNTER | S5490 N7490

S5490—A,F,W • N7490—A,F

DIGITAL 54/74 TTL SERIES

DESCRIPTION

The S5490/N7490 is a high-speed, monolithic decade counter consisting of four dual-rank, master-slave flip-flops internally interconnected to provide a divide-by-two counter and a divide-by-five counter. Gated direct reset lines are provided to inhibit count inputs and return all outputs to a logical "0" or to a binary coded decimal (BCD) count of 9. As the output from flip-flop A is not internally connected to the succeeding stages, the count may be separated in three independent count modes:

1. When used as a binary coded decimal decade counter, the BD input must be externally connected to the A output. The A input receives the incoming count, and a count sequence is obtained in accordance with the BCD count sequence truth table shown above. In addition to a conventional "0" reset, inputs are provided to reset a BCD 9 count for nine's complement decimal applications.

2. If a symmetrical divide-by-ten count is desired for frequency synthesizers or other applications requiring division of a binary count by a power of ten, the D output must be externally connected to the A input. The input count is then applied at the BD input and a divide-by-ten square wave is obtained at output A.

3. For operation as a divide-by-two counter and divide-by-five counter, no external interconnections are required. Flip-flop A is used as a binary element for the divide-by-two function. The BD input is used to obtain binary divide-by-five operation at the B, C, and D outputs. In this mode, the two counters operate independently; however, all four flip-flops are reset simultaneously.

The 5490/7490 is completely compatible with Series 54 and Series 74 logic families. Average power dissipation is 160mW.

PIN CONFIGURATIONS

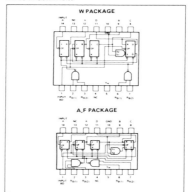

W PACKAGE

A,F PACKAGE

LOGIC TRUTH TABLES

BCD COUNT SEQUENCE (See Note 1)

COUNT	OUTPUT			
	D	C	B	A
0	0	0	0	0
1	0	0	0	1
2	0	0	1	0
3	0	0	1	1
4	0	1	0	0
5	0	1	0	1
6	0	1	1	0
7	0	1	1	1
8	1	0	0	0
9	1	0	0	1

RESET/COUNT (See Note 2)

RESET INPUTS				OUTPUT			
$R_{0(1)}$	$R_{0(2)}$	$R_{9(1)}$	$R_{9(2)}$	D	C	B	A
1	1	0	X	0	0	0	0
1	1	X	0	0	0	0	0
X	X	1	1	1	0	0	1
X	0	X	0	COUNT			
0	X	0	X	COUNT			
0	X	X	0	COUNT			
X	0	0	X	COUNT			

NOTES:
1. Output A connected to input BD for BCD count.
2. X indicates that either a logical 1 of a logical 0 may be present.
3. Fanout from output A to input BD and to 10 additional Series 54/74 loads is permitted

SCHEMATIC DIAGRAM

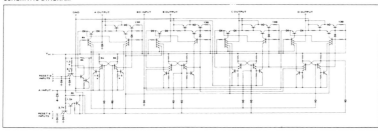

Solid State Division

Digital Integrated Circuits
Monolithic Silicon

CD4000A Series

COS/MOS IC's for
Low-Voltage (3-15V) Applications

Gates
Flip-Flops
Latches
Multivibrators
Shift Registers
Counters

Display Counter/
   Decoder/Drivers
Multiplexers
Arithmetic Circuits
Memories (RAM's)
Phase-Locked Loop

The CD4000A series of COS/MOS devices described in this DATABOOK are a comprehensive line of digital integrated circuits intended for a wide variety of applications in logic systems.

The features shown at the right plus the inherent advantages of COS/MOS IC's over other logic devices permit the logic-system design engineer to achieve outstanding electrical performance, high reliability, and simplified circuitry in a wide variety of equipment designs.

COS/MOS features:

- Supply voltage compatibility with both DTL and T^2L bipolar logic families —
  eliminates need for 2nd and 3rd power supplies for clock driver circuits, and insures voltage level compatibility when interfacing.
  Provides direct connection between COS/MOS and bipolar device in most applications; in other applications, only a single pull-up resistor may be required.
- Low power —
  10 nW typ. for gates
  10 µW typ. for MSI circuits
- Wide voltage range — 3 to 15 volts

- High noise immunity —
  45% of supply voltage (typ.)

- High speed —
  to 10 MHz for gates and flip flops
  to 5 MHz for counters and registers

- Logic compatibility T^2L & DTL interfacing
  (see ICAN 6602)

- High fanout

- Excellent temperature stability — ±1.5% shift in transfer characteristics over −55 to +125°C

- Inputs fully protected

- High input impedance — $10^{12}\Omega$ typ. ——
  Input current typically 10 pA

- Low " 1 " and " 0 " —level output impedance

- Clock voltage = supply voltage

- Single phase clock

- Wide operating temperature range —
  ceramic package types — −55°C to +125°C
  plastic package types — −40°C to +85°C

Applications:

- Automotive
- Data Terminals
- Instrumentation
- Medical
  Electronics
- Alarm Systems

- Appliances
- Watches, Clocks
- Industrial Controls
- Remote Metering
- Computers
- Calculators

Communications —
Pocket Pagers
Hand-Held Radios
Remote Control
Telemetry
Synthesizers

***** Complementary symmetry metal - oxide semiconductor

File No. 479

Digital Integrated Circuits
Monolithic Silicon

CD4000A, CD4001A
CD4002A, CD4025A
Types

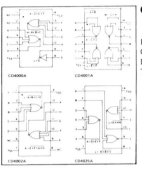

CD4000A   CD4001A

CD4002A   CD4025A

COS/MOS NOR Gates
(Positive Logic)

Dual 3 Input plus Inverter   CD4000AD, CD4000AE, CD4000AK
Quad 2 Input   CD4001AD, CD4001AE, CD4001AK
Dual 4 Input   CD4002AD, CD4002AE, CD4002AK
Triple 3 Input   CD4025AD, CD4025AE, CD4025AK

Special Features

- Medium speed operation. . . . . . $t_{PHL} = t_{PLH} = 25$ ns (typ.)
  at $C_L = 15$ pF

- Low "high"- and "low"-level output impedance. 500Ω
  and 200Ω (typ), respectively at $V_{DD} - V_{SS} = 10$ V

The combination of these devices and the RCA NAND
positive logic gate types CD4011A, CD4012A, and CD4023A
can account for appreciable package-count savings in various
logic function configurations.

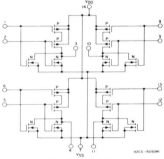

Fig.1.2—Schematic diagram for
type CD4001A.

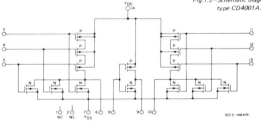

Fig.1.1—Schematic diagram for type CD4000A.   For maximum ratings, see page 20.

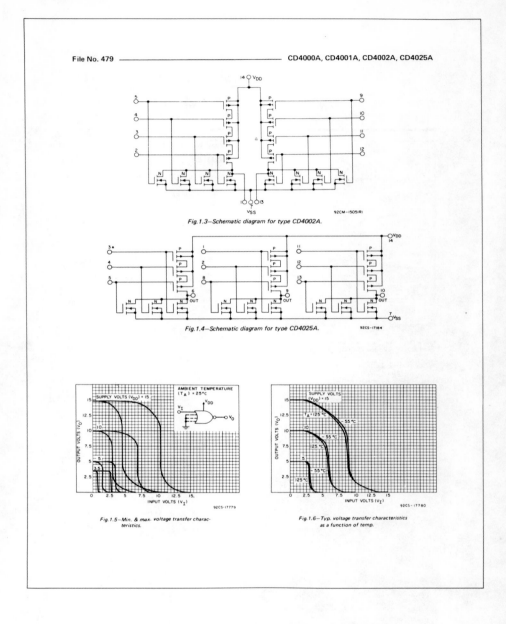

File No. 479 ————————————————————— CD4000A, CD4001A, CD4002A, CD4025A

Fig. 1.3—Schematic diagram for type CD4002A.

Fig. 1.4—Schematic diagram for type CD4025A.

Fig. 1.5—Min. & max. voltage transfer characteristics.

Fig. 1.6—Typ. voltage transfer characteristics
as a function of temp.

CD4000A, CD4001A, CD4002A, CD4025A ———————————————— File No. 479

STATIC ELECTRICAL CHARACTERISTICS (All inputs .... $V_{SS} \leq V_1 \leq V_{DD}$)
(Recommended DC Supply Voltage ($V_{DD} - V_{SS}$) .... 3 to 15 V)

CHARACTERISTIC	SYMBOL	TEST CONDITIONS V_o Volts	V_{DD} Volts	−55°C Min.	Typ.	Max.	25°C Min.	Typ.	Max.	125°C Min.	Typ.	Max.	UNITS	CHARACTERISTIC CURVES & TEST CIRCUITS Fig. No.
Quiescent Device Current	I_L		5	–	–	0.05	–	0.001	0.05	–	–	3	µA	–
			10	–	–	0.1 •	–	0.001	0.1 •	–	–	6 •		
Quiescent Device Dissipation/Package	P_D		5	–	–	0.25	–	0.005	0.25	–	–	15	µW	–
			10	–	–	1	–	0.01	1	–	–	60		
Output Voltage Low-Level	V_{OL}	$V_I = V_{DD}$ $I_O = 0A$	5	–	–	0.01	–	0	0.01	–	–	0.05	V	1.5 1.6
			10	–	–	0.01	–	0	0.01	–	–	0.05		
High-Level	V_{OH}	$V_I = V_{SS}$ $I_O = 0A$	5	4.99	–	–	4.99	5	–	4.95	–	–	V	1.7
			10	9.99	–	–	9.99	10	–	9.95	–	–		
Threshold Voltage N-Channel	$V_{TH}N$	$I_D = 10 \mu A$		–	1.7	–	–	1.5	–	–	–	1.3	V	–
P-Channel	$V_{TH}P$	$I_D = -10 \mu A$		–	-1.7	–	–	-1.5	–	–	-1.3	–	V	–
Noise Immunity (All Inputs) *For Definition, See Appendix*	V_{NL}	$I_O = 0$	3.6 / 5	1.5	–	–	1.5	2.25	–	1.4	–	–	V	–
			7.2 / 10	3	–	–	3	4.5	–	2.9	–	–		
	V_{NH}		0.95 / 5	1.4	–	–	1.5	2.25	–	1.5	–	–	V	–
			2.9 / 10	2.9	–	–	3	4.5	–	3	–	–		
Output Drive Current: N-Channel	I_{DN}	$V_I = V_{DD}$	0.4* / 5	0.5	–	–	0.40	1	–	0.28	–	–	mA	1.8 1.10
			0.5 / 10	1.1 •	–	–	0.9	2.5	–	0.65 •	–	–		
P-Channel	I_{DP}	$V_I = V_{SS}$	2.5# / 5	-0.62	–	–	-0.5	-2	–	-0.35	–	–	mA	1.9 1.11
			9.5 / 10	-0.62 •	–	–	-0.5	-1	–	-0.35 •	–	–		
Input Current	I_I			–	–	–	–	10	–	–	–	–	pA	–

▲ Maximum noise-free saturated Bipolar output voltage.
≠ Minimum noise-free saturated Bipolar output voltage.

Values shown with black dot (•) are the data referred to in the footnote on Page 4 of RIC-102 "High-Reliability COS/MOS CD4000A Series Types".

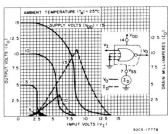

Fig. 1.7 – Typ. current & voltage transfer characteristics.

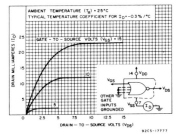

Fig. 1.8 – Typ. n-channel drain characteristics.

File No. 479 —————————————————————————— CD4000A, CD4001A, CD4002A, CD4025A

STATIC ELECTRICAL CHARACTERISTICS (All inputs .... $V_{SS} \leq V_1 \leq V_{DD}$)
(Recommended DC Supply Voltage ($V_{DD} - V_{SS}$) .... 3 to 15 V)

CHARACTERISTIC	SYMBOL	TEST CONDITIONS	V_O Volts	V_{DD} Volts	LIMITS CD4000AE, CD4001AE, CD4002AE, CD4025AE −40°C Min	Typ	Max	25°C Min	Typ	Max	85°C Min	Typ	Max	UNITS	CHARACTERISTIC CURVES & TEST CIRCUITS Fig No.
Quiescent Device Current	I_L			5	—	—	0.5	—	0.005	0.5	—	—	15	μA	
				10	—	—	5	—	0.005	5	—	—	30		
Quiescent Device Dissipation/Package	P_D			5	—	—	2.5	—	0.025	2.5	—	—	75	μW	
				10	—	—	5.0	—	0.05	5.0	—	—	300		
Output Voltage Low Level	V_{OL}	$V_I = V_{DD}$ $I_O = 0A$		5	—	—	0.01	—	0	0.01	—	—	0.05	V	1.5 1.6
				10	—	—	0.01	—	0	0.01	—	—	0.05		
High Level	V_{OH}	$V_I = V_{SS}$ $I_O = 0A$		5	4.99	—	—	4.99	5	—	4.95	—	—	V	1.7
				10	9.99	—	—	9.99	10	—	9.95	—	—		
Threshold Voltage N Channel	V_{THN}	$I_D = 10\,\mu A$			—	1.7	—	—	1.5	—	—	1.3	—	V	–
P Channel	V_{THP}	$I_D = -10\,\mu A$			—	-1.7	—	—	-1.5	—	—	-1.3	—	V	–
Noise Immunity (All Inputs) For Definition See Appendix	V_{NL}	$I_O = 0$	3.6	5	1.5	—	—	1.5	2.25	—	1.4	—	—	V	–
			7.2	10	3	—	—	3	4.5	—	2.9	—	—		
	V_{NH}		0.95	5	1.4	—	—	1.5	2.25	—	1.5	—	—	V	–
			2.9	10	2.9	—	—	3	4.5	—	3	—	—		
Output Drive Current N-Channel	I_{DN}	$V_I = V_{DD}$	0.4▲	5	0.35	—	—	0.3	1	—	0.24	—	—	mA	1.8 1.10
			0.5	10	0.72	—	—	0.6	2.5	—	0.48	—	—		
P-Channel	I_{DP}	$V_I = V_{SS}$	2.5#	5	-0.35	—	—	-0.3	-1	—	-0.24	—	—	mA	1.9 1.11
			9.5	10	-0.3	—	—	-0.25	-1	—	-0.2	—	—		
Input Current	I_I				—	—	—	—	10	—	—	—	—	pA	

▲ Maximum noise-free saturated Bipolar output voltage
Minimum noise-free saturated Bipolar output voltage

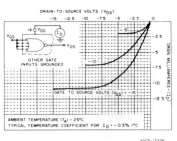

Fig. 1.9 – Typ. p-channel drain characteristics.

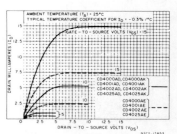

Fig. 1.10 – Min. n-channel drain characteristics.

signetics 256-BIT BIPOLAR FIELD-PROGRAMMABLE ROM [32 x 8 PROM] 8223

DIGITAL 8000 SERIES TTL/MEMORY

DESCRIPTION

The 8223 is a TTL 256-Bit Read Only Memory organized as 32 words with 8 bits per word. The words are selected by five binary address lines; full word decoding is incorporated on the chip. A chip enable input is provided for additional decoding flexibility, which causes all eight outputs to go to the high state when the chip enable input is high.

This device is fully TTL or DTL compatible. The outputs are uncommitted collectors, which permits wired AND operation with the outputs of other TTL or DTL devices. These outputs are capable of sinking twelve standard DCL loads. Propagation delay time is 50ns maximum. Power dissipation is 310 milliwatts with 400 milliwatts maximum. The 8223 may be programmed to any desired pattern by the user. (See fusing procedure.) This feature is ideal for prototype hardware and systems requiring propriety codes.

A Truth Table/Order Blank is included on page 4-43 for ordering custom patterns.

FEATURES

- BUFFERED ADDRESS LINES
- ON THE CHIP DECODING
- CHIP ENABLE CONTROL LINE
- OPEN COLLECTOR OUTPUTS
- DIODE PROTECTED INPUTS
- NO SEPARATE FUSING PINS
- BOARD LEVEL PROGRAMMABLE

APPLICATIONS

PROTOTYPING
VOLUME PRODUCTION
MICROPROGRAMMING
HARDWIRED ALGORITHMS
CONTROL STORE

LOGIC DIAGRAM

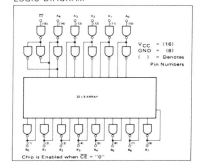

V_{CC} = (16)
GND = (8)
( ) = Denotes Pin Numbers

Chip is Enabled when $\overline{CE}$ = "0"

ELECTRICAL CHARACTERISTICS (S8223 $-55°C \leqslant T_A \leqslant +125°C$ N8223 $0°C \leqslant T_A \leqslant 75°C$; $4.75V \leqslant V_{CC} \leqslant 5.25V$)

CHARACTERISTICS	LIMITS				"0" A_n	"1" A_n	$\overline{CHIP}$ ENABLE	OUTPUTS	NOTES
	MIN.	TYP.	MAX.	UNITS					
"1" Output Leakage Current (N8223-)			100	μA			2.0V		13
(S8223-)			250	μA					
"0" Output Voltage (N8223-) (S8223-)			0.4	V	0.8V	2.0V	0.8V	9.6mA	6,10
(N8223-)			0.5	V	0.8V	2.0V	0.8V	16mA	6,10
"1" Input Current									
An, Address			40	μA		4.5V			
Chip Enable Input			80	μA			4.5V		
"0" Input Current									
An, Chip Enable	-0.1		-1.6	mA	0.4V		0.4V		

DIGITAL 8000 SERIES TTL/MEMORY ■ 8223

$T_A = 25°C$ and $V_{CC} = 5.0V$

CHARACTERISTICS	LIMITS				"0" A_n	"1" A_n	$\overline{CHIP\ ENABLE}$	OUTPUTS	NOTES
	MIN.	TYP.	MAX.	UNITS					
Propagation Delay									
$\overline{An\ to\ Bn}$		35	50	ns				DC F.O.=12	7,12
$\overline{Chip\ Enable}$ to B_n		35	50	ns		4.5V		DC F.O.=12	7,12
Power Consumption		310/62	400/77	mW/mA		4.5V	4.5V		14
Input Latch Voltage	5.5			V			10mA		11

NOTES:
1. All voltage measurements are referenced to the ground terminal. Terminals not specifically referenced are left electrically open.
2. All measurements are taken with ground pin tied to zero volts.
3. Positive current is defined as into the terminal referenced.
4. Positive logic definition: "UP" Level = "1" "DOWN" Level = "0".
5. Precautionary measures should be taken to ensure current limiting in accordance with Absolute Maximum Ratings should the isolation diodes become forward biased.
6. Output sink current is supplied through a resistor to V_{CC}.
7. One DC fan-out is defined as 0.8mA.
8. One AC fan-out is defined as 50pF.
9. Manufacturer reserves the right to make design and process changes and improvements.
10. By DC tests per the truth table, all inputs have guaranteed thresholds of 0.8V for logical "0" and 2.0V for logical "1".
11. This test guarantees operation free of input latch-up over the specified operating power supply voltage range.
12. For detailed test conditions, see AC testing.
13. Connect an external 1k resistor from V_{CC} to the output terminal for this test.
14. $V_{CC} = 5.25V$.

AC TEST FIGURE AND WAVEFORMS

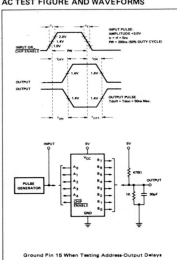

Ground Pin 15 When Testing Address-Output Delays

SCHEMATIC DIAGRAM

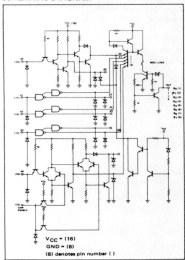

$V_{CC} = (16)$
GND = (8)
(8) denotes pin number ( )

DIGITAL 8000 SERIES TTL/MEMORY ■ 8223

8223 PROGRAMMING PROCEDURE

The 8223 may be programmed by using Curtis Electro Devices, Spectrum Dynamics or Data I/O Programmers.

The 8223 Standard part is shipped with all outputs at logical "0". To write a logical "I" proceed as follows:

Programming Procedure A

Simple Programming Procedure using "bench" Equipment

1. Start with pin 8 grounded and V_{CC} removed from pin 16.
2. Remove any load from the outputs.
3. Ground the Chip Enable.
4. Address the desired location by applying ground (i.e., 0.4V maximum) for a "0", and +5.0V (i.e., +2.8V minimum) for a "1" at the address input lines.
5. Apply +12.5V ± 0.5V to the output to be programmed through a 390 ohm ± 10% resistor. Program one output at a time.
6. Apply +12.5V to V_{CC} (pin 16) for 50 msec to 1sec (max.) with a V_{CC} rise time of 50μsec or less. If 1.0 second is exceeded, the duty cycle should be limited to a maximum of 25%. The V_{CC} overshoot should be limited to 1.0V maximum. If necessary, a clamping circuit should be used. The V_{CC} current requirement is 40 mA maximum at +12.5V. Several fuses can be programmed in sequence until 1.0 sec of high V_{CC} time is accumulated before imposing the duty cycle restriction.

   NOTE: Normal practice in test fixture layout should be followed. Lead lengths, particularly to the power supply, should be as short as possible. A capacitor of 10 microfarads minimum, connected from the +12.5V to ground, should be located close to the unit being programmed.
7. Remove the programming voltage from pin 16.
8. Open the output.
9. Proceed to the next output and repeat, or change address and repeat procedure.
10. Continue until the entire bit pattern is programmed into your custom 8223.

Fast Programming Procedure — Programming Procedure B

1. Remove V_{CC} (open or ground pin 16).
2. Remove any load from the output.
3. Ground $\overline{CE}$ (pin 15).
4. Address the word to be programmed by applying 5 volts of a "1" and ground for a "0" to the address lines. (Solid TTL logic levels are ok, but we suggest buffer drivers or Utilogic OR/NOR gates for the addressing).
5. Apply +12.5V ± 0.5V to the output to be programmed through a 390 ohm ± 10% resistor. Program one output at a time.
6. Apply +12.5V to V_{CC} (pin 16) for 25-50mS. The V_{CC} rise time must be 50μsec or less. Limit the V_{CC} overshoot to 1.0 volts max.
7. Reduce V_{CC} to ground (< 0.5V) and remove the load from the output.
8. Immediately repeat steps 5 and 6 for other outputs of the same word, or repeat 4 through 6 for a different word. Continue programming for a max of 1 second. Then remove power for 4 seconds and continue until the entire bit pattern is programmed.

After programming the 8223, the unit should be checked to insure the code is correct. If additional fuses must be opened, they may be programmed during verification.

Fast Programming Procedure — Programming Procedure C

Steps 1 through 5 are the same as in Procedure B.

6. Apply a 5mS pulse to V_{CC} (pin 16). Limit the V_{CC} overshoot.
7. Reduce V_{CC} to 5 volts for 10-15uS and verify the fuse opened (output is now a "1". If the bit programmed go on to the next bit to be programmed. If the bit did not program, then reduce V_{CC} to ground (or open) for 1-5uS and repeat step 6 and 7 until the fuse programs (1 second total time max).
8. Continue programming at this rate for 1 second. Remove all power from the device for 4 seconds then continue programming procedure.

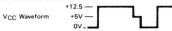

V_{CC} Waveform

BOARD LEVEL PROGRAMMING PROCEDURE FOR THE 8223

The chip select controls which 8223 is being programmed when several PROMS are collector OR'd. To program in this manner, the only changes required are:

1. The 390 ohm resistor is reduced to $\frac{200 \text{ ohm}}{N}$ where N is the number of outputs tied together ($2 \le N \le 12$).
2. Reduce max fuse pulse width from 1 second max to 0.92 sec max.

MANUAL PROGRAMMER DIAGRAM

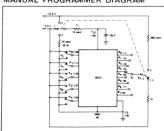

S_1 = Single pole 9 position switch
S_2 through S_6 = single pole 2 position switch
S_7 = Two pole 3 position switch with ground connected to the middle position of the section connected to V_{CC}, pin 16 to go from 5 volts to 12.5V the switch will momentarily ground V_{CC}) and positions 1 and 2 of the other section connected to 5.0V to provide the needed 5 volts to the output for verification.

NOTE: 1. The 10μf capacitor across pin 16 to ground is required to eliminate noise from V_{CC}.

2. During programming switch S_7 must be in position 2 long enough for the 1.0μf capacitor to discharge to less than 0.5 volts.

BULLETIN NO. 1191   APRIL 1974   PRELIMINARY

BAR GRAPH DISPLAY
FOR INSTRUMENTS

TYPE

BG 16101–2

The SELF-SCAN ® bar graph is a flat panel indicator displaying two separate bar graphs, each containing 100 elements for 1% resolution. The display segments are printed on 0.05-inch center spacing. This unique device is a member of a series of analog displays that should find use in such diverse applications as:

- Process Control
- Automobile Displays
- Panel Meters
- Depth Indicators
- Aircraft Displays
- Level Indicators
- Analog Indicators

Employing Burroughs' internal scanning technique, the display needs only six active drivers to operate the two independent information channels. The display elements of both bars are bussed together in a 3-phase arrangement. Operating in a scanning mode and refreshed at 70 Hz or greater, the panel presents a flicker-free display to the viewer.

Suitable logic external to the display device is used to generate a reset pulse and 100 clock pulses in a 3-phase sequence. The display anodes are switched on at reset time and are switched off at the appropriate clock count, thus determining the height of the bar. Typically, this switching is accomplished by comparing an unknown signal with a reference voltage ramp. The ramp can be linear or non-linear (such as logarithmic). A block diagram of the required circuitry is shown in Figure 2. For more detailed information regarding the theory of operation, consult Application Note BG101.

SUGGESTED OPERATING CONDITIONS

Parameter	Symbol	Value	Units
Keep Alive Anode Voltage	E_{KA}	230 ±5%	Vdc
Keep Alive Anode Resistor	R_{KA}	1	M Ω
Typical Keep Alive Anode Current	I_{KA}	100	uAdc
Display Anode Supply Voltage	E_{DS}	230 ±5%	Vdc
Display Anode Resistor	R_{DS}	18	K Ω
Typical Display Anode Current	I_{DS}	5	mA
Cathode Off Bias	E_{KO}	82	Vdc
Display Anode Off Bias Voltage	E_{bo}	100	Vdc
Scan Time Per Cathode	t_S	140	uSec
Applied Reset Pulse Width	t_{AR}	140	uSec

Burroughs Ⓑ

PIN CONNECTIONS

Pin	Connection
1	Channel No. 1 Anode
2	Phase 1 Cathode
3	Phase 3 Cathode
4	Reset Cathode
5	Keep Alive Anode
6	Keep Alive Cathode
7	Phase 2 Cathode
8	Channel No. 2 Anode

DISPLAY CHARACTERISTICS

Segment Length	0.100 inches Nominal
Segment Width	0.020 inches, Nominal
Segment Spacing	0.050 inches, Nominal
Light Output	60 ft. L, Nominal (Note 1)
Color	Neon Orange (Note 2)
Viewing Angle	150°

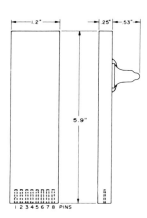

Figure 1. OUTLINE DRAWING

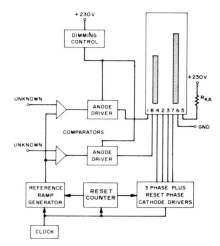

Figure 2. TYPICAL DRIVE CIRCUIT

NOTES

1. Luminance is measured using an ICI corrected Gamma Scientific Model 2020 Photometer mounted normal to an unfiltered panel operating under normal drive conditions. The light output is measured over a typical glowing area with an 0.006" probe.

2. Color filters may be used to alter the apparent color of the display.

3. For further information, write to Burroughs Corp., Electronic Components Division, P.O. Box 1226, Plainfield, New Jersey 07061; or call our special sales/applications assistance number, (201) 757-3400 or (714) 835-7335.

MANUFACTURER'S SPECIFICATION SHEETS— DISCRETE DEVICES

APPENDIX 2

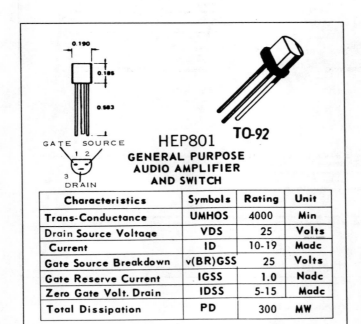

HEP801 TO-92

GENERAL PURPOSE AUDIO AMPLIFIER AND SWITCH

Characteristics	Symbols	Rating	Unit
Trans-Conductance	UMHOS	4000	Min
Drain Source Voltage	VDS	25	Volts
Current	ID	10-19	Madc
Gate Source Breakdown	v(BR)GSS	25	Volts
Gate Reserve Current	IGSS	1.0	Nadc
Zero Gate Volt. Drain	IDSS	5-15	Madc
Total Dissipation	PD	300	MW

2N2646 (SILICON)

Silicon annular PN unijunction transistors designed for use in pulse and timing circuits, sensing circuits and thyristor trigger circuits.

CASE 22 A
(TO-18 Modified)
(Lead 3 connected to case)

MAXIMUM RATINGS (T_A = 25°C unless otherwise noted)

Rating	Symbol	Value	Unit
RMS Power Dissipation*	P_D	300*	mW
RMS Emitter Current	I_e	50	mA
Peak Pulse Emitter Current**	i_e	2**	Amp
Emitter Reverse Voltage	V_{B2E}	30	Volts
Interbase Voltage	V_{B2B1}	35	Volts
Operating Junction Temperature Range	T_J	-65 to +125	°C
Storage Temperature Range	T_{stg}	-65 to +150	°C

* Derate 3.0 mW/°C increase in ambient temperature. The total power dissipation (available power to Emitter and Base-Two) must be limited by the external circuitry.

** Capacitor discharge — 10 µF or less, 30 volts or less.

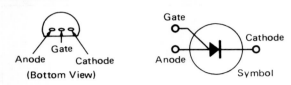

HEP S9001 — Programmable Unijunction Transistor

Care should be taken not to exceed device ratings.

MAXIMUM RATINGS ($T_A = 25^\circ C$)

Characteristic	Symbol	Rating	Unit
Power Dissipation	P_D	375	mW
Forward Anode Current	I_T	200	mA
Gate Current	I_G	±20	mA
Forward Current, Repetitive	I_{TRM}		
1% duty cycle, 100µs pulse		1.0	Amp
1% duty cycle, 20µs pulse		2.0	Amp
Forward Current, Non-Repetitive	I_{TSM}	5.0	Amp
10µs pulse			
Gate — Cathode Forward Voltage	V_{GKF}	40	Vdc
Gate — Cathode Reverse Voltage	V_{GKR}	5.0	Vdc
Gate — Anode Reverse Voltage	V_{GAR}	40	Vdc
Anode — Cathode Voltage	V_{AK}	40	Vdc

For detailed cross-reference of HEP devices with JEDEC and manufacturer's numbers refer to the latest edition of the HEP Cross-Reference Guide.

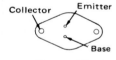

SILICON NPN

(Bottom View)

HEP 703 — Power Transistor

Care should be taken not to exceed the device ratings.

MAXIMUM RATINGS ($T_A = 25^\circ C$)

Characteristics	Symbol	Rating	Unit
Collector-Base Voltage	BV_{CBO}	70	Vdc
Collector-Emitter Voltage	BV_{CEO}	60	Vdc
Emitter-Base Voltage	BV_{EBO}	5.0	Vdc
Collector Current	I_C	5.0	Amp
Total Dissipation @ $25^\circ C$	P_D	25	Watts
Small-signal Cut-off Freq.	f_t	6	MHz
Current Gain (beta)	h_{FE}	85 Typical	

For detailed cross-reference of HEP devices with JEDEC and manufacturer's numbers refer to the latest edition of the HEP Cross-Reference Guide.

NOTE: This device has the collector connected to the metal case. For best results, use HEP 452 mounting kit (sold as accessory item).

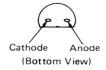

Cathode   Anode
(Bottom View)

HEP Z0212 — Zener Diode

Care should be taken not to exceed the device ratings.
MAXIMUM RATINGS (T_A = 25°C)

Characteristics	Symbol	Rating	Unit
Nominal Zener Voltage @ I_{ZT}	V_Z	5.6	Volts
Test Current	I_{ZT}	20	mA
Zener Impedance @ I_{ZT}	Z_{ZT}	15	ohms
Operating and Storage Junction Temperature Range	T_J, T_{Stg}	-65 to +150	°C

For detailed cross-reference of HEP devices with JEDEC and manufacturer's numbers refer to the latest edition of the HEP Cross-Reference Guide.

CASE 29(1)
(TO–92)

NPN silicon annular transistors designed for general purpose switching and amplifier applications and for complementary circuitry with PNP types 2N4402 and 2N4403. Features one-piece, injection-molded plastic package for high reliability.

Maximum Ratings (T_A = 25°C unless otherwise noted)

Rating	Symbol	Value	Unit
Collector-Emitter Voltage	V_{CEO}	40	Vdc
Collector-Base Voltage	V_{CB}	60	Vdc
Emitter-Base Voltage	V_{EB}	6.0	Vdc
Collector Current—Continuous	I_C	600	mAdc
Total Device Dissipation T_A = 25 °C Derate above 25 °C	P_D	310 2.81	mW mW/°C
Operating and Storage Junction Temperature Range	T_J, T_{stg}	−55 to + 135	°C

Thermal Characteristics

Characteristic	Symbol	Max	Unit
Thermal Resistance, Junction to Case	θ_{JC}	0.137	°C/mW
Thermal Resistance, Junction to Ambient	θ_{JA}	0.357	°C/mW

ELECTRICAL CHARACTERISTICS ($T_A = 25$ °C unless otherwise noted)

Characteristic	Fig. No.	Symbol	Min	Max	Unit
OFF CHARACTERISTICS					
Collector-Emitter Breakdown Voltage* ($I_C = 1$ mAdc, $I_B = 0$)		BV_{CEO}*	40	—	Vdc
Collector-Base Breakdown Voltage ($I_C = 0.1$ mAdc, $I_E = 0$)		BV_{CBO}	60	—	Vdc
Emitter-Base Breakdown Voltage ($I_E = 0.1$ mAdc, $I_C = 0$)		BV_{EBO}	6.0	—	Vdc
Collector Cutoff Current ($V_{CE} = 35$ Vdc, $V_{EB(off)} = 0.4$ Vdc)		I_{CEX}	—	0.1	μ Adc
Base Cutoff Current ($V_{CE} = 35$ Vdc, $V_{EB(off)} = 0.4$ Vdc)		I_{BL}	—	0.1	μ Adc
ON CHARACTERISTICS					
DC Current Gain	15	h_{FE}			—
($I_C = 0.1$ mAdc, $V_{CE} = 1$ Vdc) 2N4401			20	—	
($I_C = 1$ mAdc, $V_{CE} = 1$ Vdc) 2N4400			20	—	
2N4401			40	—	
($I_C = 10$ mAdc, $V_{CE} = 1$ Vdc) 2N4400			40	—	
2N4401			80	—	
($I_C = 150$ mAdc, $V_{CE} = 1$ Vdc)* 2N4400			50	150	
2N4401			100	300	
($I_C = 500$ mAdc, $V_{CE} = 2$ Vdc)* 2N4400			20	—	
2N4401			40	—	
Collector-Emitter Saturation Voltage*	16, 17, 18	$V_{CE(sat)}$			Vdc
($I_C = 150$ mAdc, $I_B = 15$ mAdc)			—	0.4	
($I_C = 500$ mAdc, $I_B = 50$ mAdc)			—	0.75	
Base-Emitter Saturation Voltage*	17, 18	$V_{BE(sat)}$			Vdc
($I_C = 150$ mAdc, $I_B = 15$ mAdc)			0.75	0.95	
($I_C = 500$ mAdc, $I_B = 50$ mAdc)				1.2	
SMALL SIGNAL CHARACTERISTICS					
Current-Gain—Bandwidth Product		f_T			MHz
($I_C = 20$ mAdc, $V_{CE} = 10$ Vdc, f = 100 MHz) 2N4400			200	—	
2N4401			250	—	
Collector-Base Capacitance ($V_{CB} = 5$ Vdc, $I_E = 0$, f = 100 kHz, emitter guarded)	3	C_{cb}	—	6.5	pF
Emitter-Base Capacitance ($V_{BE} = 0.5$ Vdc, $I_C = 0$, f = 100 kHz, collector guarded)	3	C_{eb}	—	30	pF
Input Impedance ($I_C = 1$ mAdc, $V_{CE} = 10$ Vdc, f = 1kHz) 2N4400	12	h_{ie}	0.5	7.5	k ohms
2N4401			1.0	15	
Voltage Feedback Ratio ($I_C = 1$ mAdc, $V_{CE} = 10$ Vdc, f = 1 kHz)	13	h_{re}	0.1	8.0	X 10^{-4}
Small-Signal Current Gain ($I_C = 1$ mAdc, $V_{CE} = 10$ Vdc, f = kHz) 2N4400	11	h_{fe}	20	250	—
2N4401			40	500	
Output Admittance ($I_C = 1$ mAdc, $V_{CE} = 10$ Vdc, f = 1 kHz)	14	h_{oe}	1.0	30	μ ohms
SWITCHING CHARACTERISTICS					
Delay Time $V_{CC} = 30$ Vdc, $V_{EB(off)} = 2$ Vdc,	1, 5	t_d	—	15	ns
Rise Time $I_C = 150$ mAdc, $I_{B1} = 15$ mAdc	1, 5, 6	t_r	—	20	ns
Storage Time $V_{CC} = 30$ Vdc, $I_C = 150$ mAdc	2, 7	t_s	—	225	ns
Fall Time $I_{B1} = I_{B2} = 15$ mAdc	2, 8	t_f	—	30	ns

* Pulse Test: Pulse Width ≤ 300 μs, Duty Cycle $\leq 2\%$.

Switching time equivalent test circuits

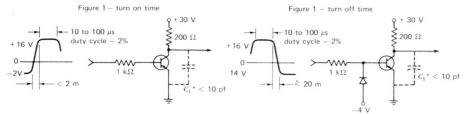

Figure 1 – turn on time   Figure 1 – turn off time

ELECTRICAL CHARACTERISTICS (T_A = 25 °C unless otherwise noted)

Characteristic		Symbol	Min	Typ	Max	Unit
Intrinsic Standoff Ratio (V_{B2B1} = 10 V) (Note 1)	2N2646	η	0.56	—	0.75	—
	2N2647		0.68	—	0.82	
Interbase Resistance (V_{B2B1} = 3 V, I_E = 0)		R_{BB}	4.7	7.0	9.1	K ohms
Interbase Resistance Temperature Coefficient (V_{B2B1} = 3 V, I_E = 0, T_A = −55 °C to +125 °C)		αR_{BB}	0.1	—	0.9	%/°C
Emitter Saturation Voltage (V_{B2B1} = 10 V, I_E = 50 mA) (Note 2)		$V_{EB(sat)}$	—	3.5	—	Volts
Modulated Interbase Current (V_{B2B1} = 10 V, I_E = 50 mA)		$I_{B2(mod)}$	—	15	—	mA
Emitter Reverse Current (V_{B2E} = 30 V, I_{B1} = 0)	2N2646	I_{EO}	—	0.005	12	μA
	2N2647		—	0.005	0.2	
Peak Point Emitter Current (V_{B2B1} = 25 V)	2N2646	I_p	—	1.0	5.0	μA
	2N2647		—	1.0	2.0	
Valley Point Current (V_{B2B1} = 20 V, R_{B2} = 100 ohms) (Note 2)	2N2646	I_V	4.0	6.0	—	mA
	2N2647		8.0	10	18	
Base-One Peak Pulse Voltage (Note 3, Figure 3)	2N2646	V_{OB1}	3.0	5.0	—	Volts
	2N2647		6.0	7.0	—	

Notes:
1. Intrinsic standoff ratio, η, is defined by equation:

$$\eta = \frac{V_p - V_{(EB1)}}{V_{B2B1}}$$

Where V_p = Peak Point Emitter Voltage
V_{B2B1} = Interbase Voltage
$V_{(EB1)}$ = Emitter to Base One Junction Diode Drop (−0.5 V @ 10 μA)

2. Use pulse techniques: PW ≈ 300 μs duty cycle ≤ 2% to avoid internal heating due to interbase modulation which may result in erroneous readings.

3. Base-One Peak Pulse Voltage is measured in circuit of Figure 3. This specification is used to ensure minimum pulse amplitude for applications in SCR firing circuits and other types of pulse circuits.

Figure 1 – Unijunction transistor symbol and nomenclature

Figure 2 – Static emitter characteristic curves

(Exaggerated to show details)

Figure 3 – V_{OB1} test circuit

(Typical relaxation oscillator)

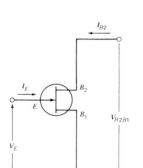

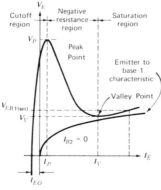

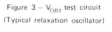

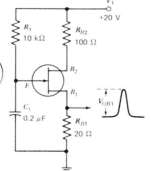

COUNT-RATE METER DERIVATION

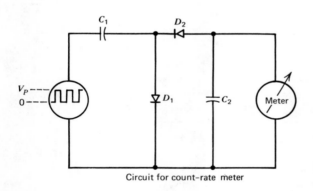

Circuit for count-rate meter

Circuit for Count-Rate Meter

Case 1: D_1 on, D_2 off.

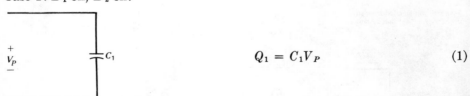

$$Q_1 = C_1 V_P \qquad (1)$$

Case 2: D_1 off, D_2 on

$$Q_1 = (C_1 + C_2) \, V_0 \qquad (2)$$

Case 3: D_1 on, D_2 off.

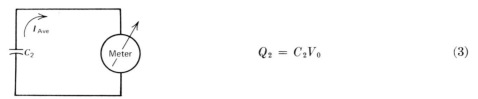

$$Q_2 = C_2 V_0 \qquad (3)$$

Substituting Equation 3 into Equation 2, we have

$$Q_1 = \left(\frac{C_1}{C_2} + 1\right) Q_2$$

If $C_2 \gg C_1$, then $Q_1 \approx Q_2$. The current through the meter is I_{ave} or

$$I_{ave} = \frac{Q_2}{t} = \frac{Q_1}{t} = \frac{C_1 V_P}{t}$$

but $1/t = f$ or

$$I_{ave} = C_1 V_P f$$

DERIVATION OF THE VOLTAGE ACROSS A CAPACITOR

Writing Kirchhoff's voltage loop equation for the figure, we have

$$V_f = iR + \frac{q}{C} = \frac{dq}{dt} + \frac{q}{C}$$

but

$$q = Cv_c \qquad \text{and} \qquad dq = Cdv_c$$

or

$$V_f = RC \frac{dv_c}{dt} + v_c$$

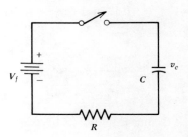

Separating variables and integrating, we have

$$-\frac{dt}{RC} = \frac{dv_c}{(v_c - V_f)}$$

$$-\frac{t}{RC} = \ln (v_c - V_f) + K$$

But at $t = 0$ sec, $v_c = V_i$ (some initial voltage across C). Then it follows that

$$K = -\ln (V_i - V_f)$$

or

$$-\frac{t}{RC} = \ln (v_c - V_f) - \ln (V_i - V_f) = \ln \left[\frac{(v_c - V_f)}{(V_i - V_f)} \right] \qquad (1)$$

Then in exponential form we have

$$e^{-t/RC} = \frac{(v_c - V_f)}{(V_i - V_f)}$$

472

Now solving for v_c, we have

$$v_c = V_f + (V_i - V_f)\, e^{-t/RC} \tag{2}$$

If $V_i = 0$ V, then Equation 2 would become

$$v_c = V_f(1 - e^{-t/RC})$$

SYMBOLS FOR SOLID-STATE DEVICES

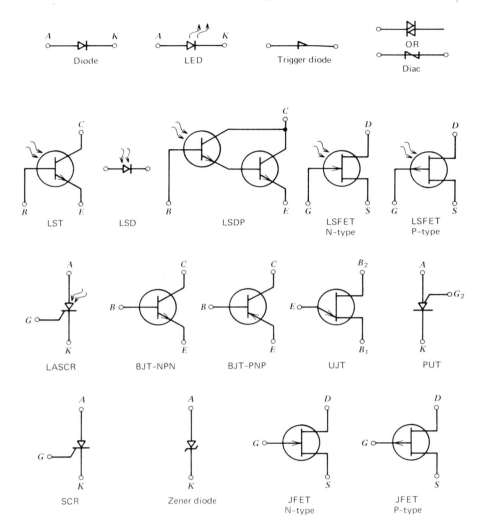

Diode

LED

Trigger diode

Diac

OR

LST

LSD

LSDP

LSFET
N-type

LSFET
P-type

LASCR

BJT-NPN

BJT-PNP

UJT

PUT

SCR

Zener diode

JFET
N-type

JFET
P-type

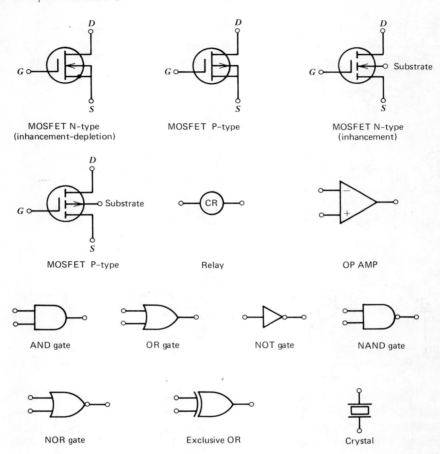

MOSFET N-type
(inhancement-depletion)

MOSFET P-type

MOSFET N-type
(inhancement)

MOSFET P-type

Relay

OP AMP

AND gate

OR gate

NOT gate

NAND gate

NOR gate

Exclusive OR

Crystal

6 DECIBEL TABLE

+dB	Voltage Ratio ($R_{in} = R_{out}$)	Power Ratio	−dB	Voltage Ratio ($R_{in} = R_{out}$)	Power Ratio
0	1.000	1.000	0	1.000	1.000
0.1	1.012	1.023	0.1	0.989	0.977
0.2	1.023	1.047	0.2	0.977	0.955
0.3	1.035	1.072	0.3	0.966	0.933
0.4	1.047	1.096	0.4	0.955	0.913
0.5	1.059	1.122	0.5	0.944	0.891
1.0	1.122	1.259	1.0	0.891	0.794
2.0	1.259	1.585	2.0	0.794	0.631
3.0	1.413	1.995	3.0	0.708	0.501
4.0	1.585	2.512	4.0	0.631	0.398
5.0	1.778	3.162	5.0	0.562	0.316
10	3.162	10.00	10	0.316	0.100
15	5.620	31.60	15	0.178	0.037
20	10.00	100.0	20	0.100	0.010
30	31.60	1000	30	3.16×10^{-2}	10^{-3}
40	100.0	10^{+4}	40	10^{-2}	10^{-4}
50	316.0	10^{+5}	50	3.16×10^{-3}	10^{-5}
100	10^{+5}	10^{+10}	100	10^{-5}	10^{-10}

COMMONLY USED ABBREVIATIONS

A/D	analog-to-digital converter
BJT	bipolar junction transistor
CB	common base
CC	common collector
CE	common emitter
CG	common gate
CMOS	complementary metal oxide semiconductor
CS	common source
D/A	digital-to-analog converter
dB	decibel
DIP	dual in-line package
DTL	diode-transistor logic
IC	integrated circuit
JFET	junction field effect transistor
LASCR	light-activated, silicon controlled rectifier
LCD	liquid crystal display
LED	light-emitting diode
LSD	light-sensitive diode
LSDP	light-sensitive Darlington pair
LSI	large-scale integration
LST	light-sensitive transistor
MOSFET	metal oxide, semiconductor field effect transistor
MSI	medium-scale integration
OP AMP	operational amplifier
PC	photocell
PIV	peak inverse voltage
PLL	phase lock loop

pROM	programmable read-only memory
PUT	programmable unijunction transistor
RAM	random access memory
ROM	read-only memory
RTL	resistor-transistor logic
SCR	silicon controlled rectifier
SCS	silicon controlled switch
SIP	single in-line package
SSI	small-scale ingtegration
UJT	unijunction transistor
TTL	transistor-transistor logic

GLOSSARY

Acceptor: Also called acceptor impurity. An impurity lacking sufficient valence electrons to complete the bonding arrangement in the crystal structure. When added to a semiconductor crystal, it accepts an electron from a neighboring atom and thus creates a hole in the lattice structure of the crystal.

Active component: A device the output of which is dependent on a source of power other than the main input signal.

Alpha: Emitter-to-collector gain of a transistor connected as a common-base amplifier. For a junction transistor, alpha is less than unity of one.

Amplifier: A device that draws power from a source other than the input signal and that produces an output, an enlarged reproduction of the essential features of its input.

Analog switch: A device that either transmits an analog signal without distortion or completely blocks it.

Barrier voltage (potential): The voltage necessary to cause electrical conduction in a junction of two dissimilar materials, such as a PN junction diode.

BCD code: A system of representing the numerals 0 to 9 in binary.

Beta: The current gain of a transistor connected as a common-emitter amplifier. It is the ratio of a small change in collector current to the corresponding change in base current, with the collector voltage constant.

Beta cutoff frequency $(f\beta)$: The frequency at which the beta of the transistor is 3 dB below the low-frequency value.

Bipolar: Describing a device in which both electrons and holes make up the current flow through the device.

All definitions except those with an asterisk can be found in *Modern Dictionary of Electronics*, Howard W. Sams & Co., Inc.

480

BJT: Abbreviation for the bipolar junction transitor. A BJT is a device that uses both positive and negative charge carriers.

Boolean algebra: A system of mathematical logic dealing with classes, propositions, on-off circuit elements, etc., associated by operators as AND, OR, NOT, EXCEPT, IF . . . THEN, etc., thereby permitting computation and demonstration as in any mathematical system.

Carbon microphone: A microphone that depends for its operation on the variation in resistance of carbon contacts.

Characteristic curves: A graph plotted to show the relationships between changing values. An example is a curve showing collector current changes as the base current varies.

*Class A amplifier: A circuit configuration in which the output current flows for a full 360°.

*Class B amplifier: A circuit configuration in which the output current flows for 180°.

*Class C amplifier: A circuit configuration in which the output current flows for less than 180°.

Clipper: A device the output of which is zero or a fixed value for instantaneous input amplitudes up to a certain value, but is a function of the input for amplitudes exceeding the critical value.

*Collector feedback biasing: A biasing configuration for a BJT in which the base resistor is connected to the collector terminal of the transistor. The equations for this arrangement are

$$I_B = \frac{V_{CE} - V_{BE}}{R_B}$$

$$V_{CE} = V_{CC} - (I_C + I_B) R_L$$

Common-base amplifier: A transistor amplifier in which the base element is common to both the input and output circuits.

Common-collector amplifier: Also called an emitter follower. A transistor amplifier in which the collector element is common to both the input and output circuits.

*Common-drain amplifier: A FET amplifier in which the drain element is common to both the input and output circuits.

Common-emitter amplifier: A transistor amplifier in which the emitter element is common to both the input and output circuits.

*Common-source amplifier: A FET amplifier in which the source element is common to both the input and output circuits.

*Common-gate amplifier: A FET amplifier in which the gate element is common to both the input and output circuits.

Complementary transistor amplifier: An amplifier that utilizes the complementary symmetry of NPN and PNP transistors.

Complementary transistors: Two transistors of opposite conductivity (PNP and NPN) that are operated in the same functional unit.

Conduction band: A partially filled energy band in which electrons can move freely, allowing the material to carry an electric current with electrons as the charge carriers.

*Constant-base current biasing: A biasing configuration for a BJT in which only a base resistor is used. The equations for this arrangement are

$$I_B = \frac{V_{CC} - V_{BE}}{R_B}$$

$$V_{CE} = V_{CC} - I_C R_L$$

Converter: A unit that changes an input dc voltage to an output dc voltage, usually at different voltage levels.

Covalent bond: A type of linkage between atoms. Each atom contributes one electron to a shared pair that constitutes an ordinary chemical bond.

Crystal: A thin slab or plate of quartz ground to a thickness that acuses it to vibrate at a specific frequency when energy is supplied.

Crystal oscillator: Also called a crystal-controlled oscillator. An oscillator in which the ferquency of oscillation is controlled by a piezoelectric crystal.

Current amplifier: A device designed to deliver a greater output current than its input current.

Current gain: The ratio of the current produced in the output circuit of an amplifier to the current supplied to the input circuit.

Current limiter: An overload-protection device that limits the maximum output current of a power supply to a preset value and automatically restores the output when the overload is removed.

Current pump: A circuit that drives, through an external load circuit, an adjustable, variable, or constant value of current regardles of the reaction of the load to the current, within rated limits of current, voltage, and load impedance.

Darlington pair: A transistor circuit that in its original form consists of two transistors in which the collectors are tied together and the emitter of the first transistor is directly coupled to the base of the second transistor. This connection of the two transistors can be regarded as a compound transistor with three terminals.

dB: Abbreviation for decibel. It is the standard unit for expressing transmission gain or loss and relative power levels.

Decoder: A device that is used to convert one type of code into another.

Depletion region: The region, extending on both sides of the reverse-biased semiconductor junction, in which all carriers are swept from the vicinity of the junction: that is, the region is depleted of carriers. This region takes on insulating characteristics.

DIAC: Two-lead ac switch semiconductor. When the voltage across the device reaches a specific value, it conducts, providing a low resistance path for current.

Diffused planar transistor: A transistor made by two gaseous diffusions, but in which the collector-base junction is defined by oxide masking. Junctions are formed beneath this protective oxide layer with the result that the device has lower reverse currents and good dc gain at low currents.

Direct coupled amplifier: An amplifier in which the output of one stage is connected to the input of the next stage without the use of intervening coupling components.

*Display: Visual presentation of alphanumeric characters. Popular displays include the LED, LCD, incandescent, fluorescent, and gas discharge types.

Donor: Also called donor impurity. An impurity atom that tends to give up an electron and thereby affects the electrical conductivity of a crystal. It is used to produce N-type semiconductor material.

Dopant: An impurity added to a semiconductor to improve its electrical conductivity.

Dual slope A/D converter: A cirucit configuration that transforms an analog quantity into its digital equivalent.

Dynamic microphone: A moving coil microphone in which the diaphragm is attached to a coil positioned in a fixed magnetic field. The sound waves strike the diaphragm, moving it and, hence, the coil back and forth. An audio frequency current is induced in the moving coil in the magnetic field and coupled to the amplifier.

Efficiency: The ratio of the output power to input power. Efficiency is usually expressed as a percentage (i.e., 0.50 equals 50%).

*Emitter-feedback biasing: A biasing configuration for a BJT in which base and emitter resistors are used. The equations for this arrangement are

$$I_B = \frac{V_{CC} - V_{BE}}{R_B + (h_{fe} + 1) R_E}$$

$$V_{CE} \approx V_{CC} - I_C(R_L + R_E)$$

Energy gap: The energy range between the bottom of the conduction band and the top of the valence band of a semiconductor.

*Equations:

A_v: Voltage gain of a common-emitter amplifier (emitter-resistor bypassed) and is equal to $\approx -R_L/h_{ib}$.

$A_{v(CB)}$:   Voltage gain for a common-base amplifier and is equal to $\approx R_L/h_{ib}$.

$A_{v(CC)}$:   Voltage gain for a common-collector amplifier and is equal to $R_L/(R_L + h_{ib})$.

A_{vf}:   Voltage gain of a common-emitter amplifier (un-bypassed emitter-resistor) and is equal to $\approx -R_L/R_E$.

h_{fe}:   ac beta and is equal to i_c/i_b.

h_{FE}:   dc beta and is equal to $(I_C - I_{CO})/(I_B + I_{CO})$.

h_{ib}:   Input resistance for the common-base amplifier and is equal to $\approx 26 \text{ mV}/I_E(\text{mA})$.

h_{ie}:   Input resistance for the common-emitter amplifier and is equal to $\approx h_{ib}(h_{fe} + 1)$.

I_{CEO}:   Collector current in a common-emitter amplifier without base current and is equal to $I_{CO}(h_{fe} + 1)$.

I_{CO}:   Leakage current for a transistor. It is a function of temperature.

$R_{in(CB)}$:   Input resistance for the common-base amplifier and is equal to $\approx h_{ib}\|R_E$.

$R_{in(CC)}$:   Input resistance for the common-collector amplifier and is equal to $(h_{fe} + 1) (h_{ib} + R_L)$.

$R_{0(CB)}$:   Output resistance for the common-base amplifier and is equal to $\approx R_L$.

$R_{0(CC)}$:   Output resistance for the common-collector amplifier and is equal to $R_L \| [h_{ib} + R_g \| R_B / (h_{fe} + 1)]$.

V_c: The voltage across a capacitor and is equal to $(I/C)\, t$ if a constant current is pumped into the capacitor.

Extrinsic semiconductor: A semiconductor the electrical properties of which depend on its impurities.

Feedback amplifier: An amplifier that uses a passive network to return a portion of the output signal back to the input to modify the performance of the amplifier.

Fiber optics: An assemblage of transparent glass fibers all bundled together parallel to one another. The length of each fiber is much greater than its diameter. This bundle of fibers has the ability to transmit a picture from one of its surfaces to the other around curves and into òtherwise inaccessible places with an extremely low loss of definition and light by a process of total reflection.

Filter: A selective network of resistors, inductors, or capacitors that offers comparatively little opposition to certain frequencies or to dc while blocking or attenuating other frequencies.

Forward bias: An external voltage applied in the conducting direction of a PN junction. The positive terminal is connected to the P-type region, and the negative terminal to the N-type region.

Gain bandwidth product (f_T):   The frequency at which the h_{fe} for the transistor equals one.

Germanium: A brittle, grayish-white metallic element having semiconductor properties. It is widely used in transistors and crystal diodes. Its atomic number is 32 and its symbol is Ge.

Heat sink: A mounting base, usually metalic, that dissipates, carries away, or radiates into the surrounding atmosphere the heat generated within a semiconductor device.

Holding current I_H: That value of average forward current (with the gate open) below which a silicon-controlled rectifier returns to the forward blocking state after having been in forward conduction.

Hybrid circuit model: A circuit configuration having a voltage generator at the input of the model and a current generator at the output of the model.

Hybrid parameters: Also called h parameters. The resultant parameters of an equivalent transistor circuit when the input current and output voltage are selected as independent variables.

IC: Abbreviation for integrated circuit. A combination of interconnected circuit elements inseparably associated on or within acontinuous substrate.

***I_{DSS}:**   Maximum drain-source current with the gate-source voltage equal to zero.

Ideal diode: A diode. When forward-biased it has zero ohms of resistance and when reverse-biased it has infinite ohms of resistance.

Intrinsic material: A semiconductor material in which there are equal numbers of holes and electrons, that is, no impurities.

Inverter: A device that converts direct current into alternating current.

JFET: Abbreviation for junction field effect transistor.

Leakage current: An undesirably small value of current that flows through the reverse biased PN junction because of minority current carriers. The magnitude of the leakage current is a function of temperature.

LED: Abbreviation for light-emitting diode. A diode of special construction that when forward-biased emits light.

Light: Radiant energy within the wavelength limits perceptible by the average human eye (roughly between 400 and 700 millimicrons).

Light-emitting diode: Abbreviated LED. A PN junction that emits light when biased in the forward direction.

Light-sensitive devices: A device exhibiting a photoelectric effect when irradiated. Examples include light-sensitive transistors and light-sensitive diodes.

Load line: A line drawn on the collector characteristic curves of a transistor on which the operating point of the transistor moves as the collector current changes. It is called a load line because the slope of the line depends on the value of the collector load resistance.

Logic circuit: A circuit that provides an input-output relationship corresponding to a Boolean algebra logic function.

Logic element: A device that performs a logic function; a gate or flip-flop, or in some cases a combintion of these devices, treated as a single entity.

Miller effect: The increase in the effective base-to-collector capacitance of a BJT in a common-emitter circuit configuration.

Monolithic IC: An integrated circuit the elements of which are formed in situ upon or within a semiconductor substrate with at least one of the elements formed within the substrate.

MOSFET: Abbreviation for metal oxide semiconductor field effect transistor, sometimes called an insulated gate FET. There are two types of MOSFETs: the enhancement type and the enhancement-depletion type.

*Negative feedback: A portion of the output signal is fed back to and 180° out of phase with the input signal. It is also called degenerative feedback. There are four types of negative feeedback:

(a) If the feedback current i_f is a function of the output voltage $i_f = f(v_0)$, the amplifier is called a transresistance amplifier. R_{inf} is low; R_{0f} is low.

(b) If the feedback current i_f is a function of the output current $i_f = f(i_0)$, the amplifier is called a current amplifier. R_{inf} is low; R_{0f} is high.

(c) If the feedback voltage v_f is a function of the output current $v_f = f(i_0)$, the amplifier is called a transconductance amplifier. R_{inf} is high; R_{0f} is high.

(d) If the feedback voltage v_f is a function of the output voltage $v_f = f(v_0)$, the amplifier is called a voltage amplifier. R_{inf} is high; R_{0f} is low.

Negative logic: A form of logic in which the more positive voltage level represents the logic level 0 and the more negative level represents the logic level 1.

Nonsinusoidal wave: Any wave that is not a sine wave. It therefore contains harmonics.

N-Type semiconductor: An extrinsic-type semiconductor in which the conduction electron density exceeds the hole density. By implication, the net ionized impurity concentration is a donor type.

Optical coupler: A light, tight container that holds a light-emitting device called the transmitter and a light-sensitive device called the receiver.

Oscillator: An oscillator may be considered an amplifier with positive feedback, with circuit parameters that restrict the oscillations of the device to a single frequency. Circuits include the Hartley, Colpitts, Armstrong, and crystal.

Oxide isolation: Electrical isolation of a circuit element by a layer of silicon oxide formed between the element and the substrate.

Passive component: A nonpowered component generally presenting some loss (expressed in decibels) to a system.

Phase shift oscillator: An oscillator in which a network having a phase shift of an odd multiple of 180° (per stage) at the oscillation frequency is connected between the output and input of an amplifier. When the phase shift is obtained by resistance-capacitance elements, the circuit is called a RC phase shift oscillator.

Photocell: A cell, such as a photovoltaic or photoconductive cell, the electrical properties of which are affected by illumination.

Photon: A quantum of electromagnetic energy. The equation is h, where h is Planck's constant and ν is the frequency associated with the photon.

PIV: Abbreviation for peak inverse voltage.

PN Junction: The region of transition between P-type and N-type material in a single semiconductor crystal.

PNPN switch: A semiconductor device that may be regarded as a two-transistor structure with two separate emitters feeding a common collector. This combination constitutes a feedback loop that is unstable for loop gains greater than unity.

Positive feedback: A portion of the output signal is fed back to and in phase with the input signal. It is also called regenerative feedback.

Positive logic: A form of logic in which the more positive voltage level represents the logic level 1 and the more negative level represents the logic level 0.

Power amplifier stage: An audio-frequency amplifier stage capable of handling considerable audio-frequency power without distortion.

Power derating: Use of computed curves to determine the correct power rating of a device or component to be used above its reference ambient temperature.

Power gain: Also called power amplification. Of an amplifying device, the ratio of the power delivered to a specified load impedance to the power absorbed by its input.

Power supply: A unit that supplies electrical power to another unit. It changes ac to dc and maintains a constant voltage output within limits.

P-Type semiconductor: An extrinsic semiconductor in which the hole density exceeds the conduction electron density. By implication, the net ionized impurity concentration is an acceptor type.

PUT: Abbreviation for programmable unijunction transistor. A solid-state switch that can be programmed. The device can be turned on at a specific anode voltage by the selection of external components.

Quiescent point: On the characteristic curves of an amplifier, the point representing the conditions existing when there is no input signal.

RAM: Abbreviation for random access memory.

RC coupling: Coupling between two stages (or more) by a combination of resistive and capacitive elements.

Rectifier: A device which, by virtue of its asymmetrical conduction characteristic, converts an alternating current into a unidirectional current.

Regulated power supply: A unit that maintains a constant output voltage or current for changes in line voltage, output load, ambient temperature, or time.

Regulator: A device, the function of which is to maintain a designated characteristic at a predetermined value or to vary it according to a predetermined plan.

Reverse bias: Also called back bias. An external voltage applied to a semiconductor PN junction to reduce the flow of current across the junction and thereby widen the depletion region. It is the opposite of forward bias.

ROM: Abbreviation for read only memory.

Saturation voltage: The voltage drop across a transistor that is saturated or fully turned on and is labeled $V_{CE(sat)}$.

SCR: Abbreviation for silicon-controlled rectifier. A four-layer PNPN semiconductor that has a gate. The gate serves the function of turning the device on with positive anode voltages. When the SCT is turned on, it continues to conduct even with the control signal removed until the anode voltage is removed, reduced, or reversed.

SCS: Abbreviation for the silicon-controlled switch. A four terminal PNPN semiconductor switching device; it can be triggered into conduction by the application of either a positive or a negative control pulse.

Semiconductor: A solid or liquid electronic conductor, with resistivity between that of metals and that of insulators, in which the electrical charge carriers concentration increases with increasing temperature. Over most of the practical temperature range, the resistance has a negative temperature coefficient. Certain semiconductors possess two types of carriers: negative electrons and positive holes.

Sinusoidal wave: A wave the displacement of which varies as the sine (or cosine) of an angle that is proportional to time and distance or both.

Silicon: A metallic element often mixed with impurities and used in the construction of solid-state devices. Its symbol is Si.

Shell: A group of electrons having a common energy level that forms part of the outer structure of an atom.

Signal generator: A device that supplies a standard voltage of known amplitude, frequency, and waveform for measuring purposes.

*Stability factors: Factors that relate the magnitude in collector current changes with respect to change in leakage current, base-emitter voltage, and beta. The total change in collector current is the sum of the individual changes or

$$\Delta I_C = S_{ICO}\Delta I_{CO} + S_{VBE}\Delta V_{BE} + S_{hFE}\Delta h_{FE}$$

where

$$S_{ICO} \approx \frac{\Delta I_C}{\Delta I_{CO}} = \frac{(h_{FE} + 1)(R_B + R_E)}{R_B + (h_{FE} + 1)\, R_E}$$

$$S_{VBE} \approx \frac{\Delta I_C}{\Delta V_{BE}} = \frac{-h_{FE}}{R_B + (h_{FE} + 1)\, R_E}$$

$$S_{hFE} \approx \frac{\Delta I_C}{\Delta h_{FE}} = \frac{(I_C - I_{CO})}{h_{FE}} \left(\frac{R_B + R_E}{R_B + (h_{FE} + 1)\, R_E} \right)$$

Substrate: The supporting material on or in which the parts of an integrated circuit are attached or made. The substrate may be passive or active.

*Switching regulator: A device that regulates by switching the series active element between cutoff and saturation for predetermined time periods. This switching action increases the efficiency of the regulator.

Thermal resistance: The ratio of the temperature rise to the rate at which heat is generated within a device under steady-state conditions.

Thyristor: A bistable device comprised of three of more junctions. At least one of the junctions can switch between reverse- and forward-voltage polarity within a single quadrant of the anode-to-cathode voltage-current characteristics. Used in a generic sense to include silicon-controlled rectifiers and gate-controlled switches as well as multilayer two-terminal devices.

Transconductance amplifier: An amplifier that supplies an output current proportional to its input voltage. From its output terminals the amplifier appears to be a current source with a high output resistance R_0.

Transducer: A device that converts energy from one form to another.

Transformer coupling: Use of a transformer to couple stages of an amplifier.

Transresistance amplifier: An amplifier that supplies an output voltage that is proportional to an input current. The transfer function of the amplifier is $v_0/i_{in} = R_m$ where R_m is the transresistance.

TRIAC: A five layer NPNPN device that is equivalent to two SCRs connected in antiparallel with a common gate. It provides switching action for either polarity of applied voltage and can be controlled in either polarity from the single gate electrode.

Truth table: A tabulation that shows the relationship between input logic levels of a digital circuit and its output logic levels.

Unipolar: A transistor in which charge carriers are of only one polarity or type (electrons or holes). Devices include field effect transistors and unijunction transitors.

*Universal biasing: A biasing configuration for a BJT in which a voltage divider network and an emitter-resistor are used. The equations for this arrangement are

$$V_{BB} = I_B(R_B + \{h_{FE} + 1\} R_E) + V_{BE}$$

$$V_{CE} = V_{CC} - I_C(R_L + R_E)$$

$$R_1 = \frac{V_{CC}}{V_{BB}} R_B$$

$$R_2 = \frac{R_1 R_B}{R_1 - R_B}$$

with the Q point given by $I_{CQ} = V_{CC}/(2R_L + R_E)$ and $V_{CEQ} = V_{CC}/(2 + R_E/R_L)$.

Valance band: In the spectrum of the solid crystal, the range of energy states containing the energy of the valance electrons that bind the crystal together. In a semiconductor material, it is just below the conduction band.

Variable capacitance diode: Also called variable voltage capacitor or VVC. A semiconductor diode in which the junction capacitance present in all semiconductor diodes has been accentuated. An appreciable change in the thickness of the junction depletion layer and a corresponding change in the capacitance occurs when the dc voltage applied to the diode is changed.

Voltage amplifier: An amplifier used specifically to increase a voltage. It is usually capable of delivering only a small output current.

Voltage doubler: A device, through the use of diodes and capacitors, that has a dc output voltage that is twice the peak amplitude of the ac input voltage.

Voltage gain: The ratio of the voltage across a specified load impedance connected to a transducer to the voltage across the input of the transducer

V_P: Pinch off voltage. This is the value of the gate voltage of a FET that blocks the drain current for all source-drain voltage below the junction breakdown value. Pinch off occurs when the depletion zone completely fills the area of the device.

y Parameters: The resultant parameters of an equivalent transistor circuit when the input and output currents are selected as independent variables.

Zener diode: A two-layer device that, above a certain reverse voltage (the zener value) has a sudden rise in current. If forward-biased, the diode is an ordinary rectifier. But, when reverse-biased, the diode exhibits a typical knee or sharp break in its current voltage graph. The voltage across the device remains essentially constant for any further increase of reverse current up to the allowable dissipation rating. The zener diode is a good voltage regulator, overvoltage protector, voltage reference, level shifter, and so on. True zener breakdown occurs at less than 6 V.

z Parameters: The resultant parameters of an equivalent transistor circuit when the input and output voltages are selected as independent variables.

INDEX

2
Sprachbrücke

Deutsch als Fremdsprache

Arbeitsheft Lektionen 6–10

von
Eva-Maria Jenkins

mit Beiträgen von
Franz Buchetmann
Siglinde Gruber
Andreas Kabisch

Klett Edition Deutsch

Sprachbrücke 2

Arbeitsheft Lektionen 6-10

Redaktion: Eva-Maria Jenkins
Illustrationen: Joachim Schreiber, Frankfurt
Zeichnungen S. 17, 28/29, 95, 104/105: Christa Janik, Leinfelden-Echterdingen

1. Auflage   1 ^{5} ^{4} ^{3} ^{2} | 1996 95 94
Alle Drucke dieser Auflage können im Unterricht nebeneinander benutzt werden, sie sind untereinander unverändert. Die letzte Zahl bezeichnet das Jahr dieses Druckes.

© Verlag Klett Edition Deutsch GmbH, München 1992.
Alle Rechte vorbehalten.
Druck: Hans Buchwieser GmbH, München · Printed in Germany

ISBN 3-12-**557255**-X

Inhalt

Verweise und Piktogramme:

A1   Ü zum Schritt A1 des Lehrbuchs

$\frac{6}{W}$ Ü zum Wortschatz von Lektion 6

$\begin{smallmatrix}A1\\\downarrow\\A3\end{smallmatrix}$ Ü zu den Schritten A1, A2, A3

$\frac{6}{G}$ Ü zur Grammatik von Lektion 6

$\begin{smallmatrix}A1\\+\\A3\end{smallmatrix}$ Ü zu den Schritten A1 und A3

$\overset{6}{\boxed{\textbf{LV}}}$ Ü zum Leseverstehen

A   Ü zum Lernstoff von Baustein A

$\frac{6}{L+S}$ Lesen und Schreiben

$\frac{A}{W}$ Ü zum Wortschatz von Baustein A

$\frac{6}{K}$ Kontrollaufgaben zu Lektion 6

$\frac{A}{G}$ Ü zur Grammatik von Baustein A

(Ü = Übung)

 Rollenspiel

 Einsatz eines Wörterbuchs ist erforderlich.

 Projekt

 Reihenübung mit Drillcharakter

 Schreiben eines zusammenhängenden Textes

Orientierung für Lehrerinnen und Lehrer

Das einsprachig deutsche Arbeitsbuch zu Sprachbrücke 2 erscheint in zwei Teilen. Der erste Teil umfaßt die Lektionen 1–5, der zweite Teil die Lektionen 6–10. Auf diese Weise kann das Übungsmaterial flexibel in verschiedenen Kursstufen eingesetzt werden. Das Arbeitsbuch zu Sprachbrücke 2 führt die Konzeption des Arbeitsbuches zu Sprachbrücke 1 weiter, setzt aber auch neue Akzente.

Generelles Ziel des Arbeitsbuches ist es, in Übungen und Aufgaben die impliziten und expliziten Lernziele von Sprachbrücke 2 aufzugreifen, sie auf vielfältige Weise einzuüben, zu variieren und zu festigen, zusammenzufassen und zu wiederholen. In seinem inneren Aufbau folgt das Arbeitsbuch der Wortschatz- und Grammatikprogression sowie dem thematisch/interkulturellen und didaktischen Ansatz des Lehrwerks und entwickelt ihn weiter. In seinem äußeren Aufbau folgt es dem Aufbau des Kursbuchs in Unterrichtsschritten (A 1, A 2, A 3 ...), Bausteinen (A, B, C ...) und Lektionen (1, 2, 3 ...).

Die Übungspalette reicht von drillähnlichen Reihenübungen bis hin zu komplexen Arbeits- und Schreibaufträgen. Neben dem Üben und Anwenden von grammatischen Strukturen liegt ein besonderer Schwerpunkt auf der Wortschatzarbeit: Wortbildung, Wörterlernen und Erweiterung des Wortschatzes sind Gegenstand zahlreicher Übungen. Die Spiel- und Rätselecke lädt immer wieder zum spielerischen Umgang mit dem Erlernten ein.

Im folgenden sei noch auf einige weitere Aspekte hingewiesen:

Interaktive/kultur-kontrastive Übungen und Aufgaben: In Partnerarbeit oder in Kleingruppen werden Aufgaben und Problemstellungen bearbeitet, bei denen es weniger auf das richtige Ergebnis ankommt, sondern auf den Austausch von Gedanken, Meinungen und Vorstellungen, die in einem nächsten Schritt dann noch einmal mündlich oder schriftlich zusammengefaßt im Plenum präsentiert werden. Häufig handelt es sich bei diesem Übungstyp um kultur-kontrastive Frage- und Aufgabenstellungen. (Beispiele: L. 1, Ü 9; L 2, Ü 7; L. 4, Ü 16 u. a.)

Landeskunde: Dem Interesse an den Veränderungen und Entwicklungen in Deutschland nach der Vereinigung der beiden deutschen Staaten im Oktober 1990 wird in einer ganzen Reihe von Übungsinhalten und Materialien Rechnung getragen. Beispiele: Erste Eindrücke nach der Maueröffnung (L. 1, Ü 9), Erwartungen der Menschen (L. 3, Ü 15), Rückblick auf die Ereignisse in den Jahren 1989/1990 (L. 9, Ü 22), die alten und neuen Bundesländer (L. 4, Ü 18), Deutschland und die europäischen Nachbarn (L. 9, Ü 24).

Wahlmöglichkeiten und/oder Binnendifferenzierung: Motivation und Interesse an Inhalten und Übungsformen sind wichtige Faktoren für das Lerninteresse und damit für den Lernerfolg. Beides aber wird aus ganz persönlichen, individuellen Quellen gespeist und ist – besonders, wenn es sich um ein Deutschlehrbuch handelt, das in vielen Ländern verwendet wird – von den Lehrbuchautoren nur äußerst begrenzt oder gar nicht vorhersehbar und planbar. Ein Ausweg aus diesem Dilemma wären Alternativangebote und Wahlmöglichkeiten. Dagegen sprechen jedoch Kosten- und Umfangsgründe. Um wenigstens die Richtung anzudeuten, werden im Arbeitsbuch bei einigen Aufgabenstellungen alternative Wege oder Texte angeboten. Manche Texte und Aufgaben sind auch fakultativ, d. h. sie können bei Zeitmangel oder fehlendem Interesse ganz weggelassen werden. Je nach Unterrichtssituation können diese Materialien jedoch auch zur Binnendifferenzierung genutzt werden.

Lesen und Leseverstehen: Sowohl in den Kursbüchern Sprachbrücke 1 und 2, als auch ganz besonders im Arbeitsbuch zu Sprachbrücke 1 ist die gezielte Vermittlung von Lesestrategien zum Aufbau des Leseverstehens ein wichtiges Anliegen. In einem fortgeschrittenen Stadium des Deutschlernens – so die Überlegung – ist es dann an der Zeit, fremdsprachige Texte „freiwillig", also aus Interesse am Textinhalt, zu lesen. Deshalb wurden zu einigen Themen der Lektionen verschiedene Lesetexte für die „Leselust zum Thema" ausgewählt. Anstelle eines großen Aufgaben- und Fragenkatalogs gibt es jeweils nur einen kleinen, die Leserichtung bestimmenden Anreiz für ein mögliches Leseinteresse. Diese Lesetexte sind natürlich fakultativ.

Nicht fakultativ dagegen sind Aufgaben, die einige Lesestrategien aufgreifen, die bisher zu kurz gekommen sind oder weiter aufgebaut werden müssen, wie z. B. „Wörter aus dem Kontext erschließen" (L. 2, Ü 18) oder die bewußte Analyse von Textkonnektoren, wobei letzteres auch für das Schreiben (Textaufbau) eine wichtige Rolle spielt.

Um Lehrenden und Lernenden die Möglichkeit zu geben, an bestimmte Lesestrategien zu erinnern und diese auch konsequent zu nutzen, wurden auf Seite 117 die leicht veränderten „Spielregeln zum Knacken deutscher Texte" aus dem Arbeitsbuch zu Sprachbrücke 1 in Form eines „Schaltplans" noch einmal abgedruckt.

Schreiben und Lesen und Schreiben: Die Rede ist hier nicht von der schriftlichen Fixierung von Grammatikübungen o. ä., sondern vom Schreiben längerer Texte, vom Ausprobieren verschiedener Textsorten (Briefe, Geschichten, Anzeigen, Bewerbungsschreiben, Lebenslauf, Gedicht, Hörspiel usw.). Das so verstandene Schreiben nimmt im fortgeschrittenen Deutschunterricht auch im Hinblick auf die Prüfung zum „Zertifikat Deutsch als Fremdsprache" des Deutschen Volkshochschulverbandes/Goethe-Instituts einen immer größeren Raum ein. Dabei darf es nicht bei der Aufforderung „Schreiben Sie bitte" bleiben. Der Textaufbau, die Präsentation und das Verknüpfen von Inhalten über die Satzebene hinaus, muß textsortenspezifisch immer wieder geübt werden. Natürlich kann das im Grundstufenbereich nur in begrenztem Rahmen geschehen. Das Arbeitsbuch stellt immer wieder entsprechende Schreibaufgaben auch in Verbindung mit vorangegangenen Lesetexten.

Damit auch das Schreiben als eine Tätigkeit erfahren wird, die Spaß machen kann, sind Vorbereitung/Planung des Textes und Umgang mit dem geschriebenen Ergebnis wichtig. Text/Schreibaufträge sollten in der Klasse vorgeplant und besprochen werden. Dabei sollte deutlich werden, daß geschriebene Texte für Leser geschrieben werden, denen der Schreibende etwas vermitteln möchte. Deshalb sollten die fertigen Texte dann in der Klasse (evtl. auch außerhalb des engen Klassenrahmens) ihre Leserinnen und Leser finden, gemeinsam „redigiert" (nicht verbessert!) werden und in einer Textmappe (Klassenmappe?), die immer wieder zum Lesen hervorgeholt werden kann, als deutlich sichtbares Ergebnis der individuellen und gemeinsamen Arbeit gesammelt werden.

Hören und Hörverstehen: Zu dem Arbeitsbuch zu Sprachbrücke 2 gibt es keine Hörkassette. Es gibt hin und wieder Aufgaben zu den auf Kassette angebotenen Texten des Kursbuches.

Hinweise zu einzelnen Übungen und Aufgaben befinden sich im Anschluß an jede Lektionsbeschreibung im Handbuch für den Unterricht.

Zum Einsatz des Arbeitsheftes im Unterricht

Die Zuordnung der Übungen zu den einzelnen Unterrichtsschritten des Lehrbuchs sagt nichts darüber aus, ob eine Übung sinnvollerweise vor oder nach der Durchnahme des Arbeitsschritts im Lehrbuch gemacht werden soll. Bei einigen Übungen wird allerdings explizit darauf hingewiesen. Zahlreiche Übungen eignen sich für die häusliche Eigenarbeit mit der Möglichkeit der Selbstkontrolle anhand des Lösungsschlüssels (ab Seite 111). Im übrigen wird man nicht nur aus Zeitgründen, sondern auch entsprechend den Unterrichtsbedingungen vor Ort eine Auswahl treffen und/oder den einen oder anderen Aspekt stärker betonen müssen.

A 1 Heimatsonne: Wortfamilie „Heimat-"

Die Begriffe (S. 6) und Definitionen (S. 7) stammen „aus dem Wahrig" (Gerhard Wahrig: Deutsches Wörterbuch, 1986).

1. Was gehört zusammen? Ordnen Sie bitte die Definitionen den passenden Begriffen zu!

1	2	3	4	5	6	7	8	9	10	11	12	13	14	15	16	17	18	19	20	21	22	23	24
d																							

2. Notieren Sie hier die beiden Adjektive aus der „Heimatsonne"!

_____ , das bedeutet: _____

_____ , das bedeutet: _____

3. Einer der Begriffe ist eng mit einer Epoche der deutschen Geschichte verknüpft. Welcher? Erklären Sie bitte den Begriff!

4. Für welche Begriffe gibt es in Ihrer Sprache keine Entsprechung?

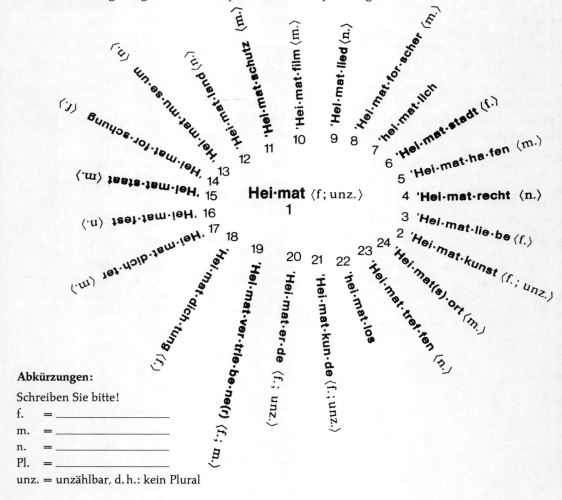

Abkürzungen:

Schreiben Sie bitte!

f. = _____

m. = _____

n. = _____

Pl. = _____

unz. = unzählbar, d. h.: kein Plural

a) Hafen, in dessen Schiffsregister ein Schiff eingetragen ist

b) Wohn- und Niederlassungsrecht

c) Stadt, in der jmd. geboren bzw. aufgewachsen ist

d) Ort, an dem man zu Hause ist, Geburts-, Wohnort; (i.w.S.) Vaterland; die ~ dieser Pflanze, dieses Tiers ist Südamerika; die alte ~ wieder einmal besuchen; keine ~ mehr haben; die ewige ~ (poet.) das Jenseits; diese Stadt ist meine zweite ~ geworden; in meiner ~ ist es Brauch ...

e) (Adj.; i.w.S.) keine Heimat besitzend; (i.e.S.) staatenlos; heimatvertrieben

f) Bewahrung der Heimat in ihrer natürl. u. künstler. Eigenart, z.B. Bewahrung von Bräuchen, Trachten, Schutz wertvoller Baudenkmäler, Verhinderung baulichen Kitsches; Sy Heimatpflege

g) (Adj.) zur Heimat gehörig, aus ihr stammend, an die Heimat erinnernd, so ähnl. wie zu Hause; wieder ~ en Boden betreten; ~ e Bräuche, Laute, Sitten, Sprache; es berührte mich ~

h) jmd., der Heimatforschung treibt; Sy Lokalforscher

i) Staat, in dem jmd. geboren ist bzw. für den er die Staatsangehörigkeit besitzt

j) die literarische Form der Heimatkunst

k) kleines Museum mit heimatkundlichen Sammlungen

l) jmd., der 1945 seinen Wohnsitz in einem der heute zur ČSFR, zu Polen u. zur ehemaligen UdSSR gehörenden Gebiete hatte u. daraus vertrieben wurde

m) Lehre von der Geschichte, Erd- u. Naturkunde der engeren Heimat (bes. als Unterrichtsfach an Grundschulen)

n) Kunst, die das Gepräge der Heimat des Künstlers trägt, z.B. Mundartdichtung

o) Liebe zur Heimat

p) Ort, in dem jmd. geboren od. aufgewachsen ist; (schweiz.) Ort, dessen Bürgerrecht jmd. besitzt

r) Land, in dem jmd. geboren od. aufgewachsen ist

s) Dichter, dessen Werk zur Heimatdichtung gehört

t) die engere Heimat, ihre Sitten u. Bräuche herausstellendes Fest

u) Erde aus der Heimat; (fig.) Heimatboden, Heimat; die ~ wieder betreten

v) die Heimat besingendes Lied

w) im ländl. Milieu spielender, dessen Sitten u. Bräuche, oft auch Trachten herausstellender Film

x) Erforschung von Natur u. Geschichte der näheren Heimat; Sy Lokalforschung

y) Treffen der Heimatvertriebenen zum Gedenken an die verlorene Heimat

Abkürzungen:

Schreiben Sie bitte!

jmd. = _____

Adj. = _____

bes. = _____

od. = _____

u. = _____

bzw. = _____

i.w.S. = im weiteren Sinne

i.e.S. = im engeren Sinne

poet. = _____

literar. = _____

z.B. = _____

fig. = figürlich, d.h.: im übertragenen Sinne

Sy = Synonym

B 1 2 **Textverständnis**

Lesen Sie bitte den Text von B 1 im Kursbuch Abschnitt für Abschnitt! Welche der folgenden Aussagen stimmen?
Kreuzen Sie die richtigen Aussagen an!

Zeile 1 – 9:

☐ Die Zeitungen nannten Hilde Domin 1987 die bedeutendste lebende Lyrikerin Deutschlands.

☐ Sie hat erst mit 40 Jahren angefangen, Gedichte zu schreiben.

☐ Zu diesem Zeitpunkt lebte sie in Deutschland.

Zeile 10 – 16:

☐ Nach 22 Jahren Exil kehrte sie nach Deutschland zurück.

☐ Ihr erster Gedichtband erschien fünf Jahre nach ihrer Rückkehr.

☐ Während der Jahre des Exils war die deutsche Sprache ihre innere Stütze.

Zeile 17 – 28:

☐ Während sie im Exil lebte, mußte Hilde Domin immer wieder eine neue Sprache lernen.

☐ Nach dem Studium übersetzte sie wissenschaftliche Texte, um Geld zu verdienen.

☐ In England durfte sie nicht Italienisch sprechen.

Zeile 29 – 36:

☐ Von Italien aus ging sie direkt in die Dominikanische Republik.

☐ Dort mußte sie auch noch Spanisch lernen.

☐ An der Universität von Santo Domingo unterrichtete sie Deutsch.

Zeile 37 – 48:

☐ Als junges Mädchen hatte Hilde Domin gern in Deutschland gelebt.

☐ Seit Hilde Domin dichtete, fühlte sie sich in Deutschland zu Hause.

☐ Die deutsche Sprache war ihre wahre Heimat. Deshalb war sie zurückgekommen.

B 1
+
B 2 3 **Hilde Domin – Eine Sprachodyssee: Warum?**

Hilde Domin war Jüdin.
Setzen Sie bitte ihren Lebenslauf in Beziehung zu den Daten und Angaben in der folgenden
Zeittafel! Benutzen Sie bitte das Wörterbuch, wenn erforderlich.

1932	Bei den Wahlen zum Reichstag erhält die Partei Adolf Hitlers (NSDAP – Nationalsozialistische Deutsche Arbeiterpartei) die meisten Stimmen.
30. Januar 1933	Adolf Hitler wird Reichskanzler.
Juli 1933	Verbot aller Parteien. Beginn der Diktatur. Beginn der Judenverfolgung.
1935	„Nürnberger Gesetze": Die Juden werden per Gesetz zu Menschen zweiter Klasse gemacht. Hunderttausende gehen ins Exil.
12. März 1938	Anschluß Österreichs
15. März 1938	Überfall auf die Tschechoslowakei
8./9. November 1939	„Reichskristallnacht": Die jüdischen Synagogen werden von SA-Leuten in Brand gesteckt. Jüdische Geschäfte werden zerstört. Die Deportation der Juden in Konzentrationslager beginnt.
1. September 1939	Überfall der deutschen Armee (Wehrmacht) auf Polen. Beginn des Zweiten Weltkriegs.
3.9.1939	Großbritannien und Frankreich erklären Deutschland den Krieg.
Mai 1940	Deutscher Angriff auf Belgien, die Niederlande, Luxemburg und Frankreich.
Juni 1940	Mussolini führt Italien auf deutscher Seite in den Krieg.

22.6.1940	Waffenstillstand zwischen Deutschland und Frankreich
22.6.1941	Überfall auf die Sowjetunion
11.12.1941	Kriegserklärung Deutschlands an die USA
20. Juli 1944	Gescheitertes Attentat auf Adolf Hitler
30. April 1945	Selbstmord Hitlers
8. Mai 1945	Kapitulation der deutschen Wehrmacht
	Besetzung Deutschlands durch die Siegermächte USA, Sowjetunion, Groß-britannien, Frankreich; Teilung in vier Besatzungszonen; Hauptstadt Berlin: vier Sektoren
Mai 1949	Gründung der Bundesrepublik Deutschland (die drei Westzonen), Hauptstadt: Bonn
Oktober 1949	Gründung der Deutschen Demokratischen Republik (sowjetisch besetzte Zone), Hauptstadt: Berlin (Ost)

4   Wie alles gekommen war ... Plusquamperfekt
B 4

30.6.1992. Der Hochzeitstag! Als sie am Morgen aufwachte, schloß sie noch einmal die Augen und erinnerte sich daran, wie alles gekommen war:

Erzählen oder schreiben Sie bitte die Geschichte!
Die Angaben dazu finden Sie unten (im Bettkasten).

Sie war damals 17 Jahre alt. Bei einem Popkonzert hatte

Sie - damals - 17 Jahre alt /
ihren Mann - bei einem Popkonzert -
kennenlernen / neben ihr - sitzen /
nach dem Konzert - ihn - einfach fragen: nach Hause fahren? / zu
ihrer Überraschung - sie anlächeln - und ja sagen / zusammen - in
seinem Auto - nach Hause fahren / vor ihrem Haus - nicht aussteigen - sondern sitzenbleiben - warten / Er: nicht aussteigen? /
Sie: wiedertreffen! / In den Arm nehmen / So - alles anfangen.

B 4
+
B 6

5 Bildgeschichte mit „als, bevor, nachdem": Textaufbau

Arbeiten Sie bitte zu zweit oder zu dritt!

1. Schreiben Sie zu jedem Bild einen oder mehrere Sätze.

2. Bringen Sie die Bilder in eine sinnvolle chronologische Reihenfolge.

3. Schreiben Sie nun eine passende Geschichte zu den Bildern. Verbinden Sie die Sätze, die Sie bei Aufgabe 1. geschrieben haben, und benutzen Sie auch die Subjunktoren „als", „bevor", „nachdem".

4. Vergleichen Sie Ihre Ergebnisse im Plenum.

6 Kleine Kopfgymnastik: bevor-vorher; nachdem-danach

1. Formen Sie bitte um! Bevor → Vorher
   Vorher → Bevor

Kairo, Stockholm und zurück

1. Bevor sie 1982 nach Kairo ging, lebte sie in Bonn:

→ *1982 ging sie nach Kairo. Vorher hatte sie in Bonn gelebt.*

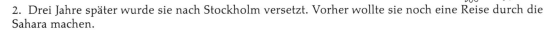

2. Drei Jahre später wurde sie nach Stockholm versetzt. Vorher wollte sie noch eine Reise durch die Sahara machen.

→

3. Bevor sie dann nach Stockholm abreiste, lernte sie ihren späteren Mann kennen.

→

4. Er konnte schon bald zu ihr nach Stockholm kommen. Aber vorher mußte er sein Studium ab-schließen.

→

5. Schließlich heirateten sie. Vorher hatten sie ein Jahr in Stockholm zusammengelebt.

→

6. Bevor sie 1989 wieder nach Kairo zurückkehrten, kam der Sohn Thomas auf die Welt.

→

2. Formen Sie bitte um! Nachdem → Danach
   Danach → Nachdem

Ein schöner Sonntag!

1. Ich frühstückte gemütlich. Danach las ich die Zeitung.

→ *Nachdem ich gemütlich gefrühstückt hatte, las ich die Zeitung.*

2. Nachdem ich zwei Stunden lang im Wald spazieren gegangen war, kochte ich das Mittagessen.

→

3. Nachdem ich sehr viel und gut gegessen hatte, legte ich mich eine Stunde lang ins Bett.

→

4. Schließlich stand ich wieder auf. Danach klingelte es an der Tür, und meine ganze Verwandtschaft kam zu Besuch.

→

5. Wir aßen den ganzen Kuchen auf. Danach schauten wir uns im Fernsehen ein Fußballspiel an.

→

6. Nachdem unsere Mannschaft gewonnen hatte, saßen wir noch bis spät in die Nacht zusammen.

→

B 4
+
B 6

7   **Vergangenheit-Gegenwart-Zukunft: Zeitenfolge**

Lesen Sie bitte!

1. Man spricht über **Vergangenes**:

Gleiche Verbform in beiden Sätzen:

(Dauer)

| **Während** sie verheiratet waren, versorgte sie den Haushalt und die Kinder, er verdiente das Geld. | **Seitdem** sie verheiratet waren, verstanden sie sich nicht mehr so gut. | Präteritum (gleichzeitig) |

Gleiche Verbform in beiden Sätzen:

| Plusquamperfekt | **Bevor** sie geheiratet hatten, hatten sie sich gut verstanden. |

Verschiedene Verbformen in Haupt- und Nebensatz:

| **Nachdem** sie geheiratet hatten, verstanden sie sich nicht mehr so gut. | (vorher) (nachher) | Plusquamp. Präteritum |

Plusquamperfekt: eher schriftsprachlich

Gleiche Verbform in beiden Sätzen:

| Perfekt | **Bevor** sie geheiratet haben, haben sie sich gut verstanden. | **Nachdem** sie geheiratet haben, haben sie sich nicht mehr so gut verstanden. | Perfekt |

Perfekt: eher umgangssprachlich

2. Man spricht über **Gegenwärtiges und Zukünftiges**:

Bitte schreiben Sie die folgenden Sätze (passend zu der Geschichte oben) zu Ende!

1. Während sie die Kinder versorgt, _____

2. Seitdem sie verheiratet sind, _____
    (beide Situationen/Handlungen dauern noch)

3. Seitdem sie das dritte Kind bekommen hat, _____ oft müde.
    (die erste Handlung ist zwar im Perfekt, aber sie hat das Kind immer noch)

4. Nachdem die Kinder groß geworden sind, _____ wieder in ihrer alten Firma.
    (die erste Handlung ist abgeschlossen)

5. Seitdem die Kinder aus dem Haus sind, _____

6. Bevor sie sich scheiden lassen, _____

> Tempusfolge
>
> nachdem er
> in Untersuchungshaft gesessen hatte
> beging er
> das Verbrechen das man
> von ihm erwartet hatte
>
> *Gerhard Sellin*

8 Ihr Text: Auskunft zur Person

B 4
+
B 6

Wählen Sie einige Sätze aus, und schreiben Sie die Sätze mit Angaben aus Ihrem eigenen Leben weiter.
Ändern Sie die Satzanfänge, wenn sie nicht passen, oder schreiben Sie andere Sätze!

1. Vergangenheit:

Bevor ich heute morgen aufgestanden bin, ...
Bevor ich aus dem Haus ging, ...
Nachdem ich aus dem Haus gegangen war, ...
Nachdem ich ... angekommen war, ...
Nachdem ich beschlossen hatte,
    Deutsch zu lernen, ...
Nachdem meine Eltern ...
Bevor ich Deutsch lernte, ...
Bevor ich in diese Stadt gekommen bin, ...
Nachdem es in meinem letzten Urlaub ...
Seitdem ich angefangen habe,
    Deutsch zu lernen ...
Seitdem ich (mit dem Rauchen?)
    aufgehört habe ...
Bevor ich mich für diesen Kurs hier
    eingeschrieben habe, ...
Nachdem ich mich von meiner Freundin/
    meinem Freund verabschiedet/getrennt hatte ...
Nachdem ich die letzte Prüfung bestanden
    hatte, ...
Seitdem man mir gesagt hat, ...

2. Gegenwart und Zukunft:

Bevor ich nachher nach Hause gehe, ...
Seitdem ich Deutsch lerne, ...
Während ich hier sitze, ...
Bevor ich Urlaub machen kann, ...
Bevor ich nach Deutschland fahren kann, ...
Bevor ich die Prüfung machen kann, ...
Während wir hier Deutsch lernen, ...
Während viele Menschen nicht genug zu
    essen haben, ...
Während es bei uns (Sommer/Winter ...) ist,
    ist es in ...
Seitdem ich weiß, wie ...
Seitdem ich (verheiratet/geschieden ...) bin, ...
Seitdem ich (Vater/Mutter ...) bin, ...
Bevor ich mit ... aufhöre, ...
Bevor ich sterbe, möchte ich ...

9 Verbal-Nominal: vor, nach, bei

C 1
↓
C 4

1. Lesen Sie bitte noch einmal die Übung 6 (1.+2.), Seite 11. Machen Sie die Probe: In welchen Fällen könnten Sie den temporalen Nebensatz mit Subjunktor durch eine entsprechende Temporalangabe ersetzen? Voraussetzung dafür ist, daß es ein passendes Nomen gibt. Machen Sie die Übung mündlich und/oder schriftlich!

Achtung:
Nebensatz <u>mit</u> Komma,
Temporalangabe <u>ohne</u> Komma!

Zum Beispiel:
Bevor sie 1982 nach Kairo ging, lebte sie in Bonn. ⟶

(Das Nomen „Gang" paßt nicht in diesem Zusammenhang, aber z.B.
das Nomen „Übersiedlung" von „übersiedeln" = woanders hinziehen, oder das Nomen „Umzug" von „umziehen".)

 Vor ihrer Übersiedlung (Vor ihrem Umzug) nach Kairo im Jahr 1982 lebte sie in Bonn.

2. Formen Sie bitte die folgenden Nebensätze mit Subjunktor in entsprechende Temporalangaben (Präposition mit Nominalgruppe) um!

1. Bevor sie geheiratet haben, haben sie sich gut verstanden.

_____

2. Während sie verheiratet waren, versorgte sie die Kinder, er verdiente das Geld.

_____

3. Seitdem sie verheiratet waren, stritten sie oft miteinander.

_____

4. Seitdem sie das dritte Kind bekommen hatte, war sie oft schlecht gelaunt.

_____

5. Nachdem die Kinder ausgezogen waren, konnte sie endlich mal an sich selber denken.

_____

6. Bevor sie sich scheiden ließen, hatten sie alles Finanzielle geregelt.

_____

7. Nachdem sie dann geschieden waren, wurden sie endlich gute Freunde.

_____

3. Temporalangaben mit „bei": Gleichzeitigkeit

Schreiben Sie bitte wie im Beispiel!

Eine Weltreise

1. Als der Zug in Stuttgart abfuhr, vergaß ich alle Sorgen.

Bei der Abfahrt des Zuges vergaß ich alle Sorgen.

2. Als wir in Frankfurt (ins Flugzeug) einstiegen, wurden wir noch einmal kontrolliert.

_____

3. Wenn ich fliege, wird mir immer schlecht.

4. Während wir zu Mittag aßen, sahen wir Singapur unter uns.

5. Als wir in Jakarta ankamen, nahm uns die heiße Luft fast den Atem.

_____

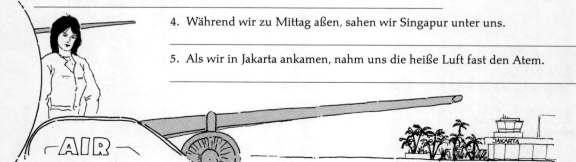

10 Auskunft zur Person

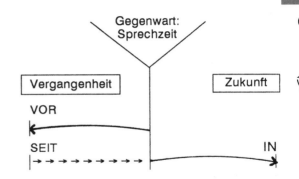

C 3

1. Reihenübung: vor, seit, in

Fragen Sie bitte reihum in der Klasse oder:
Jede/r schreibt ein paar Fragen auf einen Zettel
und schickt den Zettel an jemanden in der
Klasse.
Beispiel: Was haben Sie vor ... gemacht?
           Vor wieviel Tagen/Jahren ...
           Seit wann sind Sie/machen Sie ...?
           Was machen Sie in ...?

(Benutzen Sie alle möglichen Zeitangaben mit: Stunde, Tag, Jahr, Woche, Monat)

2. Reihenübung wie 1. Benutzen Sie jetzt Zeitangaben wie:
letztes/vorletzten (Jahr/Monat ...)
diese/nächste/übernächste (Woche/...)
ab morgen/übermorgen/nächsten Monat ...
bis morgen/übermorgen usw.

11 Geschichtlicher Rückblick und Vorausblick mit Temporalangaben

**C 3
+
C 6**

1. Schreiben Sie bitte passende Präpositionen und Artikel in die Lücken, und ergänzen Sie die
Kasusendungen!
(Der Text enthält einige historische Begriffe, die Sie vielleicht nicht kennen. Klären Sie vorher, um
welche Ereignisse es sich bei den folgenden Begriffen handelt: der Kalte Krieg, die Ostpolitik, die
Währungsunion, die Vereinigung der beiden deutschen Staaten.)

1. _____ mehr als fünfzig Jahr_____ begann der Zweite Weltkrieg. _____ achten Mai 1945
endete er mit der Kapitulation des Deutschen Reiches. _____ _____ dies_____ Tag mußten
Millionen Menschen sterben. _____ 1945 _____ 1989 bestimmte der „Kalte Krieg" das
politische Klima nicht nur in Europa.
2. _____ mehr als 40 Jahren wurde Deutschland als Folge des Krieges geteilt. _____
der folgend_____ Jahr_____ standen sich die beiden deutschen Staaten feindlich gegenüber.
Erst _____ _____ Ostpolitik Willy Brandts _____ _____ 70er Jahr_____ kam Bewegung in
die deutsch-deutschen Beziehungen.
3. Und ein_____ Tag_____ geschah es: _____ _____ Nacht _____ neunten _____
den zehnten November 1989 wurde die Mauer, die mitten durch Berlin führte, geöffnet.
4. Schon ein halb_____ Jahr _____ _____ Maueröffnung kam die Währungsunion, ein Jahr
später, _____ dritten Oktober, kam dann schon die Vereinigung der beiden deutschen Staaten.
5. _____ _____ Ende des Kalten Krieges hofften die Menschen darauf, daß es _____
einig_____ Jahr_____ oder vielleicht _____ ein paar Jahrzehnten keine Kriege und keinen Hunger mehr auf der Welt geben wird. _____ _____ dies_____ Zeitpunkt wird es leider noch
viele Rückschläge* und Enttäuschungen geben.

2. Ihr Text:

Arbeiten Sie bitte in Gruppen!
Schreiben Sie einen kurzen Informationstext über die Geschichte Ihres Landes in den letzten Jahren
oder Jahrzehnten.

6

12   Christa Pereiras Lebenslauf: Auf Jahr und Tag

Ergänzen Sie bitte den Lebenslauf von Christa Pereira mit passenden Temporalangaben im richtigen Kausus! Arbeiten Sie bei dieser Übung bitte zu zweit!
Lösen Sie die Aufgabe mit Hilfe der Angaben auf Seite 56 im Kursbuch!

1952, sechs Jahre, 28 Jahre, zehn Jahre lang,
weitere zwei Jahre, 16 Jahre, 1971-1977, ein Tag, 1978,
zwei Monate, folgende drei Jahre, diese Zeit, 1975,
dieser Tag, dasselbe Jahr, dasselbe Jahr,
29. bis 36. Lebensjahr, drei Jahre

Wann war das alles?

Christa wurde ____1952____ in Essen geboren. Mit _____ kam sie in die Schule.

_____ lernte sie, mit _____ machte sie dann ihren Schulabschluß.

Während der _____ wurde sie als Zahntechnikerin ausgebildet. Von

_____ arbeitete sie beim Zahnlabor Müller & Co. Während _____ hat

sie gut verdient. _____ beschloß sie, doch das Abitur zu machen. Deshalb besuchte

sie _____ lang ein Abendgymnasium; _____ bestand sie das Abitur.

In _____ fing sie mit dem Studium der Germanistik und Geschichte an.

Mit _____ heiratete sie Miguel Pereira.

Zu ihrer Hochzeit lud sie alle ihre Freunde ein. Von _____ an hat sich ihr Leben sehr

verändert: noch im _____ nämlich, zu Weihnachten, kam ihre erste Tochter auf die

Welt. Nach _____ bekam sie noch eine Tochter. Wegen der Kinder ist sie von

_____ zu Hause geblieben. Aber danach wurde ihr klar, daß sie wieder in die Arbeits-

welt zurückkehren wollte.

Da sie keine Stelle fand, besuchte sie _____ lang einen Fortbildungskurs in der Hoff-

nung, _____ wieder eine Arbeitsstelle zu finden.

D   13   Trauriger Wortschatz

1. Menschen verlieren ihre Heimat. Ergänzen Sie bitte die fehlenden Bezeichnungen:

Nomen	Personenbezeichnungen (Pl.)	Verben
die Flucht	_____	_____ , _____
das Exil	*Exilanten*	*ins Exil gehen*
die Auswanderung	_____	_____
die Emigration	_____	_____
die Vertreibung	*Vertriebene*	_____
das Asyl →	_____	*Asyl beantragen*

→ Asylbewerber: Sie haben Asyl beantragt, aber über ihren Antrag ist noch nicht
entschieden.

2. Welche Wörter bezeichnen – unfreiwilliges Verlassen der Heimat?
                              – freiwilliges Verlassen der Heimat?

1. **Exotische Tiere** für die Deutschen. Wie heißen sie auf deutsch?
Des Rätsels Lösung: In Berlin findet man sie fast gegenüber vom B...

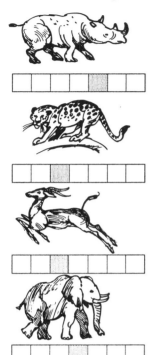

B _ _ _ _ _ _ _ O

2. **Redensarten:**

Deutsch:

Sie ist vor Freude
aus dem Häuschen.

Türkisch:
Etekleri zil çalıyor.
(Ihre Röcke klingeln.)

Ihre Sprache:

_____

_____

Er ist vom Regen in die Traufe
gekommen.

Englisch:
He jumped out of the frying
pan into the fire.

_____

_____

17

6   15   Eine Biographie schreiben

1. Wählen Sie gemeinsam einige für Ihr Land wichtige Persönlichkeiten (aus Geschichte oder Politik oder Literatur oder ...).

2. Bilden Sie bitte Gruppen! Jede Gruppe erarbeitet und schreibt auf deutsch die Biographie einer Persönlichkeit. Erstellen Sie auch eine Zeittafel!
Verteilen Sie Aufgaben an die Gruppenmitglieder, z.B.:
– Lebensdaten aufstellen
– Liste der Werke (Auswahl)
– Weitere Dokumente suchen
– Bildmaterial
Zur Vorbereitung auf das Schreiben kann aus sämtlichen Materialien erst einmal eine Collage zusammengestellt werden. Während die Collage alle Elemente gleichzeitig und im Überblick darstellt, müssen in der schriftlichen Biographie alle Daten und Fakten in eine chronologische Reihenfolge gebracht werden. Hängen Sie die Biographien im Klassenraum auf, und besprechen Sie Ihre Arbeiten!

6   16   Wortschatz:
W   Mehrsprachigkeit (-sprachig, (-)sprachlich)

> ʼMut·ter·spra·che ⟨f.⟩ die Sprache, die man von Kind auf gelernt hat; Deutsch ist meine ~
>
> ʼFremd·spra·che ⟨f.⟩ Sprache, die nicht die Muttersprache ist; mehrere ~n beherrschen
>
> ʼfremd·spra·chig ⟨Adj.⟩ eine fremde Sprache sprechend, schreibend; ~e Bücher in fremden Sprachen; ~er Unterricht in einer fremden Sprache gehaltener Unterricht
>
> ʼfremd·sprach·lich ⟨Adj.⟩ auf eine fremde Sprache bezüglich; ~er Unterricht Unterricht über eine fremde Sprache; ein ~es Buch ein Buch über eine fremde Sprache (z. B. Lehrbuch)
>
> **(Wahrig: Deutsches Wörterbuch)**

1. Schreiben Sie bitte **Erklärungen** oder **Definitionen**!

Es gibt drei deutschsprachige Länder.

D.h. *Es gibt drei Länder,* _____

_____

Er ist zweisprachig aufgewachsen.

D.h. _____

Mehrsprachigkeit ist _____

_____

muttersprachlicher Unterricht – das ist _____

fremdsprachlicher Unterricht – das ist _____

Das ist ein sprachliches Problem – das bedeutet: _____

2. Ergänzen Sie bitte!

1. Hilde Domin hatte von 1940 – 1954 in einem *spanisch* _____ Land gelebt.

2. Aussage einer Türkin, die in der Bundesrepublik Deutschland lebt und Deutsch und Türkisch spricht: „Ich empfinde meine _____ als etwas sehr Positives.

Ein _____ Mensch ist ein Mensch, der wie zwei Menschen denken kann."

3. Im vereinten Europa müssen alle Kinder mehrere Sprachen lernen.

_____ ist dann nichts Besonderes mehr.

4. In der Bundesrepublik Deutschland haben viele Kinder anderer Nationalität neben dem normalen Unterricht an einer deutschen Schule nachmittags auch noch _____ Unterricht.

5. In den normalen Buchhandlungen in der Bundesrepublik findet man leider kaum

_____ Bücher.

6. Jennys Mutter ist Spanierin, ihr Vater ist Engländer, sie leben in Deutschland. Jenny ist von Anfang

an _____

aufgewachsen.

7. (zu der nebenstehenden Statistik): In vielen Schulen der Bundesrepublik gibt

es _____

Klassen.

Diese _____

sollte im Unterricht besser genutzt werden.

(aus: Die Grundschulzeitschrift 43/1991)

Schule L.
*Grundschule,
September 1990*

165 Schüler aus
15 Staaten

79	aus Deutschland
47	aus der Türkei
10	aus Jugoslawien
6	aus Griechenland
4	aus Portugal
4	aus Pakistan
3	aus Italien
3	aus Polen
2	aus Ghana
2	aus Tunesien
1	aus Marokko
1	aus Spanien
1	aus Peru
1	aus dem Iran
1	aus Afghanistan

Schule T-H.
*Grund, Haupt- und
Realschule, April 1989*

440 Schüler aus
23 Staaten

213	aus Deutschland
137	aus der Türkei
26	aus Jugoslawien
12	aus Griechenland
10	aus Italien
7	aus Portugal
4	aus Spanien
2	aus Afghanistan
2	aus Albanien
2	aus Ecuador
2	aus Gambia
2	aus Ghana
2	aus dem Iran
2	aus Korea
2	aus dem Libanon
2	aus Marokko
2	aus Pakistan
2	aus Sri Lanka
2	aus Taiwan
2	aus Thailand
2	aus Tunesien
2	aus Vietnam
1	aus Frankreich

Schülerschaft an zwei Hamburger Schulen

17 Diskussion: Zwei- oder Mehrsprachigkeit? Bei uns ist das die Regel!

1. Lesen Sie bitte zuerst die folgenden Aussagen über Zweisprachigkeit!

2. Diskutieren Sie die Aussagen in der Klasse!

Was ist denn eigentlich Zweisprachigkeit?

Verschiedene Ansichten von Lehrerinnen und Lehrern

„Jemand ist zweisprachig, wenn er sich auch noch in einer anderen als der Muttersprache in Alltagssituationen zurechtfinden und ausdrücken kann, seine Grundbedürfnisse verbal darlegen kann und es schafft, sich aus Notsituationen zu befreien."

„Ich sehe Zweisprachigkeit als die Fähigkeit an, zwei Sprachen fehlerfrei sprechen zu können."

„Zweisprachigkeit bedeutet die Fähigkeit, sich in zwei Sprachen durchs Leben zu schlagen."

„Zweisprachigkeit ist die Fähigkeit eines Menschen, sich in zwei Sprachen gleichermaßen differenziert – besonders auch spontan – so auszudrücken, daß er von seinem Gesprächspartner vollkommen verstanden wird und dabei selbst das Gefühl hat, den von ihm beabsichtigten Inhalt optimal vermittelt zu haben."

(Aus: Die Grundschulzeitschrift 43/1991)

3. **Ihr Text:**

Wie ist es in Ihrem Land? Ist dort Zwei- oder Mehrsprachigkeit die Regel?
Welche Sprachen müssen die Menschen in Ihrem Land beherrschen, um sich in verschiedenen Bereichen zu behaupten? Wie lernen sie diese Sprachen? In welchen Bereichen braucht man diese Sprachen?

Und: Welche Sprachen werden Sie in Zukunft noch lernen (müssen)? Warum?

Schreiben Sie bitte in die Tabelle!

Sprachen, die man bei uns beherrschen muß:	Für diese Bereiche:	So lernt man sie:

Sprachen, die ich noch lernen will/muß:	Gründe:

18 Irgendwie – Irgendwo – Irgendwann

1. Ergänzen Sie bitte!

Seltsamer Vortrag!

A: Wie war der Vortrag?

B: _Irgendwie_____ seltsam!

A: Was heißt „_____ seltsam"?

B: Na, komisch halt, ich kann auch nicht
genau sagen, was und wie, er war
halt _____ seltsam!

Unbestimmtes Ziel:

A: Wo wollt ihr denn hin?

B: Ich weiß nicht, _____.

A: Was heißt hier „_____"?

B: Na, _____ halt.
Ist doch egal, wohin wir fahren!

Egal wer!

A: Wer soll dann das machen?

B: _____ halt!

A: Was heißt hier „_____"?

B: Na, _____ halt, ist doch egal, wer!

Irgend_____, Irgend_____, Irgend_____,
fängt dann die nächste Liebe an.

2. Ergänzen Sie bitte!

Irgendwann hat
irgend_____
irgend_____ mal
irgend_____ gesagt.
Das hat der aber
irgend_____
falsch verstanden.
Seitdem sprechen die
nicht mehr miteinander.

(Aus: Sichtwechsel, Arbeitsbuch)

3. Denken Sie sich bitte ähnliche Situationen aus, auch mit „nirgendwo/nirgendwohin/irgendwelche"
usw. Spielen Sie die Situationen!

19 Wortschatz: Wer schreibt was?

1. Wann spricht man vom Dichter, wann vom Autor? Kreuzen Sie bitte an! Zu einigen der Perso-
nenbezeichnungen gibt es auch entsprechende Bezeichnungen für „das Produkt". Schreiben Sie diese
Bezeichnungen bitte an den linken Rand!

		Gedichte	Romane/ Erzählungen	Zeitungs- artikel	Sach- bücher	Theater- stücke	Lehr- bücher
die Dichtung	Dichter/in	☐	☐	☐	☐	☐	☐
_____	Lyriker/in	☐	☐	☐	☐	☐	☐
_____	Autor/in	☐	☐	☐	☐	☐	☐
_____	Schriftsteller/in	☐	☐	☐	☐	☐	☐
_____	Dramatiker/in	☐	☐	☐	☐	☐	☐
_____	Poet/Poetin	☐	☐	☐	☐	☐	☐

6

2. Überall ein anderer

1. Was Sie alles sein können:

Kunde	im Supermarkt
	beim Arzt
	im Hotel
	im Konzert
	im Theater
	in einem fremden Land
	im Flugzeug
	im Deutschkurs
	in Ihrer Familie

2. Für welche Bezeichnungen gibt es weibliche Formen?

$\frac{6}{W}$ **20 Zu, zu, zu ...**

Auf deutsch „zu ...". – Und in Ihrer Sprache? Vergleichen Sie bitte! Notieren Sie bitte den entsprechenden Ausdruck in Ihrer Muttersprache!

1. Das Buch wurde rechtzeitig zu ihrem 50. Geburtstag fertig. _____
2. Die Muttersprache wurde ihr zur Heimat. _____
3. Trotz seiner französischen Herkunft wurde er zum deutschen Dichter. _____
4. Finden Sie zu jedem Abschnitt eine Überschrift. _____
5. Sie lebte bis zu ihrem Tod in Köln. _____
6. Sie verdienen zum Leben zu wenig, zum Sterben zu viel. _____ _____
7. Sie fühlte sich in vielen Ländern der Welt zu Hause. _____
8. Zu Beginn seines Vortrags sprach er über die Jahre des Exils. _____
9. Niemand kam ihr zu Hilfe. _____
10. Setz dich doch zu mir! _____
11. Sie geht noch zur Schule. _____
12. Leg das Buch zu den anderen Büchern! _____
13. Von Tag zu Tag, von Woche zu Woche ging es ihm schlechter. _____
14. Wir brauchen Gesetze zum Schutz der Natur. _____
15. Das ist zu schön, um wahr zu sein! _____ _____
16. Es gibt noch so viel zu tun. _____
17. Heute sind alle Läden zu. _____
18. Zu diesem Buch gibt es eine Cassette. _____
19. Laß das Auto stehen! Wir gehen zu Fuß. _____
20. Gehören Sie auch zur Familie? Oder sind Sie nur zu Besuch? _____ _____

---

6 21 Leselust zum Thema (fakultativ)

Lesen Sie bitte die Überschriften der beiden folgenden Texte, und überfliegen Sie nur die ersten neun, zehn Zeilen jedes Textes!
Überlegen Sie dann, wovon der Text wohl handelt, und entscheiden Sie danach, welchen der beiden Texte Sie lesen wollen.

Fremd in der Schweiz

Eigentlich fühle ich mich in der Schweiz wohl.

Schon 14 Jahre sind seit damals vergangen; ich war eine junge Braut und wollte mein Bestes tun, damit recht bald eine echte „Schweizerehefrau"
5 aus mir werden konnte, zur Freude meines Mannes und seiner Verwandten.

Ich wusste* damals nicht, dass das alles so viele Tränen, so viel Entmutigungen und Wut kosten würde.
10 Heute würde ich allen, die in die Schweiz kommen wollen, raten, einen Vorbereitungskurs zu nehmen. Diese hypothetische Person darf erwarten, dass die Leute sehr nett und freundlich sind, aber nicht auf die Dauer; sie muss lernen,
15 immer dankbar dafür zu sein, wenn sie mit Schweizern in Kontakt kommt, dass sie in einem so sauberen, reichen und ordnungsliebenden Land leben darf.

Sie soll versuchen, kein Heimweh zu zeigen.
20 Ich habe gelernt, Schweizer Rezepte zu kochen wie „Rösti und Bratwürste", „Gschnätzlets Zürcherart" etc., und Fondue? Klar! Heute lade ich unsere Nachbarn zu einer Fondue-Party ein und preise die kulinarischen Fähigkeiten meines
25 Mannes sehr hoch, eine so herrliche nach Fendant duftende Fondue zu kochen.

Ich habe gelernt, in der Wohnung mit einem leichten Schritt zu laufen, um die Nachbarn vom unteren Stock nicht zu stören; ich versuche täg-
30 lich mit leiser Stimme zu sprechen, sonst kann es vorkommen (oh wehe), dass unsere Nachbarn auf der linken Seite zu viel über uns wissen. Heute ist es mir völlig klar, warum mein Mann, wenn er um 12 Uhr nachts auf die Toilette geht,
35 kein Wasser braucht, es könnte die Nachbarn wecken!

Ich habe gelernt, nicht mehr so offen und temperamentvoll zu reden; es ist für mich immer noch ein Rätsel, wie die Schweizer reagie-
40 ren können. Wenn ich einmal richtig wütend bin, rede ich lieber mit meinem eigenen Wandspiegel.

Ich erinnere mich an die ersten Tage in Hergiswil in meiner Wohnung am Stadtrand mit
45 Blick auf den Pilatus... es war so atemberaubend!

Ich träumte schon davon, alle meine italienischen Verwandten einzuladen, um zu zeigen, wie schön meine neue Welt war mit dem mit Geranien geschmückten Balkon und rundum die
50 Wiesen, die so grün und duftig waren.

Am Abend kam mein Mann nach Hause und ich in seinen Armen glaubte, dass die ganze Welt in Ordnung sei. Ich glaubte sogar, im Paradies zu sein. Alle waren so nett, so hilfsbereit, so höf-
55 lich; es war ein Glück für mich, in der Schweiz zu leben. Die Leute, die ich sah, waren wohlhabend, und ich war überzeugt, dass die Schweiz ein reiches Land sein müsste. Es war sicher nichts mehr zu wünschen übrig.
60 Ich kann einfach nicht begreifen, warum diese schönen Tage für immer vergangen sind. Innert kurzer Zeit merkte ich, dass ich einsam war, ich hörte immer öfter sagen „wir sind nicht in Italien". Die Leute waren so distanziert und kühl
65 geworden. Wie kleine Kinder, die von einem Tag zum anderen kein Interesse mehr an ihrem Spielzeug haben.

Mein Mann sagte oft „Ich will meine Ruhe haben" oder „Ich habe keine Zeit". Ich war auch den
70 ganzen Tag im Büro, neun lange Stunden in Stille, ohne jeglichen Kontakt mit anderen Kollegen.

Das schöne Paradies war ein Paradies aus Plastik geworden, bis der Tag kam, an dem ich diese Stille und Kälte nicht mehr ertragen konnte.
75 Ich nahm einen Zug nach Süden, zurück nach Hause in meine Heimat, wo die Leute laut auf den Strassen singen, wo die Küchen nach schmackhaften Mahlzeiten duften, wo jeder sein Herz öffnen kann, ohne Angst, gehänselt zu werden.
80 Ich bin reifer geworden, die vielen Jahre haben etwas Positives gebracht. Ich habe gelernt, mit Schweizern umzugehen, d. h. ich erwarte nicht viel von ihnen, so kann ich mich um so mehr freuen, wenn ich einmal eine Person treffe, die
85 warmherzig ist. (Meistens waren solche Leute viel im Ausland.) Aber eine Schweizer Freundin werde ich nie haben, glaube ich.

Und ich weiss auch warum, ich bin hier wie ein Baum ohne Wurzeln, ich fühle mich so unsi-
90 cher; die Schweizer können auch nichts dafür.

Marisa Zago: Ich habe gelernt ... Aus: Fremd in der Schweiz.

Diskussion und Bericht:

1. Was mußte die Autorin in der Schweiz lernen, und warum? Haben Sie schon vergleichbare Erfahrungen beim Aufenthalt in einem fremden Land gemacht? Erzählen Sie in der Gruppe!

2. Berichten Sie den Kursteilnehmern, die den anderen Text gelesen haben, über die Erfahrungen der Autorin in der Schweiz.

---

* In der Schweiz wird ß als ss geschrieben.

Die Suche nach den Deutschen

Von João Ubaldo Ribeiro Frankfurter Rundschau 8. Juni 1991

Am Anfang schien es leicht. Schließlich sind wir in Deutschland und einen Deutschen zu treffen, sollte nicht schwer sein, wir hatten sogar gedacht, wir würden schon eine ganze
5 Reihe kennen. Jetzt nicht mehr. Jetzt wissen wir, daß das so einfach nicht ist, und ich habe gewisse Befürchtungen, daß wir nach Brasilien zurückkehren, ohne einen einzigen Deutschen gesehen zu haben. Das habe ich zufällig
10 entdeckt, als ich mit meinem Freund Dieter sprach, den ich für einen Deutschen gehalten hatte.

„Jetzt bin ich doch wahrhaftig schon ein Jahr in Deutschland, wie die Zeit vergeht",
15 sagte ich, als wir in einer Kneipe am Savigny-Platz ein Bierchen tranken.

„Ja", sagte er. „Die Zeit vergeht schnell, und du hast Deutschland nun gar nicht kennengelernt."
20 „Was heißt das, nicht kennengelernt? Ich bin doch die ganze Zeit über kaum fort gewesen."

„Na eben. Berlin ist nicht Deutschland. Das hier hat mit dem wirklichen Deutschland
25 überhaupt nichts zu tun."

„Darauf war ich nicht gefaßt. Wenn Berlin nicht Deutschland ist, dann weiß ich nicht mehr, was ich denken soll, dann ist alles, was ich bis heute über Deutschland gelernt habe,
30 falsch."

„Glaubst du etwa, daß eine Stadt wie Berlin, voller Menschen aus aller Herren Ländern, wo nichts so schwierig ist wie ein Restaurant zu finden, das nicht italienisch, jugo-
35 slawisch, chinesisch oder griechisch ist – alles, nur nicht deutsch – und wo das Mittagessen für neunzig Prozent der Bevölkerung aus Döner Kebab besteht, wo du dein ganzes Leben zubringen kannst, ohne ein einziges Wort
40 Deutsch zu sprechen, wo alle sich wie Verrückte anziehen und mit Frisuren herumlaufen, die aussehen, wie ein Modell der Berliner Philharmonie, da glaubst du, das sei Deutschland?"
45 „Na ja, also ich dachte immer, ist doch so, oder? Schließlich ist Berlin..."

„Da irrst du dich aber gewaltig. Berlin ist nicht Deutschland. Deutschland, das ist z. B. die Gegend, aus der ich komme."
50 „Vielleicht hast du recht. Schließlich bist du Deutscher und mußt wissen, wovon du redest."

„Ich bin kein Deutscher."

„Wie bitte? Entweder bin ich verrückt oder du machst mich erst verrückt. Hast du nicht
55 gerade gesagt, du seist in einer wirklich deutschen Gegend geboren?"

„Ja, aber das will in diesem Fall nichts heißen. Die Gegend ist deutsch, aber ich fühle mich nicht als Deutscher. Ich finde, die Deut-
60 schen sind ein düsteres, unbeholfenes, verschlossenes Volk... Nein, ich bin kein Deutscher, ich identifiziere mich viel mehr mit Völkern wie dieser, das sind fröhliche, entspannte, lachende Menschen, die offen sind...
65 Nein, ich bin kein Deutscher."

„Also laß mal gut sein, Dieter, natürlich bist du Deutscher, bist in Deutschland geboren, siehst aus wie ein Deutscher, deine Muttersprache ist Deutsch..."
70 „Meine Sprache ist nicht Deutsch. Ich spreche zwar deutsch, aber in Wahrheit ist meine Muttersprache der Dialekt aus meiner Heimat, der ähnelt dem Deutschen, ist aber keins."

„Halt mal, du bringst mich ja völlig durch-
75 einander. Erst sagst du, deine Heimat sei wirklich deutsch und jetzt sagst du, dort spricht man nicht die Sprache Deutschlands. Das verstehe ich nicht."

„Ganz einfach. Was du die Sprache
80 Deutschlands nennst, ist Hochdeutsch, und das gibt es nicht, es ist eine Erfindung, etwas Abstraktes. Niemand spricht Hochdeutsch, nur im Fernsehen und in den Kursen vom Goethe-Institut, alles gelogen. Der wirkliche
85 Deutsche spricht zu Hause kein Hochdeutsch, die ganze Familie würde denken, er sei verrückt geworden. Nicht einmal die Regierenden sprechen Hochdeutsch, ganz im Gegenteil, du brauchst dir nur ein paar Reden anzu-
90 hören. Es wird immer deutlicher, daß du die Deutschen wirklich nicht kennst."

Nach dieser Entdeckung unternahmen wir verschiedene Versuche, einen Deutschen kennenzulernen, aber alle, auch wenn wir uns
95 noch so anstrengten, schlugen unweigerlich fehl. Unter unseren Freunden in Berlin gibt es nicht einen einzigen Deutschen. In Zahlen ausgedrückt ist das etwa so: 40% halten sich für Berliner und meinen, die Deutschen seien
100 ein exotisches Volk, das weit weg wohnt; 30% fühlen sich durch die Frage beleidigt und wollen wissen, ob wir auf irgend etwas anspielen und rufen zu einer Versammlung gegen den Nationalismus auf; 15% sind Ex-Ossis, die
105 sich nicht daran gewöhnen können, daß sie

keine Ossis mehr sein sollen; und die rest-
lichen 15 % fühlen sich nicht als Deutsche, dieses düstere, unbeholfene, verschlossene Volk
usw. usw.

Da uns hier nicht mehr viel Zeit bleibt, wird
es langsam ernst. Wir beschlossen also, be-
scheiden in einige Reisen zu investieren. Zu-
nächst wählten wir München und freuten uns
schon alle über die Aussicht, endlich einige
Deutsche kennenzulernen, als Dieter uns besuchte und uns voller Verachtung erklärte, in
München würden wir keine Deutschen finden,
sondern Bayern – eine Sache sei Deutschland,
eine andere Bayern, es gebe keine größeren
Unterschiede auf dieser Welt. Leicht ent-
täuscht fuhren wir dennoch hin, es gefiel uns
sehr, aber wir kamen mit diesem dummen
Eindruck zurück, daß wir Deutschland nicht
gesehen hatten – es ist nicht leicht, das zu be- 125
werkstelligen. Noch weiß ich nicht recht, wie
ich der Schande entgehen kann, daß wir nach
unserer Rückkehr aus Deutschland in Brasilien gestehen müssen, wir hätten Deutschland
nicht kennengelernt. Eins ist jedoch sicher: ich 130
werde mich beim DAAD wegen falscher Versprechungen beschweren und deutlich machen, daß sie mich beim nächsten Mal gefäl-
ligst nach Deutschland bringen sollen, sonst
sind wir geschiedene Leute. 135

Übersetzung: Ray-Güde Mertin

Das ist doch seltsam!

Dieter, der deutsche (?) Freund des brasilianischen DAAD*-Stipendiaten João Ubaldo Ribeiro, sagt
über sich selbst: „Ich fühle mich nicht als Deutscher. Ich finde, die Deutschen sind ein düsteres,
unbeholfenes, verschlossenes Volk ... Nein, ich bin kein Deutscher ...“

Und was sagt er über ...

Berlin: _____

Deutschland: _____

seine Muttersprache: _____

das Hochdeutsche: _____

München: _____

João Ubaldo Ribeiro fragte seine Freunde in Berlin: Fühlt ihr euch als Deutsche?

Und das waren die Antworten:

40 % _____

30 % _____

15 % _____

15 % _____

Wie erklären Sie sich das? Glauben Sie, daß João Ubaldo Ribeiro übertreibt*?

Fragen Sie Deutsche, die Sie kennen: Fühlen Sie sich/Fühlst du dich als (richtiger) Deutscher/(richtige) Deutsche? Berichten Sie dann in der Klasse, was für Antworten Sie bekommen haben.

Als was fühlen Sie sich? Diskutieren Sie in der Klasse.

**2
K**

22 Kontrollaufgaben (Wiederholung)

1. Formen Sie bitte den Bericht vom **gestreßten Manager** mit Hilfe von „bevor" und „nachdem" um:

1. Herr Ernst schlief ein. Vorher las er bis drei Uhr morgens die neuen Statistiken.

Bevor Herr Ernst einschlief, las er bis drei Uhr morgens die neuen Statistiken.

2. Er wachte auf. Danach stand er sofort auf. _____

3. Zuerst trank er eine Tasse Kaffee. Dann duschte und rasierte er sich. _____

4. Er zog sich an. Danach ging er aus dem Haus. _____

5. In der Firma las er die Zeitung. Danach begann er sofort mit seiner Arbeit. _____

6. Er schrieb seinen Bericht über die Situation der Firma. Vorher diskutierte er darüber mit seinen Mitarbeitern. _____

7. Er nahm an einer Konferenz teil. Vorher telefonierte er mit mehreren Kunden. _____

8. In der Konferenz stellte er die neue Werbestrategie vor. Vorher hörte er sich an, was die anderen zu sagen hatten. _____

9. Um 12 Uhr nahm er das Flugzeug nach Straßburg. Vorher plante er mit seiner Sekretärin die Termine für den nächsten Tag. _____

10. In Straßburg trank er schnell eine Tasse Kaffee. Danach führte er Gespräche mit Abgeordneten des Europa-Parlaments. _____

11. Er wollte um 21 Uhr zurückfliegen. Vorher brach er zusammen. _____

2. **Sprachkurs als Heiratsvermittler**

Setzen Sie bitte die richtigen Zeitangaben ein, und achten Sie auf die richtigen Endungen!

_____*Am*_____ 16.5.1955 wurde Lorenz Lang in der Nähe von Kiel geboren. _____ sein____ 6. _____ sein____ 18. Lebensjahr ging er in die Schule und machte _____ Jahr____ 1974 das Abitur. 18 Monate _____ war er dann beim Militär. _____ Frühjahr 1976 fuhr er für sechs Monate in die USA. _____ d____ erst____ drei Monate besuchte er einen Englisch-Sprachkurs. Ein____ Tag____ lernte er dort eine sympathische Mexikanerin kennen. _____ sein____ Geburtstag organisierte er ein großes Fest und lud auch sie ein. _____ dies____ Abend waren beide oft zusammen. _____ d____ Sprachkurs machten sie _____ 28. Juni _____ 18. Juli eine Reise durch Mexiko. _____ dies____ Reise verliebten sie sich ineinander, und ein____ Abend____ fragte er sie, ob sie ihn heiraten wolle. Sie brauchte ziemlich lange für ihre Antwort. _____ ein____ halb____ Stunde antwortete sie dann: „Ja".

1   Auftaktseite: Liebe

1. Es gibt **viele Arten von Liebe,** z. B.:

Liebe zwischen Mann und Frau

Liebe der Eltern zu den Kindern

Liebe _____

Liebe _____

_____

_____

_____

Seid umschlungen, Millionen, diesen Kuß der ganzen Welt.
Schiller: Hymne an die Freude

2. Liebe zwischen **Mann und Frau.** Schreiben Sie bitte, was Ihnen dazu einfällt, in das Assozio-gramm!

ist nur ein Wort
?

Liebe

reiner Egoismus
zu zweit
?

das einzige, wofür es
sich lohnt, zu leben
?

2   Erster Kontakt im fremden Land: Briefe schreiben (Parallel zu A 4 oder vor A 4)   **A 1**

1. Für die **Kursteilnehmerinnen:**

In einem Brief an ihre Schwester schreibt Eva auch von dem netten jungen Mann, der sie im Flug-hafenrestaurant angesprochen hat. Schreiben Sie den Brief bitte weiter!

(Zur Erinnerung: In Briefen benutzt man für die Wiedergabe von Gesprächen in der Regel keine direkte Rede; in der indirekten Rede wird in einem Brief in familiärem Stil meist der Indikativ benutzt. Die Pronomen für die Anrede, „Du, Dich" usw., schreibt man groß. Natürlich schreibt Eva im Brief an ihre Schwester auch, was sie bei dieser Begegnung gefühlt und gedacht hat.)

> *Lilastadt, den ...*
>
> *Liebes Schwesterherz,*
>
> *nun bin ich schon seit ein paar Tagen hier in Lilaland, und ich muß sagen, alles, was ich bisher gesehen habe, gefällt mir sehr gut.*
> *Lilastadt selbst ist eine anregende Stadt: interessante, moderne Architektur, mit viel Grün und kleinen Seen mitten drin. Die Atmosphäre ist bunt und international; es wimmelt nur so von Menschen aus allen Ländern dieser Welt.*
> *Gleich am ersten Tag nach meiner Ankunft habe ich einen netten jungen Lilaländer kennengelernt. Ich saß im Flughafenrestaurant und wartete auf Beate, die mich abholen wollte. Ich fing schon an nervös zu werden, da stand plötzlich ...*

2. Für die **Kursteilnehmer:**

Schreiben Sie bitte einen Brief an eine(n) deutsche(n) Bekannte(n)/Brieffreund(in), und erzählen Sie ihm (ihr), wie Sie im Flughafenrestaurant von Lilastadt eine deutsche Touristin kennengelernt haben.

(Zur Erinnerung: In Briefen benutzt man in der Regel keine direkte Rede. Schreiben Sie auch über Ihre Gefühle und Unsicherheit bei dieser Kontaktaufnahme.)

Lilastadt, den ...

Lieb...

Du weißt doch, daß ich seit ... Deutsch lerne. Bis jetzt fehlte mir einfach die Gelegenheit, mein Deutsch auch mal außerhalb des Klassenzimmers auszuprobieren. Aber gestern, als ich im Flughafenrestaurant wieder mal einen Kaffee getrunken hatte und gehen wollte (Du weißt ja, ich mag diese Flughafenatmosphäre), da sah ich am Nebentisch ...

---

3. Lesen Sie sich bitte Ihre Briefe vor! Gibt es Unterschiede zwischen den weiblichen und männlichen Briefeschreibern?

A 1 + A 2 3 „Anbandeln": Wie macht man das?

Hat Koto Kana nun mit Eva „angebandelt" oder hat er nicht?
Das können Sie entscheiden, wenn Sie die Geschichte weiterschreiben (A 5).

Gibt es bei Ihnen Gelegenheiten „anzubandeln"? Wo? Wie macht man das? Was kann man da sagen?
Und was wird geantwortet?
So machten es Faust und Gretchen in Goethes berühmtem Schauspiel „Faust! Der Tragödie erster Teil" (1808).

'an|ban·deln, 'an|bän·deln ⟨V. i.⟩ eine Liebesbeziehung anknüpfen, einen Annäherungsversuch machen; (selten) Streit anfangen; ⟨Nebenform von⟩ anbinden; sie hat mit ihm angebändelt sich in eine Liebelei eingelassen

Wahrig: Deutsches Wörterbuch

Straße

FAUST. MARGARETE *vorübergehend.*

FAUST. Mein schönes Fräulein, darf ich wagen,
Meinen Arm und Geleit Ihr anzutragen?
MARGARETE.
Bin weder Fräulein, weder schön,
Kann ungeleitet nach Hause gehn.
(Sie macht sich los und ab.)

So macht es Koto Kana:
Schreiben Sie bitte!

Die Antwort:

Was könnte man in anderen Situationen sagen?

Vor einem Schaufenster: In einer Ausstellung: In der Mensa:

Nach einem Vortrag: Beim Reifenwechsel:

Andere Situationen:

A 3   4   Attribut: Partizip I und Partizip II

1. Lesen Sie bitte die Tabelle!

Partizip I	**die liebende Frau** = eine Aktion = ein Prozeß = dauert noch an	= die Frau, die liebt	Aktiv
Partizip II	**die geliebte Frau** = ein Zustand (dauert noch an)	= die Frau, die geliebt wird	Passiv
	die gekochten Eier	= die Eier sind gekocht worden = die Eier sind gekocht	Passiv Zustands- passiv
	der eingefahrene Zug = ein Resultat = ein Zustand = ein abgeschlossener Vorgang	= der Zug ist eingefahren	Vergangen- heit

2. **Wann kann man Partizip I benutzen, wann kann man Partizip II benutzen?**

Machen Sie die Probe!

Zum Beispiel: verletzen – Worte/Kritik/Bemerkungen/Kind/Hund usw.
Aktiv: Worte können verletzen ⟶ verletzende Worte (Partizip I)
Passiv: (Worte können nicht verletzt werden)
         Aber: Ein Kind/Jedes Lebewesen kann verletzt werden. ⟶ ein verletztes Kind (Partizip II)

Zum Beispiel: steigen – Kosten/Unruhe/Mann...
Aktiv: Kosten können steigen. ⟶ steigende Kosten
         Ein Mensch kann zum Beispiel im Beruf aufsteigen. ⟶ ein aufsteigender junger Mann
Zustand/Resultat: Die Kosten sind gestiegen. ⟶ gestiegene Kosten
(Bei den intransitiven Verben, die kein Vorgangspassiv und kein Zustandspassiv bilden, kann man das Partizip II nicht als Attribut benutzen, Beispiel: duftende Rose (~~geduftete Rose~~), rennende Kinder (~~gerannte Kinder~~)...)

Probieren Sie selbst: **Was ist möglich?**
Beispiel: steigende Unruhe
          gestiegene Unruhe   beides ist möglich

steigen – Unruhe/Preise/Einkommen/Wasser
retten – Idee/Gedanke/Pilot/Hund/Ufer
faszinieren – Stimme/Persönlichkeit/Zuschauer
zerstören – Wirkung/Stadt/Hoffnungen/Struk-
   turen

übertreiben – Worte/Hoffnungen/Pessimismus
singen – Kinder/Lied/Vogel
fahren – Zug/Kilometer
starten/landen – Flugzeug

3. **Reihenübung** in der Klasse:

a) Bilden Sie bitte Partizip I und/oder Partizip II! Sprechen Sie bitte reihum!
Beispiel: Die Schuhe sind geputzt ⟶ die geputzten Schuhe

die Hausaufgaben sind korrigiert – die Waren wurden importiert – die Nachfrage steigt – der Export steigt – das Formular ist ausgefüllt – die Ölquellen brennen – die Bilder sind gestohlen – die Kinder schreien – die Dichter sind vergessen worden – der Bericht ist veröffentlich worden – der Arbeiter liest – das Wasser fließt – die Stadt ist geschmückt – die Menschen schlafen – die Ferien beginnen – der Vertrag ist unterschrieben worden – der Schnee ist gefallen – die Wände sind bemalt – die Kinder malen – das Wasser kocht

b) Lösen Sie das Partizip bitte auf:
Beispiel: der gekürzte Text ⟶ der Text ist gekürzt (worden)

die fallenden Regentropfen
der verlorene Krieg
die gefallenen Soldaten
die drohende Arbeitslosigkeit
die zu spät gekommenen Gäste
geliehenes Geld

die wiedergefundene Freiheit
ein nie geküßter Mund
die selbstklebende Briefmarke
schön klingende Worte
der aufgehende Mond
eine schmerzende Wunde

ein leuchtender Stern
eine beeindruckende Leistung
die sich selbst erfüllende*
　　Prophezeiung
ein brennendes Problem
ein ungelernter Arbeiter

c) Verstehen Sie die **Wendungen**?
Sie ist ein gebranntes Kind.
Schlafende Hunde soll man nicht wecken.

5 (Kitschige) Liebe auf den ersten Blick – mit Partizip   A 3

Bitte schreiben Sie alles mit Partizipialattributen!

1. Als sie den Raum betrat, wußte er, daß sie es war, die er schon immer gesucht hatte!
Sie – die große Liebe seines Lebens

mit ihren _____*duftenden*_____ Haaren,

ihren _____ ,

ihren _____ ,

_____

_____

_____

_____

| Ihre Haare dufteten. |
| Ihre Augen leuchteten. |
| Ihre Wangen waren gerötet. |
| Ihr Mund war zart geschminkt. |
| Ihre Stimme zitterte. |
| Ihr Rock schwang hin und her. |
| Ihre Beine waren schön geformt. |

2. Und ihr ging es nicht anders. Sie sah nur noch ihn:

| Seine Haare waren frisch gewaschen. |
| Sein Mund lachte. |
| Seine Haut war leicht gebräunt. |
| Seine Zähne blitzten. |
| Sein Hemd war frisch gebügelt. |
| Seine Hände wirkten kräftig. |

Seine frisch _____

_____

_____

_____

_____

_____

Alle konnten es sehen, es war wirklich Liebe auf den ersten Blick.

* erfüllen

A 3   6   Sommergewitter mit Partizip

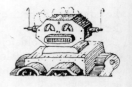

1. Lesen Sie bitte!

> Wortstellung bei erweitertem Partizip:
>
> **Die Wolken** ziehen über den Himmel.
>
> **Die** über den Himmel ziehenden **Wolken**

2. Schreiben Sie den folgenden Text bitte neu: Machen Sie mit Hilfe des Partizips aus zwei Sätzen einen Satz, oder verwandeln Sie den Relativsatz in ein Partizip.

Beginnen Sie bitte so:
Die untergehende Sonne schickt ihre letzten Strahlen ...

1. Die Sonne geht unter. Sie schickt ihre letzten Strahlen über die Erde.
2. Ein Gewitter droht. Es verdunkelt den ganzen Himmel.
3. Die Blitze, die über den Himmel zucken, erschrecken die kleinen Kinder.
4. Blätter tanzen. Sie werden vom Wind gejagt.
5. Die Menschen, die in die Häuser geflüchtet sind, hoffen, daß das Gewitter bald losbricht.
6. Hinter den Fenstern sieht man ängstliche Gesichter. Die Fenster sind geschlossen.
7. Die Wassermassen, die auf die Erde stürzen, bringen endlich die Abkühlung, die von allen erhofft wurde.
8. Kaum hat der Regen aufgehört, sieht man schon wieder Kinder, die spielen, und Menschen, die geschäftig hin- und hereilen*.

B
W   **7   Beziehungswortschatz** (fakultativ)

1. Liebesgeschichten haben ganz verschiedene Formen.

Welche der Wörter unten bezeichnen
– eine dauerhafte Beziehung, bei der ein Mann und eine Frau zusammenleben, ohne verheiratet zu sein
– den Zustand „verheiratet"
– eine eher oberflächliche Beziehung von kürzerer Dauer. Das ist auf jeden Fall die Meinung des Sprechers, der das Wort benutzt.

Tragen Sie bitte die Ziffern in die entsprechenden Spalten unten ein! Welche Wörter sind vom Sprecher eher negativ gemeint? Kennzeichnen Sie sie mit einem Negativstrich!

Die Wörter:

1 Ehe	7 Affaire
2 Ehe ohne Trauschein	8 Abenteuer
3 Liebesbeziehung	9 Onkelehe
4 Verhältnis	10 Bratkartoffelverhältnis
5 Romanze	11 Liebelei
6 wilde Ehe	12 Scheinehe

dauerhaft, nicht verheiratet						
verheiratet						
oberflächlich/kürzer						

*eilen = schnell laufen

Onkelehe: Eine Witwe mit Kindern lebt mit einem Mann zusammen. Sie heiratet ihn nicht, um ihre Rente aus der ersten Ehe nicht zu verlieren.

Bratkartoffelverhältnis: Ein Mann hat ein Verhältnis mit einer Frau, weil es für ihn praktisch und angenehm ist. Er hat aber keine ernsthafteren Absichten.

2. Sammeln Sie den „Beziehungswortschatz" in Ihrer Sprache. Welche Beziehungen werden wie bezeichnet? Machen Sie auch eine Tabelle!

8 Kleiner Wortschatz: „selber" oder auch „selbst" C 1

Ergänzen Sie bitte!

Ich möchte ich _____*selber*_____ sein!

Du sollst _____ _____ sein!

Er will ganz _____ _____ sein.

Sie ist ganz _____ _____ geblieben.

Wer kann schon _____ _____ verstehen?

Wir können _____ _____ nicht verstehen.

Ihr müßt _____ _____ entscheiden.

'Lie·be ⟨f. 19⟩ I ⟨unz.; i. w. S.⟩ **1** *starke Zuneigung, starkes Gefühl des Hingezogenseins, opferbereite Gefühlsbindung* (Menschen ~, Mutter ~, Nächsten ~, Tier ~, Vaterlands ~); ⟨i. e. S.⟩ *starke geschlechtsgebundene, opferbereite Geschlechtsbeziehung; Ggs Haß; heftiger Drang, heftiges Verlangen, Streben nach etwas* (Freiheits ~, Gerechtigkeits ~, Wahrheits ~); ⟨umg.⟩ *Gefälligkeit, Freundlichkeit* **2** *Glaube, Hoffnung und* ~ (nach 1. Korinther, 13,13); *ein (Kind der ~ K. aus einer ganz bes. glücklichen Ehe; uneheliches K.; **Lust** und ~ zu einer Sache haben eine S. gern tun; **Werke** der ~ tun der Barmherzigkeit; auf ein **Wort** der ~ warten* **3** *seine ~ war erkaltet, erloschen, gestorben; jmdm. eine* ~ **erweisen**; *jmds.* ~ (nicht) **erwidern**; *jmdm. seine* ~ **gestehen**; **tun** *Sie mir die* ~! ⟨umg.⟩ *ich bitte sehr darum!* **4** *alte ~ rostet nicht* ⟨Sprichw.⟩ *Jugendliebe od. -freundschaft ist von langer Dauer;*

Brennende ~ ⟨Bot.⟩ — *Herzblume;* **brüderliche,** *kindliche, schwesterliche,* **mütterliche, väterliche ~; eheliche,** *geschlechtliche ~;* **erbarmende ~; göttliche ~; große,** *heiße, herzliche, innige,* *leidenschaftliche, treue ~;* **heimliche,** *stille ~; eine ~ ist der anderen* **wert** *jmdm., der gefällig ist, tut man auch* **gern** *einen Gefallen* **5** *jmdn.* **aus** ~ *heiraten; etwas aus ~ zu jmdm. tun;* ~ **für** *jmdn. empfinden, fühlen; (kein) Glück* **in** *der* ~ *haben; in* ~ *entbrannt sein; etwas* **mit** *viel* ~ *tun mit viel Geduld u. Mühe; die ~* **zu** *den Eltern, zu den Kindern;* ~ *zur Musik, zur Kunst, zur Natur;* ~ *zur Wahrheit;* ~ **zwischen** *Mann und Frau, Eltern und Kindern* II ⟨zählb.; umg.⟩ **6** *Liebschaft; jmd., mit dem man eine Liebschaft hat od. hatte, jmd., den man liebt od. geliebt hat* **7** *alle seine* ~n; *meine erste ~; sie war meine große ~; seine vielen ~n* [< ahd. *liubi; zu lieb*]

(Wahrig: Deutsches Wörterbuch)

9 Liebesmetaphorik international (Erste Textauswertung) D 1

1. Lesen Sie bitte den Text D1 im Kursbuch! Suchen und notieren Sie drei zentrale Begriffe, um die es in diesem Text geht.

1. _____

2. _____

3. _____

2. Fassen Sie mündlich kurz zusammen, was über diese Begriffe gesagt wird.

* die Rente

D 1
+
D 2

10 Wortschatz: lieben, mögen, gern haben

- **jemanden/etwas lieben:** starkes Gefühl für **Personen** oder **Sachen.** Wenn die Person oder Sache nicht da ist, hat man Sehnsucht.
- **jemanden lieb haben:** starkes Gefühl für **Personen,** etwas vorsichtiger ausgedrückt
- **jemanden gern haben:** positives, freundliches Gefühl, meist für **Personen**
- **jemanden/etwas mögen:** positives Gefühl für **Personen** und **Sachen**

1. Ergänzen (verändern) Sie bitte jede/r für sich die Assoziogramme mit Ihren ganz persönlichen „Lieben" und „Vorlieben". Vergleichen Sie in der Klasse, und sprechen Sie darüber!

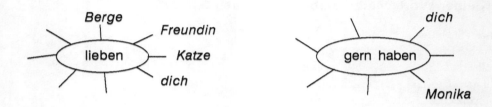

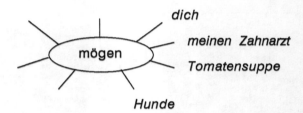

liebe

Ich liebe meine Mutter,
weil sie mit mir kuschelt.

Ich liebe meinen Hamster,
weil er mit mir spielt.

Ich liebe die Stille,
weil ich dann
besser arbeiten kann.

Ich liebe meinen Papa,
weil ich ihn halt liebe.

Ich liebe meine Uroma,
weil sie 100 Jahre alt
ist.

Lynn, 8 Jahre

„Ich mag dich, aber ich liebe dich nicht."
Warum ist das ein schlimmer Satz für einen Liebenden?
Übersetzen Sie bitte den Satz in Ihre Muttersprache!

2. Ihr Text!

Schreiben Sie auch ein Gedicht:
Ich liebe ...
oder
Ich mag ...

3. Man sagt auch:

Ich mag es, wenn ... (ein Mann mich auf der Straße anlächelt), oder:
Ich mag es nicht, wenn ...

Ich habe es gern, wenn ... (eine Frau mir nachschaut), oder:
Ich habe es nicht gern, wenn ...

Ihr Text:

Was haben Sie (nicht) gern? Was mögen Sie (nicht) gern? Schreiben Sie bitte!

Ich habe es nicht gern, wenn ⎯⎯⎯⎯⎯⎯⎯⎯⎯⎯⎯⎯⎯⎯⎯⎯⎯⎯⎯

Ich mag es nicht, wenn ⎯⎯⎯⎯⎯⎯⎯⎯⎯⎯⎯⎯⎯⎯⎯⎯⎯⎯⎯⎯⎯

Ich mag es, wenn ⎯⎯⎯⎯⎯⎯⎯⎯⎯⎯⎯⎯⎯⎯⎯⎯⎯⎯⎯⎯⎯⎯⎯⎯

Ich hab' es gern, wenn ⎯⎯⎯⎯⎯⎯⎯⎯⎯⎯⎯⎯⎯⎯⎯⎯⎯⎯⎯⎯⎯⎯

4. **etwas gern(e) machen** oder **etwas gern(e) tun** – (weil es Spaß macht)
etwas gern(e) essen – (weil es gut schmeckt)

Lesen und sprechen Sie bitte!

Ich	treibe esse gehe mache lebe	(nicht)	gern	Spaziergang Sport Gemüse spazieren hier

Sprechen Sie bitte reihum in der Klasse! Fragen und antworten Sie bitte:
Was machen Sie gern? oder: (Tanzen/Singen/... Sie gern?)
Was essen Sie gern? oder: Essen Sie auch gern (Eis, Schokolade, ...)?

5. **Ihr Text!**

Wenn Sie jetzt noch Lust haben, das „Mögen, Lieben, Gerne haben, Gerne essen" zusammen auszuprobieren, dann schreiben Sie einen Text oder ein Gedicht über Ihre positiven Gefühle für Menschen, Dinge, Aktivitäten ...

Zum Beispiel: Ich liebe meinen Mann.
Ich gehe gern im Wald spazieren.
Ich mag es, wenn es draußen stürmt und ich im warmen Zimmer sitze ...

Sammeln Sie die Texte ein, und raten Sie, wer welchen Text geschrieben hat.

11   Linksattribut, Rechtsattribut

D 3

(Nach Aufgabe D 3, 1.)

1. Formen Sie bitte die von Ihnen gebildeten Appositionen um in Adjektive, die (links) vor dem Nomen stehen!
Bilden Sie jeweils Sätze.
Beispiel:

Adjektive:		**Apposition:**	
Schwarzer, duftender	Kaffee, Kaffee	schwarz und duftend,	stand auf dem Tisch.

12   Bücherwurm: Appositionen

Erweitern Sie bitte die folgenden Sätze durch Appositionen!
Achten Sie auf den Kasus.

Gestern habe ich Hans Knapp,

_____, in einem Buch-   alter Bekannter

geschäft wiedergetroffen. Dieser Bekannte,

_____, wohnt erst seit   interessanter Mann

einigen Monaten hier und lebt sehr zurückgezogen. Er interessiert
sich sehr für Kunst und schreibt auch selbst Bücher darüber. Ich
kenne ihn, weil er zusammen mit meinem Mann,

_____,   sein Studienkollege

an der Universität gearbeitet hat. Im Buchgeschäft Wurm,

_____,   größtes der Stadt

suchte er ein ganz besonderes Buch und bat mich, ihm zu helfen.
Dieses Buch über die Kunst des 19. Jahrhunderts,

_____, wollte er seinem   schön gedruckt und farbig

Chef, _____,   Leiter der Abteilung

zum Geburtstag schenken.

D 3   13   **Attribute rechts vom Nomen**

1. **Liebeskummer** (Attribut mit Präposition)
Ergänzen Sie bitte die fehlenden Angaben!

Der Apotheker
hat Tabletten   gegen Kopfschmerzen

                gegen _____

                _____

hat Säfte   gegen _____

                gegen _____

hat Salben*   gegen _____

                _____

und
sieht ratlos aus
als ich frage
was er gegen meine Herzschmerzen
aufgrund chronischer Sehnsucht
empfehlen kann

* die Salbe

2. Bilden Sie bitte Attribute mit Präposition!

Beispiel: ein Kosename für die Liebste

Dank	Wahrheit
Demonstration	Liebste
Politik	gestern
Brot	Welt
ein Gedicht	Arme
ein Leben	Kinder
Hoffnung	Geschenk
Kampf	Ungerechtigkeit
ein Kosename	Krieg
Schnee	Umwelt
Haß	Hunger
Liebe	ein besseres Leben
ein Herz	Hesse

3. Bilden Sie bitte Genitivattribute!

Beispiel: die Einsamkeit der Dichterin

Herr Müller	der Drache
das Geld	die Verliebten
das Werk	die Natur
das Jahr	seine Frau
die Schönheit	der Autofahrer
die Angst	dieser Planet
die Einsamkeit	der Torhüter
die Vertrautheit	die Reichen
die Hoffnung	die Ärmsten
das Leben	die Flüchtlinge
der Schutz	die Dichterin

Ein Herz für Kinder

14   Welche? Was für ein? Links- und Rechtsattribute   D 3
(Diese Aufgabe zeigt die Rolle der Attribute im Text.)

Auf der nächsten Seite finden Sie einen Textausschnitt aus dem Kursbuch abgedruckt. Aber es fehlen alle Attribute.

1. Lesen Sie bitte den Text, wie er jetzt ist, laut vor. Was fällt Ihnen auf?

2. Arbeiten Sie zuerst bitte allein:
Stellen Sie (still) Fragen nach den unterstrichenen Wörtern mit den Fragepronomen „Welcher/Welche...“? und „Was für ein?“ Schreiben Sie die Fragen in die rechte Spalte.

3. Arbeiten Sie nun bitte zu dritt:
A liest einen Satz. B stellt die Fragen. C nimmt den Text im Lehrbuch und antwortet mit den fehlenden Attributen.

Beispiel:
A: Wo Beate bloß bleibt? In der Woche hat sie die Zeit doch noch bestätigt.
B: In welcher Woche? –C: In der vergangenen Woche.
B: Welche Zeit? –C: Die Ankunftszeit.

Wechseln Sie bitte auch die Rollen.

4. Machen Sie bitte die Aufgabe zu Hause noch einmal schriftlich: Schreiben Sie die Antworten ins Arbeitsbuch!

Treffpunkt für zwei

Aus Text A 1, S. 98:	**Fragen:**	**Antworten:**
Wo Beate bloß bleibt? In der ? <u>Woche</u> hat sie die ? <u>zeit</u> doch noch bestätigt! Eva Lehmann blättert <u>mit</u> ? <u>Unruhe</u> im ? Bordbuch.	In welcher Woche? Welche Zeit? Was für eine Unruhe? In welchem Bordbuch?	*In der vergangenen Woche.* *Die Ankunftszeit.* *Mit steigender Unruhe.* *Im vor ihr*
Eine halbe Stunde sitzt sie jetzt schon <u>im</u> ? <u>Restaurant</u> ? und wartet darauf, daß ihre Freundin sie abholt. Wahrscheinlich gibt es auch hier <u>einen</u> ? <u>verkehr</u>, ? , denkt sie und ißt noch etwas von <u>der</u> ? <u>torte</u>.	*Was für*	
„Entschuldigen Sie bitte ..." hört Eva plötzlich jemand sagen. Sie blickt auf. Vor ihr steht <u>ein</u> ? <u>Mann</u>. „Entschuldigen Sie bitte", sagt <u>der</u> ? <u>Mann</u> noch einmal, „daß ich Sie so einfach anspreche, aber..." Eva ist überrascht, hier <u>auf dem</u> <u>Flughafen</u> ? auf deutsch angesprochen zu werden. „... aber äh, ich ... äh." Er schweigt einen Augenblick. „Darf ich mich vielleicht zu Ihnen setzen?" fragt er dann schnell.	*Was für* *Welcher*	
Eva blickt <u>durch das</u> ? <u>Fenster,</u> und dann sieht sie <u>dem</u> ? <u>Mann</u> direkt <u>ins</u> ? <u>Gesicht</u>. Er sieht gar nicht so unsympathisch aus. „Hm, na ja, bitte", antwortet sie dann.	Wie ist das Gesicht?	

15 Textaufbau: Eine Geschichte ausschmücken und weiterschreiben

Alles, was die folgende Geschichte ein bißchen lebendig und farbig machen könnte, und viele zusätzliche Informationen wurden weggelassen.
Sie können nun die Geschichte neu schreiben, eventuell weiterschreiben und mit Attributen schmücken.

Das Mädchen und der Wolf

Eines Nachmittags saß ein Wolf in einem Wald und wartete, daß ein Mädchen mit einem Korb des Weges käme. Endlich kam auch ein Mädchen des Weges, und sie trug einen Korb. „Bringst du den Korb zu deiner Großmutter?" fragte der Wolf. Das Mädchen sagte ja, und nun erkundigte sich der Wolf, wo die Großmutter wohne. Das Mädchen gab ihm Auskunft, und er verschwand im Wald.
Als das Mädchen das Haus betrat, sah sie, daß jemand im Bett lag. Sie war noch keine drei Schritte auf das Bett zugegangen, da merkte sie, daß es nicht ihre Großmutter war, sondern der Wolf, ...

(Weitere Informationen und Einzelheiten:
Es geschah an einem Sonntag. Das Mädchen war klein und lieb. Sie hatte eine rote Mütze. Die Mütze hatte ihr ihre Großmutter geschenkt. Der Wolf war groß und böse. Der Korb war voller Lebensmittel für die Großmutter. Die Großmutter war krank. Die Großmutter wohnte in einem Haus im Wald. Das Haus lag mitten im Wald. Der Wald war dunkel. Die Person im Bett trug ein Nachthemd und eine Nachthaube* ...)

---

* die Haube = die Mütze

G 16 Telefonitis

1. Was paßt wo?

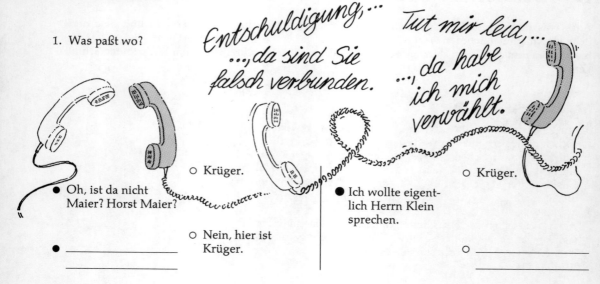

○ Krüger.

● Oh, ist da nicht
Maier? Horst Maier?

● _____

_____

○ Nein, hier ist
Krüger.

● Ich wollte eigent-
lich Herrn Klein
sprechen.

○ Krüger.

○ _____

2. **Rollenspiele:**

> Sie möchten heute abend mit einem deut-
> schen Freund ins Kino. Seine Schwester ist
> am Telefon.

> Sie haben um 17 Uhr einen Termin beim
> deutschen Botschafter, können aber erst um
> 18 Uhr da sein. Rufen Sie auf der Botschaft
> an!

> Um 23 Uhr nachts ruft ein/e Unbekannte/r
> an und macht Ihnen (natürlich auf deutsch)
> Komplimente am Telefon.

> Sie rufen Ihren Deutschlehrer/Ihre Deutsch-
> lehrerin an. Sie wollen ihn/sie zu einem Fest
> in ihrem Haus einladen – eine fremde Per-
> son meldet sich am Telefon.

> Koto Kana ruft Eva am selben Abend noch
> einmal an (sie hatte ihm ihre Telefonnum-
> mer gegeben) und wiederholt seine Einla-
> dung zum Fest an der deutschen Abteilung
> (Uhrzeit, er will sie abholen usw.).

> Sie kommen müde nach Hause, wollen ein
> gemütliches Bad nehmen: In der Badewanne
> liegt ein Krokodil. Das Tier hat offensicht-
> lich Hunger. Sie rufen den Notdienst an.

Erfinden Sie selbst weitere „Telefonsituationen"!

1. Liebesrätsel

Bilden Sie bitte aus den Silben Wörter!

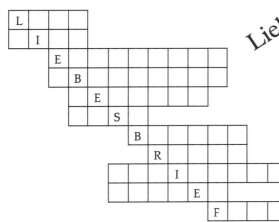

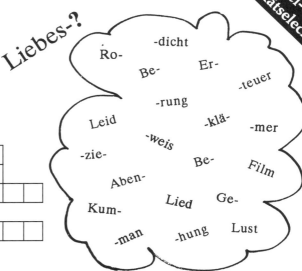

Liebes-?

Ro- -dicht
Be- Er-
-teuer
-rung
Leid -klä- -mer
-weis
-zie- Be- Film
Aben- Lied Ge-
Kum-
-man -hung Lust

2. Traummann und Traumfrau

Ergänzen Sie bitte die Grußanzeigen aus der Zeitung!

Liebe SABINE

es tut mir leid, wenn ich _____ weh getan habe, bitte verzeih. Auf _____ „Anrufe" konnte ich nicht reagieren, weil ich _____ neue Anschrift/Tel.-Nr. nicht kenne. Seit Monaten bin ich auf der Suche nach _____, leider ohne Erfolg. Bitte zeig mir den Weg zu _____. In Liebe

„ODYSSEUS" R.

TRAUMMANN
4. 6. - 4. 12. 1991

Vor 6 Monaten hat _____ große Liebe begonnen.
Dafür, _____ geliebter Schatz, danke ich _____ von ganzem Herzen.
Mögen _____ Träume einmal wahr werden.
In tiefer Liebe
_____ STERNTALER

_____ Herz weint, weil es das Feuer der Liebe löschen möchte, aber vergebens brennt mein Herz weiter voll Liebe zu _____ -, SUSANNE!!!

TRAUMMANN

Geliebte INGE!

Auch wenn ich es nicht sage jeden Tag - so weißt ____ doch, wie lieb ich _____ hab'. Bin ich auch in der Ferne, sind heute getrennt auch wir -, ____ hörst mein Herz, wie es ruft nur nach ____! Und ich schau' den weißen Wolken nach und fange an zu träumen ...

Ewig - ____ DUSCHA

Liebster DUSCHA _____,
auch ich denke sehnsüchtig täglich an _____.
Warum nur läßt _____ mich so lange auf ein Zeichen von _____ warten; wo und wann ich _____ wiedersehen kann?
In Liebe _____ Schatz

ALICE

3. Was sagen die **Sprichwörter?** Gibt es ähnliche oder andere Sprichwörter in Ihrer Sprache?

Keine Liebe ohne Leid.
Was sich liebt, das neckt sich.
Gegen die Liebe ist kein Kraut gewachsen.
Liebe macht blind.
Freunde in der Not gehen tausend auf ein Lot.
Um den Freund zu erkennen, mußt du erst ein Pfund Salz mit ihm gegessen haben.

4. Redewendungen

Versuchen Sie bitte herauszufinden, was hier dargestellt wird, was die entsprechende Redewendung bedeuten könnte, und wie sie lautet. Finden Sie bitte Situationen, in denen die Redewendungen passen. Was sagt man in Ihrer Sprache in dieser Situation?

_____   _____   _____   _____

_____   _____   _____   _____

7/D 18 Umgangsformen zwischen Männern und Frauen

1. Arbeiten Sie bitte zu zweit oder zu dritt!

Welche der folgenden Umgangsformen und Regeln sind für Sie in Ihrem Land eher üblich? Kreuzen Sie das übliche Verhalten bitte am linken Rand an! Wenn keine dieser Verhaltensweisen in Ihrem Land üblich sind, dann schreiben Sie bitte die übliche (oder Ihre) Verhaltensweise in die freie Linie. Am rechten Rand finden Sie angekreuzt, was heute in der Bundesrepublik Deutschland aus der Sicht einer etwa 40jährigen weiblichen Person möglich oder eher üblich ist. Natürlich kommt es dabei auch noch auf die Situation an. [?] bedeutet: kann schon mal sein.

2. Vergleichen Sie, und diskutieren Sie in der Klasse!

Bei Ihnen auch üblich: **Deutschland**   **Bei Ihnen auch üblich:**   **Deutschland**

Ein Mann und eine Frau gehen zusammen in ein Lokal:
☐ Er hält ihr die Tür auf. ☒
☐ Sie hält ihm die Tür auf. ☐
☐ Er geht im Gastraum voraus, sucht einen Tisch. ☒
☐ Sie geht im Gastraum voraus, sucht einen Tisch. ☐
☐ _____
☐ Er bestellt das Essen für beide. ☒

☐ Sie bestellt das Essen für beide. [?]
   (bei Geschäftsessen)
☐ Jeder bestellt für sich. ☒
☐ _____

Beim Mantelanziehen:
☐ Er hilft ihr in den Mantel. ☒
☐ Sie hilft ihm in den Mantel. ☐
☐ Jeder zieht seinen Mantel alleine an. ☒
☐ _____

Bei Ihnen auch üblich: Deutschland

Einkaufen:
- [ ] Er kauft ein. [x]
- [ ] Sie kauft ein. [x]
- [ ] Sie kaufen zusammen ein: [x]
- [ ] Er trägt die schwere Einkaufstasche. [x]
- [ ] Sie trägt die schwere Einkaufstasche. [?]
- [ ]

Autofahren:
- [ ] Er fährt, öffnet ihr aber vorher den Beifahrersitz und läßt sie einsteigen. [x]
- [ ] Sie fährt, öffnet ihm aber vorher den Beifahrersitz und läßt ihn einsteigen. [ ]
- [ ] Er fährt immer, auch wenn sie dabei ist. [?]
- [ ] Sie fährt nur, wenn er müde ist. [x]
- [ ]

Beim Tanzen (in der Disco, bei Festen ...):
- [ ] Nur Männer dürfen Frauen auffordern. [ ]
- [ ] Auch Damenwahl ist möglich. [x]
- [ ] Wenn zu wenig Männer da sind, tanzen auch Frauen miteinander. [?]
- [ ] Wenn zu wenig Frauen da sind, tanzen auch Männer miteinander. [ ]
- [ ] Man darf keine Frau alleine am Tisch sitzen lassen. [ ]
- [ ] Es gibt nur nach Geschlechtern getrennte Tanzveranstaltungen. [ ]
- [ ]

Bei Ihnen auch üblich: Deutschland

Der Dame fällt etwas auf den Boden (Taschentuch, Kamm, Handtasche):
- [ ] Er bückt sich schnell und hebt es auf. [x]
- [ ] Sie bückt sich und hebt es auf. [x]
- [ ] Sie wartet darauf, daß er sich bückt und es aufhebt. [?]
- [ ]

Kontaktaufnahme:
- [ ] Nur er kann mit einer Frau Kontakt aufnehmen (sie zum Beispiel einfach ansprechen). [ ]
- [ ] Sie kann genausogut einen Mann ansprechen. [x]
- [ ] Er ist empört, wenn sie ablehnt. [?]
  (manchmal)
- [ ] Sie ist empört, wenn er ablehnt. [ ]
- [ ]

Trennung nach längerer Beziehung:
- [ ] Man gibt die Geschenke zurück. [ ]
- [ ] Beide behalten ihre Geschenke. [x]
- [ ]

Heiraten:
- [ ] Sie wartet darauf, daß er ihr einen Heiratsantrag macht. [x]
- [ ] Es ist normal, daß sie ihm, wenn sie will, einen Heiratsantrag macht. [?]
- [ ] Er muß bei den Eltern um ihre Hand anhalten. [ ]
- [ ] Die Eltern werden nicht gefragt, nur informiert. [x]
- [ ]

Komm tanzen!

Kommentar: In der Bundesrepublik Deutschland gibt es keine strengen Vorschriften mehr für Umgangsformen zwischen Männern und Frauen. Je nach Generationszugehörigkeit (z. B. Sechzigjährige, Vierzigjährige oder Zwanzigjährige und Jüngere) gibt es unterschiedliche Erfahrungen, Erwartungen und Verhaltensweisen. Wahrscheinlich gibt es aufgrund der unterschiedlichen Entwicklung der Gesellschaften in den alten und neuen Bundesländern auch Unterschiede zwischen West- und Ostdeutschland.

7/D 19 Diskussion

1. Hier sind ein paar Sprichwörter auseinandergebrochen. Können Sie sie wieder zusammensetzen?

Jung gefreit

Drum prüfe

das Ende der Liebe

wer sich ewig bindet

nie gereut

Die Ehe ist

2. Diskutieren Sie Ihre Ergebnisse! Welcher Aussage würden Sie zustimmen, welcher nicht? Bilden Sie Pro- und Contra-Gruppen.
Jede Gruppe sammelt und notiert Argumente, um ihren Standpunkt zu vertreten. Dann kann die Diskussion beginnen. Bestimmen Sie einen Diskussionsleiter (siehe dazu Arbeitsbuch Lektion 3, Seite 51, Ü 2).

7 20 Leselust zum Thema (fakultativ)

Wählen Sie bitte einen der drei Texte!
Informieren Sie sich! Amüsieren Sie sich! Diskutieren Sie!

1. **Duzen** oder **Siezen** in deutschen Firmen? Informieren Sie sich!

Du, Chef! Neuer Trend im Büro

dpa **München** –Wie ist es richtig: „Guten Tag, Herr Direktor" oder „Hallo, Hans"?

Noch können sich fast alle Chefs auf das „Sie" verlassen. Aber womöglich nicht mehr lange.

Für den Wirtschaftspsychologen Professor Dr. Lutz von Rosenstiel (Universität München) steht fest: In den deutschen Büros ist das „Du" mächtig auf dem Vormarsch! Die These des Wissenschaftlers: Die Manager von morgen wachsen in einer Du-Kultur auf – und werden deshalb diese Gewohnheit auch im Job übernehmen.

Die ersten Unternehmen haben schon auf den neuen Trend reagiert. Beim schwedischen Möbelhauskonzern Ikea müssen sich alle Angestellten bis hinauf zum Boß duzen lassen – auch in deutschen Filialen. Sprecher Dierk Gronholz: „Natürlich ist es für Vorgesetzte leichter, aus der Distanz des ‚Sie' Kritik zu vermitteln. Aber als nicht gerade ultra-seriöseste Firma haben wir mit dem ‚Du' bisher kaum Probleme gehabt."

Ähnlich sieht es die Düsseldorfer Modehaus-Kette „Esprit", die vor 16 Jahren von einer Gruppe junger Leute aufgebaut wurde. Hier duzt jeder jeden, und der Erfolg ist groß. „Die familiäre Atmosphäre ohne lästige Formalitäten steigert sogar das Arbeitstempo", berichtet „Esprit"-Sprecher Detlev Krause.

Ein Vorbild für andere Firmen? Bei Siemens in München ist man noch skeptisch: „Als 150jähriges Unternehmen können wir nicht von heute auf morgen so tun, als wären wir eine Firma von Gleichaltrigen ..." Auch bei der Deutschen Bank bleibt es bis auf weiteres beim „Sie": „Es gibt gewachsene Traditionen, die einer besonderen Diskussion nicht bedürfen!"

Vielleicht kommt der Kompromiß aus Amerika. In der Deutschland-Niederlassung des US-Computerkonzerns Hewlett-Packard in Böblingen reden sich die Mitarbeiter mit Vornamen an – aber ganz höflich mit einem „Sie".

Hamburger Abendblatt 12.1.1991

2. Die letzte Sprechblase ist <u>noch</u> leer.

3.

Neue Studie über die Liebe
Alle wollen Geld

Das kann ja nicht gutgehen! Deutsche Männer wünschen sich Frauen, die zweieinhalb Jahre jünger sind als sie. Deutsche Frauen sind erst glücklich, wenn der Geliebte ihnen vier Jahre Lebenserfahrung voraus hat. Eine neue US-Studie enthüllt jetzt, was Männer und Frauen wirklich von ihren Partnern wollen. Die Wissenschaftler befragten Heiratskandidaten in 33 Ländern.

Von Anja Malanowski

München – Schlechte Zeiten für die Liebe. Denn was Frauen und Männer weltweit voneinander erwarten, klafft ganz schön auseinander – mit einer Ausnahme. Wenn es ums Geld geht, greifen (fast) alle gerne zu.

US-Professor David Buss, Psychologe und Leiter des Projekts: „Das war für uns sehr überraschend. Frauen aller Kontinente halten immer noch nach dem zuverlässigen Beschützer Ausschau – vor allem aber schauen sie aufs Geld. Die Männer suchen bei ihren Bräuten erstmal Schönheit und Jugend, doch dann folgt gleich der Blick ins Portemonnaie."

Dieses für Feministinnen und Soziologen niederschmetternde Ergebnis habe jede einzelne Testperson bestätigt. Weitere interessante Einzelheiten:

Keuschheit

Hier bestehen zwischen Nord und Süd die größten Unterschiede. Daß eine Frau unberührt in die Ehe geht, spielt für Deutsche, Schweden, Ozeanier und Nordamerikaner kaum noch eine Rolle. Für Osteuropäer und Afrikaner schon eher. Als fast unverzichtbares Gut müssen Iranerinnen ihre „Unschuld" mit in die Ehe bringen. Auch für die Inder ist die Keuschheit ihrer Ehefrauen sehr wichtig. Mehr noch als ihre Männer haben die Chinesinnen dieses Sittlichkeitsideal verinnerlicht. Für sie ist Jungfräulichkeit die wertvollste Mitgift. Umgekehrt schätzen die deutschen Frauen durchaus Männer mit Erfahrung.

Reichtum

Spitzenreiter der „Geldrangliste" sind die Indonesierinnen. Für sie ist Reichtum ein unverzichtbares Männer-Accessoire. Deutsche Frauen stufen Geld als „wichtig" ein. Nur den Holländerinnen ist es egal, was der Zukünftige auf der hohen Kante hat. Bei den Männern ist der Latin Lover in erster Linie scharf aufs Geld: Venezolaner und Kolumbianer klären vorher gern das Finanzielle. Deutsche Männer empfinden vermögende Frauen als „angenehm".

Beruf

Ein Geschlechterproblem scheint weltweit gelöst zu sein. Denn überall haben Frauen und Männer akzeptiert, daß ihre Partner ehrgeizig und arbeitsam sind. Deutsche Frauen halten Emsigkeit sogar für unverzichtbar. Einzige Bedingung: Der Mann darf nicht mit seinem Beruf verheiratet sein. Eine Ausnahme bildet die schwarze Bevölkerung Südafrikas. Nur dort erwarten Männer mehr Einsatz von den Frauen als diese von ihren Gatten.

Altersunterschied

Klar ist nur: Die Braut sollte jünger sein. In Nigeria sechs, in Kolumbien viereinhalb, in Schweden mindestens drei Jahre. Bundesrepublik siehe oben. Auch Frauen bestehen auf diesem kleinen Unterschied. In den USA muß der ideale Ehemann zweieinhalb, in Iran sogar fünf Jahre älter sein.

Schönheit

Hübsches Gesicht, aufregende Figur – für Männer in aller Welt immer noch der Hauptgrund, eine Frau zu heiraten. Besonders Palästinenser, Bulgaren und die deutschen Männer lassen sich bei ihrer Partnerwahl von äußeren Reizen beeinflussen. Frauen stellen laut US-Studie weniger Ansprüche an ihren Traummann, wenn die Kasse stimmt. Gutes Aussehen ist nicht entscheidend. Trotz aller Emanzipation – ein Beschützer ist erwünscht.

Hamburger Abendblatt, 9.5.89

1. Überraschen Sie einige Ergebnisse dieser Studie? Warum?
2. Wie stehen Sie zu den fünf genannten Punkten? Machen Sie eine Umfrage in der Klasse (in Ihrem Sprachinstitut?).

21 Wiederholung: Präpositionen und Adjektivendungen

1. **Zufällige Bekanntschaft**: Präpositionen
Setzen Sie bitte die fehlenden Präpositionen und Artikelwörter ein!

Liebe Ellen,

verzeih mir, daß ich so lange nicht mehr geschrieben habe, aber ich war in letzter Zeit sehr beschäftigt. Ich muß Dir nun einige Neuigkeiten erzählen!
In diesem Sommer hatte ich sechs Wochen Urlaub. Also ging ich _____ Reisebüro, das sich gleich _____ Haus _____ Ecke befindet. Ob ich lieber _____ Gebirge, _____ Meer, _____ Insel oder _____ Stadt fahren möchte, fragte mich die nette Angestellte. Das war gar nicht so einfach, denn all die Orte, die _____ Prospekten zu sehen waren, schienen wunderschön. Wollte ich lieber _____ Norden oder _____ Süden fahren? Sollte ich ganz allein _____ Land oder _____ Bergen wandern? Ich überlegte auch, ob ich nicht zuerst _____ Dir und dann weiter _____ meine Geburtsstadt direkt _____ Meer fahren sollte. Doch es kam alles anders. Als ich _____ Büro wieder _____ die Straße trat, kam ein starker Wind, und alle Prospekte fielen _____ Boden. Ein Mann, der _____ mir stand, half mir beim Einsammeln meiner Prospekte. Ich blickte ihn an, weil ich ihm danken wollte, und schaute direkt _____ zwei große dunkle Augen. Du kannst Dir vorstellen, wie peinlich mir diese Situation war. Er aber lächelte mich freundlich an und fragte mich, ob ich nicht Lust hätte, mit ihm einen Kaffee zu trinken. Wir gingen _____ Marktplatz und setzten uns _____ gemütliches Café. Seitdem haben wir gemeinsam viel erlebt. Ich bin nämlich gar nicht weggefahren, sondern in der Stadt geblieben. Und ich kann Dir sagen, auch ein Sommer _____ Stadt kann schön sein: mal _____ Kino, dann _____ Theater, _____ Ausstellung oder _____ Freunden gehen, _____ Park herumlaufen oder einfach _____ Hause bleiben und – sich erholen.
Laß Dir einen guten Rat geben: plane Deinen Urlaub nicht mehr!

Viele liebe Grüße – Deine Susanne

2. **Partnersuche**
Die Adjektivendungen sind verlorengegangen.
Bitte ergänzen Sie!
Schreiben Sie eigene Partneranzeigen.

34j. HORST, ledig
kathol., ein sympathisch___ jung___ Mann mit natürlich___ Wesen, humorvoll, kinderlieb und naturliebend, spielt gern Schach, mag Spaziergänge zu zweit und wünscht sich ein___ treu___, einfühlsam___ Partnerin. Suchst Du auch eine ehrlich___, verständnisvoll___ Partnerschaft, dann ruf doch einfach mal an unter ...

Mit 40 fängt das Leben erst an ...
Welcher humorvoll-ausgeglichen___, zärtlich___ und emanzipiert___ Mann, der seine Träume und seine Pläne noch nicht zu den Akten gelegt hat, hat Lust, seine Zukunft gemeinsam mit mir zu gestalten? Ich bin eine sehr lebendig___, aufgeschlossen___, hübsch___ Frau (Studienrätin, 41/1,68/55/NR, im Raum 6), mag fremd___ Sprachen und Länder, Menschen, Natur und Großstadt und freue mich auf die Zuschrift eines Mannes (mit Bild), der auch glaubt, daß nicht alles im Leben Zufall ist.

Neuanfang (Raum 4 oder 5)
Lebensfroh___, junggeblieben___ Frau, 53 J., gesch., 3 erwachsen___ Kinder, psychologisch___ Tätigk., wünscht sich lebendig___ Partnerschaft mit warmherzig___, offen___, neugierig___ Mann (auch jünger), der wie sie Mut zu einer neu___, reifer___ Beziehung mitbringt, in der jeder dem anderen viel persönlich___ Raum lassen kann. Ich freue mich auf Ihre Bildzuschrift.

1 Auftaktseite: Beschreibung einer Landschaft

Betrachten Sie bitte das Bild im
Kursbuch, S. 113!
Wählen Sie eine der beiden
folgenden Aufgaben!

hinten
Im Hintergrund
links
In der Mitte
rechts
Im Vordergrund
vorn(e)

1. Wenn Sie sich sicher fühlen: **Freie Bildbeschreibung**

Beschreiben Sie, was Sie auf dem Bild sehen. Gliedern Sie Ihre Bildbeschreibung:
Im Vordergrund .../In der Mitte des Bildes .../Im Hintergrund ...
Vergleichen Sie Ihre Bildbeschreibung mit der Bildbeschreibung in 2.

2. Wenn Sie sich nicht so sicher fühlen: **Bildbeschreibung mit Lückentext**

Ergänzen Sie bitte die vorgegebene Bildbeschreibung! Jedes dritte Wort fehlt.

In der _____ des Bildes _____ ein alter _____ , der nur _____ wenige
Äste _____ . Aber er _____ viele Blätter. _____ Spitze des _____ ist wahrscheinlich
_____ einem Sturm _____ . Der Baum _____ auf einer _____ Wiese.
Unter _____ Baum weiden _____ oder Schafe, _____ kann es _____ genau
erkennen.
_____ Baum steht _____ sehr weit _____ Ufer eines _____ entfernt. Der
_____ führt ziemlich _____ Wasser, einige _____ stehen mitten _____ Wasser,
auch _____ dem Baum _____ der Mitte _____ Bildes ist _____ . Wahrscheinlich hat
_____ in letzter _____ viel geregnet.
_____ gegenüberliegenden Ufer _____ Flusses sieht _____ die Spitze _____ Kirchturms.
Dort _____ ein Dorf. _____ Hintergrund des _____ erheben sich _____ Berge.
Man _____ nicht erkennen, _____ sie bewaldet _____ oder nicht. _____ Himmel, der
_____ obere Drittel _____ Bildes ausfüllt, _____ ziemlich bewölkt.

_____ vorn, vor _____ eigentlichen Bild, _____ noch eine _____ gezeichnet.
Die _____ zeigen fünf _____ vor zwölf. _____ der Innenseite _____ Deckels der
_____ sieht man einen Baum, _____ Äste nach _____ hängen und
_____ keine Blätter _____ trägt. Derjenige, _____ die Taschenuhr _____ hat, will
_____ sagen: Wenn _____ nichts für _____ Natur tut, _____ es bald _____ spät.

3. Welche Jahreszeit zeigt das Bild Ihrer Meinung nach? Welche Farben hat diese Landschaft?

A 2 „Es war eine Mutter, die hatte vier Kinder den Frühling, den Sommer, den Herbst und den Winter"

(Text eines Liedes)

Welche Bilder auf der nächsten Seite sind Frühlingsbilder (F), Sommerbilder (S), Herbstbilder (H) aus
Deutschland? Ein Bild ist ein Winterbild. Welches?
Diskutieren und begründen Sie bitte Ihre Entscheidungen!

2

3

4

7

6

B 1   **3**   **Frühlingsreise: Wortschatz**

1. Lesen Sie bitte den Text von B 1 noch einmal Abschnitt für Abschnitt, und ergänzen Sie die fehlenden Wörter und Ausdrücke in der folgenden Aufgabe mit Ausdrücken aus dem Text!

Évora:

Die Mandelbäume _____ schon im Januar.

Die Orangen _____ an den Bäumen.

Ihm fehlt das _____ Natur.

Córdoba:

Am 31. März beträgt die _____ 16 Grad,

die _____ ist minimal, das Wetter ist _____ und _____ .

San Sebastián:

Die herrliche, grüne _____ mit ihren _____ , _____ Bergen erinnert

ihn ans Allgäu. Er war _____ und _____ .

Oberrotweil:

(Drei Verben dafür, daß er von den Meteorologen etwas gelernt hat:)

Er schreibt: Sie _____ mich, daß ... (Beamtendeutsch)

Ich _____ , daß ...

Es _____ , daß ... (Beamtendeutsch)

(Drei Verben für „eilige Fortbewegung“:)

Der Frühling _____ nicht, _____ nicht, _____ nicht.

Helsinki:

Als das Flugzeug _____ , sah er Apfelbäume auf dem Vorfeld

des Flughafens.

2. Welche **Synonyme** stehen im Text für:

der Wechsel der Jahreszeiten: _____

Ich war „begeistert“: _____

Als das Flugzeug landete ...: _____

3. **Undeutliche oder blasse Farben.** Ergänzen Sie bitte!

1. Die Farbe der Hügel ist zwischen blau und grün, aber sie geht doch eher ins Blaue:

sie ist _____ . 2. Am frühen Morgen ist der Himmel manchmal ganz leicht rot: er ist

_____ . 3. Nach der ersten Wäsche schon war aus der leuchtend gelben Bluse ein trauriges

_____ Etwas geworden.

4. Wie heißen die Adjektive? Ergänzen Sie bitte!

die sommer_____ Hitze   die winter_____ Landschaft

die herbst_____ Kühle   das frühlings_____ Wetter

4 Wortbildung: Zusammengesetzte Wörter

1. **Substantive** (siehe auch: Arbeitsbuch Lektion 3, Übung 21)

Wie heißen die Substantive? Schreiben Sie bitte!

die Feuchtigkeit in der Luft: _____

ein Baum, der Mandeln trägt: _____

Waren für den Export: _____ *Exportwaren* _____

der Export von Waren: _____

eine Reise im Frühling: _____

> Bestimmungs-
> wort auf -heit,
> -keit, -ing und
> -ung → + **s**:
> Frühling**s**sonne

Was bedeuten die Substantive? Schreiben Sie vorne auch den Artikel des Substantivs!

_____ Kindheitserinnerung = _____ *Erinnerung an die Kindheit* _____

_____ Dolmetscherprüfung = _____

_____ Flüchtlingslager = _____

_____ Apfelblüte = _____

_____ Hügellandschaft = _____

_____ Seenlandschaft = _____

_____ Entspannungsübung = _____

_____ Bildungssystem = _____

_____ Kleiderordnung = _____

_____ Karrierefrau = _____

2. **Adjektive** (siehe auch: Arbeitsbuch Lektion 3, Übung 22)

Wie heißen die zusammengesetzten Adjektive?

Sie ist eine sehr _____ Frau. (Sie hat Freude an Kontakten)

Er liebte ihre _____ Augen. (Ihre Augen haben die Form von Mandeln.)

Er war sehr _____ . (Er wollte immer Recht haben.)

Sie macht eine _____ Berufsausbildung (nah an der Praxis)

Er verstand sogar die _____ Nachrichten. (Nachrichten in deutscher Sprache)

Was bedeuten die Adjektive?

eine abwechslungsreiche Tätigkeit
ein dreijähriges Kind
ein weltberühmter Professor
eine kinderfreundliche Gesellschaft
ein liebenswerter Mensch
eine leidvolle Vergangenheit
eine lebenslange Arbeit

eine kinderfeindliche Gesellschaft
frauenfeindliche Werbung
eine herzförmige Tomate
eine familienfreundliche Politik
ein sternklarer Himmel
ein buntkariertes Hemd
ein todkranker Mann

3. Partizip:

Ergänzen Sie bitte das passende Partizip!

1. Er ist ein sehr _____ Mensch.
(Er ist sich seiner Pflicht bewußt.)

2. Die Regierung beschloß _____ Maßnahmen.
(Die Maßnahmen sollten die Wirtschaft fördern.)

3. In diesem Land gab es schon vor 2000 Jahren eine _____ Zivilisation.
(Die Zivilisation war hoch entwickelt.)

4. Viele Abiturienten machen zuerst eine _____ Ausbildung.
(Die Ausbildung bezieht sich auf einen Beruf = ist auf einen Beruf bezogen.)

5. In der chemischen Industrie gibt es viele _____ Arbeitsplätze.
(Die Arbeitsplätze machen krank.)

6. Mit den _____ Haaren habe ich sie fast nicht mehr erkannt.

7. Leider gibt es auf der Welt noch viele _____ Regime.
(In diesen Regimen werden die Menschen verachtet.)

8. Wann bezahlen Sie die noch _____ Rechnungen?
(Die Rechnungen stehen noch offen; d.h.: sie sind noch nicht bezahlt.)

5. Superlative:

Zusammengesetzte Adjektive mit „hoch-", „tief-", „super-", „blitz-" und „stock-":

hochkompliziert = _____ *sehr kompliziert* _____

hochliterarische Texte = _____

ein hochtechnisiertes Land = _____

Anfang einer Rede: „Hochverehrte Gäste ..." = _____

Sie schaute ihn mit tieftraurigen Augen an. = mit _____

In den Straßen sah man nur tiefverschleierte Frauen. = _____

Der Himmel war tiefblau. = _____

Er hielt sich selbst für superschlau. = _____

ein superschneller Wagen = _____

Draußen war es stockdunkel. = _____

So ein stockdummer Kerl! = _____

In dem Häuschen war es blitzsauber. = _____

Die ganze Sache ging blitzschnell. = _____

Superlative in der Jugendsprache von 1992:
etwas ist <u>voll gut</u> = sehr gut, prima
etwas ist <u>affengeil</u> = etwas ist ganz toll
Das ist <u>megaout</u> = Das ist abolut vorbei/unmodern/nicht mehr aktuell.

---

5   Landschaft – Natur – Kultur

(Vor dem Hören/Lesen von C 1:)

1. „Eine herrliche Landschaft!"

Wann sagen oder denken Sie das? Sammeln und diskutieren Sie bitte in der Klasse! Benutzen Sie das Wörterbuch, wenn Ihnen Wörter fehlen. (Wenn die Vorstellungen in der Klasse zu unterschiedlich sind, sollte jede/r für sich auf einem Blatt Papier einen „Wortigel" machen. Die Ergebnisse werden im Klassenraum aufgehängt und besprochen.)

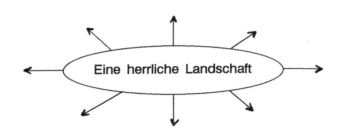

2. Was stellen Sie sich unter den beiden folgenden Begriffen vor? Schreiben Sie bitte!

                   unberührte Natur

„Naturlandschaft" _____

                        _____

                        _____

                                _____

                                _____ **„Kulturlandschaft"**

                                _____

                                _____

Gibt es in Ihrer Sprache vergleichbare Begriffe, die diese Unterschiede ausdrücken?

3. Wie sieht die Landschaft aus, in der Sie leben? Sammeln Sie gemeinsam an der Tafel!

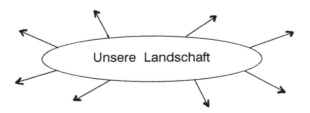

4. Schreiben

Schreiben Sie bitte einen Text für den Reisekatalog eines deutschen Reisebüros: Beschreibung der Landschaft, in der Sie leben, und was daran für Besucher (Touristen) besonders interessant und reizvoll ist. Ihr Ziel: Sie wollen das Reisebüro überzeugen, Reisen in Ihre Region zu planen. (Sie können natürlich auch einen Brief an einen deutschen Bekannten/eine Bekannte schreiben.)

8

C 1
+
D 1

6   Landschaften: Wortschatz

Benutzen Sie bei diesen Aufgaben das Wörterbuch, wenn Sie nicht weiterwissen!

1. Wo ist was in dieser „Landschaft"? Schreiben Sie bitte die Wörter an die richtige Stelle!

**das Gebirge – die Ebene/das Flachland – der Hügel – die Hochebene – die Anhöhe/die Erhebung –
das Tal – der Gipfel – der Berg**

2. **der Bach – der Fluß – der Kanal – der Strom – das Meer**

– Suchen Sie bitte einen Oberbegriff für diese Wörter: _____

– Ein Wort paßt nicht in die Reihe. Welches? Warum? _____

– Sehen Sie einen Grund für die Anordnung der Wörter oben? _____

– Was haben Bach, Fluß und Strom, aber Kanal und Meer nicht? _____

– Welche der fünf haben „ein Ufer" oder „eine Küste" oder „einen Strand"? _____

3. **der See – die See**
Was ist der Unterschied zwischen den beiden? Finden Sie Beispiele!

4. **der Garten – die Wiese – das Feld – der Wald – die Savanne – der Park – die Wüste – die
Pampa**
Was gibt es nur durch Menschenhand?

5. **Landschaft – Region – Umwelt – Gegend – Gebiet**

Einige dieser Begriffe können in verschiedenen Kontexten teilweise synonym gebraucht werden,
einige nicht.
Probieren Sie es an den folgenden Sätzen aus!

1. Was für ein... herrliche... _____ !

2. Dies... _____ ist typisch für Deutschland.

3. Die _____ muß durch bessere Gesetze geschützt werden.

4. In d... _____ gibt es keine Wälder mehr.

5. Wir machen eine Rundreise durch die schönsten _____ des Landes.

6. Was ist der Unterschied zwischen „Landschaft" und „Umwelt"? Versuchen Sie, ihn zu beschreiben!

7. **der Regenwald – der Urwald – der Dschungel**
In welchen Ländern gibt es das?

7 *sein + zu* : Passiversatz

1. Wo findet man solche Anweisungen?
Z. B. a) An Tankstellen.

b) _____

_____

c) _____

_____

d) _____

_____

e) _____

_____

f) _____

_____

g) _____

_____

h) _____

_____

i) _____

_____

| Passiv: | Das Rauchen muß unter-lassen werden. |
| Passiversatz: | Das Rauchen ist zu unter-lassen. |

a) Der Motor ist abzustellen.

b) Türen und Fenster sind zu schließen.

c) Die Fahrkarte ist zu entwerten.

d) Die Sicherheitsgurte sind anzulegen.

e) Ein- und Ausgänge sind freizuhalten.

f) Die Treppen sind einmal wöchentlich zu reinigen.

g) Das Klavierspielen ist nach 22 Uhr zu unterlassen.

h) Die Haustüre ist jeden Abend abzuschließen.

i) Die Wege sind sauber zu halten.

2. Schreiben Sie nette Anweisungen!

Beispiel:

Ausländer sind freundlich zu empfangen!

3. Eine schlechte Rede

Die Rede wurde auf einer Fortbildungsveranstaltung gehalten. Sie schreiben einen Bericht über die Rede. Schreiben Sie die Sätze bitte mit *sein + zu*!

1. Man konnte den Redner kaum sehen.
Der Redner war _____

2. Seine Stimme konnte man auch kaum hören.

_____

3. Was er sagte, konnte man kaum verstehen.

_____

4. Man muß die Ergebnisse der Untersuchung abwarten.

_____

5. Man muß die Richtigkeit seiner Aussagen in Frage stellen.

_____

6. Der Schlußsatz muß überprüft werden.

_____

C 2   8   _haben + zu_ + Infinitiv = _müssen_

> Unpersönlich: Das Rauchen **ist zu unterlassen**.
>
> Persönlich: Die Passagiere **haben das Rauchen zu unterlassen**.
>
> = Die Passagiere müssen/sollen das Rauchen unterlassen.

1. Schreiben Sie bitte die Sätze neu mit persönlichem Subjekt! (Zwei Möglichkeiten)

Das Singen ist zu unterlassen. ⟶ Die Mieter _____

⟶ Die Mieter _____

Mitgeführte Tiere sind zu melden. ⟶ Die Reisenden _____

⟶ _____

Fehler sind zu vermeiden. ⟶ Die Lernenden _____

⟶ _____

2. Unterstreichen Sie bitte
die _haben + zu_ + Infinitiv-Formen
im folgenden Ausschnitt
aus einer Hausordnung.

---

§ 3 Treppenhausreinigung

Wenn nicht die Reinigung des Treppenhauses vom Vermieter übernommen ist, haben die Bewohner des Erdgeschosses den Erdgeschoßflur, Haustür, Haustreppe und die Kellertreppe zu säubern, die Bewohner der anderen Stockwerke haben für die Reinigung des vor ihrer Wohnung liegenden Vorplatzes und der nach dem nächsten unteren Stockwerk führenden Treppe sowie des Treppenhausfensters zu sorgen. Die Bewohner des obersten Stocks sind außerdem verpflichtet, für die Sauberhaltung der Bodentreppe und des Vorplatzes auf dem Boden zu sorgen. Mehrere auf demselben Flur wohnende Parteien haben die Reinigung abwechselnd auszuführen. Wird bei dem Transport von Gegenständen Schmutz verursacht, so hat der Mieter, für den es geschieht, Hauseingang, Treppen usw. sofort selbst zu säubern.

---

3. Was haben die Bewohner des Hauses zu tun? Schreiben Sie bitte!

Die Bewohner des Erdgeschosses _____

_____

Die Bewohner der anderen Stockwerke _____

_____

Die Bewohner des obersten Stockwerks _____

_____

Mehrere auf demselben Flur wohnende Parteien _____

---

4. Ersetzen Sie die _haben + zu_ + Infinitiv-Formen durch _müssen_ + Infinitiv. Wie klingt das? Was finden Sie besser?

9 Attribute links vom Substantiv: Partizip I mit *zu* C 2

Gebrauch vor allem in offiziellen Redesituationen (Politikerreden, Nachrichten, Vorträge) und in der Schriftsprache. Diese Form hat passivische Bedeutung.

Kleine Drillübung

Sprechen Sie bitte reihum wie im Beispiel:
Was ist ...
– eine kaum zu bezahlende Reise = eine Reise, die kaum zu bezahlen ist
　　　　　　　　　　　　　　　= eine Reise, die kaum bezahlt werden kann
　　　　　　　　　　　　　　　= eine Reise, die man kaum bezahlen kann

Und nun Sie bitte!

Was ist/sind ...
– eine leicht zu treffende Wahl – ein schwer zu durchschauender Charakter – ein leicht zu übersetzender Text – eine schwer zu erlernende Sprache – eine gut zu verstehende Aussprache – nicht einzuhaltende Versprechungen – schwer zu erfüllende Hoffnungen – kaum zu beeinflussende Faktoren – gut zu lösende Probleme – sofort zu erledigende Arbeiten – schwer zu fassende Phänomene – leicht zu erklärende Unfallzahlen –

10 Drängende Probleme unserer Zeit. (Erweiterte) Attribute links vom Substantiv: Partizip I und Partizip II mit *zu*

C 2
↓
C 3

1. Schreiben Sie bitte die Sätze neu mit aus der rechten Spalte gebildeten Linksattributen!

Beispiel:

Die ▼ Kriminalität in den Großstädten ist ein ▲ Problem. | Die K. wächst.
Das P. ist ernst zu nehmen.

Die **wachsende** Kriminalität in den Großstädten ist ein **ernst zu nehmendes** Problem.

1. Wie die ▼ Generationen auf unserem Planeten leben werden, das ist eine ▼ Frage.

Die G. kommen nach (uns).
Die F. ist schwer zu beantworten.

_____

_____

2. Zwischen der ▼ Anzahl der Autos und der ▼ Umweltverschmutzung besteht ein ▼ Zusammenhang.

Die A. der Autos steigt.
Die U. wird immer größer.
Der Z. ist leicht zu erklären.

_____

_____

3. Den Militärs immer noch mehr Geld zu geben, wäre ein Fehler.

Die M. drängen auf mehr Waffen.
Der F. ist nicht wieder gut zu machen.

_____

4. Die Erde für <u>unsere</u> <u>Kinder</u> wieder schöner zu machen, ist <u>eine</u> <u>große</u> <u>Aufgabe</u>.

Die K. wachsen heran.
Die A. ist nicht leicht zu lösen.

_____

_____

_____

5. ↓ <u>Vorurteile</u> gegen Fremde und Andersartige stören das friedliche Zusammenleben der Menschen.

Die V. sind durch nichts zu ent-schuldigen.

_____

_____

6. <u>Die</u> <u>Menschheit</u> muß sich entscheiden zwischen einem vernünftigen Umgang mit ihrem Planeten oder dem Ende der Menschheit auf <u>einem</u> <u>Planeten</u>. <u>Eine</u> <u>Wahl</u> könnte man meinen.

Die M kämpft um ihr Überleben.
Der P. stirbt.
Die W. ist leicht zu treffen.

_____

_____

_____

2. Unterstreichen Sie die Partizipialkonstruktionen in den folgenden Sätzen. Drücken Sie sie dann anders aus!

Thema: Scheidung
1. In vielen Ländern ist die Ehe eine nicht leicht zu lösende Verbindung.
2. Ob die Ehe allerdings die Menschen glücklich macht, ist eine nur schwer zu beantwortende Frage.
3. Eine Scheidung bedeutet oft kaum zu bezahlende Kosten für die Ehepartner.
4. Manchmal beginnt mit der Scheidungsabsicht ein nicht enden wollender Kampf um die Kinder.
5. Für manche ist das ein durch nichts zu widerlegendes Argument gegen die Scheidung.

C 3   11   (Erweitertes) Attribut links vom Substantiv: Partizip II

Schreiben Sie bitte wie im Beispiel:

Natürlich ...

ist eine Frage, die viel diskutiert wird, _____ *eine viel diskutierte Frage.* _____

ist eine Antwort, die häufig gegeben wird, _____

ist eine Prüfung, die gut vorbereitet worden ist, _____

ist ein Buch, das viel und gern gelesen wird, _____

ist ein Theater, das vom Staat subventioniert wird, _____

sind Arbeiter, die schlecht bezahlt werden, _____

ist eine Autobahn, die (von) viel(en) Autos befahren wird, _____

Suchen Sie bitte weitere Beispiele!

12   Verstehbar und Verständlich (1)

1. Manchmal gibt es zu demselben Verbstamm ein Adjektiv auf *-bar* und eins auf *-lich*. Häufig gibt es nur eine Form von beiden. Manchmal gibt es deutliche Bedeutungsunterschiede zwischen den beiden Formen, manchmal sind die Bedeutungsunterschiede kaum wahrnehmbar.
Die Adjektive auf *-lich* gibt es manchmal nur mit der Vorsilbe *un-* (Negation).

Beispiele:

-bar			**-lich**
eßbar	nicht eßbar		
vermeidbar	– unvermeidbar		unvermeidlich
vergleichbar	– unvergleichbar		unvergleichlich
		vergeßlich –	unvergeßlich
denkbar	– undenkbar		
bewohnbar	– unbewohnbar		
verletzbar	– unverletzbar	verletzlich –	unverletzlich
(Körper)		(Psyche)	
lesbar	– unlesbar	leserlich –	unleserlich
trennbar	– untrennbar		unzertrennlich
beschreibbar	– unbeschreibbar		unbeschreiblich

2. Ergänzen Sie bitte die folgende Tabelle mit den entsprechenden Adjektiven. Benutzen Sie bitte ein Wörterbuch!

	-bar	**-lich**
voraussehen		
belehren		
entbehren		
verwechseln		
erklären		
verständigen		
verstehen		
bewohnen		
bewegen		
mißverstehen		
definieren		
kritisieren		
brennen		
lösen		
danken		
erziehen		

C 4   **13**   **Verstehbar und verständlich (2)**

1. **Kleine Drillübung**

Antworten Sie bitte reihum wie im Beispiel:

Was ist ...   Das ist ...

– eine unvergleichliche Schönheit?   – eine Schönheit, die man mit nichts vergleichen kann.

– ein unbelehrbarer Mensch – unentbehrliche Hilfe – ein unverwechselbares Aussehen – ein unerklärlicher Vorgang – ein unbewohnbarer Planet – eine denkbare Lösung – bewegliche Figuren – eine mißverständliche Erklärung – eine unlösbare Aufgabe – eine undankbare Aufgabe – ein dankbarer Mensch – ein trennbares Verb – ein undefinierbares Phänomen – eine unglaubliche Geschichte – ein unvergeßliches Erlebnis – eine unsachliche Behauptung

2. Bilden Sie bitte aus den folgenden Sätzen **Nominalgruppen mit** dem entsprechenden **Adjektiv!** Schreiben Sie die Nominalgruppen in eine Tabelle!

Diesen Vorgang kann man nicht erklären.
Das Wasser kann man nicht trinken.
Diese Geschichte ist nicht zu glauben.
Diesen Fehler kann niemand verzeihen.
So ein Erlebnis kann man nicht wiederholen.
Diese Entwicklung ist vorauszusehen.
Die Schönheit dieser Landschaft kann man
  nicht beschreiben.

Dieses Kind ist schwer zu erziehen.
Das Material läßt sich gut formen.
Ihren Mut kann man mit nichts vergleichen.
Die Handschrift kann man gut lesen.
Diesen Text kann niemand verstehen.
Ihr Verhalten ist nicht zu begreifen.

– bar	– lich
untrinkbares Wasser	*ein unerklärlicher Vorgang*

E 1   **14**   **Wortschatz: Synonyme**

In der linken Spalte auf der gegenüberliegenden Seite finden Sie den Text E 1 aus dem Kursbuch, S. 124. Einige Ausdrücke aus dem Text sind jedoch durch synonyme Ausdrücke ersetzt. Vergleichen Sie bitte den hier abgedruckten Text mit dem Text im Kursbuch, unterstreichen Sie die geänderten Ausdrücke, und schreiben Sie die synonymen Ausdrücke aus dem Kursbuchtext in die rechte Spalte.

„Und sonntags flüchtet der Mensch aus der Welt der Arbeit in die Natur."

Was heißt hier Natur?

„Natürlich gibt es Natur", sagte neulich jemand. Natürlich –das heißt in der Umgangssprache soviel wie selbstverständlich. Also kann man den Satz auch so übersetzen: Selbstverständlich gibt es das Selbstverständliche. Aber: Wenn man mit Natur das meint, was unabhängig von der Tätigkeit des Menschen existiert, dann kann man Garten, Wiese, Wald und Feld nicht als Natur bezeichnen. Und nicht einmal die abgelegenste Wüsten- oder Hochgebirgsregion ist ganz frei von menschlicher Aktivität.

Im Zeitalter der Romantik entstand das Ideal der vom Menschen unberührten Natur. Es entwickelte sich als Gegenbild zu einer die Natur immer mehr ausbeutenden Ökonomie, die die Produktion ohne Beschränkung ausweiten wollte. Die nicht beherrschte, „wilde" Natur wurde zum Zufluchtsort für den sonntags aus der Welt der Arbeit flüchtenden Menschen.

Mit der Trennung der Welt der Arbeit von der Freizeitwelt entwickelten sich im 19. Jahrhundert gleichzeitig zwei unterschiedliche Naturvorstellungen. Auf der einen Seite gab es den Naturbegriff der exakten Wissenschaften. Er verstand sich selbst als objektiv, d.h. als wahr. Auf der anderen Seite steht ein idealer Begriff von Natur, der die Tradition von Aufklärung und Romantik fortführt. In dessen Mittelpunkt stehen die Begriffe Emanzipation und Zivilisationskritik. Diese Naturvorstellung gilt als stärker subjektiv.

Allerdings gibt es in jüngster Zeit immer mehr politisch denkende Naturwissenschaftler, die den Natur- und Umweltschutz mit aller Macht unterstützen. Sie vertreten damit auch den subjektiven Begriff von Natur. Natur hier und heute? Natürlich gibt es sie auch heute. Nur, was man darunter versteht, ist alles andere als selbstverständlich.

Synonyme Ausdrücke (im Kursbuch)

_____

_____

_____

_____

_____

_____

_____

_____

_____

_____

_____

_____

_____

_____

_____

_____

_____

_____

_____

_____

_____

_____

_____

_____

_____

_____

_____

_____

_____

_____

_____

_____

_____

_____

_____

_____

E 2
+
E 4

15   Hier geht's ums Attribut

1. Sammeln Sie bitte durch Zuruf an der Tafel 15 bis 20 Substantive, die Ihnen gerade einfallen!

2. Jede/r wählt 3 bis 4 dieser Substantive und bildet so viele Attribute (links und/oder rechts) wie möglich zu jedem Substantiv. Lesen Sie die Ergebnisse im Plenum vor. Korrigieren Sie gemeinsam.

Beispiel:

> **eine Schule**
> **eine Schule** ohne Angst
> **eine Schule** ohne Angst für unsere Kinder
> **eine** Grund**schule** mit Englischunterricht
> **eine** gute **Schule**
> **eine** kinderfreundliche **Schule**
> **die** noch zu schaffende **Schule** der Zukunft
> **eine** gern besuchte **Schule**
> **eine** von Kindern, Eltern und
> Lehrern gemeinsam gestaltete **Schule**
> **eine Schule**, die von Kindern, Lehrern und Eltern gemeinsam gestaltet
> wird.

3. **Schreiben**

Schreiben Sie nun eine Geschichte, in der diese Begriffe und einige der von Ihnen erarbeiteten Attribute vorkommen. (Wenn es Ihnen trotz größter Mühe nicht gelingt, die Begriffe zu einer Geschichte zu verbinden, dürfen Sie maximal zwei weitere Begriffe aus der Sammlung an der Tafel auswählen und in Ihre Geschichte einbauen.)

Lesen Sie sich Ihre Geschichten vor!

16   Spiel- und Rätselecke

1. Wie wird aus den einzelnen Strophen wieder ein ganzes **Gedicht**? Numerieren Sie die Strophen bitte in der richtigen Reihenfolge!

Apfel-Kantate

Und was bei Sonn' und Himmel war,
erquickt nun Mund und Magen
und macht die Augen hell und klar.
So rundet sich das Apfeljahr.
Und mehr ist nicht zu sagen.

Dann waren Blätter grün an grün
und grün an grün nur Blätter.
Die Amsel nach des Tages Mühn,
sie sang ihr Abendlied gar kühn -
und auch bei Regenwetter.

Der Herbst, der macht die Blätter steif.
Der Sommer muß sich packen.
Hei! daß ich auf dem Finger pfeif':
da sind die ersten Äpfel reif
und haben rote Backen!

Der Apfel ist nicht gleich am Baum.
Da war erst lauter Blüte.
Da war erst lauter Blütenschaum.
Da war erst lauter Frühlingstraum
und lauter Lieb und Güte.

Hermann Claudius

2. Wie heißt wohl das deutsche **Sprichwort**,
das hier gezeichnet ist?

3. **Das Stadt-Land-Fluß-Spiel**

Stadt	Land	Fluß	Berg/Gebirge	Berufe	Namen	Punkte
Mannheim	Marokko	Main	Mount Everest	Maler	Monika	

In diesem Spiel geht es darum, Namen von Städten, Ländern, Flüssen usw. zu finden. Wenn die Klasse sich darauf einigt, so oft wie möglich geographische Angaben aus den deutschsprachigen Ländern zu benutzen (dafür gibt es Pluspunkte!), sollten mindestens eine Landkarte, am besten aber mehrere Atlanten im Klassenraum sein.

Das Spiel wird in Gruppen mit maximal 5 Teilnehmern gespielt. Spieler A sagt „A" und beginnt in Gedanken, das deutsche Alphabet aufzusagen (A, B, C, D, E, ...). Wenn Spieler B „Stop!" sagt, nennt A den Buchstaben, den er gerade gedacht hat, z. B.: „M": Jetzt schreibt jede/r so schnell wie möglich die entsprechenden Namen in die Tabelle.

Wer zuerst fertig ist, sagt „Stop!" Nun wird verglichen, was in den Spalten steht.
– Wenn zwei denselben Namen (Beruf) geschrieben haben, bekommt jede/r 5 Punkte.
– Für verschiedene Namen (Berufe) bekommt jede/r 10 Punkte.
– Wenn in einer Spalte nur eine/r etwas gefunden hat, bekommt er/sie 20 Punkte.
– Für jede deutschsprachige Lösung gibt es 10 zusätzliche Punkte.

Nach fünf Durchgängen werden die Punkte zusammengezählt. Wer die meisten Punkte hat, hat gewonnen.

Man kann das Spiel auch mit anderen Begriffen spielen, z. B.: mit deutschen Wörtern verschiedener Wortart:

Substantiv	Verb	Adjektiv	Zusammengesetztes Substantiv/Adjektiv	Namen	

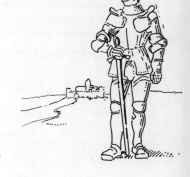

17 „Natürliche" Redewendungen

1. Setzen Sie bitte in den folgenden Minidialogen die passenden Redewendungen ein.
Wie lauten die entsprechenden Wendungen in Ihrer Muttersprache?

über den Berg sein - Bis dahin fließt noch viel Wasser den Berg hinunter. -
den Wald vor lauter Bäumen nicht sehen - mit der Wahrheit hinterm Berg halten -
Das paßt jetzt nicht in die (politische) Landschaft! - über Berg und Tal sein -
sich wie die Axt im Wald benehmen

1. ● Wie geht es Herrn Müller?
   ○ Er war lange krank, aber jetzt _____ .

2. ● Für Familien mit vielen Kindern sollte es bald Steuersenkungen geben.
   ○ Das _____ . Bei den Staatsschulden!

3. Die deutsche Vereinigung wird wahrscheinlich Milliarden kosten? Aber weil die Politiker
   wiedergewählt werden wollen, _____

4. ● Ich werde einfach mit meiner Doktorarbeit nicht fertig. Soviel Material, soviel Fakten!
   ○ _____ Du mußt Dich auf das Wesentliche konzentrieren!

5. Wenn du erreichen willst, was du möchtest, dann kannst du dich nicht _____
   _____ , dann mußt du ganz ruhig und vorsichtig an die Sache herangehen.

6. Als die Polizei kam, war der Bankräuber schon längst _____ .

7. ● Wart' nur, eines Tages wirst du auch heiraten und Kinder haben.
   ○ Du meine Güte, _____ .

2. Verstehen Sie die folgenden Wendungen? Erklären Sie sie bitte mit anderen Worten!
Gibt es diese oder ähnliche Redewendungen in Ihrer Muttersprache?

Er hat eine eiserne Natur.
Das liegt in der Natur der Sache.
Das sind Fragen grundsätzlicher Natur.
Sie ist von Natur aus ängstlich.
Er ist trotz seines Erfolges ganz natürlich geblieben.

„Er hat eine eiserne Natur."

18 Textumformung/Wörter aus dem Kontext erschließen

1. Was bedeutet das Wort „Naturmuffel"? Benutzen Sie bitte kein Wörterbuch, sondern versuchen Sie, das Wort aus dem Text selbst zu erschließen. Welche anderen Wörter helfen Ihnen dabei? Bestimmen Sie auch die Wortart!

2. Warum werden die Deutschen hier als „Naturmuffel" bezeichnet?

Deutsche sind Naturmuffel

dpa **Allensbach** - Viele Bundesbürger sind Wald-, Feld- und Wiesenmuffel. Nur 13 Prozent unternehmen „sehr oft" Ausflüge ins Grüne, 19 Prozent raffen sich immerhin noch „ziemlich oft" zu Spaziergängen oder Wanderungen auf. 26 Prozent dagegen kommen „kaum oder nie" dazu, zehn Prozent „ziemlich selten" und 32 Prozent „manchmal". Das ergab eine Umfrage des Instituts für Demoskopie Allensbach.

Hamburger Abendblatt, 29. 4. 1991

---

3. Welche Frage hat man bei der Umfrage gestellt? Könnte man bei Ihnen auch eine Umfrage zu diesem Thema machen?

4. Stellen Sie bitte die statistischen Angaben in dem Text in Form einer Tabelle dar!

5. Beurteilen Sie bitte die Ergebnisse der Umfrage aus Ihrer Sicht: Würden Sie auch sagen „Die Deutschen sind Naturmuffel", wenn Sie diese Statistik betrachten?

6. Überlegen Sie: Was bedeutet „Morgenmuffel"?
Versuchen Sie, andere Zusammensetzungen mit „Muffel" zu bilden.

19 Kontrollaufgaben (Wiederholung)

1. Was paßt?
Ergänzen Sie bitte, und notieren Sie danach in der rechten Spalte die entsprechenden Adjektive!

a) zu lösende b) zu übersetzende c) zu gebrauchende d) zu realisierende e) zu beantwortende

Beispiel:		Adjektive
Was haben alle gern?	Leicht zu bedienende Geräte	_____
1. Was macht man im Urlaub gern?	Leicht _____ Kreuzworträtsel.	_____
2. Was haben Deutschlernende gern?	Leicht _____ Texte.	_____
3. Worauf hoffen viele Menschen?	Auf bald _____ Zukunftspläne.	_____
4. Was sollten Prüfer vermeiden?	Schwer _____ Fragen.	_____
5. Was sollten die Lehrer schnell abschaffen?	Nicht mehr _____ Deutschlehrbücher.	_____

2. Streichen Sie bitte in diesem Text aus der Zeitung alle Attribute und Angaben weg.

Was bleibt übrig?

Frau stürzte aus dem Fenster

Eine etwa 27 Jahre alte Frau stürzte am Mittwoch abend gegen 20.30 Uhr aus bislang ungeklärter Ursache aus einem Fenster im dritten Stock eines Mehrfamilienhauses an der Straße Kleine Freiheit auf St. Pauli. Die Schwerverletzte wurde in das Krankenhaus Altona eingeliefert.

---

3. In einigen der folgenden Sätze fehlt das kleine Wörtchen „zu". In welchen? Schreiben Sie es bitte an die richtige Stelle:

a) Der Vertrag ist so schnell wie möglich unterschreiben.

b) Der Vertrag ist schnell unterschrieben worden.

c) Diese Geschichte ist unglaublich.

d) Diese Geschichte ist nicht glauben.

e) Die Bewohner des Hauses haben die Türen abends schließen.

f) Die Bewohner des Hauses haben die Türen abends geschlossen.

g) Ich brauche dringend einen guten Rat von dir.

h) Du brauchst mir nur einen guten Rat geben.

i) Ich habe vergessen, dich fragen, was wir tun sollen.

j) Ich frage dich, was wir tun sollen.

k) Deutsch ist eine schwer lernende Sprache.

l) Deutschlernen ist schwer.

m) Es ist schwer, Deutsch lernen.

n) Hunde sind an der Leine führen.

o) Hunde werden an der Leine geführt.

p) Wir dürfen nie aufhören Fragen stellen.

q) Wir müssen immer wieder Fragen stellen.

r) Hier sind noch einige Fragen stellen.

s) Hier sind noch einige Fragen offen.

t) Ich habe heute einiges erledigt.

u) Ich habe heute noch einiges erledigen.

v) Heute ist noch einiges erledigen.

4. **Schlagzeilen in der Zeitung!** Was bedeuten sie?

Frühling in Osteuropa!

Zweiter Frühling des beliebten Schauspielers!

Endlich: Konjunkturfrühling in Ostdeutschland!

Frühling in den Schaufenstern!

Prinzessin Di und Charles: Ein neuer Ehefrühling?

Frühlingserwachen: auch in der Textilindustrie

Boris und Karen: Frühlingsgefühle!

Schrecklicher Unfall riß vier junge Leute aus dem Frühling ihres Lebens!

Wissenschaftlich bewiesen: Der dritte Frühling ist der schönste!

1 Auftaktseite: Geschichte und Geschichten

1. Betrachten Sie bitte den Buch- und den Zeitschriftentitel (links oben und rechts unten) auf der Auftaktseite, und lesen Sie den Text oben rechts. Bearbeiten Sie dann die Aufgaben!

a) Worum geht es hier? Kreuzen Sie bitte an!

Es geht um ☐ ein Buch
☐ einen Film
☐ die Geschichte einer Familie
☐ deutsche Geschichte am Beispiel einer Familie

b) Was ist eine „Chronik"?

c) In welcher Epoche beginnt und endet die „Geschichtschronik", von der hier die Rede ist?

d) Auf der Zeitschiene unten fehlt noch ein wichtiges historisches Ereignis. Welches? Schreiben Sie den Namen dieses Ereignisses bitte an die entsprechende Stelle (auch auf S. 68 im Arbeitsbuch)!

2. „Geschichtsschreibung von oben" – „Geschichtsschreibung von unten"

a) Lesen Sie dazu bitte den Text links unten auf der Auftaktseite, und notieren Sie Stichwörter und Angaben zu diesen Begriffen!

„Geschichte von oben"

_____

_____

_____

_____

_____

„Geschichte von unten"

_____

_____

_____

_____

_____

Sammeln Sie bitte weitere Stichwörter zu dem einen oder anderen Begriff!

b) Erinnern Sie sich bitte an Ihren Geschichtsunterricht in der Schule! Haben Sie dort eher „Geschichte von oben" oder „Geschichte von unten" gelernt? Notieren Sie bitte Beispiele!

_____   _____

_____   _____

_____   _____

c) Können Sie Beispiele von Büchern, Filmen usw. zur Geschichte Ihres Landes nennen, die mehr in die eine oder die andere Richtung gehen?

_____   _____

_____   _____

_____   _____

_____   _____

d) „Sehnsucht nach Heimat" – nennt der SPIEGEL seine Titelgeschichte zu diesem Film und nicht „Sehnsucht nach der Heimat". Sehen Sie einen Unterschied? (Zum Begriff „Heimat" siehe im Kursbuch S. 82 und im Arbeitsbuch Lektion 6, Übung 1.)

A 1
B 1
C 1
D 1

2 Bilder aus einem Film

Auf den Seiten 128, 132, 136, 140 im Kursbuch finden Sie vier Filmstreifen aus dem Film „Heimat" abgedruckt. Jeder Filmstreifen zeigt einen Ausschnitt zu einer bestimmten historischen Epoche zwischen 1919 und 1982. Betrachten Sie bitte die vier Filmstreifen, überlegen und diskutieren Sie:

Welche Zeit zeigen die Bilder? (Indikatoren sind zum Beispiel die Kleidung und das Aussehen der Leute, bestimmte Gegenstände usw.). Welche Situationen/Orte können Sie erkennen (glauben Sie zu erkennen)? Notieren Sie bitte! Sprechen Sie bitte miteinander über Ihre Ergebnisse!

Seite	Zeit (vermutlich)	Indikatoren	Ort/Situation (vermutlich)
128	_____	_____	_____
132	_____	_____	_____
136	_____	_____	_____
140	_____	_____	_____

A 1 3 Bild und Text

1. Betrachten Sie jetzt bitte die Bilder von A 1, Kursbuch S. 128! Decken Sie dabei den Text rechts ab!

Überlegen Sie:
– Was passiert hier?
– Was für Leute sieht man auf den Bildern?
– Was machen die Leute?
– Was für eine Stimmung haben die Leute Ihrer Meinung nach?

2. Betrachten Sie die Bilder, und hören Sie den Text auf der Kassette! (Lassen Sie den Text im Buch immer noch abgedeckt!) Versuchen Sie dann, die Fragen zu beantworten!

– Wer spricht? Zu wem?
– Was empfinden Sie bei dieser Rede?
– Passen die Bilder und die Rede zusammen oder gibt es einen Widerspruch?
  Suchen Sie Erklärungen!

4 Eine kämpferische Rede (fakultativ)

(Zu Aufgabe A 1., 2. im Kursbuch)

1. Lesen Sie nun bitte den abgedruckten Text aus dem Drehbuch, und suchen Sie Elemente in der Rede (Begriffe/Begriffspaare), die darauf hinweisen, daß der Redner

☐ die Nationalsozialisten (Nazis) eher ablehnt,

☐ eher mit der Ideologie der Nationalsozialisten sympathisiert.

Unterstreichen Sie bitte die entsprechenden Begriffe!

2. Eine Regieanweisung lautet: „Die Dorfbewohner haben bei diesen Worten völlig leere Gesichter." – Was soll damit ausgedrückt werden?

5 Je länger die Tage, desto (um so) kürzer die Nächte

1. Bilden Sie bitte Aussagen mit „je … desto" oder „je … um so"!
Beispiele: starker Kaffee = guter Kaffee: je stärker der Kaffee, desto besser
gutes Essen → große Liebe: je besser das Essen, desto größer die Liebe

Und nun Sie bitte!
Machen Sie die Übung bitte reihum, sprechen Sie ziemlich schnell! Finden Sie weitere Beispiele!

heiße Liebe – kurze Liebe; große Wohnung – teure Wohnung; reiche Männer – junge Frauen; kleine Kinder – gestreßte Eltern; gute Lehrer – zufriedene Schüler; frisches Obst – gesundes Obst; schnelles Auto – gefährliches Auto; alter Wein – teurer Wein; …

2. Ergänzen Sie bitte, und/oder finden Sie selbst Beispiele!

Je selbstbewußter die Frau, um so …
Je reicher die Leute, …
Je schöner der Abend, …
Je friedlicher die Demonstration, …
Je intensiver der Gedankenaustausch, …
Je verrückter die Mode, …
Je größer die Liebe, …

Je … die Reichen, desto … die Armen
Je … deine Fragen, … … meine Antworten
Je … die Gäste, … … die Party
Je … die Brautleute, … … die Ehe
Je … das Thema, … … die Diskussion
Je … die Lehrerin, … … der Deutschunterricht
Je … die Hoffnung, … …

6 Je … desto/Je … um so: Satzgliedstellung

Der Nebensatz mit „je …" steht immer vorn:

Nebensatz		V	Hauptsatz	V	
Je dunkler	die Nacht	(ist),	desto heller	leuchten	die Sterne.
(Wenn	die Nacht	dunkel ist,		leuchten	die Sterne heller.)

Bilden Sie bitte die Argumentationskette mit „je … desto"!

1. *Je benachteiligter die Frauen sind, desto unzufriedener sind sie.*

(Wenn die Frauen benachteiligt sind, werden sie unzufrieden.)

2. _____

(Wenn die Frauen unzufrieden sind, treten sie kämpferisch für ihre Rechte ein.)

3. _____

(Wenn die Frauen selbstbewußt auftreten, werden viele Männer unsicher.)

4. _____

(Wenn die Männer im Haushalt viel mitarbeiten, können die Frauen ihre Karriere bewußt planen.)

5. _____

(Wenn Väter häufig mit ihren Kindern spielen, ist es gut für die ganze Familie.)

6. _____

(Wenn Frauen ihre Ziele energisch verfolgen, werden sie erfolgreich sein.)

7. _____

8. _____

A/W   7   Die goldenen zwanziger Jahre (fakultativ)

Lesen Sie bitte den Text, und ergänzen Sie dann die Tabelle auf S. 71 oben! Suchen Sie bitte die Angaben im Text!

Die goldenen zwanziger Jahre

Nach den wirtschaftlichen und politischen Wirren aus den Anfangsjahren der Weimarer Republik wird die Lage ab der Mitte der 20er Jahre relativ stabil. Das kulturelle Leben und die glänzende Unterhaltungswelt nehmen einen großen Aufschwung.

Gleichzeitig bleibt jedoch die Erinnerung an die Inflationsjahre bestehen, neue Unruhe kündigt sich
5 an: Die Arbeitslosenzahlen steigen langsam aber stetig, die Kabinette wechseln immer häufiger. Die Zeit ist von Hektik und Schnellebigkeit gekennzeichnet. Moden, Hits und Stars verschwinden genauso schnell von der Bühne, wie sie dorthin aufgestiegen sind.

Typische Formen der Unterhaltung in den 20er Jahren sind:

Operette: Nach wie vor gehören Operetten zu den beliebten Unterhaltungen. Die „Kleine Schwester
10 der Oper" zeigt meist eine lockere, heitere Handlung. Gesprochene Dialoge, Gesang- und Tanzeinlagen wechseln einander ab. In Berlin feiert vor allem der österreichische Komponist Franz Léhar mit seinen Stücken große Erfolge.

Revue: In den Revuen werden verschiedene Programmnummern meist ohne inhaltlichen Zusammenhang aneinandergereiht. Dabei wechseln Gesangs-, Sprech- und Tanzszenen. Die Revuen wer-
15 den mit prachtvollen Kostümen und Dekorationen und einer aufwendigen Bühnentechnik ausgestattet. Hauptattraktion der Revuen sind die attraktiven Tänzerinnen. Gezeigt wird viel Bein.

Varieté: Im Varietétheater werden vor allem artistische und akrobatische Nummern zu einem bunten Unterhaltungsprogramm zusammengestellt.

Kabarett: Eine andere Form der Aneinanderreihung von Kleinkunstformen stellt das Kabarett dar.
20 Hier werden in Chansons und Sketchen politische und gesellschaftliche Mißstände karikiert. Das Publikum amüsiert sich jedoch eher über die Kritik und begreift sie nur selten als Aufforderung zum Handeln. Die Texte zu den Chansons stammen vor allem von Walter Mehring, Klabund, Kurt Tucholsky, Erich Kästner und Joachim Ringelnatz.

Film: Das noch relativ junge Medium wird schnell zu einer Attraktion für breite Schichten. Riesige
25 Filmpaläste mit komfortabler Ausstattung entstehen, in denen in- und ausländische Produktionen gezeigt werden.

Schlager: Wie schon zu Beginn der 20er Jahre – während der Inflationszeit war etwa der Schlager „Pleite" große Mode – nehmen die Schlagertexte häufig bezug auf mehr oder weniger aktuelle Ereignisse und Entwicklungen. Seit etwa 1925 kommen Schlager mit Nonsense-Texten auf, in denen man
30 sich in erster Linie an Wortspielen und grotesken Situationen erfreut.

Typische Formen der Unterhaltung in den 20er Jahren:

Varieté	Operette	Kabarett	Film	Schlager	Revue
	lockere, heitere Handlung; Dialoge, Gesang- und Tanzeinlagen				

8 Eine Hochzeit im Dritten Reich   B 1

1. In diesem Abschnitt des Films geht es um eine Hochzeit. Betrachten Sie bitte die drei Bilder des Filmstreifens! Versuchen Sie, die Bilder (Personen, Ort, vermutliche Situation/Handlung) in bezug auf das Thema „Hochzeit" zu interpretieren und zu beschreiben.

2. Hören Sie den Text von der Kassette, und betrachten Sie dabei die Bilder! Was für eine Hochzeit findet hier statt? Welchen Moment der Hochzeit zeigen Bild 2 und Bild 3?

3. Das Dritte Reich begann mit der Machtübernahme der Nationalsozialisten am 30. Januar 1933 und endete mit der bedingungslosen Kapitulation der deutschen Armee am 8./9. Mai 1945 (siehe auch Zeittafel im Arbeitsbuch S. 68).

a) Zu welchem Zeitabschnitt des Dritten Reichs gehört dieser Filmausschnitt?

b) Warum kommt Anton nicht zur Hochzeit nach Hause?

4. **Eilfried – Katharina – Martha:** Wer von den dreien steht den Nationalsozialisten nahe? Woraus kann man das schließen?

5. **Schreiben** (fakultativ)

Wählen Sie eine der drei folgenden Schreibaufgaben!

a) Brief von Martha an Anton: Martha schreibt, wie sie die Hochzeit zu Hause „gefeiert" haben.

b) Anton schreibt in sein Tagebuch: Heute war mein Hochzeitstag...

c) Schreiben Sie die Kurzbeschreibung der Szene für die Regie! Sie können so beginnen: Szene „Ferntrauung". Die Kamera zeigt...

9 „Zeitworte"   B 1

Viele Wörter einer Sprache sind „historisch", d.h. sie gehören in eine ganz bestimmte Epoche. Sie bezeichnen z.B. bestimmte historische Ereignisse, bestimmte Situationen, Verhältnisse oder Gegenstände aus einer bestimmten Zeit. Die Bedeutung dieser Wörter versteht man nur, wenn man weiß, in welche Zeit sie gehören, und wenn man etwas über diese Zeit weiß.
Mit dem Wort „Machtübernahme" ist in Deutschland zum Beispiel meistens die „Machtübernahme durch die Nazis" gemeint.

Welche anderen Wörter im Text sind Ihrer Meinung nach solche „Zeitworte"?

B 3
+
B 4

10 Konjunktiv II/Irrealis: Was wäre, wenn ...

> Beachten Sie:
> 1. Konjunktiv II in der Umgangssprache: meist „würde" + Infinitiv
> Ausnahmen:
> – Alle Modalverben: dürfte, müßte, könnte, sollte, wollte
> – „sein" und „haben": wäre, hätte
> – Konjunktivformen einiger Verben, die besonders häufig vorkommen, z.B.:
> käme, ginge, ließe, gäbe, säße, bliebe, wüßte
> 2. In literarischen Texten findet man häufiger Konjunktiv-II-Formen von regelmäßigen und unregelmäßigen Verben als in der Umgangssprache.

1. Ergänzen Sie bitte die Endungen:

Ich würd_____ es machen,

wenn du es auch machen würd_____.

Aber er würd_____ es nur machen,

wenn ihr alle mitmachen würd_____.

2. **Wenn jeder eine Blume pflanzen würde**

Drücken Sie bitte die folgenden Zeilen des Gedichts von Peter Härtling umgangssprachlich aus!

3 *Wenn jeder* _____

4 *Wenn jeder* _____

5 _____

7 _____

10 _____ *ihre Sorgen* _____

3. **Zukunftschancen**

Formulieren Sie bitte Zukunftschancen wie im Beispiel! Finden Sie weitere oder andere Beispiele für Ihr Land/aus Ihrem Land!

Beispiel:
Frieden und soziale Gerechtigkeit schaffen

→ Wenn es den Menschen gelänge (gelingen würde), Frieden und soziale Gerechtigkeit auf der Welt zu schaffen, dann gäbe es weniger Hunger auf der Welt.

Und nun Sie bitte!
Soldaten: nicht in Krieg ziehen/niemand: Auto fahren/die reichen Länder: von ihrem Reichtum abgeben/die Umwelt: überall schützen/keine Atomkraftwerke bauen/alle Politiker: selbstlos/alle Menschen: einander helfen/die Kinder: weniger Fernsehen – mehr lesen/alle Grenzen öffnen/...

4. **Schlechte Aussichten! Was wäre, wenn ...**
die Dinosaurier kommen zurück/die Erde hört auf, sich zu drehen/die Menschen werden nicht klüger/das Ozonloch wird immer größer/Außerirdische greifen die Erde an/es gibt kein Fernsehen mehr/ alle Computer stürzen ab/...

5. **Ihr Text:**
Was würden Sie (anders) machen, wenn ...
– Sie die Macht in ihrem Land hätten ...
– Sie die Welt verändern könnten ...
– Sie tun und lassen könnten, was Sie wollen, ...
– Sie eines Morgens plötzlich als Mann/als Frau aufwachen würden ...
– Sie eine Million DM gewinnen würden ...
– eines Morgens ein Dinosaurier vor Ihrer Haustür stünde ...
...

11 So etwas tun Sie nicht! Oder vielleicht doch?

B 3 + B 4

1. In welchen Situationen würden Sie das tun? Schreiben Sie bitte!

– ein Geheimnis verraten
– einem Kind eine Ohrfeige geben
– Geld zum Fenster rauswerfen
– Ihre Seele dem Teufel verkaufen
– ein Buch wegwerfen
– Ihren besten Freund/Ihre beste Freundin anlügen
– auf das Dach Ihres Hauses steigen
– ein Fahrrad „stehlen"
...

2. Jede/Jeder schreibt 3 bis 5 ähnliche Situationen auf und legt sie anderen Kursteilnehmern vor.

12 Konjunktiv II: Was wäre gewesen, wenn ... (Vergangenheit)

B 4

1. Wählen Sie aus, und formulieren Sie bitte Ihre Annahmen!
Finden Sie andere Beispiele!

Was wäre gewesen, wenn ...
... der Müller den König nicht angelogen hätte
... Rumpelstilzchen der Müllerstochter nicht geholfen hätte
... der Wolf Rotkäppchen nicht gefressen hätte
... die Menschen nicht aus dem Paradies vertrieben worden wären
... die Deutschen Hitler nicht gefolgt wären
...

2. **Ihr Text:**
Wenn ich das gewußt hätte!

Wie oft denken wir, daß manches in unserem Leben vielleicht anders gelaufen wäre, wenn wir uns anders entschieden hätten, wenn wir manches vorher gewußt hätten, wenn ...

Erinnern Sie sich: Was wäre in Ihrem Leben (wahrscheinlich) anders gelaufen, wenn ...

B 3
+
B 6

13   Wenn ich das nur (auch) könnte!

Was sagen oder denken Sie? Sprechen/Schreiben Sie bitte!

Beispiel:
Ihr Freund kann wunderbar Klavier spielen. Sie möchten auch so gut spielen können.

→   *Wenn ich nur auch so gut (Klavier)spielen könnte!*

Und nun Sie bitte!

1. Sie möchten mehr Zeit für Ihre Hobbys haben.
2. Sie wünschen sich, daß Ihre Freundin/Ihr Freund endlich kommt.
3. Sie müssen **leider** eine Prüfung machen.
4. Sie sind in der Stadt, mit wenig Geld in der Tasche. In einem Schaufenster sehen Sie Ihre Traum-
schuhe. Das Geld reicht nicht, Schecks haben Sie nicht dabei ...
5. Ein Bekannter war schwer krank. Aber Sie haben es nicht gewußt.
6. Ihre Nachbarin hat einen tollen Sportwagen. Sie möchten so
gerne einmal damit fahren, aber sie läßt Sie nicht ans Steuer.
7. ...

Wiederholen Sie alle Wünsche, aber diesmal ohne „wenn".
Z. B.: Könnte ich nur auch so gut (Klavier)spielen!

B 6   **14   Ach hätte ich, ach wäre ich ...! Auch eine Liebesgeschichte**

Lesen Sie bitte! Schreiben Sie bitte!
Er sah sie in der U-Bahn wieder.
Fünfzehn Jahre waren vergangen.
Er war damals nach Amerika ausgewandert. ...
Sie hatte ihn zum Flughafen begleitet. ...
Er hatte ihr nie mehr geschrieben ...

Und plötzlich fühlte er, daß er sie all die Jahre
vermißt hatte.
Und während er sie heimlich ansah, dachte er:

Ach, hätte ich sie doch nicht verlassen!

Ach wäre ich _____

_____

_____

_____

_____

(Warum habe ich sie verlassen?)

(Wie dumm ich damals war!)

(Wir haben nicht über unsere Gefühle
gesprochen.)

(Ich habe sie nicht in den Arm genommen.)

(Ich habe sie nicht geküßt.)

(Ich bin so schüchtern gewesen.)

(Nun ist es zu spät!)

In diesem Augenblick trafen sich ihre Blicke, und sie erkannte ihn. Wie ein Blitz durchfuhren sie
tausend Gedanken. Was für Gedanken? Was meinen Sie?

_____   _____

_____   _____

_____   _____

15 Amizigarett'

1. Nach Aufgabe 1. und 2. von C 1:
Wie passen Bild zwei und drei des Filmstreifens zusammen?
Erzählen Sie bitte die Geschichte!

2. Hören und lesen Sie bitte den kleinen Dialog von C 3, und antworten Sie dann bitte!
– Was machen die beiden Sprecher?
– Wo und wann findet das Gespräch statt?

3. **Lebensmittelmarkt – Schwarzmarkt – Marktwirtschaft:**
Welches der drei Wörter paßt zu diesem Dialog?

16 Ich wäre Ihnen sehr dankbar, wenn ...

1. Nein, so geht es wirklich nicht! Sagen Sie es bitte höflich!

Tür schließen!   Hörer abnehmen!   Buch schicken!
Nicht immer das Benehmen kritisieren!   Dem Kollegen etwas ausrichten!
Das Formular ausfüllen!   Den Text zusammenfassen!
Den Koffer tragen helfen!   Mich an den Termin erinnern!

2. Was sagen Sie?

1. Sie brauchen ein Buch für Ihre Diplomarbeit. Sie fragen die deutsche Lektorin, ob sie es Ihnen besorgen kann.

2. Sie arbeiten in einer deutschen Firma. Sie wollen mit Ihrem Vorgesetzten über eine Gehalts-erhöhung/den Termin für Ihren Jahresurlaub/ein paar Tage Sonderurlaub sprechen.
a) Bitten Sie um einen Termin.
b) Wie beginnen Sie das Gespräch für die Gehaltserhöhung/für den Urlaub?

3. Was sagen Sie, wenn Sie in Deutschland sind?

1. Sie wollen einen Passanten auf der Straße nach der Uhrzeit/nach dem Weg zum Rathaus/zum Heimatmuseum/zur Kongreßhalle/zum Bahnhof ... fragen.

2. Sie sind zum ersten Mal bei Familie X. Sie müssen dringend auf die Toilette.

3. In der Apotheke: Der Apotheker soll Ihnen ein Mittel gegen Kopfschmerzen empfehlen.

4. Anruf bei einem Arzt. Die Sprechstundenhilfe ist am Apparat. Bitten Sie um einen Termin.

5. Im Supermarkt: Sie haben eingekauft. An der Kasse warten viele Leute. Sie haben es aber sehr eilig und haben keine Zeit zu warten.

6. Sie fahren im Schlafwagen von Hamburg nach Wien. Sie sind in einem Abteil mit zwei anderen Personen. Sie wissen aber, daß es Schlafwagenabteile gibt, die nicht belegt sind. Fragen Sie den Schlafwagenschaffner, ob ...

4. Schreiben
Schreiben Sie bitte!

Sie wollen im Sommer einen Sommerkurs an einer deutschen Universität besuchen. Schreiben Sie bitte an den DAAD (Kennedyallee 50, D 5300 Bonn 2), und bitten Sie um die Informationsbroschüre „Sommerkurse an deutschen Hochschulen".

C 7
+
C 8

17 Als hätten sie nichts gewußt!

1. Lesen Sie bitte!

Und als endlich alles vorbei war, da wollte es niemand gewesen sein. Viele Leute sagten:

„Wir haben ja nicht geahnt, was alles passieren würde."

„Das hat man doch nicht wissen können!"

„Wir haben nichts gesehen und nichts gehört."

„Wir waren nicht dabei."

„Wir haben nicht mitgemacht."

„Wir haben nichts damit zu tun."

„Wir haben nichts Böses getan."

„Wir sind ganz und gar unschuldig."

„Es geht uns nichts an, was passiert ist."

2. Schreiben Sie bitte weiter mit den Aussagen unter 1.!

Und als endlich alles vorbei war, da wollte es niemand gewesen sein. Die meisten taten,

als hätten sie nicht geahnt, was alles passieren würde,

als hätte man das nicht

_____

_____

_____

_____

_____

_____

_____

3. Aber zwanzig Jahre später, da kamen die Kinder und fragten: „Was habt ihr damals gemacht?"
Und als die Eltern sagten, sie hätten nichts gewußt, da wurden die Kinder böse und sagten:

Tut doch nicht so, als ob ihr nichts gewußt hättet!

Tut doch

_____

_____

_____

_____

_____

_____

18   Wer höflich ist, tut manchmal so, als ob ...

Sie sind ein höflicher Mensch. Deshalb müssen Sie manchmal so tun, als ob.
Hier ein paar Beispiele für solche Situationen:

Schreiben und ergänzen Sie bitte!
(Finden und bearbeiten
Sie andere Situationen.)

Doch Sie tun, als ob ...

Sie denken:

1. Doch Sie tun, *als ob Ihnen*
   das Essen _____ ,

und sagen:

Die Situation

1. Sie sind eingeladen,
aber das Essen schmeckt
Ihnen nicht.

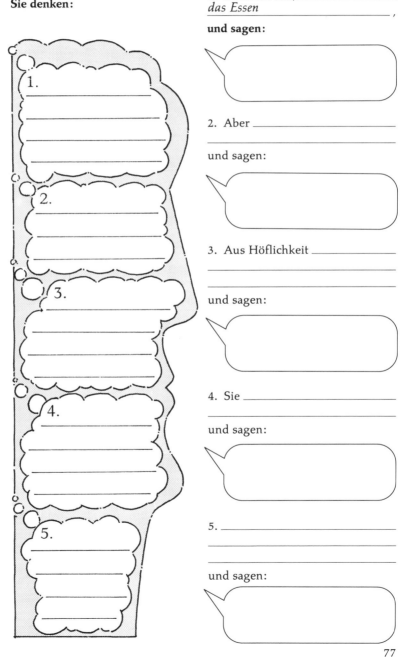

2. Aber _____

und sagen:

2. Zu Ihrem Geburtstag
bekommen Sie eine Vase
geschenkt. Die Vase ge-
fällt Ihnen nicht.

3. Ein unangemeldeter
Besuch steht vor Ihrer
Tür. Es paßt Ihnen gar
nicht, denn Sie müssen
für den nächsten Tag
eine Arbeit fertig-
schreiben.

3. Aus Höflichkeit _____

und sagen:

4. Sie hören etwas, was
Sie eigentlich nicht hören
sollten. Die Gesprächs-
partner schauen Sie fra-
gend an.

4. Sie _____

und sagen:

5. Ein Bekannter, den Sie
nicht besonders mögen,
möchte mit Ihnen ausge-
hen. Sie wollen nicht mit
ihm ausgehen.

5. _____

und sagen:

9

D 2 19 Die fünfziger Jahre

1. „Wunderwörter":

Fräuleinwunder –Wirtschaftswunder – Das Wunder von Bern

Verstehen Sie diese Ausdrücke?

2. Welche Schlagworte aus der Collage von Kursbuch, Seite 141, passen zu den unterstrichenen Ausdrücken im folgenden Text? Schreiben Sie die Schlagworte bitte in die rechte Spalte!

Die Anfänge der Wohlstandsgesellschaft

1 Etwa seit 1952/53 erreicht der	
2 <u>wirtschaftliche Aufschwung</u> auch breitere	*Wiederaufbau*
3 Schichten der Bevölkerung. Die allmähliche	
4 Steigerung des Lebensstandards zeigt sich	
5 in regelrechten <u>Konsumwellen</u>: Einer „Eß-",	
6 „Kleidungs-" und dann „Wohnungswelle"	
7 folgt gegen Ende der fünfziger Jahre	
8 schließlich der <u>Trend zu Reise und Urlaub</u>.	
9 Parallel dazu bekommt <u>das Auto</u> mehr und	
10 mehr <u>die Rolle des zentralen Statussymbols</u>.	
11 Der <u>sich ausbreitende Wohlstand</u> bestimmt	
12 die Grundeinstellung vieler Menschen: Die	
13 <u>Konzentration auf materielle Verbesserungen</u>,	
14 <u>auf Familie und häusliches Leben</u> und der	
15 <u>Stolz auf das Erreichte</u> drängen das Inter-	
16 esse an der Politik und an gesellschaft-	
17 lichen Veränderungen in den Hintergrund.	

D 2 20 Es sieht so aus/Es scheint, als ob ...

1. Lesen Sie bitte!

In den fünfziger Jahren waren die meisten Westdeutschen vor allem mit dem Wiederaufbau ihres Landes beschäftigt:
– Sie wollten nichts mehr von den Verbrechen der Nazi-Zeit hören.
– Sie erinnerten sich nicht mehr an die Leiden des Krieges.
– Sie vergaßen die Opfer.
– Sie hatten nur noch materielle Interessen.
– Sie dachten nur ans Essen, Kaufen, Reisen.
– Ein eigenes Auto, ein eigenes Haus waren ihre wichtigsten Ziele.
– Sie verstanden die heranwachsenden Jugendlichen nicht.
– Sie hörten am liebsten kitschige Schlagermusik.
– Sie interessierten sich nicht besonders für die Politik.
– Sie wollten endlich anfangen zu leben.

2. War es wirklich so, oder sah es nur so aus? Wahrscheinlich hatten viele die Erinnerungen nur verdrängt*.
Verändern Sie bitte die Aussagen oben, und benutzen Sie die Ausdrücke „Es schien, als (ob)…/Es sah aus, als (ob)…

Beginnen Sie bitte so:
In den fünfziger Jahren schienen die Westdeutschen vor allem mit dem Wiederaufbau ihres Landes beschäftigt zu sein. Es schien, als ob sie nichts mehr von den Verbrechen der Nazi-Zeit hören wollten. Es sah aus, …

21 Es war einmal… D 3

Was fällt Ihnen zu diesem Bild ein?
Notieren Sie bitte Stichwörter, und sprechen Sie darüber!

---

* verdrängen

D 22 Wende und Vereinigung (1989 + 1990)

Welcher Text zu welchem Bild? Ordnen Sie bitte zu!

1

2

4

3

5

Bild-Nr.

	Ungarn öffnet seine Grenze zu Österreich.
	40. Jahrestag der Staatsgründung der DDR. Gorbatschow zu Honecker: „Wer zu spät kommt, den bestraft das Leben."
	9./10. November 1989: Öffnung der Mauer in Berlin
	Freundlicher Empfang
	In Leipzig demonstrieren 150 000 für die Vereinigung beider deutscher Staaten.

Hamburg – Aquarell von Hans Nordmann, mit freundlicher Genehmigung des Künstlers und der Poster-Galerie in Hamburg.

Bild-Nr.

1.7.1990: Die Währungsunion tritt in Kraft. Die Bürger der DDR stehen vor den Banken Schlange, um ihr Geld umzutauschen.

Die „Mauerspechte" beginnen ihr Werk.

12.9.1990: Die vier Siegermächte des Zweiten Weltkriegs und die beiden deutschen Staaten unterzeichnen einen Vertrag, in dem Deutschland die volle Souveränität zurückgegeben wird.

3.10.1990: Beitritt der DDR zur Bundesrepublik Deutschland

2.12.1990: Erste gesamtdeutsche Bundestagswahl

D 6   23   Das Grundgesetz der Bundesrepublik Deutschland

1. Lesen Sie bitte!

Das Grundgesetz garantiert

für alle Deutschen:

8	– das Recht auf Freizügigkeit
☐	– das Recht auf freie Berufsausübung
☐	– die freie Wahl des Arbeitsplatzes
☐	– das aktive Wahlrecht
☐	– das passive Wahlrecht

für alle Menschen, die in seinem Geltungsbereich* leben:

☐	– freie Entfaltung und Unverletzlichkeit der Person
☐	– Gleichheit vor dem Gesetz
☐	– Glaubens- und Religionsfreiheit
☐	– Meinungsfreiheit
☐	– Versammlungsfreiheit
☐	– Unverletzlichkeit des Brief-, Post- und Fernmeldegeheimnisses
☐	– Unverletzlichkeit der Wohnung

2. Ordnen Sie diese Rechte und Freiheiten bitte den Angaben im Kursbuch, Seite 144, zu!
Schreiben Sie die entsprechenden Nummern in die Kästchen!

GRUNDGESETZ

für die Bundesrepublik Deutschland

I. Die Grundrechte

Artikel 1

[Menschenwürde, Grundrechtsbindung der staatlichen Gewalt]

(1) Die Würde des Menschen ist unantastbar. Sie zu achten und zu schützen ist Verpflichtung aller staatlichen Gewalt.
(2) Das Deutsche Volk bekennt sich darum zu unverletzlichen und unveräußerlichen Menschenrechten als Grundlage jeder menschlichen Gemeinschaft, des Friedens und der Gerechtigkeit in der Welt.
(3) Die nachfolgenden Grundrechte binden Gesetzgebung, vollziehende Gewalt und Rechtsprechung als unmittelbar geltendes Recht.

Artikel 2

* der Geltungsbereich des Grundgesetzes = Gebiet, in dem das Grundgesetz gilt, also: Die Bundesrepublik Deutschland

24 Deutschland – In der Mitte Europas

D

1. Deutschland und seine Grenzen

Schreiben Sie bitte die Jahreszahlen in die entsprechenden Karten:

1990
1949–1990
1871–1918
1919–1939

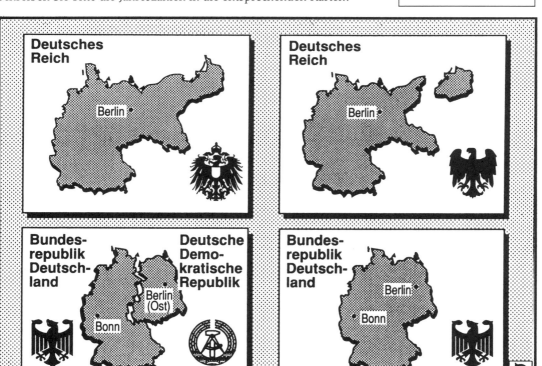

2. Deutschland und seine Nachbarn

Die Bundesrepublik Deutschland hat neun Nachbarländer. Wie heißen sie?
Hier sind die Autokennzeichen dieser Länder:

CH A NL F B
L DK CZ PL

Wie viele Nachbarländer hat Ihr Land? Wie heißen sie auf deutsch?

3. Drei große europäische Ströme

Wie heißen sie? Durch welche Länder fließen sie?

_____ : CH — D + F — NL → Nordsee

_____ : D — A — H — YU — R → Schwarzes Meer

_____ : CZ — D → Nordsee

25 Geschichte in „Zeitsätzen"

Das Gedicht von Rudolf Otto Wiemer erfaßt das Leben einer ganzen Generation in Deutschland seit dem Beginn des zwanzigsten Jahrhunderts.

Notieren Sie bitte a) die entsprechende Jahreszahl
b) den historischen Begriff oder ein passendes Schlagwort für die Epoche!

Jahreszahl	Schlagwort für die Epoche	Zeitsätze
_____	_____	Als wir sechs waren, hatten wir Masern.
1914	*Erster Weltkrieg*	Als wir vierzehn waren, hatten wir Krieg.
_____	_____	Als wir zwanzig waren, hatten wir Liebeskummer.
_____	_____	Als wir dreißig waren, hatten wir Kinder.
_____	_____	Als wir dreiunddreißig waren, hatten wir Adolf.
_____	_____	Als wir vierzig waren, hatten wir Feindeinflüge.
_____	_____	Als wir fünfundvierzig waren, hatten wir Schutt.
_____	_____	Als wir achtundvierzig waren, hatten wir Kopfgeld.
_____	_____	Als wir fünfzig waren, hatten wir Oberwasser.
_____	_____	Als wir neunundfünfzig waren, hatten wir Wohlstand.
_____	_____	Als wir sechzig waren, hatten wir Gallensteine.
_____	_____	Als wir siebzig waren, hatten wir gelebt.

Rudolf Otto Wiemer

9 26 Leselust zum Thema (fakultativ)

Wie haben Kinder die deutsche Einheit erlebt?

Die Journalistin Annegret Hofmann war zwischen August 1990 und Dezember 1991 mit Kassettenrecorder und Notizblock unterwegs und hat mit Kindern gesprochen. Daraus ist ein Buch entstanden: Annegret Hofmann: Unterwegs nach Deutschland. Kinder im Niemandsland. (Aufbau Verlag, Berlin und Weimar 1992.) Im folgenden zwei Berichte aus diesem Buch.

Vergleichen Sie: früher – jetzt; Tomi – Maria.

Tomi, 12, ist 1990 mit seinen Eltern – Vater Handwerker, Mutter Kassiererin – von Ost- nach Westberlin übergesiedelt.

Ich bin jetzt der dritte Schüler aus dem Osten in unserer Klasse, ich denk, da gibt's jetzt keine Unterschiede mehr. Wir sind doch alle Berliner. Manche haben schon vergessen, daß ich aus'm Osten bin.

Am Anfang hatte ich manche Worte nicht gekannt. Affengeil zum Beispiel. Das benutz ich aber jetzt auch, und auch mein Freund in Hohenschönhausen* kennt das jetzt.

Die Schule hier gefällt mir besser, da ist es irgendwie nicht so streng. Wir hatten drüben mehr Disziplin. Wenn Mathe war, durften eben nur die Mathebücher auf dem Tisch liegen. Hier ist alles wild durcheinander.

Wir wohnen hier auch im Neubau, aber der ist besser als der in Hohenschönhausen. Mein Zimmer ist größer, und es ist nicht immer so überheizt. Die Küche hat ein Fenster, das war dort auch nicht so.

Meine Mutter hat erst gedacht, wir dürften hier unseren Hund, den Asko, nicht halten, ging aber. Der rennt immer mit mir los, wenn wir radfahren. Meistens in der Königsheide. Im Osten. Da war ja früher mal die Grenze. Manchmal sieht man noch was davon. Auf den Grenztürmen kann man schön spielen.

Mein Vater hat viel Arbeit, er hat wenig Zeit für uns. Mutti macht auch oft Überstunden. Das war früher nicht so, da war sie immer um drei zu Hause.

Ich hab viele Freunde, nicht nur einen. Was man so mit denen macht: radfahren, Fußball spielen, Comics austauschen und Gameboyspiele. Ich hab nicht sehr viele, manche aber doch.

Urlaubmachen ist hier schöner. Man kann weit wegfahren. Wir waren in Spanien. Das war gar nicht teuer. Das hat die Einheit gebracht.

Früher war man eingesperrt.

Man Vater sagt, das wichtigste ist jetzt erst mal, viel Geld zu verdienen. Dann kann man sich was leisten. Er war schon immer Handwerker, aber er hatte keinen Betrieb. Jetzt auch nicht, aber er will sich selbständig machen. Vielleicht in Ostberlin, da gibt's viel Arbeit, sagt er.

---

* Hohenschönhausen: Stadtteil in Ostberlin

Meine Schwester hat im Osten eine Wohnung. Sie ist nicht mit umgezogen. Die wollte nicht. Mein Vater redet jetzt nicht mehr mit ihr. Ich kann sie auch nicht besonders leiden. Ich glaube, die ist jetzt arbeitslos.

Alles, was ich da schon im Kinderzimmer hatte, hab ich mitgebracht. Jetzt hab ich aber noch viel mehr, mehr Spiele, mehr Comics und so. Alles voll. Sonst hab ich nichts mitgebracht, die Schulbücher nicht. Die hab ich in den Mülleimer geworfen. Das Pionierbuch auch. Ich bin da nie gern hingegangen. Das war mehr so Zwang. Hier ist man freier, ehrlich. Ich find's echt affengeil hier.

Maria, 12, besuchte in Ostberlin eine Sprachspezialklasse für Russisch, ihre Eltern sind Wissenschaftler, sie hat einen behinderten Bruder.

Mein Zimmer hab ich jetzt ganz anders eingerichtet als früher. Die Puppen sind weg, auch viele Spiele hab ich verschenkt. Ich hab jetzt eine DDR-Fahne, so eine mit Emblem, an der Wand. Auch mein Pionierbuch hab ich noch und die Abzeichen. Behalt ich alles. Bestimmt für immer.

Viele Menschen erwarten jetzt von einem, daß man sagt – mir gefällt es jetzt besser. Das kann ich aber nicht sagen. Ich hab mich früher wohler gefühlt. Sicherer. Ich hatte alles, was ich brauchte, und die anderen Sachen, die von drüben, die hab ich nicht gebraucht.

Als ich das erstemal drüben war, fand ich eigentlich alles genauso wie bei uns. Häuser gleich, Bäume gleich, nur eben die Geschäfte. Viele haben das schau* gefunden. Ich geh da jetzt nicht mehr rüber.

Was soll ich da?

Meinem Vater ist zum Ende des Jahres gekündigt worden. Bei meiner Mutter ist der Arbeitsplatz auch unsicher. Vieles zu Hause ist anders als früher. Wir müssen sparen. Dabei haben wir früher auch nicht üppig gelebt. Aber jetzt weiß man nicht, wie es weitergehen soll.

---

* schau: ugs für „toll", „großartig"

27 Kontrollaufgaben (Wiederholung)

1. Wortschatz: Welche Definitionen passen?

1 Sie ist eine geborene Bäcker.
2 Sie ist eine geborene Lehrerin.
3 Er ist gelernter Bäcker.
4 Er hat Bäcker gelernt.
5 Man wird nicht als Bäcker geboren.
6 Sie ist ungelernte Arbeiterin.
7 Sie wurde im Betrieb angelernt.

Das ist die Definition für:

☐ Sie ist eine sehr gute Lehrerin.
☐ Sie ist schon bei der Geburt wie eine Lehrerin gewesen.
☐ Ihr Mädchenname ist Bäcker.
☐4 Er hat das Bäckerhandwerk gelernt.
☐ Er hat den Beruf Bäcker gelernt.
☐ Den Bäckerberuf muß man lernen.
☐ Sie hat die Tätigkeit, die sie im Betrieb ausübt, dort gelernt.
☐ Sie hat keinen Beruf gelernt.
☐ Sie hat im Betrieb einen Beruf gelernt.

2. Konjunktiv II

Der Konjunktiv II steht

a) als Ausdruck von Höflichkeit
b) als Ausdruck einer Möglichkeit
c) als Hinweis auf ein Geschehen in der Vergangenheit, das aber nicht passiert ist.

Welche Bedeutung trifft zu?
Kreuzen Sie bitte an!

	a	b	c
1. Könnten Sie mir bitte die Speisekarte geben?			
2. Wir hätten nicht mitmachen dürfen.			
3. Ich hätte noch eine Idee.			
4. Könnten Sie mir bitte beim Ausfüllen des Formulars helfen?			
5. Ihr könntet ihn fragen.			
6. Wären wir nur klüger gewesen!			
7. Da wäre ich vorsichtig.			
8. Das hätten Sie einfacher haben können!			
9. Hätten Sie heute Zeit für mich?			
10. Das wäre besser gewesen.			
11. Bitte, ich wüßte gerne, wann die Kasse geöffnet wird!			
12. Wir könnten dir dabei helfen.			
13. Würden Sie mir bitte noch einen anderen Pullover zeigen?			

3. Eine undurchsichtige Geschichte: Irrealer Vergleich. Verbinden Sie bitte die beiden Sätze!

Klappt es? – Es sieht nicht so aus. _____

Ist er krank? – Es scheint so. _____

Interessiert es ihn überhaupt? – Er tut so. _____

Weiß seine Frau nicht Bescheid? – Sie verhält sich so. _____

Kommt da jemand? – Es klingt so. _____

Wird es ein Erfolg? – Wir tun so. _____

Müssen wir die anderen informieren. – Es sieht so aus. _____

1 Auftaktseite: Worum geht es hier?

1. Auf der Auftaktseite sind die Umschlagseiten* von vier Büchern abgedruckt. Lesen Sie bitte die Texte, und überlegen Sie gemeinsam, worum es in dieser Lektion wohl geht.
Schreiben Sie dann bitte:

Hier geht es ganz generell um _____

2. Sagen Sie es jetzt bitte präziser!

Im „Knigge" geht es um _____

Im „Duden" _____

Im Buch unten links _____

Im BGB _____

3. Kennen Sie andere (Lebens-)Bereiche, die durch Vorschriften geregelt sind?

2 Normen u. ä.*: Einstieg in den Text

 A 1

**das Gesetz – die Regel – der Maßstab –
die Vorschrift – die Konvention – die Norm – die Sitte**

Alle diese Wörter finden Sie im Text von A 1, Kursbuch S. 146.

1. Überlegen Sie bitte: Welches Wort könnte der Oberbegriff für alle anderen sein?

2. Lesen Sie bitte den Text von A 1, und unterstreichen Sie diese Begriffe jedesmal, wenn sie vorkommen! Welcher Begriff wird im Text als Oberbegriff gebraucht?

3. Lesen Sie den Text bitte noch einmal, und ergänzen Sie die Tabelle:

a) Welche Anwendungsbereiche/Beispiele werden **im Text** für die einzelnen Begriffe genannt?
b) Diskutieren und finden Sie bitte **weitere Anwendungsbereiche**/andere Beispiele!

Normen					
	techn. Normen	Regel(n)	Vorschrift(en)	Konvention/Sitte	Gesetz(e)
Beispiele aus dem Text:	*DIN*				
Andere Beispiele:		*Fußball*			

---

4. Überlegen Sie bitte!

a) Wie nennt man Verstöße gegen die Norm in den verschiedenen Bereichen?

Zum Beispiel: Fußball: Foul

              Sprache: _____

b) Wer achtet darauf, daß die Regeln eingehalten werden?

Zum Beispiel: Fußball: Schiedsrichter

              Straßenverkehrsordnung: _____

              Sprache: _____

5. **Diskussion:**

> Die Normverstösse von heute
> sind die Normen von morgen!

Suchen Sie bitte Beispiele für diese Aussage aus der Gegenwart und Vergangenheit Ihres Landes!

A 3   3   Ein Telegramm für Petra

1. Lesen Sie bitte den Telegrammtext! Was fällt Ihnen auf? Erklären Sie bitte!

| TEL |

DEUTSCHE BUNDESPOST   **Telegramm**

| Datum | Uhrzeit | Empfangen von | | Vermerke/Verzögerungsvermerke | Datum | Uhrzeit |

Platz | Empfangen Namenszeichen | | | Aufgabe-Nr. | Wortzahl | | Gesendet Platz | Namenszeichen | Aufgabetag | Uhrzeit |

Bezeichnung der Aufgabe-TSt
aus **Hamburg/90**

▼ Gebührenpflichtige Dienstvermerke

▼ Name des Empfängers, Straße, Hausnummer usw.

Petra Böhlmann, Am Weiher 17

▼ Bestimmungsort – Bestimmungs-TSt

2100 Hamburg 90

Abfliege Mailand 15.30 Uhr mit LH 503.
Ankomme Hamburg 17.45 Uhr. Vorschlag
treffen Hotel Elysée. Abendessen
19.30 Uhr. Dr. Müller

Die stark umrahmten Teile sind vom Absender auszufüllen. Bitte Rückseite beachten.

| | DM | Pf | Auf ungenügende Anschrift / Besonderheiten / Dienstzeit hingewiesen | Absender (Name und Anschrift, ggf. Ortsnetzkennzahl und Telefonnummer, diese Angaben werden nicht mittelegrafiert) |

Feste Gebühr | | |

Wort | | Angenommen |

2. Wie würden Sie die Information formulieren, wenn Sie sie auf einer Postkarte oder in einem Brief schreiben würden?

88

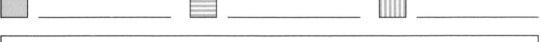

4 Värschtahtöpperkaimundaart? Deutsch in der Schweiz   **B**

1. Schreiben Sie bitte, um welches Sprachgebiet es sich handelt:

☐ _____   ▤ _____   ▥ _____

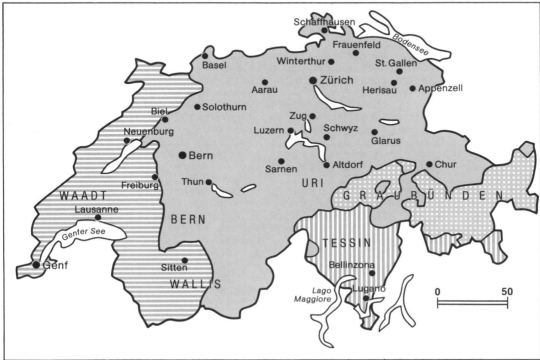

2. Suchen Sie bitte die **Information im Text auf Seite 90!**   ▦ rätoromanisch [LV]

a) „Värschtahtöpperkaimundaart?" heißt auf Hochdeutsch: _____

_____

b) Ergänzen Sie bitte mit **Informationen aus dem Text!**

Die deutsche Hochsprache wird in der Schweiz _____ genannt.

Sie wird nur _____ gebraucht. Wenn

die Schweizer miteinander reden, dann sprechen sie _____, und

zwar den _____ ihrer Region.

Obwohl die Bundesverfassung als Landessprache _____ vor-

schreibt, wird auch in der Schule, in der Wirtschaft, an der Universität und in

den meisten Fernseh- und Rundfunksendungen _____ gesprochen.

Es gibt kein einheitliches Schweizerdeutsch, es gibt viele verschiedene

_____. Für die meisten Schweizer gilt, was Dürrenmatt so aus-

gedrückt hat: Der Dialekt ist meine _____, Hochdeutsch ist meine

_____.

Deutsch in der Schweiz

Von Irmgard Locher

„Ich habe mir vorgenommen, jetzt endlich Schweizerdeutsch zu lernen", sagt der deutsche Tourist, der jedes Jahr seinen Urlaub in der Schweiz verbringt. Das „Grüezi mitenand" beherrscht er schon nahezu perfekt und amüsiert sich königlich, wenn er beim Wandern in den Schweizer Bergen von seinen eigenen Landsleuten für einen Eidgenossen gehalten wird. „Die antworten dann mit ‚Tach' ", strahlt er.

Die Sache mit seinem Lerneifer hat nur einen Haken: Es gibt gar kein Schweizerdeutsch, sondern nur eine Vielzahl teils sehr unterschiedlicher Dialekte. So kann es einem Ostschweizer durchaus passieren, daß man ihn in einem abgelegenen Tal des Berner Oberlands für einen Ausländer hält. ...

Der Dialekt hat in der deutschsprachigen Schweiz seit Jahren schon Konjunktur. Die Mundartwelle rollt und rollt und verdrängt die Hochsprache immer mehr. Wenn ein Heimattheater, ein Liedermacher oder ein Kabarettist dem Volk aufs Maul schaut, wenn die Werbung sich durch Dialekt beim Konsumenten einzuschmeicheln versucht, ist das noch zu verstehen, wenn aber auch in den elektronischen Medien vorwiegend Dialekt gesprochen wird, dann kann das die gesamtschweizerische Kommunikation doch ernstlich beeinträchtigen, ganz abgesehen von den Ausländern diesseits und jenseits der Grenze, die mit den Programmen ihre liebe Mühe haben. Vor vier Jahren wurde in Funk und Fernsehen bereits zu 60 Prozent Dialekt geredet. Inzwischen dürfte die Prozentzahl merklich in die Höhe geschnellt sein, denn seither kamen an die 30 Lokalradios dazu, in denen kein einziges Wort Hochdeutsch mehr zu hören ist. ...

Auch in Schule, Kirche, Wirtschaft, Politik und Militär macht sich die Mundart breit und breiter. Dies, obwohl die Bundesverfassung Deutsch als Landessprache vorschreibt, nicht aber Schweizerdeutsch. Schriftdeutsch, wie die Hochsprache in der Schweiz genannt wird, ist in der Praxis nur eine Sprache zum Schreiben – und natürlich zum Lesen –, aber nicht zum Sprechen. Bei Veranstaltungen, an denen auch Eidgenossen aus den anderen Sprachgebieten oder Ausländer teilnehmen, passiert es immer wieder, daß ein Vortragender fragt: „Värschtahtöpperkaimundaart?", bevor er im vertrauten Dialekt weiterredet. Übersetzt heißt das: „Versteht jemand keine Mundart?" Die Französisch- oder Italienisch-Sprechenden verstehen die Frage gar nicht und können sich schon allein deshalb nicht wehren. Sie, die schon mit dem für sie schweren und komplizierten Hochdeutsch Mühe genug haben, fühlen sich bei solchen Gelegenheiten ausgestoßen und kommen sich im eigenen Land wie Bürger zweiter Klasse vor.

Noch vor einem Jahrzehnt galt es als selbstverständliche Höflichkeit unter den Deutschschweizern, sofort vom Dialekt auf Hochdeutsch umzuschalten, sobald sich Eidgenossen anderer Zunge oder Ausländer zu ihnen gesellten. Doch das ist heute nicht mehr Mode. Lieber weichen sie ins Englische aus, statt sich der ungewohnten Schriftsprache zu bedienen; am deutlichsten zeichnet sich dieser Trend bei Technikern und Wissenschaftlern ab. Sogar an den Universitäten und Hochschulen wird mehr und mehr Dialekt gesprochen, von den Professoren und Assistenten in den Seminaren ebenso wie von den Studenten. „Sie möchten wohl Hochdeutsch mit mir reden", sagt eine Genfer Studentin in Bern, „aber ihnen fehlen dann einfach die Worte; sie sind wie blockiert." ...

Von dieser Schwierigkeit bleiben allerdings nicht einmal die Dichter und Denker der Schweiz verschont; auch für die kommt der Dialekt aus dem Bauch, das Hochdeutsche dagegen aus dem Kopf. Der Berner Friedrich Dürrenmatt* drückt das so aus: „Berndeutsch ist für mich die Muttersprache, Hochdeutsch die Vatersprache."

Stuttgarter Zeitung 8.1.1988 (gekürzt)

3. Wie werden die Schweizer im Text auch genannt? Fragen Sie Ihre Lehrerin/Ihren Lehrer, was dieser Name bedeutet.

---

* Friedrich Dürrenmatt, deutschsprachiger Schriftsteller aus der Schweiz (1921–1990).

5 Information: In Österreich heißt der Blumenkohl Karfiol   B

Das Deutsch, das in Österreich gesprochen wird, unterscheidet sich nicht so sehr vom Hochdeutschen. Die Aussprache klingt ein bißchen wie Bayerisch.
Unterschiede gibt es vor allem bei den Lebensmitteln und auf der Speisekarte.

Kleines Glossar für Interessierte:

Deutsch	Österreichisch
der Blumenkohl	der Karfiol
die Tomate	der Paradeiser
das Hörnchen	das Kipferl
das Brötchen	die Semmel
die Sahne	Obers
die Süßspeise	die Mehlspeise
die Kartoffel	der Erdapfel
der Quark	der Topfen
das Abendessen	das Nachtmahl
die Aprikose	die Marille
das Bonbon	das Zuckerl

6 Begrüßung und Verabschiedung in D, A, CH   B

Schreiben Sie die Grußformeln bitte in die Tabelle!

Grüezi mitenand!   Tag wohl!   Ciao!   Tach!

Tschüß!   Grüezi!   Guten Tag!

Griaß Sie!   Auf Wiedersehn!   Grüß Gott!   Salü!

Uf Wiederluega!   Servus!   Pfiat Di!   (Auf) Wiederschaun!   Läb wohl!

	(Hoch)Deutsch D	Österreichisches Deutsch A	Schweizer Deutsch CH
Begrüßung			
Verab-schiedung			

C 1   7   Was man tun und lassen soll: soziale Konventionen

In einem neueren deutschen „Benimmbuch" (aus dem Jahr 1987) findet man einige „Goldene Regeln"
für höfliches und korrektes Benehmen in der Öffentlichkeit aufgelistet.

1. Überlegen Sie bitte: Welche sprachliche Form haben die meisten Regeln? Welche sprachliche
Form könnte man noch benutzen? In welcher sprachlichen Form werden solche Regeln in Ihrer
Sprache formuliert?

2. **Diskussion**
Finden Sie die Regeln sinnvoll? Benehmen sich die Deutschen, die Sie kennen, entsprechend?
Welche Regeln sind vielleicht ein bißchen altmodisch und gelten möglicherweise nur für ältere
Leute? Gibt es ähnliche Regeln bei Ihnen? Welche anderen Regeln gibt es bei Ihnen? Gibt es bei
Ihnen auch „Benimm-Bücher"?

1979

1987

 3. Fragen Sie Deutsche, wie sie die Regeln finden?

Goldene Regeln

* *Man denkt nicht an sich selbst zuerst,*
  sondern man nimmt auf andere Rücksicht.

* *Man läßt anderen Leuten den Vortritt, im Haus,*
  im Bus, auch in der Warteschlange,
  wenn man merkt, jemand hat es eilig.

* *Man drängelt sich nicht vor.*
  In einer Schlange stellt man sich hinten an.

* *Ein Herr redet mit einer Dame nicht mit*
  den Händen in den Hosentaschen, nicht mit
  einer brennenden Zigarette im Mund.

* *Ein Herr steht auf,*
  wenn eine Dame das Zimmer betritt.

* *Wenn man mit jemand spricht,*
  schaut man ihm in die Augen.

* *Man sitzt nicht mit weit gespreizten Beinen,*
  auch nicht als Mann oder Mädchen in Hosen.

* *Man sitzt gerade und liegt nicht halbwegs auf dem*
  Rücken im Stuhl.

* *In der Öffentlichkeit kämmt und schminkt man sich*
  nicht.

* *Man fragt nicht „Was?" sondern „Wie bitte?",*
  wenn man etwas nicht verstanden hat.

* *Man fällt keinem ins Wort, sondern wartet höflich,*
  bis der Redner/die Rednerin ausgeredet hat.

* *Man spricht nicht mit vollem Mund.*

* *Was auf keinen Fall erlaubt ist:*
  Lautes Lachen mit aufgerissenem Mund; angeben;
  Neugier zeigen und nach dem Einkommen fragen; lange
  Krankengeschichten erzählen; falsch verwendete Fremd-
  wörter; jemand lange anstarren; schmatzen und
  schlürfen bei Tisch; aufstoßen nach dem Essen, ...

(Nach: Sybil Gräfin Schönfeldt: 1×1 des guten Tons. Das neue
Benimmbuch. rororo Sachbuch 8887. Reinbek 1987)

4. Machen Sie eine Liste:
Wie benehmen sich Leute in Deutschland, die bei vielen als unhöflich gelten?
Zum Beispiel: Sie nehmen nicht auf andere Rücksicht.

5. Oder: Machen Sie **Rollenspiele**:
Der eine/Die eine verhält sich unkorrekt (z. B. kämmt und schminkt sich in der Klasse).

a) Reagieren Sie höflich!
b) Reagieren Sie empört!

6. **Ihr Text:** Setzen Sie sich in kleinen Gruppen zusammen, und stellen Sie „Zehn goldene Regeln"
für gutes Benehmen in Ihrem Land auf!

Stellen Sie dann in der Klasse Ihre Regeln vor! Haben alle ähnliche Regeln? Diskutieren Sie bitte,
und einigen Sie sich in der Klasse auf zehn bis maximal fünfzehn Regeln.
Schreiben Sie das Ergebnis auf ein großes Papier und hängen Sie es an die Wand!

Zehn goldene Regeln für gutes Benehmen in ...

7. Sie können Übung 4. und/oder 5. auch für Ihr Land machen. Z. B.: Ein Deutscher/Eine Deutsche
verhält sich unkorrekt. Wie reagieren Sie? Was sagen Sie?

C 2 8 Der Ton macht die Musik und:
Auf die Situation kommt es an

Arbeiten Sie bitte in kleinen Gruppen! Überlegen Sie:

Was drücken die folgenden Sätze aus oder:
Was können sie in bestimmten Situationen ausdrücken?
Sprechen Sie sich die Sätze laut vor
(probieren Sie ruhig verschiedene Intonationsmöglichkeiten
aus), und stellen Sie sich bestimmte Situationen vor,
in denen man diese Sätze sagen könnte. Kreuzen Sie dann an!

a) Aufforderung an den Hörer
b) Aufforderung, die auch für den Sprecher gilt
c) Begrüßungsform
d) Verabschiedung
e) Drohung
f) Erlaubnis
g) Redeeinleitung

Die Sätze

Was die Sätze ausdrücken:

Satz	a	b	c	d	e	f	g
Gehen Sie jetzt bitte!							
Gute Nacht!							
Fassen Sie mich ja nicht an!							
Liebe Zuhörer!							
Setzen Sie sich doch!							
Komm rein!							
Verschwinde! Sonst...							
Gehen wir!							
Hallo!							
Lassen Sie mich jetzt bitte allein!							
Guten Abend!							
Lassen Sie mich in Ruhe!							
Liebe Anwesende!							
Guten Tag!							
Bis bald!							
Laß uns anfangen!							
Welch eine Überraschung!							
Hör mal zu!							
Treten Sie bitte ein!							
Sprechen Sie bitte etwas lauter!							
Auf Wiedersehen!							
Darüber müssen wir noch diskutieren!							
Tschüß!							
Sag mal! Was soll das!?							
Schließ bitte das Fenster!							
Ciao! (Tschau!)							
Sehr geehrte Damen und Herren!							
Halten Sie mich bloß nicht für so dumm!							
Werte Kolleginnen und Kollegen!							
Kommt, machen wir Schluß!							
Grüß dich!							
Lieber Hans!							

9 Aufforderungen: Auf das Wie kommt es an!

Wie erreiche ich, was ich will?
Schreiben Sie bitte die nach Ihrer Meinung erfolgreichsten Aufforderungen in die Sprechblasen!

10 Liebesgeschichte für [s]: Rechtschreibung

Ergänzen Sie bitte – s, – ss oder ß!

Als ich in das Zimmer trat, sah ich sie. Sie sa_____ in einem hohen, schönen Se_____el und la_____.
Ich war sofort wieder hingeri_____en, so sü_____ sah sie au_____. Gewi_____ liebt sie mich noch
nach all den Jahren, dachte ich, aber ich wollte es genau wi_____en. Ich fiel vor ihr nieder, kü_____te
ihre Fü_____e und sprach: „Bewei_____e, da_____ du mich lieb_____t und kü_____e mich." Sie gab
mir einen Ku_____, und wir beschlo_____en, zu_____ammen e_____en zu gehen.

In einem Landga_____thaus a_____en wir Gemü_____e und Nu_____kuchen aus feinen Nü_____en
und tranken Wein aus hohen Glä_____ern. Plötzlich fing sie heftig an zu weinen, bi_____ ihre
Na_____e ganz na_____ war und sprach: „La_____ mein Herz, o la_____ mich sein, ich mu_____ dich
verla_____en." „Was, du willst mich versto_____en? Blo_____ nicht!" Sie sprach: „Doch, es mu_____
sein, wir mü_____en uns trennen. Auch mir zerrei_____t es das Herz. Du mu_____t jetzt mutig sein:
Du wei_____t es noch nicht, aber ich soll einen König heiraten." „Wie hätte ich da_____ wi_____en
sollen, ich war doch bis vor kurzem noch auf Rei_____en. Jetzt fühl' ich mich ganz zerri_____en!
Sei wenigstens gro_____zügig und gib mir deine Adre_____e!"

So geschah es. Von Zeit zu Zeit sehen wir uns
bei einem Fe_____te_____en auf einem der
Schlö_____er des Königs. Dann grü_____en
wir uns heimlich mit den Augen.

D 11 Sonntag ist Ruhetag: Satzzeichen

Setzen Sie bitte alle fehlenden Satzzeichen ein!

Aufstehen es ist schon spät hörte ich meine Mutter rufen Ich antwortete Heute
ist doch Sonntag warum muß ich auch heute so früh aufstehen Susanne hat
recht rief mein Vater wir bleiben heute ja zu Hause Meine Mutter meinte
dann aber Das ist kein Grund einfach faul im Bett liegen zu bleiben Wollen
wir spazierengehen fragte ich die Großmutter Ich weiß nicht antwortete sie es
ist heute so kalt Wenn du willst sagte mein Bruder gehen wir heute abend
ins Konzert.

' ! , . ? : ; - „ "

12 Nicht ganz ernstgemeinter Brief an die Autoren von „Sprachbrücke": Satzzeichen

Diesen Beschwerdebrief hat eine Gruppe von Deutschlernenden geschrieben, die Schwierigkeiten mit dem Komma hatte.

1. Lesen Sie bitte den Brief, und setzen Sie die Kommazeichen!

_____ den _____

An die Autoren des Lehrwerks „Sprachbrücke"!

Sehr geehrte Damen und Herren

nachdem wir nun schon fast zwei Jahre lang mit Ihrem Buch Deutsch gelernt haben stellen wir fest daß Sie uns das Wichtigste erst am Schluß erklärt haben nämlich die Zeichensetzung und ganz besonders die Sache mit dem Komma.

Nun ist es fast schon zu spät wie Sie an diesem Schreiben sehen können denn „Was Hänschen nicht lernt lernt Hans nimmermehr". Diesen Brief den Sie hoffentlich ohne allzu große Mühe lesen können auch wenn sämtliche Kommas fehlen schreiben wir damit Sie das nächste Mal wenn Sie ein Lehrwerk schreiben schon viel früher damit anfangen auf die Bedeutung des Kommas hinzuweisen. Einige von uns haben nämlich diesen kleinen Strich bisher kaum beachtet und sie fragen sich ob man ihnen diesen Fehler jemals verzeihen wird. Andere haben ihn zwar von Zeit zu Zeit gesetzt aber wenn man nicht weiß wie und wo und wann dann nützt es auch nichts.

Es ist ja nicht nur eine Frage der Rechtschreibung und der Bedeutung sondern auch eine Frage der Ästhetik. So ein Text sieht einfach mit Kommas an der richtigen Stelle besser aus als ohne Kommas oder etwa nicht? Wir hoffen nun daß sich jemand findet der uns zeigen kann wo wir die Kommas in diesem Brief setzen müssen.

Zum Schluß bleibt uns nur noch zu sagen daß wir im übrigen mit Ihrem Buch sehr zufrieden waren und daß es anstrengend aber auch schön war damit zu arbeiten.

In der Hoffnung Ihnen eines Tages selbst zu begegnen grüßt Sie recht herzlich Ihre Deutschgruppe aus ...

---

2. Zählen und vergleichen Sie bitte: 30 Kommas müssen es sein.

D 13 Und noch einmal: Komma, Komma

1. Wortschlangen

Bestimmen Sie bitte die Wortgrenzen, und setzen Sie die Satzzeichen!

1. UMTEXTEVERSTEHENZUKÖNNENMÜSSENNICHTNURDIEWÖRTER
BEKANNTSEINSONDERNAUCHDIEPUNKTEUNDKOMMASRICHTIGGESETZT
SEINWEILSONSTALLESWIEEINGROSSERWORTSALATAUSSIEHT

2. ZUMVERSTÄNDNISDERBEDEUTUNGEINESSATZESISTDERPLATZDES
KOMMASWICHTIGWEILSEHROFTDERPLATZDESKOMMASNICHTNURDAS
LESENERLEICHTERTSONDERNAUCHDIEBEDEUTUNGEINESSATZES
ÄNDERNKANN

2. Schachtelsätze

Lesen Sie bitte die folgenden Schachtelsätze!

a) Wenden Sie bitte die Merksätze von 1. an, und setzen Sie Kommas!

b) Drücken Sie die Inhalte dieser drei Sätze in kürzeren, einfacheren Sätzen aus.

1. Der Brief von mir geschrieben wurde nachdem ich ihn einem Bekannten mit der Bitte ihn sofort aufzugeben mitgegeben hatte leider erst nach Tagen zur Post gebracht.

2. Mein Onkel ein reicher Mann schenkte noch kurz vor seinem Tod den größten Teil seines hart erarbeiteten Vermögens nicht seinen Verwandten sondern einer älteren Dame die immer für ihn gesorgt hatte.

3. Alles was sie hatte mußte verkauft werden weil sie ohne überhaupt zu überlegen immer Karten spielte und einmal als sie wieder verloren hatte sehr große Schulden machen mußte.

Schachtelsätze sind Sätze, in die (zu) viele Inhalte mit immer neuen Ein schüben und Nebensätzen ineinander ge schach telt werden. oder aneinandergereiht

14 (Wort)s(ch)atzrätsel: Erkennen Sie mich?

1. Ich trenne, was getrennt werden muß.
2. Ich markiere eine Pause oder einen eingeschobenen Gedanken.
3. Ich möchte, daß jetzt gleich etwas getan wird.
4. Ich zeige, daß ich etwas nicht weiß, aber daß ich glaube, daß Sie es wissen.
5. Was zwischen mir steht, ist nicht so wichtig.
6. Ich bin weder Fisch noch Fleisch, bin mehr als 1. und weniger als 10.
7. Ich bin immer vorn und hinten mit dabei, wenn jemand etwas sagt.
8. Ich bringe vieles zusammen, z. B. daß-Sätze.
9. Hoppla, nach mir kommt was Wichtiges, passen Sie gut auf!
10. Ich bin der Schluß.

11. Wir alle zusammen sind das Lösungswort.

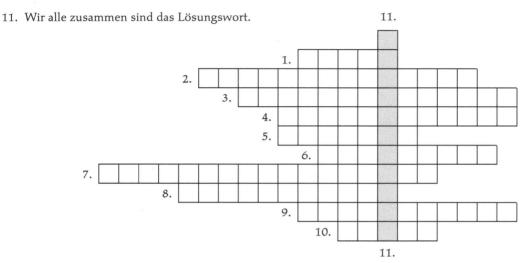

15 Hägar der Schreckliche

Übertragen Sie bitte die Texte des Cartoons in Ihre Muttersprache!

1. Lautmalende Wiedergabe von Geräuschen – nur eine Konvention?

a) Wie werden Geräusche wahrgenommen und schriftlich wiedergegeben? Ordnen Sie bitte die Wiedergabe der Geräusche „auf deutsch" den abgebildeten Gegenständen und Handlungen zu! Schreiben Sie die passenden Nummern in die Kästchen!

b) Wie werden diese Geräusche in Ihrer Sprache schriftlich wiedergegeben?

Auf deutsch

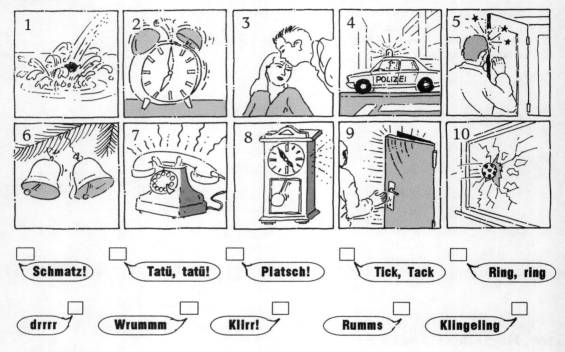

Schmatz!   Tatü, tatü!   Platsch!   Tick, Tack   Ring, ring

drrrr   Wrummm   Klirr!   Rumms   Klingeling

In Ihrer Sprache

2. Wann sagen (schreiben) die Deutschen **was?**

Schreiben Sie bitte den „sprechenden" Ausdruck in die Sprechblasen!
Was sagen Sie in Ihrer Sprache?

> **Au! Aua! Pssst!**
> **Ha ha ha!**
> **brrrr Igittigitt!**

17   Es, es, es und es

Das kleine Wörtchen „es" macht häufig Schwierigkeiten. Deshalb sollen hier einige Anwendungen
noch einmal gezeigt und geübt werden.

1. „es"-Pronomen

Ergänzen Sie bitte!
Hast du mein Buch gesehen? – Hier liegt _____ doch.

Wo liegt denn Feuerland? – Ich weiß _____ nicht.

2. Wetter: Schreiben Sie bitte Wetterverben und Wetterausdrücke!

Es regnet.   *Es*_____   _____   _____

Es ist kalt.   *Es*_____   _____   _____

3. Uhrzeit, Tages- und Jahreszeiten

Wie spät ist _____?   – _____ _____ schon ziemlich spät.

Wieviel Uhr ist _____? – _____ _____ erst ein Uhr.

_____ ist dunkel.   _____ _____ Frühling.   _____ _____ schon Abend.

4. Wendungen

Wie geht _____ Ihnen? – _____ _____ mir schon besser.

_____ _____ höchste Zeit!

_____ geht vorwärts.   _____ _____ _____ bergauf.   _____ gibt nichts, was _____ _____ gibt.

5. Wendungen bestimmter Verben. Formulieren Sie bitte eine Antwort!

Worum geht es hier? – _____

Worum handelt es sich denn? – _____

Worauf kommt es Ihnen im Leben an? – _____

6. „es" – Korrelat für einen „daß"-Satz oder einen Infinitiv mit „zu"

Bilden Sie bitte Sätze mit und ohne „es" wie in den Beispielen a) und b):

Beispiel:
a) Frauen haben heute mehr Chancen als früher. Das stimmt schon.

⟶ *Es stimmt schon, daß Frauen heute mehr Chancen haben als früher.*

⟶ *Daß Frauen heute mehr Chancen haben als früher, stimmt schon.*

Und nun Sie bitte:

Frauen verdienen trotzdem immer noch weniger als Männer. Das ist eine Tatsache.

_____

_____

Viele Männer helfen im Haushalt. Das ist erfreulich.

_____

Männer in Führungspositionen können Beruf und Familie vereinen. Das ist normal.

_____

Frauen denken zuerst an die Familie. Das wird immer noch für selbstverständlich gehalten.

_____

Die Altersarmut ist bei Frauen besonders weit verbreitet. Das ist bekannt.

_____

_____

b) Schreiben Sie die Sätze bitte weiter!

Beispiel: Es war vernünftig, *das Studium nicht abzubrechen.* _____

Und nun Sie bitte:

Es war schön, _____

Es empfiehlt sich, _____

Es lohnt sich, _____

Es ist sinnvoll, _____

Es ist unklug, _____

Es gelingt mir selten, _____

18 Diskussion

1. **Verstehen Sie die Deutschen richtig?**

a) Ein Deutscher/Eine Deutsche sagt: „Komm doch mal vorbei!"

Ist das ☐ eine ernstgemeinte Einladung?

☐ eine Einladung zum Essen?

☐ nur eine höfliche Floskel?

Wie reagieren Sie?

b) Sie haben ein Problem. Ein Deutscher/Eine Deutsche, den Sie/die Sie um Rat und Hilfe bitten, sagt: „Mal sehen, was ich für Sie tun kann."

Wird diese Person ☐ sich wirklich für Sie einsetzen?

oder: ☐ Ist es nur eine höfliche Floskel?

oder: ☐ Ist es ein Mittel, um Sie loszuwerden?

c) Ein Deutscher/Eine Deutsche sagt: „Meld dich doch mal!"/„Ich melde mich dann mal!"

Ist das ☐ eine ernstgemeinte Aufforderung/ein ernstgemeintes Versprechen?

oder: ☐ eine Möglichkeit, ein Gespräch unverbindlich, aber freundlich, zu beenden?

Wie reagieren Sie?

d) In der Stadt: Ein Deutscher/Eine Deutsche sagt: „Kommst du mit essen?"

Ist das ☐ eine Einladung nach Hause?

☐ eine Einladung (d.h. Sie müssen nicht bezahlen) ins Restaurant?

☐ nur eine „Aufforderung" (Bitte), gemeinsam essen zu gehen; aber jeder zahlt für sich?

2. **Wie ist es bei Ihnen?**

– Sie laden für den Abend Freunde ein: Ist das automatisch eine Einladung zum Essen?
– Blumen als Gastgeschenk bei einer Einladung: Ist das auch bei Ihnen üblich?
– Viele Deutsche essen heutzutage lieber italienische, griechische, asiatische Küche als traditionelle deutsche Küche? Gibt es ähnliche Erscheinungen bei Ihnen?
– Eine Einladung nach Hause bedeutet häufig viele Stunden Vorbereitung für die Gastgeber. Warum machen Sie das? Wäre es nicht einfacher, in ein Restaurant einzuladen?
– Die Vorbereitungen für eine Einladung bleiben meist den Frauen überlassen, vor allem das Kochen. Männer übernehmen vielleicht das Einkaufen, häufig kümmern sie sich nur um die Getränke. Wie ist die Arbeitsverteilung bei Ihnen?

19 Textaufbau: Abschied vom Rumpelstilzchen

1. Erinnern Sie sich noch an den Anfang des Märchens aus Lektion 1? Wenn nicht, dann lesen Sie bitte in Lektion 1, Übung 18, noch einmal nach!
Sie haben die Geschichte damals weitererzählt, wie Sie dachten, daß sie weitergehen könnte.
Wie aber ging die Geschichte bei den Gebrüdern Grimm weiter? Die Zeichnungen (S. 104/105) zeigen es. Schreiben Sie bitte die Grimmsche Version mit Hilfe der Zeichnungen zu Ende!
Und so machen Sie es: Schreiben Sie bitte zuerst zu jedem Bild einen oder mehrere Sätze! Bringen Sie diese Textabschnitte dann in eine chronologische Reihenfolge! Schreiben Sie Überleitungen zwischen den einzelnen Textteilen, und verändern Sie die Textteile so, daß sie nun zueinander passen. Schreiben Sie eine passende Überleitung von dem in Lektion 1 abgedruckten Textende zu Ihrem Textanfang hier! Lesen Sie Ihren Text noch einmal und redigieren Sie ihn!
Der originale Grimm-Text befindet sich im Lehrerhandbuch. Lassen Sie ihn sich von Ihrer Lehrerin/Ihrem Lehrer vorlesen, und vergleichen Sie mit Ihren Texten.

Bei der Redaktion Ihres Textes sollten Sie folgende Punkte überprüfen:

- die Stellung der Satzglieder und Verbklammer
  (Hinweise: Kursbuch S. 71, 74, 124, 135, 155)
- die Kongruenz von Subjekt und Verb
- die Verwendung der Zeiten (Hinweise:
  Kursbuch S. 87)
- die Funktion und Form der verschiedenen Er-
  gänzungen (Verwendung von Akk., Dat.,
  Gen.) und Angaben

- die Verwendung der Präpositionen (Hinwei-
  se: Kursbuch S. 88/89)
- die Adjektivendungen in der Nominalgruppe
  (Hinweise: Kursbuch S. 18, S. 122)
- die Nebensätze und die Stellung des Verbs im
  Nebensatz (Hinweise: Kursbuch S. 87)
- die Textkonnektoren (Satzanschlüsse)
- Rückverweisende und vorausweisende
  Pronomen
- Satzzeichen (Hinweise: Kursbuch S. 153)

2. Alte Märchen werden von Schriftstellern unserer Zeit **umgeschrieben**. Rosemarie Künzler schrieb das Märchen vom Rumpelstilzchen neu mit neuem Ausgang.

Lesen Sie bitte!
Welche häufig in Märchen anzutreffende Norm wird hier durchbrochen?

... Als das Mädchen allein war, kam das Männchen zum dritten Mal. Es fragte: „Was gibst du mir, wenn ich dir helfe? Aber die Müllerstochter hatte nichts mehr zu verschenken. „So versprich mir, wenn du Königin wirst, dein erstes Kind!" Da fiel es dem Mädchen wie Schuppen von den Augen. „Du spinnst!" rief das Mädchen dem Männchen zu. „Niemals werde ich diesen abscheulichen König heiraten! Niemals würde ich mein Kind hergeben!" „Ich spinne nicht! Ich spinne nie mehr!" schrie das Männlein wütend. „Ich habe umsonst gesponnen!" Das Männlein stieß mit dem rechten Fuß vor Zorn so tief in die Erde, daß die Kammertür aufsprang. Da lief die Müllerstochter in die weite Welt hinaus und war gerettet.

20 Testen Sie Ihr Deutsch!

Mit der folgenden Übung können Sie Ihre Grammatikkenntnisse überprüfen und erkennen, welche grammatischen Strukturen Sie noch nicht so gut beherrschen. Sie ist gleichzeitig eine Vorbereitung auf den Grammatiktest in der Prüfung zum „Zertifikat Deutsch als Fremdsprache", denn mit diesem Aufgabentyp werden in der Zertifikatsprüfung Ihre Grammatikkenntnisse getestet.

Jede mit A, B, C usw. gekennzeichnete Übungsfolge bildet einen zusammenhängenden Kontext, der Ihnen hilft, die richtigen Lösungen zu finden. Überlegen Sie bitte, welche der vier angegebenen Alternativen (a, b, c, d) in die Lücke des jeweiligen Satzes paßt! Markieren Sie die richtige Lösung! (Es gibt immer nur **eine richtige** Lösung.)

Die richtigen Lösungen finden Sie wie immer im Lösungsschlüssel. Wenn Sie die richtige Lösung nicht gewußt haben, finden Sie bei jeder Aufgabe einen Hinweis (→) auf die entsprechende Seite in Sprachbrücke 1 oder Sprachbrücke 2, wo Sie noch einmal nachlesen können. Wiederholen Sie dann auch die entsprechenden Übungen und Aufgaben im Arbeitsbuch.

Beispiel:

Wenn du _____, wie lange ich schon hier sitze!
a) gewußt hast
b) weißt
X   c) wüßtest
d) wußtest   → **1**, S. 135

A   1. _____ beschäftigst du dich denn gerade? –
Ich lese gerade die vielen Bewerbungsschreiben.
a) wo
b) womit
c) wozu
d) wofür   → **1**, S. 161, 165

2. Unsere Stellenangebote _____ viele Leute zu interessieren. –
a) können
b) scheinen
c) müssen
d) werden   → **2**, S. 34

3. Ich _____ dir gerne geholfen, diese Bewerbungsschreiben zu beantworten, aber ich habe keine Zeit.
a) könnte
b) würde
c) hätte
d) sollte   → **2**, S. 61, 134

B   4. Dieser Brief hätte schon vor zwei Tagen geschrieben _____ . –
a) werden müssen
b) werden
c) müssen   → **1**, S. 66, 179
d) haben müssen   **2**, S. 41, 43

5. Ich _____ das, aber ich hatte wirklich keine Zeit.
a) kenne
b) mag
c) will
d) weiß   → **2**, S. 58

6. Es waren so viele andere wichtige Dinge zu _____ . –
   a) schreiben
   b) erledigen
   c) telefonieren
   d) kaufen

7. Gut, sprechen wir nicht mehr darüber, denn je mehr Zeit vergeht,
   _____ größer wird das Problem.
   a) aber
   b) doch
   c) so
   d) desto   → **2**, S. 129

C   8. Wir wollten gerade _____ Haus gehen, als du anriefst.
   a) ausgehen
   b) hinaus
   c) aus dem
   d) in dem

9. Du hast eigentlich _____ . Wir hätten uns früher melden müssen. –
   a) recht
   b) richtig
   c) sicher
   d) gern

10. Wenn ich nicht zuhause bin, dann kannst du auf jeden Fall _____ ,
    wann du kommst.
    a) ausrichten
    b) ausgerichtet lassen
    c) ausrichten lassen
    d) ausrichten gelassen   → **2**, S. 30

D 11. Kannst du mir bitte Äpfel mitbringen? Ich möchte einen Apfelkuchen backen
     und brauche noch _____ . –
     a) einen
     b) anderes
     c) solche
     d) welche   → **1**, S. 95

12. Warum sagst du mir das erst jetzt! Die hättest du schon gestern kaufen müssen.
    Heute _____ die Geschäfte _____ . –
    a) werden ... schließen
    b) werden ... geschlossen
    c) sind ... geschlossen
    d) müssen ... schließen   → **2**, S. 43

13. Ganz _____ zu dir, bin ich eben vergeßlich. –
    a) im Gegenteil
    b) im Vergleich
    c) im Gegensatz
    d) ähnlich

14. Du _____ dir alles aufschreiben, dann würde dir das nicht passieren.
    a) darfst
    b) solltest
    c) willst
    d) kannst

E 15. Erinnern Sie sich an die Freunde, mit _____ ich in Argentinien war? –
  a) den
  b) ihnen
  c) sie
  d) denen

→ 1, S. 169

16. Nein, ich erinnere mich nicht mehr _____ . –
  a) dazu
  b) daran
  c) sie
  d) an sie

→ 1, S. 165

17. _____ drei Wochen wollen sie eine Reise durch Deutschland machen. –
  a) in
  b) bis
  c) vor
  d) nach

→ 2, S. 42

18. Kommen sie dann auch _____ Ihnen? –
  a) von
  b) mit
  c) bei
  d) zu

→ 1, S. 94

19. Ja, ich hoffe. Ich würde mich auch sehr _____ freuen.
  a) darüber
  b) darauf
  c) daran
  d) damit

→ 1, S. 165

F 20. Dieser Text ist zu schwierig, er ist _____ .
  a) unübersetzbar
  b) unübersetzlich
  c) unübergesetzt
  d) unübersetzhaft

→ 2, S. 121

21. Vielleicht ist Ihnen bei der Auswahl der Texte ein Fehler _____ . –
  a) gemacht
  b) gefunden
  c) unterlaufen
  d) entstanden

22. Das glaube ich nicht, außerdem übertreibst du.
  Dieser Text _____ ganz einfach _____ . –
  a) kann ... übersetzen
  b) ist ... übersetzen
  c) muß ... übersetzen
  d) ist ... zu übersetzen

→ 2, S. 119

23. Das werden wir ja sehen: _____ zeige ich dir meine Übersetzung.
  a) nachher
  b) vorher
  c) nach
  d) nachdem

→ 2, S. 86

G 24. In letzter Zeit verhält Rolf sich so, _____ überhaupt nichts passiert wäre. –
   a) als
   b) ob
   c) als ob
   d) obwohl   → **2,** S. 139

25. _____ , was ist denn passiert?
   a) Sag denn
   b) Sagt
   c) Sage, eben
   d) Sag' mal

26. Ich wäre dir sehr dankbar, wenn du nicht _____ fragen würdest. –
   a) darüber
   b) danach
   c) darum
   d) mich   → **1,** S. 165

27. Entschuldige, ich dachte, du wärest schon darüber _____ .
   a) hinweggekommen
   b) vergessen
   c) gestanden
   d) verstanden

H 28. Ich möchte in einer Stadt leben, _____ es keine Autos gibt. –
   a) bei der
   b) in der
   c) in den
   d) die   → **1,** S. 169

29. Vielleicht _____ es eine solche Stadt mal _____ . –
   a) werden ... geben
   b) wird ... geben
   c) wird ... gegeben
   d) ist ... gegeben   → **2,** S. 41

30. Ich zweifle _____ einer solchen Möglichkeit.
   a) von
   b) mit
   c) zu
   d) an

31. Das wäre einfach zu _____ .
   a) schöner
   b) schönsten
   c) schön
   d) schöne

I 32. Das Klima wird sich in den nächsten Jahren stark _____ . –
   a) verwechseln
   b) wechseln
   c) verändern
   d) umformen

33. _____ weißt du das?
   a) Worin
   b) Woraus
   c) Woher
   d) Wo

34. Ich lese sehr viele Zeitungen. Aber das _____ auch in vielen Büchern.
Heute schreiben ja alle über Zukunftsaussichten. –
   a) schreibt
   b) steht
   c) liest
   d) sieht   → **1**, S. 30

35. Dieses _____ Bewußtsein der Umwelt gegenüber ist sehr positiv.
   a) wachsende
   b) wachsendes
   c) wachsens
   d) gewachsene   → **2**, S. 67

K 36. Kannst du mir bitte auch mal das Buch geben, _____ du es zurückgibst? –
   a) vor
   b) vorher
   c) bevor
   d) wenn   → **2**, S. 87, 88

37. _____ , du bekommst es morgen.
   a) In Ordnung
   b) Auf Wiedersehen
   c) Im Grunde
   d) Unbedingt

L 38. _____ bist du schrecklich kompliziert! –
   a) In solcher Hinsicht
   b) In mancher Hinsicht
   c) In mancher Weise
   d) In dieser Art

39. Man braucht sich doch nicht immer alles _____ !
   a) fallen gelassen
   b) gefallen zu lassen
   c) gefallen gelassen
   d) zu gefallen   → **2**, S. 34

Lektion 6

1

1	2	3	4	5	6	7	8	9	10	11	12
d	n	o	b	a	c	g	h	v	w	f	r

13	14	15	16	17	18	19	20	21	22	23	24
k	x	i	t	s	j	l	u	m	e	y	p

(S. 6 unten): f. = femininum; m. = maskulinum; n. = neutrum; Pl. = Plural; (S. 7 unten): jmd. = jemand; Adj. = Adjektiv; bes. = besonders; od. = oder; u. = und; poet. = poetisch; literar. = literarisch; bzw. = beziehungsweise

4 Möglicher Text: Sie war damals 17 Jahre alt. Bei einem Popkonzert hatte sie ihren Mann kennengelernt, er hatte neben ihr gesessen. Nach dem Konzert hatte sie ihn einfach gefragt: „Kannst du mich nach Hause fahren?" Zu ihrer Überraschung hatte er sie angelächelt und ja gesagt. Zusammen waren sie in seinem Auto nach Hause gefahren. Vor ihrem Haus war sie nicht ausgestiegen, sondern einfach sitzen geblieben und hatte gewartet. Er hatte sie gefragt: „Willst du nicht aussteigen?" Da hatte sie ihn gefragt, ob sie sich vielleicht wiedertreffen könnten. Da hatte er sie in den Arm genommen. So hatte alles angefangen.

6 1. 2. Bevor sie nach Stockholm versetzt wurde, wollte sie noch eine Reise durch die Sahara machen. 3. Dann reiste sie nach Stockholm ab. Vorher hatte sie ihren späteren Mann kennengelernt. 4. Bevor er zu ihr nach Stockholm kommen konnte, mußte er sein Studium abschließen. 5. Bevor sie heirateten, lebten sie ein Jahr in Stockholm zusammen. 6. 1989 kehrten sie nach Kairo zurück. Vorher war der Sohn Thomas auf die Welt gekommen.
2. 2. Dann ging ich zwei Stunden lang im Wald spazieren. Danach kochte ich das Mittagessen. 3. Ich aß gut und viel. Danach legte ich mich eine Stunde lang ins Bett. 4. Nachdem ich wieder aufgestanden war, klingelte es an der Tür, und meine ganze Verwandtschaft kam zu Besuch. 5. Nachdem wir den ganzen Kuchen aufgegessen hatten, schauten wir uns im Fernsehen ein Fußballspiel an. 6. Unsere Mannschaft gewann. Danach saßen wir noch bis spät in die Nacht zusammen.

7 2. 1. Während sie die Kinder versorgt, verdient er das Geld. 2. Seitdem sie verheiratet sind, verstehen sie sich nicht mehr so gut. 3. Seitdem sie das dritte Kind bekommen hat, ist sie oft müde. 4. Nachdem die Kinder groß geworden sind, arbeitet sie wieder in ihrer alten Firma. 5. Seitdem die Kinder aus dem Haus sind, haben sie sich nichts mehr zu sagen. 6. Bevor sie sich scheiden lassen, regeln sie alles Finanzielle.

9 1. Zu 6. 1: 2. Vor ihrer Versetzung nach Stockholm wollte sie ... 3. Vor ihrer Abreise lernte

sie ... 4. Nach dem Abschluß seines Studiums kam er ... 5. Vor ihrer Heirat lebten sie ... 6. Nach der Geburt ihres Sohnes Thomas kehrten sie ... zurück / Vor ihrer Rückkehr nach Kairo kam ... auf die Welt.
Zu 6. 2: 1. Nach dem Frühstück las ich die Zeitung. 2. Nach meinem Waldspaziergang kochte ich ... 3. Nach dem Essen legte ich mich ... 4. Nach meinem Mittagsschlaf klingelte es ... 5. Vor dem Fußballspiel aßen wir ... 6. Nach dem Ende des Spiels saßen wir ...
2. 1. Vor ihrer Heirat ... 2. Während der Ehe ... 3. Seit ihrer Heirat ... 4. Seit der Geburt ihres dritten Kindes ... 5. Nach dem Auszug der Kinder ... 6. Vor der Scheidung ... 7. Nach der Scheidung ...
3. 2. Beim Einsteigen ... 3. Beim Fliegen ... 4. Beim Mittagessen ... 5. Bei der Ankunft.

11 1. Vor mehr als fünfzig Jahren / Am / Bis zu diesem Tag / Von 1945 bis 1989. 2. Vor / Während der folgenden Jahre / Erst mit der Ostpolitik Willy Brandts in den 70er Jahren. 3. Und eines Tages geschah es: In der Nacht vom neunten auf den zehnten November 4. ein halbes Jahr nach der Maueröffnung / am dritten Oktober / 5. Nach dem Ende des Kalten Krieges ... daß es in einigen Jahren oder vielleicht in ein paar Jahrzehnten / Bis zu diesem Zeitpunkt ...

13 die Flucht, Flüchtlinge, fliehen, flüchten / Auswanderer, auswandern / Emigranten, emigrieren / vertreiben / Asylanten

14 1. Das Lösungswort heißt: Bahnhof Zoo. (= Bahnhof in Berlin. Eigentlich: Bahnhof Zoologischer Garten. Direkt gegenüber liegt der Berliner Zoo [Zoologischer Garten].)

18 1. Seltsamer Vortrag: In allen Lücken „irgendwie".
Egal wer: In allen Lücken „irgendwer".
Unbestimmtes Ziel: In allen Lücken „irgendwohin".
2. Irgendwann hat
irgendjemand
irgendwem mal
irgendwas gesagt.
Das hat der aber
irgendwie
falsch verstanden. Seitdem ...
(Aus: Bachmann / Brecheisen / Gerhold / Wessling: Arbeitsbuch zu: Sichtwechsel. Elf Kapitel zur Sprachsensibilisierung. S. 9)

19 1. die Lyrik, das Drama, die Poesie
2. 1. + 2. Kunde / Kundin-Supermarkt; Patient / in-Arzt; Gast-Hotel; Zuhörer / in-Konzert; Zuschauer / in-Theater; Tourist / in-in einem fremden Land; Passagier-Flugzeug; Schüler / in, Kursteilnehmer / in-Deutschkurs

22 **1.** 2. Nachdem er aufgewacht war, stand er sofort auf. Bevor er sich duschte und rasierte, trank er eine Tasse Kaffee. 4. Nachdem er sich angezogen hatte, ging er aus dem Haus. 5. Nachdem er in der Firma die Zeitung gelesen hatte, begann er sofort mit seiner Arbeit. 6. Bevor er seinen Bericht über die Situation der Firma schrieb, diskutierte er darüber mit seinen Mitarbeitern. 7. Bevor er an einer Konferenz teilnahm, telefonierte er noch mit mehreren Kunden. 8. Nachdem er sich angehört hatte, was die anderen zu sagen hatten, stellte er die neue Werbestrategie vor. 9. Nachdem er mit seiner Sekretärin die Termine für den nächsten Tag geplant hatte, nahm er das Flugzeug nach Straßburg. 10. Bevor er dort Gespräche mit den Abgeordneten des Europa-Parlaments führte, trank er schnell noch eine Tasse Kaffee. 11. Bevor er dann um 21 Uhr zurückfliegen konnte, brach er zusammen.

2. <u>Von</u> seinem 6. <u>bis zu</u> seine<u>m</u> 18. Lebensjahr ging er in die Schule und machte <u>im</u> Jahr 1974 das Abitur. 18 Monate <u>lang</u> war er dann beim Militär. <u>Im</u> Frühjahr 1976 fuhr er für sechs Monate in die USA. <u>Während der</u> ersten drei Monate besuchte er einen Englisch-Sprachkurs. <u>Eines</u> Tages lernte er dort eine sympathische Mexikanerin kennen. <u>An</u> seine<u>m</u> Geburtstag organisierte er ein großes Fest und lud auch sie ein. <u>Seit</u> diese<u>m</u> Abend waren beide oft zusammen. <u>Nach dem</u> Sprachkurs machten sie <u>vom</u> 28. Juni <u>bis</u> 18. Juli eine Reise durch Mexiko. <u>Auf</u> dieser Reise verliebten sie sich ineinander, und <u>eines</u> Abend<u>s</u> fragte er sie, ob sie ihn heiraten wolle. Sie brauchte ziemlich lange für ihre Antwort. <u>Nach</u> ein<u>er</u> halb<u>en</u> Stunde antwortete sie dann: „Ja".

Lektion 7

5 **1.** mit ihren duftenden Haaren / ihren leuchtenden Augen / ihren geröteten Wangen / ihrem zart geschminkten Mund / ihrer zitternden Stimme / ihrem hin- und herschwingenden Rock / ihren schön geformten Beinen
2. seine frisch gewaschenen Haare / sein lachender Mund / seine leicht gebräunte Haut / seine blitzenden Zähne / sein frisch gebügeltes Hemd / seine kräftig wirkenden Hände

6 **2.** 2. <u>Ein drohendes Gewitter</u> verdunkelt den ganzen Himmel. 3. <u>Die über den Himmel zuckenden Blitze</u> erschrecken die kleinen Kinder. 4. Die <u>tanzenden Blätter</u> werden vom Wind gejagt. 5. <u>Die in die Häuser geflüchteten Menschen</u> hoffen, daß das Gewitter bald losbricht. 6. <u>Hinter den geschlossenen Fenstern</u> sieht man ängstliche Gesichter. 7. <u>Die auf die Erde stürzenden Wassermassen</u> bringen endlich <u>die von allen erhoffte Abkühlung</u>. 8. Kaum hat der Regen aufgehört, sieht man schon wieder spie-

<u>lende Kinder</u> und geschäftig <u>hin- und hereilende Menschen</u>.

8 ich selber / du selber / er selber / sie selber / sich selber / uns selber / euch selber

12 Hans Knapp, <u>einen alten Bekannten</u>, ... / Dieser Bekannte, <u>ein interessanter</u> Mann, ... / zusammen mit meinem Mann, <u>seinem Studienkollegen</u>, ... / Im Buchgeschäft Wurm, <u>dem größten der Stadt</u>, ... / Dieses Buch ..., <u>schön gedruckt und farbig</u>, ... / seinem Chef, <u>dem Leiter der Abteilung</u>, ...

17 2.

Liebe SABINE

es tut mir leid, wenn ich <u>Dir</u> weh getan habe, bitte verzeih. Auf <u>Deine</u> „Anrufe" konnte ich nicht reagieren, weil ich <u>Deine</u> neue Anschrift/Tel.-Nr. nicht kenne. Seit Monaten bin ich auf der Suche nach <u>Dir</u>, leider ohne Erfolg. Bitte zeig mir den Weg zu <u>Dir</u>. In Liebe

„ODYSSEUS" R.

TRAUMMANN
4. 6. - 4. 12. 1991

Vor 6 Monaten hat <u>unsere</u> große Liebe begonnen.
Dafür, <u>mein</u> geliebter Schatz, danke ich <u>Dir</u> von ganzem Herzen. Mögen <u>unsere</u> Träume einmal wahr werden.
In tiefer Liebe
DEIN STERNTALER

Mein Herz weint, weil es das Feuer der Lieb löschen möchte, aber vergebens brennt mein Her weiter voll Liebe zu <u>Dir</u> -, SUSANNE!!!

TRAUMMANN

Liebster DUSCHA <u>mein</u>,
auch ich denke sehnsüchtig täglich an <u>Dich</u> Warum nur läßt <u>Du</u> mich so lange auf ei Zeichen von <u>Dir</u> warten; wo und wann ic <u>Dich</u> wiedersehen kann?
In Liebe <u>Dein</u> Schatz

ALICE

> **Geliebte INGE!**
>
> *Auch wenn ich es nicht sage jeden Tag - so weißt Du doch, wie lieb ich Dich hab'. Bin ich auch in der Ferne, sind heute getrennt auch wir -, Du hörst mein Herz, wie es ruft nur nach Dir! Und ich schau' den weißen Wolken nach und fange an zu träumen ...*
>
> *Ewig - Dein DUSCHA*

4. Die Redewendungen heißen:
um den heißen Brei herumreden / mit der Tür ins Haus fallen / jemandem das Wort abschneiden / jemanden zum Fressen gern haben

19 Die Sprichwörter heißen:
Jung gefreit, nie gereut.
Drum prüfe, wer sich ewig bindet.
Die Ehe ist das Ende der Liebe.

21 **1.** ins Reisebüro / im Haus an der Ecke / ins Gebirge, ans Meer, auf eine Insel, in eine Stadt / in den Prospekten / in den Norden, in den Süden / auf dem Land, in den Bergen / zu Dir, in meine Geburtsstadt direkt am Meer / aus dem Büro, auf die Straße, auf den Boden / neben mir / in zwei große Augen / zum Marktplatz, in ein Café / in der Stadt: ins Kino, ins Theater, in eine Ausstellung, zu Freunden, im Park, zu Hause.
2.

Mit 40 fängt das Leben erst an ...
Welcher humorvoll-ausgeglichene, zärtliche und emanzipierte Mann, der seine Träume und seine Pläne noch nicht zu den Akten gelegt hat, hat Lust, seine Zukunft gemeinsam mit mir zu gestalten? Ich bin eine sehr lebendige, aufgeschlossene, hübsche Frau (Studienrätin, 41/1,68/55/NR, im Raum 6), mag fremde Sprachen und Länder, Menschen, Natur und Großstadt und freue mich auf die Zuschrift eines Mannes (mit Bild), der auch glaubt, daß nicht alles im Leben Zufall ist.

34j. HORST, ledig
kathol., ein sympathischer junger Mann mit natürlichem Wesen, humorvoll, kinderlieb und naturliebend, spielt gern Schach, mag Spaziergänge zu zweit und wünscht sich eine treue, einfühlsame Partnerin. Suchst Du auch eine ehrliche, verständnisvolle Partnerschaft, dann ruf doch einfach mal an unter ...

Neuanfang (Raum 4 oder 5)
Lebensfrohe, junggebliebene Frau, 53 J., gesch., 3 erwachsene Kinder, psychologische Tätigk., wünscht sich lebendige Partnerschaft mit warmherzigem, offenem, neugierigem Mann (auch jünger), der wie sie Mut zu einer neuen, reiferen Beziehung mitbringt, in der jeder dem anderen viel persönlichen Raum lassen kann. Ich freue mich auf Ihre Bildzuschrift.

Lektion 8

1 **2.** In der Mitte des Bildes steht ein alter Baum, der nur noch wenige Äste hat. Aber er hat viele Blätter. Die Spitze des Baumes ist wahrscheinlich bei einem Sturm abgebrochen. Der Baum steht auf einer großen Wiese. Unter dem Baum weiden Kühe oder Schafe, man kann es nicht genau erkennen.
Der Baum steht nicht sehr weit vom Ufer eines Flusses entfernt. Der Fluß führt ziemlich viel Wasser, einige Bäume stehen mitten im Wasser, auch unter dem Baum in der Mitte des Bildes ist Wasser. Wahrscheinlich hat es in letzter Zeit viel geregnet.
Am gegenüberliegenden Ufer des Flusses sieht man die Spitze eines Kirchturms. Dort ist ein Dorf. Im Hintergrund des Bildes erheben sich hohe Berge. Man kann nicht erkennen, ob sie bewaldet sind oder nicht. Der Himmel, der das obere Drittel des Bildes ausfüllt, ist ziemlich bewölkt.
Jemand hat eine Taschenuhr vor das Bild gemalt. Die Zeiger zeigen fünf Minuten vor zwölf. Auf der Innenseite des Deckels der Uhr sieht man einen Baum, dessen Äste nach unten hängen und der keine Blätter mehr trägt. Derjenige, der die Taschenuhr gemalt hat, will wahrscheinlich sagen: Wenn man nichts für die Natur tut, ist es bald zu spät.

3 **3.** 1. bläulich 2. rötlich 3. ein gelbliches Etwas
4. sommerlich / herbstlich / winterlich / frühlingshaft

4 **1.** die Luftfeuchtigkeit / der Mandelbaum / der Warenexport / die Frühlingsreise.
Prüfung für Dolmetscher / Lager für Flüchtlinge / Blüte der Apfelbäume / Landschaft mit Hügeln / Landschaft mit Seen / Übung zur Entspannung / System für die Bildung und Ausbildung der Menschen in einem Land / Wie man sich in welchen Situationen korrekt kleidet / eine Frau, die in ihrem Beruf Erfolg hat und Karriere macht
2. eine kontaktfreudige Frau / mandelförmige Augen / rechthaberisch / eine praxisnahe Berufsausbildung / deutschsprachigen Nachrichten
3. ein pflichtbewußter Mensch / wirtschaftsfördernde Maßnahmen / eine hochentwickelte Zivilisation / eine berufsbezogene Ausbildung / krankmachende Arbeitsplätze / menschenverachtende Regime / offenstehende Rechnungen

6 **2.** Oberbegriff: Gewässer
„Kanal" paßt nicht (ist nicht natürlich)
Grund für die Anordnung: Größe
Bach, Fluß, Strom haben „ein Bett"
Ufer: Bach, Fluß, Strom, Meer
Küste: Meer
Strand: Meer, eventuell Fluß

113

3. der See: Binnengewässer
die See = das Meer (die Nordsee, die Südsee)
4. Nur durch Menschenhand: Garten, Feld, Park

7 3. Der Redner war kaum zu sehen. 2. Seine Stimme war kaum zu hören. 3. Was er sagte, war kaum zu verstehen. 4. Die Ergebnisse der Untersuchung sind abzuwarten. 5. Die Richtigkeit seiner Aussagen ist in Frage zu stellen. 6. Der Schlußsatz ist zu überprüfen.

8 1. Die Mieter sollen das Singen unterlassen. Die Mieter haben das Singen zu unterlassen. Die Mieter dürfen nicht singen. / Die Reisenden müssen mitgeführte Tiere melden. Die Reisenden haben mitgeführte Tiere zu melden. / Die Lernenden sollen Fehler vermeiden. Die Lernenden haben Fehler zu vermeiden.

10 1. Wie <u>die nachkommenden Generationen</u> auf unserem Planeten leben werden, das ist <u>eine schwer zu beantwortende Frage.</u>
2. <u>Zwischen der steigenden Anzahl der Autos</u> und <u>der immer größer werdenden Umweltverschmutzung</u> besteht <u>ein leicht zu erklärender Zusammenhang.</u> 3. <u>Den auf mehr Waffen drängenden Militärs</u> noch mehr Geld zu geben, wäre <u>ein nicht wieder gutzumachender Fehler.</u> 4. Die Erde für <u>unsere heranwachsenden Kinder</u> wieder schöner zu machen, ist <u>eine nicht leicht zu lösende,</u> große <u>Aufgabe.</u> 5. <u>Durch nichts zu entschuldigende Vorurteile</u> gegen Fremde und Andersartige stören das friedliche Zusammenleben der Menschen. 6. <u>Die um ihr Überleben kämpfende Menschheit</u> muß sich entscheiden zwischen ... oder dem Ende der Menschheit <u>auf einem sterbenden Planeten.</u> 7. <u>Eine leicht zu treffende Wahl</u> könnte man meinen.

13 2. -bar: unwiederholbar / voraussehbar / schwer erziehbar / gut formbar / nicht vergleichbar / nicht verstehbar / gut lesbar
-lich: unglaublich / unverzeihlich / unbeschreiblich schön / unvergleichlich / unverständlich / unbegreiflich

16 2. Das Sprichwort heißt: Wie man in den Wald hineinruft, so ruft (tönt) es wieder heraus.

17 1. 1. Er war lange krank, aber <u>jetzt ist er über den Berg.</u> 2. <u>Das paßt jetzt nicht in die politische Landschaft.</u> 3. Weil die Politiker wiedergewählt werden wollen, <u>halten sie mit der Wahrheit hinter dem Berg.</u> 4. <u>Du siehst den Wald vor lauter Bäumen nicht.</u> 5. Wenn du die erreichen willst, was du möchtest, dann <u>kannst du dich nicht wie die Axt im Wald benehmen.</u> 6. Als die Polizei kam, <u>war der Bankräuber schon längst über alle Berge.</u> 7. Du meine Güte, <u>bis dahin fließt noch viel Wasser den Berg hinunter.</u>

19 1. 1. a) 2. b) 3. d) 4. e) 5. c)
Adjektive; bedienbar / übersetzbar / realisierbar / beantwortbar / unbrauchbar

---

2. Eine Frau stürzte aus einem Fenster. Die Verletzte wurde in das Krankenhaus eingeliefert.
3. a) Der Vertrag <u>ist</u> so schnell wie möglich <u>zu unterschreiben.</u> d) Die Geschichte <u>ist nicht zu glauben.</u> e) die Bewohner des Hauses <u>haben</u> die Türen abends <u>zu schließen.</u> h) Du <u>brauchst</u> mir <u>nur</u> einen guten Rat <u>zu geben.</u> i) Ich habe <u>vergessen,</u> dich <u>zu fragen,</u> ... k) Deutsch ist <u>eine schwer zu lernende Sprache.</u> m) <u>Es ist schwer,</u> Deutsch <u>zu lernen.</u> n) Hunde <u>sind</u> an der Leine <u>zu führen.</u> p) Wir dürfen nie <u>aufhören, Fragen zu stellen.</u> r) Hier <u>sind</u> noch einige <u>Fragen zu stellen.</u> u) Ich <u>habe</u> heute noch <u>einiges zu erledigen.</u> v.) Heute <u>ist</u> noch <u>einiges zu erledigen.</u>

Lektion 9

6 2. Je unzufriedener die Frauen sind, desto kämpferischer treten sie für ihre Rechte ein. 3. Je selbstbewußter die Frauen auftreten, desto unsicherer werden viele Männer. 4. Je mehr die Männer im Haushalt mitarbeiten, desto bewußter können die Frauen ihre Karriere planen. 5. Je häufiger die Väter mit ihren Kindern spielen, desto besser ist es für die ganze Familie. 6. Je energischer die Frauen ihre Ziele verfolgen, desto erfolgreicher werden sie sein.

10 1. Ich würde es machen, wenn du es auch machen würd<u>est.</u> Aber er würd<u>e</u> es nur machen, wenn ihr alle mitmachen würd<u>et.</u>

13 1. Wenn ich nur mehr Zeit für meine Hobbys hätte! 2. Wenn nur mein Freund / meine Freundin endlich käme! 3. Wenn ich nur keine Prüfung machen müßte! 4. Wenn ich nur Geld dabei hätte! / Wenn ich wenigstens meine Schecks dabei hätte! / Wenn ich doch jetzt 200 Mark finden würde! / Wenn ... 5. Wenn ich nur gewußt hätte, daß er krank ist! 6. Wenn sie mich nur einmal mit ihrem tollen Wagen fahren lassen würde!

14 Ach wäre ich doch damals nicht so dumm gewesen! / Hätten wir doch über unsere Gefühle gesprochen! / Hätte ich sie doch in den Arm genommen! / Hätte ich sie doch geküßt! / Ach, wäre ich doch nicht so schüchtern gewesen! / Ach! Wäre es doch jetzt nicht zu spät!

17 2. Die meisten taten, als hätte man das nicht wissen können / als hätten sie nichts gesehen und nichts gehört / als wären sie nicht dabeigewesen / als hätten sie nicht mitgemacht / als hätten sie damit nichts zu tun (gehabt) / als hätten sie nichts Böses getan / als wären sie ganz und gar unschuldig / als ginge es sie nichts an, was passiert ist (als ginge sie, was passiert ist, nichts an).

3. Tut doch nicht so, als ob ihr nichts hättet wissen können / als ob ihr nichts gesehen und gehört hättet / als wärt ihr nicht dabeigewesen / als hättet ihr nicht mitgemacht / als hättet ihr damit nichts zu tun (gehabt) / als ob ihr nichts Böses getan hättet / als ginge euch alles, was passiert ist, nichts an!

24 Die Nachbarländer: siehe Lektion 4, Übung 18
Die drei Flüsse heißen: der Rhein / die Donau / die Elbe

27 **2.** 1. a) 2. c) 3. b) 4. a) 5. b) 6. c) 7. b) 8. c) 9. a) 10. c) 11. a) 12. b) 13. a)

Lektion 10

6 **(Hoch)Deutsch**
Begrüßung: Guten Tag! Tach! Grüß Gott!
Verabschiedung: Auf Wiedersehen! Tschüß! Ciao!
Österreichisches Deutsch
Begrüßung: Servus! Guten Tag! Grüß Gott! Griaß di! (= Grüß dich!)
Verabschiedung: Servus! Ciao! Pfiat di! (= (Gott) behüte dich!)
(Auf) Wiederschaun! Auf Wiedersehn! Tschüß!
Schweizer Deutsch
Begrüßung: Grüezi! Grüezi mitenand (= (Ich) grüße euch alle miteinander)! Tag wohl! Salü!
Verabschiedung: Uf Wiederluega! (= Auf Wiedersehen) Salü! Ciao! Läb (leb) wohl!

10 saß / Sessel / las / hingerissen / süß / aus / Gewiß / wissen / küßte / Füße / Beweise / daß / liebst / küsse / Kuß / beschlossen / zusammen / essen / Landgasthaus / aßen / Gemüse / Nußkuchen / Nüssen / Gläsern / bis / Nase / naß / laß / laß / muß / verlassen / verstoßen / bloß / muß / müssen / zerreißt / mußt / weißt / das / wissen / Reisen / zerrissen / großzügig / Adresse / Festessen / Schlösser / grüßen

11 „Aufstehen! Es ist schon spät!" hörte ich meine Mutter rufen. Ich antwortete: „Heute ist doch Sonntag! Warum muß ich auch heute so früh aufstehen?" „Susanne hat recht!" rief mein Vater. „Wir bleiben heute ja zu Hause!" Meine Mutter meinte dann: „Aber das ist kein Grund, einfach faul im Bett liegen zu bleiben." „Wollen wir spazierengehen?" fragte ich die Großmutter. „Ich weiß nicht", antwortete sie, „es ist heute so kalt." „Wenn du willst", sagte mein Bruder, „gehen wir heute abend ins Konzert."

12 Den vollständigen Text siehe Seite 116.

13 **1.** 1. Um Texte verstehen zu können, müssen nicht nur die Wörter bekannt sein, sondern auch die Punkte und Kommas richtig gesetzt sein, weil sonst alles wie ein großer Wortsalat aussieht. 2. Zum Verständnis der Bedeutung eines Satzes ist der Platz des Kommas wichtig, weil sehr oft der Platz des Kommas nicht nur das Lesen erleichtert, sondern auch die Bedeutung des Satzes ändern kann.

2. 1. Der Brief, von mir geschrieben, wurde, nachdem ich ihn einem Bekannten mit der Bitte, ihn sofort aufzugeben, mitgegeben hatte, leider erst nach Tagen zur Post gebracht. 2. Mein Onkel, ein reicher Mann, schenkte noch kurz vor seinem Tod den größten Teil seines hart erarbeiteten Vermögens nicht seinen Verwandten, sondern einer älteren Dame, die immer für ihn gesorgt hatte. 3. Alles, was sie hatte, mußte verkauft werden, weil sie, ohne überhaupt zu überlegen, immer Karten spielte und einmal, als sie wieder verloren hatte, sehr große Schulden machen mußte.

14 1. Komma 2. Gedankenstrich 3. Ausrufezeichen 4. Fragezeichen 5. Klammer 6. Semicolon 7. Anführungsstriche 8. Bindestrich 9. Doppelpunkt 10. Punkt

16 **1.** a) 1: Platsch! 2: drrr! 3: Schmatz! 4: Tatü, tatü! 5: Rumms! 6: Klingeling! 7: Ring, ring! 8: Tick, Tack! 9: Wrumm! 10: Klirr!
2. 1: Au! Aua! 2: brrr! 3: Igittigitt! 4: Pssst! 5: Hahaha!

17 1. Hast du mein Buch gesehen? — Hier liegt <u>es</u> doch.
Wo liegt denn Feuerland? — Ich weiß <u>es</u> nicht.
2. Es schneit. Es ist heiß. Es blitzt. Es donnert. Es hagelt. Es ist regnerisch. Es ist windig. Es ist frisch. Es ist kühl.
3. Wie spät ist <u>es</u>? — <u>Es</u> ist schon ziemlich spät. Wieviel Uhr ist <u>es</u>? — <u>Es</u> ist erst ein Uhr. <u>Es</u> ist dunkel. <u>Es ist</u> Frühling. <u>Es ist</u> Abend.
4. Wie geht <u>es</u> Ihnen? — <u>Es geht</u> mir schon besser.
<u>Es ist</u> höchste Zeit! <u>Es geht</u> vorwärts. <u>Es geht</u> bergauf. <u>Es gibt</u> nichts, was <u>es nicht</u> gibt.
6. <u>Es ist</u> eine Tatsache, <u>daß</u> Frauen trotzdem immer noch weniger verdienen als Männer. Daß Frauen trotzdem immer noch weniger verdienen als Männer, ist eine Tatsache. / <u>Es ist erfreulich,</u> <u>daß</u> viele Männer im Haushalt helfen. Daß viele Männer im Haushalt helfen, ist erfreulich. / <u>Es ist</u> <u>normal, daß</u> Männer in Führungspositionen Beruf und Familie vereinen können. Daß Männer ... ist normal. / <u>Es wird immer noch für selbstver-</u><u>ständlich gehalten, daß</u> Frauen zuerst an die Familie denken. Daß Frauen ..., wird immer noch für selbstverständlich gehalten. / <u>Es ist bekannt,</u> <u>daß</u> die Altersarmut bei Frauen besonders weit verbreitet ist. Daß die Altersarmut ..., ist bekannt.

18 a) „Komm doch mal vorbei!" — ist eher eine höfliche Floskel. Aber man darf, wenn man Lust hat, schon mal kommen. Dann darf man aber nicht erwarten, daß man etwas zu essen bekommt oder daß der Besuchte bei einem unerwarteten Besuch auch Zeit hat.

b) Die Person wird sich wirklich bemühen, Ihnen bei der Lösung Ihres Problems zu helfen.
c) Es ist eher eine Möglichkeit, ein Gespräch unverbindlich, aber freundlich, zu beenden.
d) Es ist eine freundliche Aufforderung, das Essen gemeinsam einzunehmen. Jeder zahlt im Prinzip für sich selbst.

20 1. b, 2. b, 3. c, 4. a, 5. d, 6. b, 7. d, 8. c, 9. a, 10. c, 11. d, 12. c, 13. c, 14. b, 15. d, 16. d, 17. a, 18. d, 19. a, 20. a, 21. c, 22. d, 23. a, 24. c, 25. d, 26. b, 27. a, 28. b, 29. b, 30. d, 31. c, 32. c, 33. c, 34. b, 35. a, 36. c, 37. a, 38. b, 39. b

---

Lösung zu Lektion 10, Übung 12 (Seite 97)

_____, den _____

An die Autoren des Lehrwerks „Sprachbrücke"!

Sehr geehrte Damen und Herren,

nachdem wir nun schon fast zwei Jahre lang mit Ihrem Buch Deutsch gelernt haben, stellen wir fest, daß Sie uns das Wichtigste erst am Schluß erklärt haben, nämlich die Zeichensetzung und ganz besonders die Sache mit dem Komma.

Nun ist es fast schon zu spät, wie Sie an diesem Schreiben sehen können, denn „Was Hänschen nicht lernt, lernt Hans nimmermehr". Diesen Brief, den Sie hoffentlich ohne allzu große Mühe lesen können, auch wenn sämtliche Kommas fehlen, schreiben wir, damit Sie das nächste Mal, wenn Sie ein Lehrwerk schreiben, schon viel früher damit anfangen, auf die Bedeutung des Kommas hinzuweisen. Einige von uns haben nämlich diesen kleinen Strich bisher kaum beachtet, und sie fragen sich, ob man ihnen diesen Fehler jemals verzeihen wird. Andere haben ihn zwar von Zeit zu Zeit gesetzt, aber wenn man nicht weiß, wie und wo und wann, dann nützt es auch nichts.

Es ist ja nicht nur eine Frage der Rechtschreibung und der Bedeutung, sondern auch eine Frage der Ästhetik. So ein Text sieht einfach mit Kommas an der richtigen Stelle besser aus als ohne Kommas, oder etwa nicht? Wir hoffen nun, daß sich jemand findet, der uns zeigen kann, wo wir die Kommas in diesem Brief setzen müssen.

Zum Schluß bleibt uns nur noch zu sagen, daß wir im übrigen mit Ihrem Buch sehr zufrieden waren und daß es anstrengend, aber auch schön war, damit zu arbeiten.

In der Hoffnung, Ihnen eines Tages selbst zu begegnen, grüßt Sie recht herzlich Ihre Deutschgruppe aus ...

Schaltplan

zum "Knacken" deutscher Texte

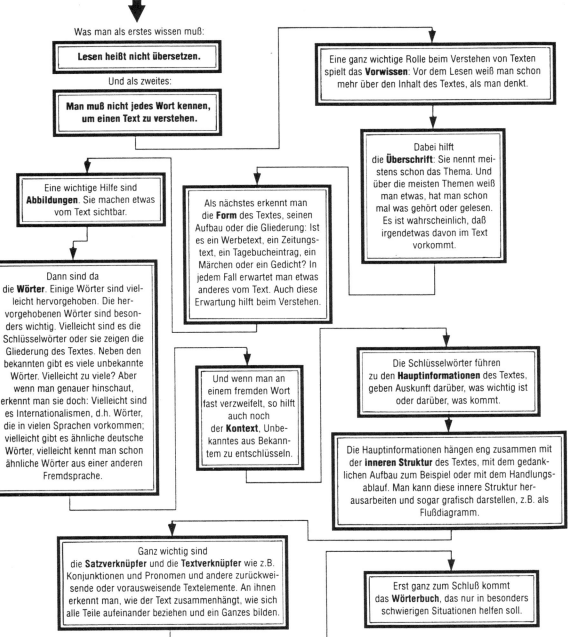

Was man als erstes wissen muß:

Lesen heißt nicht übersetzen.

Und als zweites:

Man muß nicht jedes Wort kennen, um einen Text zu verstehen.

Eine ganz wichtige Rolle beim Verstehen von Texten spielt das **Vorwissen**: Vor dem Lesen weiß man schon mehr über den Inhalt des Textes, als man denkt.

Dabei hilft die **Überschrift**: Sie nennt meistens schon das Thema. Und über die meisten Themen weiß man etwas, hat man schon mal was gehört oder gelesen. Es ist wahrscheinlich, daß irgendetwas davon im Text vorkommt.

Eine wichtige Hilfe sind **Abbildungen**. Sie machen etwas vom Text sichtbar.

Als nächstes erkennt man die **Form** des Textes, seinen Aufbau oder die Gliederung: Ist es ein Werbetext, ein Zeitungstext, ein Tagebucheintrag, ein Märchen oder ein Gedicht? In jedem Fall erwartet man etwas anderes vom Text. Auch diese Erwartung hilft beim Verstehen.

Dann sind da die **Wörter**. Einige Wörter sind vielleicht hervorgehoben. Die hervorgehobenen Wörter sind besonders wichtig. Vielleicht sind es die Schlüsselwörter oder sie zeigen die Gliederung des Textes. Neben den bekannten gibt es viele unbekannte Wörter. Vielleicht zu viele? Aber wenn man genauer hinschaut, erkennt man sie doch: Vielleicht sind es Internationalismen, d.h. Wörter, die in vielen Sprachen vorkommen; vielleicht gibt es ähnliche deutsche Wörter, vielleicht kennt man schon ähnliche Wörter aus einer anderen Fremdsprache.

Und wenn man an einem fremden Wort fast verzweifelt, so hilft auch noch der **Kontext**, Unbekanntes aus Bekanntem zu entschlüsseln.

Die Schlüsselwörter führen zu den **Hauptinformationen** des Textes, geben Auskunft darüber, was wichtig ist oder darüber, was kommt.

Die Hauptinformationen hängen eng zusammen mit der **inneren Struktur** des Textes, mit dem gedanklichen Aufbau zum Beispiel oder mit dem Handlungsablauf. Man kann diese innere Struktur herausarbeiten und sogar grafisch darstellen, z.B. als Flußdiagramm.

Ganz wichtig sind die **Satzverknüpfer** und die **Textverknüpfer** wie z.B. Konjunktionen und Pronomen und andere zurückweisende oder vorausweisende Textelemente. An ihnen erkennt man, wie der Text zusammenhängt, wie sich alle Teile aufeinander beziehen und ein Ganzes bilden.

Erst ganz zum Schluß kommt das **Wörterbuch**, das nur in besonders schwierigen Situationen helfen soll.

Aus: Fremdsprache Deutsch. Zeitschrift für die Praxis des Deutschunterrichts. Heft 2/1990: Arbeit mit Texten.

Quellennachweis: Abbildungen

Seite 10: Bildgeschichte. © 1989, Stern and Joseph Farris.
Seite 16: Christa Fischer, Waldhausen.
Seite 19: Foto: Süddeutscher Verlag/Lothar Schulz.
Seite 29: Zeichnung oben: Harald Stetzer, Schwäbisch Gmünd.
Seite 39: Zeichnung Schwangerschaftstest. Aus: Hans Ritz (Hrsg.): Bilder vom Rotkäppchen. Muriverlag, Göttingen, S. 63.
Seite 45: Koseworte. Aus: Eva H.: Küß mich, ich bin eine verzauberte Geschirrspülmaschine. Lappan Verlag, Oldenburg 1984.
Seite 49: Fotos: ANTHONY, Starnberg: Paradekissen/Jorde; Brotzeit/Pfeiffer; Martinszug/van der Kallen; Vögel/A. Gröger; Umweltbehörde Stadtreinigung Hamburg: Laubabfuhr; E.-M. Jenkins: Osterstrauch, Frühlingsspaziergang.
Seite 60: Bavaria/Images.
Seite 79: Foto: Andreas Sprenger/Kunst und Bild GmbH, Berlin.
Seite 80/81: Nr. 6: HAMBURGER ABENDBLATT; Fotos: Süddeutscher Verlag.
Seite 82: Grundgesetz für die Bundesrepublik Deutschland. Stand: Oktober 1990.
Seite 83: Schaubild 8527: Globus Kartendienst.
Seite 87: Foto: Internationaler Stecker. E.-M. Jenkins.
Seite 89: Kartographie: Oberländer.
Seite 92: Abbildung links: Erschienen im Wilhelm Heyne Verlag, München. Umschlaggestaltung Atelier Schütz, München. Abbildung rechts: Umschlaggestaltung Britta Lembke.
Seite 99: Cartoon: © 1992 KFS/Distr. BULLS.
Seite 105: Titelbild von Willi Glasauer von: Neues vom Rumpelstilzchen. Hg. von Hans-Joachim Gelberg, Beltz Verlag, Weinheim und Basel 1981, Programm Beltz und Gelberg, Weinheim.

Quellennachweis: Texte

Seite 6/7: Stichwörter zu „Heimat". Aus: Gerhard Wahrig: Deutsches Wörterbuch. © 1991 Bertelsmann Lexikon Verlag GmbH, München.
Seite 12: Gerhard Sellin: Tempusfolge. Aus: Bundesdeutsch. hg. von Rudolf Otto Wiemer. Peter Hammer Verlag, Wuppertal 1974.
Seite 18: Stichwörter: Muttersprache, Fremdsprache, fremdsprachig, fremdsprachlich. Aus: Gerhard Wahrig: Deutsches Wörterbuch. © 1991 Bertelsmann Lexikon Verlag GmbH, München.
Seite 19: Schüler an zwei Hamburger Grundschulen. Aus: Die Grundschulzeitschrift 43/1991.
Seite 20: Ansichten von Lehrerinnen. Aus: Die Grundschulzeitschrift 43/1991.
Seite 23: Fremd in der Schweiz. Texte von Ausländern. Hg. von Irmela Kummer, Elisabeth Winiger, Kurt Fendt und Roland Schärer. Cosmos Verlag, Muri bei Bern 1987, 2. Auflage 1991.
Seite 24/25: João Ubaldo Ribeiro aus: Frankfurter Rundschau vom 8.6.1991.
Seite 28/33: Stichwort: anbandeln-anbändeln/Liebe. Aus: Gerhard Wahrig: Deutsches Wörterbuch. © 1991 Bertelsmann Lexikon Verlag GmbH, München.
Seite 34: „Liebe" von Lynn. Aus: Fragen und Versuche 51/1990. Pädagogik Kooperative e.V., Goebenstr. 8, 2800 Bremen 1.
Seite 44: Du, Chef! Neuer Trend im Büro. © dpa Deutsche Presse-Agentur, Hamburg.
Seite 46: Alle wollen nur das eine — Geld: HAMBURGER ABENDBLATT vom 9.5.89.
Seite 54: Stichwort: Landschaft. Aus: Duden Bd. 8: Die Sinn- und Sachverwandten Wörter. Bibliographisches Institut & F.A. Brockhaus AG, Mannheim.
Seite 62: Hermann Claudius: Apfel-Kantate. Aus: Hermann Claudius: In meiner Mutter Garten. Bertelsmann, Gütersloh.
Seite 65: Deutsche sind Naturmuffel. Aus: HAMBURGER ABENDBLATT vom 29.4.91.
Seite 70: Die goldenen zwanziger Jahre (gekürzt). Aus: Chronik der Deutschen. Chronik Verlag, Harenberg Kommunikation, Dortmund 1983.
Seite 84: Zeitsätze. Von Rudolf Otto Wiemer. Aus: Wortwechsel. © Wolfgang Fietkau Verlag, Berlin.
Seite 85: Annegret Hofmann: Unterwegs nach Deutschland. Kinder im Niemandsland. © Aufbau-Verlag, Berlin und Weimar.
Seite 90: Irmgard Locher aus: Stuttgarter Zeitung vom 8.1.1988 (gekürzt).
Seite 92/93: Sybil Gräfin Schönfeld: 1×1 des guten Tons. Das neue Benimmbuch. rororo Sachbuch 8877.
Seite 105: Copyright Rosemarie Künzler-Behncke, Angererstr. 38, 8000 München 40.